Texts in Computational Science and Engineering

5

Editors

Timothy J. Barth
Michael Griebel
David E. Keyes
Risto M. Nieminen
Dirk Roose
Tamar Schlick

Michael Griebel
Stephan Knapek
Gerhard Zumbusch

Numerical Simulation in Molecular Dynamics

Numerics, Algorithms, Parallelization, Applications

With 180 Figures, 43 in Color, and 63 Tables

 Springer

Michael Griebel

Institut für Numerische Simulation
Universität Bonn
Wegelerstr. 6
53115 Bonn, Germany
e-mail: griebel@ins.uni-bonn.de

Stephan Knapek

e-mail: stknapek@yahoo.de

Gerhard Zumbusch

Institut für Angewandte Mathematik
Friedrich-Schiller-Universität Jena
Ernst-Abbe-Platz 2
07743 Jena, Germany
e-mail: zumbusch@mathematik.uni-jena.de

Library of Congress Control Number: 2007928345

Mathematics Subject Classification (2000): 68U20, 70F10, 70H05, 83C10, 65P10, 31C20, 65N06, 65T40, 65Y05, 68W10, 65Y20

ISSN 1611-0994
ISBN 978-3-540-68094-9 Springer Berlin Heidelberg New York

Springer is a part of Springer Science+Business Media
springer.com

© Springer-Verlag Berlin Heidelberg 2007

Typesetting: by the authors using a Springer TeX macro package
Cover design: WMX Design GmbH, Heidelberg
Production: LE-TeX Jelonek, Schmidt & Vöckler GbR, Leipzig

Printed on acid-free paper   46/3180/YL - 5 4 3 2 1 0

Preface

The rapid development of parallel computing systems made it possible to recreate and predict physical processes on computers. Nowadays, computer simulations complement and even substitute experiments. Moreover, simulations allow to study processes that cannot be analyzed directly through experiments. This accelerates the development of new products since costly physical experiments can be avoided. Additionally, the quality of products can be improved by the investigation of phenomena not previously accessible. Hence, computer simulation plays a decisive role especially in the development of new substances in the material sciences as well as in biotechnology and nanotechnology.

Many interesting processes cannot be described nor understood in their entirety in a continuum model, but have to be studied with molecular or atomic models. The numerical simulation of models on this length scale usually relies on particle methods and other methods of molecular dynamics. Areas of application range from physics, biology, and chemistry to modern material sciences.

The fundamental mathematical model here is Newton's second law. It is a system of ordinary differential equations of second order. The law describes the relationship between the force acting on a particle and the resulting acceleration of the particle. The force on each particle is caused by the interaction of that particle with all other particles. The resulting system of differential equations has to be numerically approximated in an efficient way. After an appropriate time discretization, forces on all particles have to be computed in each time step. Different fast and memory-efficient numerical methods exist which are tailored to the short-range or long-range force fields and potentials that are used. Here, especially the linked cell method, the particle mesh method, the P^3M method and its variants, as well as several tree methods such as the Barnes-Hut method or the fast multipole method are to be mentioned. The execution times of these methods can be substantially reduced by a parallel implementation on modern supercomputers. Such an approach is also of fundamental importance to reach the very large numbers of particles and long simulation times necessary for some problems. The numerical methods mentioned are already used with great success in many different implementations by physicists, chemists, and material scientists. However, without a deeper understanding of the particular numerical method applied it is hard to make changes or modifications, or to parallelize or otherwise optimize the available programs.

The aim of this book is to present the necessary numerical techniques of molecular dynamics in a compact form, to enable the reader to write a molecular dynamics program in the programming language C, to implement this program with MPI on a parallel computer with distributed memory, and to motivate the reader to set up his own experiments. For this purpose we present in all chapters a precise description of the algorithms used, we give additional hints for the implementation on modern computers and present numerical experiments in which these algorithms are employed. Further information and some programs can also be found on the internet. They are available on the web page

$$\text{http://www.ins.uni-bonn.de/info/md.}$$

After a short introduction to numerical simulation in chapter 1, we derive in chapter 2 the classical molecular dynamics of particle systems from the principles of quantum mechanics. In chapter 3 we introduce the basic modules of molecular dynamics methods for short-range potentials and force fields (linked cell implementation, Verlet time integration). Additionally, we present a first set of examples of their use. Here, the temperature is taken into account using statistical mechanics in the setting of an NVT ensemble. The Parrinello-Rahman method for the NPT ensemble is also reviewed. Subsequently we discuss in detail the parallel implementation of the linked cell method in chapter 4 and give a set of further examples. In chapter 5 we extend our methods to molecular systems and more complex potentials. Furthermore, in chapter 6 we give an overview of methods for time integration.

Different numerical methods for the efficient computation of long-range force fields are discussed in the following chapters 7 and 8. The P^3M method approximates long-range potentials on an auxiliary grid. Using this method, further examples can be studied that involve in particular Coulomb forces or, on a different length scale, also gravitational forces. We review both the sequential and the parallel implementation of the SPME technique, which is a variant of the P^3M method. In chapter 8 we introduce and discuss tree methods. Here, the emphasis is on the Barnes-Hut method, its extension to higher order, and on a method from the family of fast multipole methods. Both sequential and parallel implementations using space filling curves are presented. In chapter 9 we give examples from biochemistry that require a combination of the methods introduced before.

We thank the SFB 611 (Collaborative Research Center sponsored by the DFG - the German Research Association) "Singular Phenomena and Scaling in Mathematical Models" at the University of Bonn for its support, Barbara Hellriegel and Sebastian Reich for valuable hints and references, our colleagues and coworkers Marcel Arndt, Attila Caglar, Thomas Gerstner, Jan Hamaekers, Lukas Jager, Marc Alexander Schweitzer and Ralf Wildenhues for numerous discussions, their support in the implementation of the algorithms as well as for the programming and computation for various model problems and applications. In particular we thank Attila and also Alex, Jan, Lukas, Marcel, Ralf and Thomas for their active help. We also thank Bernhard Hientzsch for the efforts he put in translating the German version[1] of this book.

Bonn,
April 2007

Michael Griebel
Stephan Knapek
Gerhard Zumbusch

---

[1] *Numerische Simulation in der Moleküldynamik, Numerik, Algorithmen, Parallelisierung, Anwendungen*, Springer Verlag, Heidelberg, 2004.

Table of Contents

1 Computer Simulation – a Key Technology

Experiments, Modelling and Numerical Simulation. In the natural sciences one strives to model the complex processes occurring in nature as accurately as possible. The first step in this direction is the description of nature. It serves to develop an appropriate system of concepts. However, in most cases, mere observation is not enough to find the underlying principles. Most processes are too complex and can not be clearly separated from other processes that interact with them. Only in rare exceptional cases one can derive laws of nature from mere observation as it was the case when Kepler discovered the laws of planetary motion. Instead, the scientist creates (if possible) the conditions under which the process is to be observed, i.e., he conducts an experiment. This method allows to discover how the observed event depends on the chosen conditions and allows inferences about the principles underlying the behavior of the observed system. The goal is the mathematical formulation of the underlying principles, i.e. a theory of the phenomena under investigation. In it, one describes how certain variables behave in dependence of each other and how they change under certain conditions over time. This is mostly done by means of differential and integral equations. The resulting equations, which encode the description of the system or process, are referred to as a mathematical model.

A model that has been confirmed does not only permit the precise description of the observed processes, but also allows the prediction of the results of similar physical processes within certain bounds. Thereby, experimentation, the discovery of underlying principles from the results of measurements, and the translation of those principles into mathematical variables and equations go hand in hand. Theoretical and experimental approaches are therefore most intimately connected.

The phenomena that can be investigated in this way in physics and chemistry extend over very different orders of magnitudes. They can be found from the smallest to the largest observable length scales, from the investigation of matter in quantum mechanics to the study of the shape of the universe. The occurring dimensions range from the nanometer range (10^{-9} meters) in the study of properties of matter on the molecular level to 10^{23} meters in the study of galaxy clusters. Similarly, the time scales that occur in these models (that is, the typical time intervals in which the observed phenomena take place) are vastly different. They range in the mentioned examples from 10^{-12} or even 10^{-15} seconds to 10^{17} seconds, thus from picoseconds or even femtoseconds up to time intervals of several billions of years. The masses occurring in the models are just as different, ranging between 10^{-27} kilograms for single atoms to 10^{40} kilograms for entire galaxies.

The wide range of the described phenomena shows that experiments can not always be conducted in the desired manner. For example in astrophysics, there are only few possibilities to verify models by observations and experiments and to thereby confirm them, or in the opposite case to reject models, i.e., to falsify them. On the other hand, models that describe nature sufficiently well are often so complicated that no analytical solution can be found. Take for example the case of the van der Waals equation to describe dense gases or the Boltzman equation to describe the transport of rarefied gases. Therefore, one usually develops a new and simplified model that is easier to solve. However, the validity of this simplified model is in general more restricted. To derive such models one often uses techniques such as averaging methods, successive approximation methods, matching methods, asymptotic analysis and homogenization. Unfortunately, many important phenomena can only be described with more complicated models. But then these theoretical models can often only be tested and verified in a few simple cases. As an example consider again planetary motion and the gravitational force acting between the planets according to Newton's law. As is known, the orbits following from Newton's law can be derived in closed form only for the two body case. For three bodies, analytical solutions in closed form in general no longer exist. This is also true for our planetary system as well as the stars in our galaxy.

Many models, for example in materials science or in astrophysics, consist of a large number of interacting bodies (called particles), as for example stars and galaxies or atoms and molecules. In many cases the number of particles can reach several millions or more. For instance every cubic meter of gas under normal conditions (that is, at a temperature of 273.15 Kelvin and a pressure of 101.325 kilopascal) contains $2.68678 \cdot 10^{25}$ atoms (Loschmidt constant). 12 grams of the carbon isotope C_{12} contain $6.02214 \cdot 10^{23}$ atoms (Avogadro constant). But large numbers of particles do not only occur on a microscopic scale. Our galaxy, the Milky Way, consists of an estimated 200 billion stars. A glance at the starry sky in a clear night delivers the insight that in such cases there is no hope at all to determine a solution of the underlying equations with paper and pencil.

These are some of the reasons why *computer simulation* has recently emerged as a third way in science besides the experimental and theoretical approach. Over the past years, computer simulation has become an indispensable tool for the investigation and prediction of physical and chemical processes. In this context, computer simulation means the mathematical prediction of technical or physical processes on modern computer systems. The following procedure is typical in this regard: A mathematical-physical model is developed from observation. The derived equations, in most cases valid for continuous time and space, are considered at selected discrete points in time and space. For instance, when discretizing in time, the solution of equations is no longer to be computed at all (that is, infinitely many) points in time, but is only considered at selected points along the time axis. Differential operators, such as for example derivatives with respect to time, can then be approximated by difference operators. The solution of the continuous equations is computed approximately at those selected points. The more densely those points are selected, the more accurately the solution can be approximated. Here, the rapid development of computer technology, which has led to an enormous increase in the computing speed and the memory size of computing systems, now allows simulations that are more and more realistic. The results can be interpreted with the help of appropriate visualization techniques. If corresponding results of physical experiments are available, then the results of the computer simulation can be directly compared. This leads to a verification of the results of the computer simulation or to an improvement in the applied methods or the model (for instance by appropriate changes of parameters of the model or by changing the used equations). Figure 1.1 shows a schematic overview of the steps of a numerical simulation.

Altogether, also for a computer experiment, one needs a mathematical model. But the solutions are now obtained approximately by computations which are carried out by a program on a computer. This allows to study models that are significantly more complex and therefore more realistic than those accessible by analytical means. Furthermore, this allows to avoid costly experimental setups. In addition, situations can be considered that otherwise could not be realized because of technical shortcomings or because they are made impossible by their consequences. For instance, this is the case if it is hard or impossible to create the necessary conditions in the laboratory, if measurements can only be conducted under great difficulties or not at all, if experiments would take too long or would run too fast to be observable, or if the results would be difficult to interpret. In this way, computer simulation makes it possible to study phenomena not accessible before by experiment. If a reliable mathematical model is available that describes the situation at hand accurately enough, it does in general not make a difference for the computer experiment, whether an experiment is carried out at a pressure of one atmosphere or 1000 atmospheres. Obviously this is different, if the experiment would actually have to be carried out in reality. Simulations that run at room temperature or at 10000 Kelvin can in principle be treated in the same way. Computer experiments can span μ-meters or meters, studied phenomena can run within femtoseconds (10^{-15}) or several millions of years. Moreover, the parameters of the experiment can easily be changed. And the behavior of solutions of the mathematical model with respect to such parameters changes can be studied with relatively little effort.

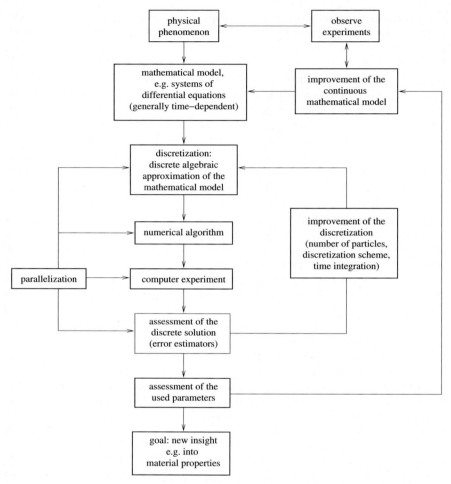

Fig. 1.1. Schematic presentation of the typical approach for numerical simulation.

By now, numerical simulation also provides clues for the correctness of models in fields such as astronomy, in which there is only a slight possibility of model verification. In nanotechnology it can help to predict properties of new materials that do not yet exist in reality. And it can help to identify the most promising or suitable materials. The trend is towards virtual laboratories in which materials are designed and studied on a computer. Moreover, simulation offers the possibility to determine mean or average properties for the macroscopic characterization of such materials. All in all, computer experiments act as a link between laboratory experiments and mathematical-physical theory.

Each of the partial steps of a computer experiment must satisfy a number of requirements. First and foremost, the mathematical model should describe reality as accurately as possible. In general, certain compromises between accuracy in the numerical solution and complexity of the mathematical model have to be accepted. In most cases, the complexity of the models leads to enormous memory and computing time requirements, especially if time-dependent phenomena are studied. Depending on the formulation of the discrete problem, several nested loops have to be executed for the time dependency, for the application of operators, or also for the treatment of nonlinearities.

Current research therefore has its focus in particular on the development of methods and algorithms that allow to compute the solutions of the discrete problem as fast as possible (multilevel and multiscale methods, multipole methods, fast Fourier transforms) and that can approximate the solution of the continuous problem with as little memory as possible (adaptivity). More realistic and therefore in general more complex models require faster and more powerful algorithms. Vice versa, better algorithms allow the use of more complex models.

Another possibility to run larger problems is the use of *vector computers* and *parallel computers*. Vector computers increase their performance by processing similar arithmetical instructions on data stored in a vector in an assembly line-like fashion. In parallel computers, several dozens to many thousands of powerful processors[1] are assembled into one computing system. These processors can compute concurrently and independently and can communicate with each other.[2] A reduction of the required computing time for a simulation is achieved by distributing the necessary computations to several processors. Up to a certain degree, the computations can then be executed concurrently. In addition, parallel computer systems in general have a substantially larger main memory than sequential computers. Hence, larger problems can be treated.

As an example we mention the ASC Red System (formerly ASCI Red) which was the first computer that reached the teraflops per second threshold in processing speed, meaning that it can process a trillion, i.e. 10^{12} floating point operations per second. This computer system consists of 9216 processors and was assembled within the framework of the Advanced Simulation and Computing Program (ASC) (formerly ASCI) of the USA. This initiative

---

[1] The processors in use today have mostly a RISC (reduced instruction set computer) processor architecture. They have fewer machine instructions compared to older processors, allowing a faster, assembly line-like execution of the instructions, see [467].

[2] To improve portability of programs among parallel computers from different manufacturers and to simplify the assembly of computers of different types to a parallel computer, uniform standards for data exchange between computers are needed. In recent years, the MPI (Message Passing Interface) platform has emerged as a de facto standard for the communication between processes, see appendix A.3.

aimed to build a series of supercomputers with processing speeds starting at
1 teraflops/s, currently in the 100 teraflops/s and into the petaflop/s range.[3]
The latest computer (2007) in this series is an IBM BlueGene/L computer
system installed at the Lawrence Livermore National Laboratory. At the moment this system is the most powerful computer in the world. It consists
of 65,536 dual processor nodes. The nodes are connected as a $32 \times 32 \times 64$
3D-torus. A total memory amount of 32 terabyte is available. BlueGene/L
achieves 280.6 teraflop/s, which is about 76 percent of its theoretical performance (367 teraflop/s). This computer occupies a floor space of 2500 square
feet. The first computer performing more than 1 petaflop/s (10^{15} floating
point operations per second) will probably be available in 2008. A computer
with a peak performance of 10 petaflop/s is planned to be realized in 2012.

Figure 1.2 shows the development of the processing speed of high performance computers from the last years measured by the parallel Linpack
benchmark.[4] The performance in flops/s is plotted versus the year, for the
fastest parallel computer in the world, and for the computers at position 100
and 500 in the list of the fastest parallel computers in the world (see [1]). Personal computers and workstations have seen a similar development of their
processing speed. Because of that, satisfactory simulations have become possible on these smaller computers.

Particle Models. An important area of numerical simulation deals with
so-called particle models. These are simulation models in which the representation of the physical system consists of discrete particles and their interactions. For instance, systems of classical mechanics can be described by
the positions, velocities, and the forces acting between the particles. In this
case the particles do not have to be very small either with respect to their
dimensions or with respect to their mass – as possibly suggested by their
name. Rather they are regarded as fundamental building blocks of an abstract model. For this reason the particles can represent atoms or molecules
as well as stars or parts of galaxies.[5] The particles carry properties of physical
objects, as for example mass, position, velocity or charge. The state and the
evolution of the physical system is represented by these properties of the par-

---

[3] These computer systems were intended for the development of software and simulation of, among other things, the production, aging, safety, reliability, testing,
and further development of the American nuclear arsenal with the aim to replace
the actual physical testing of atomic bombs. One fundamental idea was that experimental measurements at real tests of atomic bombs have relatively large
errors and that the same accuracy can be reached with computer simulations.

[4] The Linpack benchmark tests the performance of computers on the solution of
dense linear systems of equations using Gaussian elimination.

[5] Galaxies are often modeled as a few mass points which describe the average of
a large group of stars and not as a collection of billions of stars.

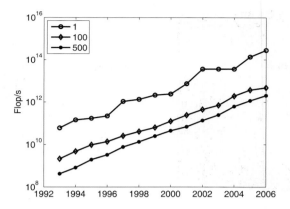

Fig. 1.2. Development of processing speed over the last years (parallel Linpack benchmark); fastest (1), 100th fastest (100) and 500th fastest (500) computer in the world; up to now the processing speed increases tenfold about every four years.

ticles and by their interactions, respectively.[6] Figure 1.3 shows the result of a simulation of the formation of the large scale structure of the universe. The model consists of 32768 particles that each represents several hundred galaxies. Figure 1.4 shows a protein called Nucleosome which consists of 12386 particles representing single atoms.

The laws of classical mechanics are used in many particle models. The use of Newton's second law results in a system of ordinary differential equations of second order describing how the acceleration of any particle depends on the force acting on it. The force results from the interaction with the other particles and depends on their position. If the positions of the particles change relative to each other, then in general also the forces between the particles change. The solution of the system of ordinary differential equations for given initial values then leads to the trajectories of the particles. This is a deterministic procedure, meaning that the trajectories of the particles are in principle uniquely determined for all times by the given initial values.

But why is it reasonable at all to use the laws of classical mechanics when at least for atomic models the laws of quantum mechanics should be used? Should not Schrödinger's equation be employed as equation of motion instead of Newton's laws? And what does the expression "interaction between particles" actually mean, exactly?

---

[6] Besides the methods considered in this book there are a number of other approaches which can be categorized as particle methods, see [448, 449], [144, 156, 380, 408], [254, 369, 394, 437] and [677, 678]. Also the so-called gridless discretization methods [55, 76, 194, 195, 270, 445] can be interpreted as particle methods.

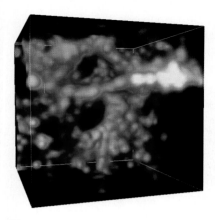

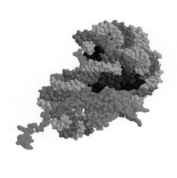

Fig. 1.3. Result of a particle simulation of the large scale structure of the universe.

Fig. 1.4. A view of the protein Nucleosome.

If one considers a system of interacting atoms which consists of nuclei and electrons, one can in principle determine its behavior by solving the Schrödinger equation with the appropriate Hamilton operator. However, an analytic or even numerical solution of the Schrödinger equation is only possible in a few simple special cases. Therefore, approximations have to be made. The most prominent approach is the Born-Oppenheimer approximation. It allows a separation of the equations of motions of the nuclei and of the electrons. The intuition behind this approximation is that the significantly smaller mass of the electrons permits them to adapt to the new position of the nuclei almost instantaneously. The Schrödinger equation for the nuclei is therefore replaced by Newton's law. The nuclei are then moved according to classical mechanics, but using potentials that result from the solution of the Schrödinger equation for the electrons. For the solution of this electronic Schrödinger equation approximations have to be employed. Such approximations are for instance derived with the Hartree-Fock approach or with density functional theory. This approach is known as ab initio molecular dynamics. However, the complexity of the model and the resulting algorithms enforces a restriction of the system size to a few thousand atoms.

A further drastic simplification is the use of parametrized analytical potentials that just depend on the position of the nuclei (classical molecular dynamics). The potential function itself is then determined by fitting it to the results of quantum mechanical electronic structure computations for a few representative model configurations and subsequent force-matching [208] or by fitting to experimentally measured data. The use of these very crude approximations to the electronic potential hypersurface allows the treatment of systems with many millions of atoms. However, in this approach quantum mechanical effects are lost to a large extent.

The following incomplete list gives some examples of physical systems that can be represented by particle systems in a meaningful way. They are therefore amenable to simulation by particle methods:

Solid State Physics: The simulation of materials on an atomic scale is primarily used in the analysis of known materials and in the development of new materials. Examples for phenomena studied in solid state physics are the structure conversion in metals induced by temperature or shock, the formation of cracks initiated by pressure, shear stresses, etc. in fracture experiments, the propagation of sound waves in materials, the impact of defects in the structure of materials on their load-bearing capacity and the analysis of plastic and elastic deformations.

Fluid Dynamics: Particle simulation can serve as a new approach in the study of hydrodynamical instabilities on the microscopic scale, as for instance, the Rayleigh-Taylor or Rayleigh-Benard instability. Furthermore, molecular dynamics simulations allow the investigation of complex fluids and fluid mixtures, as for example emulsions of oil and water, but also of crystallization and of phase transitions on the microscopic level.

Biochemistry: The dynamics of macromolecules on the atomic level is one of the most prominent applications of particle methods. With such methods it is possible to simulate molecular fluids, crystals, amorphous polymers, liquid crystals, zeolites, nuclear acids, proteins, membranes and many more biochemical materials.

Astrophysics: In this area, simulations mostly serve to test the soundness of theoretical models. In a simulation of the formation of the large-scale structure of the universe, particles correspond to entire galaxies. In a simulation of galaxies, particles represent several hundred to thousand stars. The force acting between these particles results from the gravitational potential.

Computer Simulation of Particle Models. In the computer simulation of particle models, the time evolution of a system of interacting particles is determined by the integration of the equations of motion. Here, one can follow individual particles, see how they collide, repel each other, attract each other, how several particles are bound to each other, are binding to each other, or are separating from each other. Distances, angles and similar geometric quantities between several particles can also be computed and observed over time. Such measurements allow the computation of relevant macroscopic variables such as kinetic or potential energy, pressure, diffusion constants, transport coefficients, structure factors, spectral density functions, distribution functions, and many more.

In most cases, variables of interest are not computed exactly in computer simulations, but only up to a certain accuracy. Because of that, it is desirable

- to achieve an accuracy as high as possible with a given number of operations,

- to achieve a given accuracy with as few operations as possible, or
- to achieve a ratio of effort (number of operations) to achieved accuracy which is as small as possible.

Clearly the last alternative includes the first two as special cases. A good algorithm possesses a ratio of effort (costs, number of operations, necessary memory) to benefit (achieved accuracy) that is as favorable as possible. As a measure for the ranking of algorithms one can use the quotient

$$\frac{\text{effort}}{\text{benefit}} = \frac{\#\ \text{operations}}{\text{achieved accuracy}}.$$

This is a number that allows the comparison of different algorithms. If it is known how many operations are minimally needed to achieve a certain accuracy, this number shows how far a given algorithm is from optimal. The minimal number of operations to achieve a given accuracy ε is called ε-complexity. The ε-complexity is thus a lower bound for the number of operations for any algorithm to achieve an accuracy of ε.[7]

The two principal components of the computer simulation of particle models are (in addition to the construction of appropriate interaction potentials) the time integration of Newton's equations of motion and the fast evaluation of the interactions between the individual particles.

Time Integration: In numerical time integration, the solution of the considered differential equation is only computed at a number of discrete points in time. Incrementally, approximations to the values at later points in time are computed from the values of the approximations at previous points in time. Once a specific integration scheme has been chosen, the forces acting on the individual particles have to be computed in each time step. For this, the negative gradient of the potential function of the system has to be computed. If we denote with $\mathbf{x}_i, \mathbf{v}_i$ and $\mathbf{F}_i$ the position of the ith particle, the velocity of the ith particle, and the force on the ith particle, respectively, we can write the basic algorithm 1.1 for the computation of the trajectories of N particles.[8] With given initial values for $\mathbf{x}_i$ and $\mathbf{v}_i$ for $i = 1, \ldots, N$, the time is increased by δt in each step in an outer integration loop starting at time $t = 0$ until the final time t_{end} is reached. The forces on the individual particles and their new positions and velocities are then computed in an inner loop over all particles.

Fast Evaluation of the Forces: There are N^2 interactions between particles in a system which consists of N particles. If self-interactions are excluded, this number is reduced by N. If we also consider that all other interactions

---

[7] The branch of mathematics and computer science that deals with questions in this context is called *information-based complexity*, see for instance [615].

[8] In principle, the algorithms described in this book can be implemented in many different programming languages. In the following, we give algorithms in the programming language C, see [40, 354], and higher level expressions.

Algorithm 1.1 Basic Algorithm

```
real t = t_start;
for i = 1, ..., N
  set initial conditions xᵢ (positions) and vᵢ (velocities);
while (t < t_end) {
  compute for i = 1, ..., N the new positions xᵢ and velocities vᵢ
    at time t + delta_t by an integration procedure from the
    positions xᵢ, velocities vᵢ and forces Fᵢ on the particle at
    earlier times;
  t = t + delta_t;
}
```

are counted twice,[9] we obtain in total $(N^2 - N)/2$ actions between particles that have to be determined to compute the forces between all particles. This naive approach therefore needs $\mathcal{O}(N^2)$ operations for N particles in each time step.[10] Thus, if the number of particles is doubled, the number of operations quadruples. Because of the limited performance of computers, this approach to the computation of the forces is only feasible for relatively small numbers of particles. However, if only an approximation of the forces up to a certain accuracy is required, a substantial reduction of the complexity may be possible.

The complexity of an approximative evaluation of the forces at a fixed time is obviously at least of order $\mathcal{O}(N)$ since every particle has to be "touched" at least once. Algorithms are called optimal if the complexity for the computation up to a given accuracy is $\mathcal{O}(N)$. If the complexity of the algorithm differs from the optimal by a logarithmic factor, meaning it is of the order $\mathcal{O}(N \log(N)^\alpha)$ with $\alpha > 0$, the algorithm is called quasi-optimal. Figure 1.5 shows a comparison of the time complexities using an optimal, a quasi-optimal and an $\mathcal{O}(N^2)$-algorithm. The evaluation of the interactions for 1000 particles with the $\mathcal{O}(N^2)$-algorithm needs as much time as the optimal algorithm needs for the approximate evaluation of the interactions for almost a million particles.

The goal is to find optimal algorithms and also to implement them on computers. The design of a suitable algorithm necessarily has to be adapted to the kind of interactions modeled and to other parameters, as for example changes in the density of the particles. It is clear that algorithms which are optimal for some form of interaction potentials may not be suitable for other forms of potentials. This can be demonstrated most easily by the difference between a potential that decays quickly and one that decays slowly. In this

---

[9] By Newton's third law the action of a particle i on a particle j is the same as the action of the particle j on the particle i.

[10] The relation $f(N) = \mathcal{O}(N^2)$ means for a function f that $f(N)/N^2$ is bounded for $N \to \infty$.

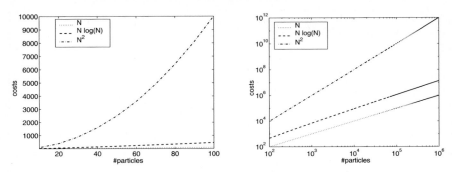

Fig. 1.5. Comparison of the complexity of an optimal, a quasi-optimal, and an $\mathcal{O}(N^2)$-algorithm (left: linear plot, right: doubly logarithmic plot).

context, for a fast decaying potential, a particle only exerts a significant force on another particle if the distance between the two particles is small. For almost uniformly distributed particles, the force evaluation can then be implemented in $\mathcal{O}(N)$ operations, since only particles in the close vicinity of a particle contribute significantly to the force acting on it. On the other hand, long-range forces such as Coulomb forces or gravitational forces decay only very slowly. Their effects can not be neglected in general even if the particles are far apart, see [219, 672].

The graphs in Figure 1.6 and 1.7 show schematically a $1/r$ behavior which is typical for long-range potentials. For very small values of the distance r the potential is very large and decreases strongly with increasing distance. The decrease slows down with increasing distance. For r small, a small change in the position of the particle has a very strong effect on the resulting potential value, compare Figure 1.6. However, a small change in the position of the particles which are farer apart (r large) only has a small effect on the resulting potential value, compare Figure 1.7. A similar statement is valid for the forces, since the force is the negative gradient of the potential. In particular, in the case of large r, one does not have to distinguish between two particles close to each other for the approximative evaluation of potentials and forces, since the resulting values of potential and force will be approximately the same. This behavior is exploited in algorithms for long-range potentials.

History. The development of computer simulations of particle models has been closely connected with the development of computers. The first article about simulation in molecular dynamics was written by Alder and Wainwright [33] in 1957. The authors studied a model of some hundred particles that interacted by elastic impact and computed the associated phase diagram.[11] Gibson, Goland, Milgram and Vineyard [252] soon afterwards inves-

---

[11] These results were a big surprise at that time since it was generally assumed that also an attractive potential was needed to generate such a phase transition.

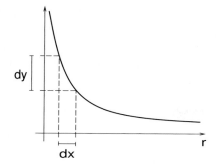

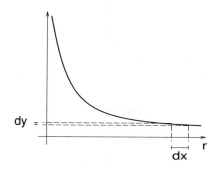

Fig. 1.6. In the near field (that is, for small values of r), a small change in position dx of the particles results in a large change dy of the potential.

Fig. 1.7. In the far field (that is, for large values of r), a small change in position dx of the particles results in a small change dy of the potential.

tigated damages caused by radioactive radiation in a molecular dynamics simulation involving 500 atoms. Rahman [498] studied properties of fluid argon in 1964. He was the first who used the Lennard-Jones potential in molecular dynamics simulations. In 1967, Verlet introduced an algorithm to efficiently manage data in a molecular dynamics simulation using neighbor-lists [645]. This paper also presented a time integration scheme[12] that serves as the standard method in molecular dynamics simulations even today. Molecules such as butane were first investigated in 1975 in [533]. Molecular dynamics simulations with constant pressure or constant temperature were described in the beginning of the eighties in the papers [42, 327, 452, 453]. Also, more complex potentials with many-body interactions were introduced quite early into simulation [57].

The potentials used in these early papers were mostly short-range potentials and the simulations were very limited because of the small capacity of the computers at that time. The simulation of models with long-range potentials, in particular for large numbers of particles, demanded further developments in computer technology and in the algorithms used. One method for the treatment of such potentials relies in its main features on Ewald [216]. There, the potential is split into its short-range and long-range part, each of which can be computed efficiently with a specific separate approach. The decisive idea is the use of fast Poisson solvers for the long-range part. Hierarchical methods, such as fast Fourier transforms or multilevel methods, are applied in these fast Poisson solvers. The variants of this so-called P^3M method [202, 324] differ in the selection of algorithms for the single components (interpolation, force evaluation, adaptivity, fast solvers, etc.). A prominent example is the so-called Particle Mesh Ewald method (PME), see [168, 215, 374], that uses

---

[12] This scheme relies on an approach presented by Störmer in 1907 [588] which can be traced back even further to Delambre in 1790 [566].

B-spline or Lagrangian interpolation together with fast Fourier transforms. See also [211, 488, 574] and [612] for a review of existing P^3M variants.

Another class of methods for long-range potentials uses an expansion (Taylor expansion, multipole expansion) of the potential functions in the distance to approximately evaluate particle interactions. The resulting data is stored in tree structures to allow efficient data management and computation. Some earlier representatives of this class that are often used in astrophysics were developed by Barnes and Hut [58] and Appel [47]. Newer variants by Greengard and Rohklin [260, 263, 525] use higher moments in the expansions.

During the last years the parallelization of algorithms for molecular dynamics simulations has attracted much attention. A description of parallel algorithms for short-range potentials can be found for example in [71, 483] and [503]. Parallel variants of the P^3M algorithm are found in [606, 613, 679] and [221], respectively. Parallel versions of the Barnes-Hut algorithm and the multipole method have been presented for example in [261, 651, 652, 654, 686]. Domain decomposition methods are employed in these versions as well as in the parallelization of algorithms for short-range potentials.

Accounts of the theoretical background together with the detailed description of different methods and application areas of molecular dynamics can be found in a number of books and edited volumes [34, 90, 147, 148, 239, 282, 324, 328, 373, 607]. By now, most of the methods have been implemented in commercial and research software packages. Examples are AL_CMD [628], Amber [471], CHARMM [125, 126, 401], DL_POLY [571], EGO [205], GROMACS [83], GROMOS [634], IMD [530], LAMMPS [482], Moldy [507], MOSCITO [463], NAMD [446], OPLS [348], ORAC [2, 491], PMD [662], SIgMA [314], SPaSM [69] and YASP [441]. Pbody [93] and DPMTA [95] provide parallel libraries for N-body problems. NEMO [604], GADGET [3] and HYDRA [4] are suitable especially for applications in astrophysics.

This book tries to build a bridge from the theory of molecular dynamics to the implementation of efficient parallel algorithms and their applications. Its primary goal is to introduce the necessary numerical techniques of molecular dynamics in a compact form, to present the necessary steps in the development of efficient algorithms, and to describe the implementation of those algorithms on sequential as well as parallel computer systems. All models and algorithms are derived in detail. Also, programs and parameters necessary for the example applications are listed. This will enable the reader to implement programs for molecular dynamics simulations, to use them on parallel computer systems, and to conduct simulations on his own.

This book primarily appeals to two audiences. On the one hand to students, teachers, and researchers in physics, chemistry, and biology that want to gain a deeper understanding of the fundamentals and the efficiency of molecular dynamics software and its applications. On the other hand it appeals to mathematicians and computer scientists by giving them the possibility to get acquainted with a number of different numerical methods from the area of molecular dynamics.

Depending on prior knowledge and interest, the reader of this book is advised to selectively choose and read particular chapters. Chapters 3 and 4 give a convenient introduction; the Sections 3.7.4 and 3.7.5 can be skipped in a first reading. The Chapters 7, 8 and 9 are intended to be self-contained, up to a few basic concepts and facts from chapters 3 and 4.

2 From the Schrödinger Equation to Molecular Dynamics

In particle methods, the laws of classical mechanics [48, 371] are used, in particular Newton's second law. In this chapter we will pursue the question why it makes sense to apply the laws of classical mechanics, even though one should use the laws of quantum mechanics. Readers that are more interested in the algorithmic details or in the implementation of algorithms in molecular dynamics can skip this chapter.

In quantum mechanics, the Schrödinger equation is taking the place of Newton's equations. But the Schrödinger equation is so complex that it can be solved analytically only for a few simple cases. Also the direct numerical solution on computers is limited to very simple systems and very small numbers of particles because of the high dimension of the space in which the Schrödinger equation is posed. Therefore, approximation procedures are used to simplify the problem. These procedures are based on the fact that the electron mass is much smaller than the mass of the nuclei. The idea is to split the Schrödinger equation, which describes the state of both the electrons and nuclei, with a separation approach into two coupled equations. The influence of the electrons on the interaction between the nuclei is then described by an effective potential. This potential results from the solution of the so-called electronic Schrödinger equation. As a further approximation the nuclei are moved according to the classical Newton's equations using either effective potentials which result from quantum mechanical computations (which include the effects of the electrons) or empirical potentials that have been fitted to the results of quantum mechanical computations or to the results of experiments.

All in all, this approach is a classical example for a hierarchy of approximation procedures and an example for the use of effective quantities. In the following, the derivation of the molecular dynamics method from the laws of quantum mechanics is presented. For further details see the large body of available literature, for example [372, 427, 554], [381, 514], and [417, 626, 627].

2.1 The Schrödinger Equation

Up to the end of the nineteenth century, classical physics could answer the most important questions using Newton's equations of motion. The Lagrange formalism and the Hamilton formalism both lead to generalized classical equations of motion that are essentially equivalent. These equations furnish how the change in time of the position of particles depends on the forces acting on them. If initial positions and initial velocities are given, the positions of the particles are determined uniquely for all later points in time. Observable quantities such as angular momentum or kinetic energy can then be represented as functions of the positions and the impulses of the particles.

In the beginning of the twentieth century the theory of quantum mechanics was developed. There, the dynamics of the particles is described by a new equation of motion, the Schrödinger equation. In contrast to Newton's equations its solution no longer provides unique trajectories, meaning uniquely determined positions and impulses of the particles, but only probabilistic statements about the positions and impulses of the particles. Furthermore, position and impulse of a single particle can no longer be measured arbitrarily accurately at the same time (Heisenberg's uncertainty principle) and certain observables, as for example the energies of bound electrons, can only assume certain discrete values. All statements that can be made about a quantum mechanical system can be derived from the state function (or wave function) Ψ which is given as the solution of the Schrödinger equation. Let us consider as an example a system consisting of N nuclei and K electrons. The time-dependent state function of such a system can be written in general as

$$\Psi = \Psi(\mathbf{R}_1, \ldots, \mathbf{R}_N, \mathbf{r}_1, \ldots, \mathbf{r}_K, t),$$

where $\mathbf{R}_i$ and $\mathbf{r}_i$ denote positions in three-dimensional space $\mathbb{R}^3$ associated to the ith nucleus and the ith electron, respectively. The variable t denotes the time-dependency of the state function. The vector space (space of configurations) in which the coordinates of the particles are given is therefore of dimension $3(N + K)$. In the following we will abbreviate $(\mathbf{R}_1, \ldots, \mathbf{R}_N)$ and $(\mathbf{r}_1, \ldots, \mathbf{r}_K)$ with the shorter notation $\mathbf{R}$ and $\mathbf{r}$, respectively.

According to the statistical interpretation of the state function, the expression

$$\Psi^*(\mathbf{R}, \mathbf{r}, t)\Psi(\mathbf{R}, \mathbf{r}, t)dV_1 \cdots dV_{N+K} \qquad (2.1)$$

describes the probability to find the system under consideration at time t in the volume element $dV_1 \cdot \ldots \cdot dV_{N+K}$ of the configuration space centered at the point $(\mathbf{R}, \mathbf{r})$. By integrating over a volume element of the configuration space one determines the probability to find the system in this domain.

We assume in the following that nuclei and electrons are charged particles. The electrostatic potential (Coulomb potential) of a point charge (with elementary charge $+e$) is $\frac{e}{4\pi\epsilon_0}\frac{1}{r}$, where r is the distance from the position of the charged particle and ϵ_0 is the dielectric constant. $1/(4\pi\epsilon_0)$ is also called Coulomb constant. An electron moving in this potential has the potential energy $V(r) = -\frac{e^2}{4\pi\epsilon_0}\frac{1}{r}$. Neglecting spin and relativistic interactions and assuming that no external forces act on the system, the Hamilton operator associated to the system of nuclei and electrons is given as the sum over the operators for the kinetic energy and the Coulomb potentials,

$$\mathcal{H}(\mathbf{R}, \mathbf{r}) := -\frac{\hbar^2}{2m_e} \sum_{k=1}^{K} \Delta_{\mathbf{r}_k} + \frac{e^2}{4\pi\epsilon_0} \sum_{k<j}^{K} \frac{1}{\| \mathbf{r}_k - \mathbf{r}_j \|} - \frac{e^2}{4\pi\epsilon_0} \sum_{k=1}^{K} \sum_{j=1}^{N} \frac{Z_j}{\| \mathbf{r}_k - \mathbf{R}_j \|}$$

$$+ \frac{e^2}{4\pi\epsilon_0} \sum_{k<j}^{N} \frac{Z_k Z_j}{\| \mathbf{R}_k - \mathbf{R}_j \|} - \frac{\hbar^2}{2} \sum_{k=1}^{N} \frac{1}{M_k} \Delta_{\mathbf{R}_k}. \tag{2.2}$$

Here, M_j and Z_j denote the mass and the atomic number of the jth nucleus, m_e is the mass of an electron and $\hbar = h/2\pi$ with h being Planck's constant. $\|\mathbf{r}_k - \mathbf{r}_j\|$ are the distances between electrons, $\|\mathbf{r}_k - \mathbf{R}_j\|$ are distances between electrons and nuclei and $\|\mathbf{R}_k - \mathbf{R}_j\|$ are distances between nuclei. The operators $\Delta_{\mathbf{R}_k}$ and $\Delta_{\mathbf{r}_k}$ stand here for the Laplace operator with respect to the nuclear coordinates $\mathbf{R}_k$ and with respect to the electronic coordinates $\mathbf{r}_k$.[1] In the following we will denote the separate parts of (2.2) in abbreviated form (written in the same order) with

$$\mathcal{H} = T_e + V_{ee} + V_{eK} + V_{KK} + T_K. \tag{2.3}$$

The meanings of the individual parts are the following: T_e and T_K are the operators of the kinetic energy of the electrons and of the nuclei, respectively. V_{ee}, V_{KK} and V_{eK} refer to the operators of the potential energy of the interactions (thus the Coulomb energy) between only the electrons, between only the nuclei, and between the electrons and the nuclei, respectively.

The state function Ψ is now given as the solution of the Schrödinger equation

$$i\hbar \frac{\partial \Psi(\mathbf{R}, \mathbf{r}, t)}{\partial t} = \mathcal{H}\Psi(\mathbf{R}, \mathbf{r}, t) \tag{2.4}$$

where i denotes the imaginary unit. The expression $\Delta_{\mathbf{R}_k}\Psi(\mathbf{R}, \mathbf{r}, t)$, which occurs in $\mathcal{H}\Psi$, stands there for $\Delta_{\mathbf{Y}}\Psi(\mathbf{R}_1, \ldots, \mathbf{R}_{k-1}, \mathbf{Y}, \mathbf{R}_{k+1}, \ldots, \mathbf{R}_N, \mathbf{r}, t)|_{\mathbf{R}_k}$, that is, the application of the Laplace operator to Ψ seen as a function of $\mathbf{Y}$ (the kth vector of coordinates) and the evaluation of the resulting function at the point $\mathbf{Y} = \mathbf{R}_k$. The operators $\Delta_{\mathbf{r}_k}$ and later $\nabla_{\mathbf{R}_k}$ and others are to be understood in an analogous way.

In the following we consider the case that the Hamilton operator $\mathcal{H}$ is not explicitly time-dependent, as we already assumed in (2.2).[2] Then, the separation approach

$$\Psi(\mathbf{R}, \mathbf{r}, t) = \psi(\mathbf{R}, \mathbf{r}) \cdot f(t) \tag{2.5}$$

of Ψ with a function $\psi = \psi(\mathbf{R}, \mathbf{r})$ that does not depend on time and a function $f = f(t)$ that depends on time when substituted into (2.4) gives rise to

---

[1] If we denote the three components of $\mathbf{R}_k$ by $(\mathbf{R}_k)_1, (\mathbf{R}_k)_2$ and $(\mathbf{R}_k)_3$, then we obtain $\Delta_{\mathbf{R}_k} = \frac{\partial^2}{\partial(\mathbf{R}_k)_1^2} + \frac{\partial^2}{\partial(\mathbf{R}_k)_2^2} + \frac{\partial^2}{\partial(\mathbf{R}_k)_3^2}$.

[2] Since the Hamilton operator $\mathcal{H}$ depends on the coordinates and impulses of the particles, it depends implicitly on time this way. If time-dependent external forces act on the system, the Hamilton operator could also explicitly depend on time. Then, one would write $\mathcal{H}(\mathbf{R}, \mathbf{r}, t)$ to reflect this dependency.

$$i\hbar\frac{df(t)}{dt}\psi(\mathbf{R},\mathbf{r}) = f(t)\mathcal{H}\psi(\mathbf{R},\mathbf{r}), \tag{2.6}$$

since $\mathcal{H}$ does not act on $f(t)$.[3] A formal division of both sides by the term $\psi(\mathbf{R},\mathbf{r}) \cdot f(t) \neq 0$ yields

$$i\hbar\frac{1}{f(t)}\frac{df(t)}{dt} = \frac{1}{\psi(\mathbf{R},\mathbf{r})}\mathcal{H}\psi(\mathbf{R},\mathbf{r}). \tag{2.7}$$

The left hand side contains only the time coordinate t, the right hand side only the coordinates in space. Therefore, both sides have to be equal to a common constant E and (2.7) can be separated. We obtain the two equations

$$i\hbar\frac{1}{f(t)}\frac{df(t)}{dt} = E \tag{2.8}$$

and

$$\mathcal{H}\psi(\mathbf{R},\mathbf{r}) = E\psi(\mathbf{R},\mathbf{r}). \tag{2.9}$$

The differential equation (2.8) describes the evolution over time of the wave function. Its general solution reads

$$f(t) = ce^{-iEt/\hbar}. \tag{2.10}$$

Equation (2.9) is an eigenvalue problem for the Hamilton operator $\mathcal{H}$ with the energy eigenvalue E. This equation is called time-independent (or stationary) Schrödinger equation. To every energy eigenvalue E_n there is one (or, in the case of degenerated states, several) associated energy eigenfunctions ψ_n. Also, for every energy eigenvalue E_n, (2.10) yields a time-dependent term f_n. The solution of the time-dependent Schrödinger equation (2.4) is then given as a linear combination of the energy eigenfunctions ψ_n and the associated time-dependent terms f_n of the form

$$\Psi(\mathbf{R},\mathbf{r},t) = \sum_n c_n e^{-iE_n t/\hbar}\psi_n(\mathbf{R},\mathbf{r}) \tag{2.11}$$

with the weights $c_n = \int \psi_n^*(\mathbf{R},\mathbf{r})\Psi(\mathbf{R},\mathbf{r},0)d\mathbf{R}d\mathbf{r}$.

Similar to the time-dependent Schrödinger equation, (2.9) is so complex that analytical solutions can only be given for a few very simple systems. The development of approximation procedures is therefore a fundamental area of research in quantum mechanics. There exists an entire hierarchy of approximations that exploit the different physical properties of nuclei and electrons [417, 626, 627]. We will consider these approximations in the following in more detail.

---

[3] One then also calls ψ state or wave function.

2.2 A Derivation of Classical Molecular Dynamics

In the following we will derive, starting from the time-dependent Schrödinger equation (2.4), the equations of classical molecular dynamics by a series of approximations. We follow [417] and [626, 627].

2.2.1 The TDSCF Approach and Ehrenfest's Molecular Dynamics

First, we decompose the Hamilton operator (2.3) as follows: We set

$$\mathcal{H} = \mathcal{H}_e + T_K \tag{2.12}$$

with the electronic Hamilton operator

$$\mathcal{H}_e := T_e + V_{ee} + V_{eK} + V_{KK}. \tag{2.13}$$

We decompose $\mathcal{H}_e$ further into its kinetic and potential part

$$\mathcal{H}_e := T_e + V_e$$

where now

$$V_e := V_{ee} + V_{eK} + V_{KK}$$

is just the operator for the potential energy of the entire system.

The wave function $\Psi(\mathbf{R}, \mathbf{r}, t)$ depends on the coordinates of the electrons and of the nuclei as well as on time. First, we separate the wave function into a simple product form[4]

$$\Psi(\mathbf{R}, \mathbf{r}, t) \approx \tilde{\Psi}(\mathbf{R}, \mathbf{r}, t) := \chi(\mathbf{R}, t)\phi(\mathbf{r}, t) \exp\left[\frac{i}{\hbar}\int_{t_0}^{t} \tilde{E}_e(t')dt'\right] \tag{2.14}$$

of the contribution of the nuclei and electrons to the full wave function Ψ. It is assumed that the nuclear wave function $\chi(\mathbf{R}, t)$ and the electronic wave function $\phi(\mathbf{r}, t)$ are normalized for any point in time t, that means that both $\int \chi^*(\mathbf{R}, t)\chi(\mathbf{R}, t)d\mathbf{R} = 1$ and $\int \phi^*(\mathbf{r}, t)\phi(\mathbf{r}, t)d\mathbf{r} = 1$ hold. The phase factor $\tilde{E}_e$ is chosen in the form

$$\tilde{E}_e(t) = \int \phi^*(\mathbf{r}, t)\chi^*(\mathbf{R}, t)\mathcal{H}_e\phi(\mathbf{r}, t)\chi(\mathbf{R}, t)d\mathbf{R}d\mathbf{r} \tag{2.15}$$

which is convenient for the following derivation of a coupled system of equations.

Now, we insert (2.14) into the time-dependent Schrödinger equation (2.4) with Hamilton operator $\mathcal{H}$, multiply from the left with $\phi^*(\mathbf{r}, t)$ and $\chi^*(\mathbf{R}, t)$

---

[4] This approximation is a so-called single determinant or single configuration ansatz for the full wave function. It can only result in a mean field description of the coupled dynamics.

and integrate over $\mathbf{R}$ and $\mathbf{r}$. Finally, we require conservation of energy, that is,

$$\frac{d}{dt} \int \tilde{\Psi}^* \mathcal{H} \tilde{\Psi} d\mathbf{R} d\mathbf{r} = 0,$$

and obtain thereby the coupled system of equations

$$i\hbar \frac{\partial \phi}{\partial t} = -\sum_k \frac{\hbar^2}{2m_e} \Delta_{\mathbf{r}_k} \phi + \left(\int \chi^*(\mathbf{R}, t) V_e(\mathbf{R}, \mathbf{r}) \chi(\mathbf{R}, t) d\mathbf{R} \right) \phi, \quad (2.16)$$

$$i\hbar \frac{\partial \chi}{\partial t} = -\sum_k \frac{\hbar^2}{2M_k} \Delta_{\mathbf{R}_k} \chi + \left(\int \phi^*(\mathbf{r}, t) \mathcal{H}_e(\mathbf{R}, \mathbf{r}) \phi(\mathbf{r}, t) d\mathbf{r} \right) \chi. \quad (2.17)$$

These equations constitute the foundation for the TDSCF approach (*time-dependent self-consistent field*) introduced by Dirac in 1930, see [181, 186]. Both unknowns again obey a Schrödinger equation, but now with a time-dependent effective operator for the potential energy which arises as an appropriate average of the other unknown. These averages can also be interpreted as quantum mechanical expectation values with respect to the operators V_e and $\mathcal{H}_e$ and give a mean field description of the coupled dynamics.

As a next step the nuclear wave function χ is to be approximated by classical point particles. For this, we first write the wave function χ as

$$\chi(\mathbf{R}, t) = A(\mathbf{R}, t) \exp\left[\frac{i}{\hbar} S(\mathbf{R}, t) \right] \quad (2.18)$$

with an amplitude $A > 0$ and a phase factor S, both real [187, 427, 536]. Substitution into the equation for the nuclei in the TDSCF system (2.17) and separating real and imaginary parts leads to the coupled system of equations

$$\frac{\partial S}{\partial t} + \sum_k^N \frac{1}{2M_k} \left(\nabla_{\mathbf{R}_k} S \right)^2 + \int \phi^* \mathcal{H}_e \phi d\mathbf{r} = \hbar^2 \sum_k^N \frac{1}{2M_k} \frac{\Delta_{\mathbf{R}_k} A}{A}, \quad (2.19)$$

$$\frac{\partial A}{\partial t} + \sum_k^N \frac{1}{M_k} \left(\nabla_{\mathbf{R}_k} A \right) \left(\nabla_{\mathbf{R}_k} S \right) + \sum_k^N \frac{1}{2M_k} A \left(\Delta_{\mathbf{R}_k} S \right) = 0. \quad (2.20)$$

Here, $\nabla_{\mathbf{R}_k} = \left(\frac{\partial}{\partial (\mathbf{R}_k)_1}, \frac{\partial}{\partial (\mathbf{R}_k)_2}, \frac{\partial}{\partial (\mathbf{R}_k)_3} \right)^T$. The abbreviation $(\nabla_{\mathbf{R}_k} S)^2$ denotes the scalar product of $\nabla_{\mathbf{R}_k} S$ with itself and $(\nabla_{\mathbf{R}_k} A)(\nabla_{\mathbf{R}_k} S)$ denotes the scalar product of the vectors $\nabla_{\mathbf{R}_k} A$ and $\nabla_{\mathbf{R}_k} S$. This system corresponds exactly to the second equation in the TDSCF system (2.17) in the new variables A and S.[5] The only term that directly depends on $\hbar$ is the right hand side of

---

[5] This is the so-called quantum fluid dynamics representation [182, 187, 427, 536, 668] which opens up another possibility to treat the time-dependent Schrödinger equation. (2.20) can be written with $|\chi|^2 \equiv A^2$ as continuity equation that locally conserves the probability density $|\chi|^2$ of the nuclei under a flow.

equation (2.19). In the limit $\hbar \to 0$ equation (2.19) gives[6]

$$\frac{\partial S}{\partial t} + \sum_{k}^{N} \frac{1}{2M_k} (\nabla_{\mathbf{R}_k} S)^2 + \int \phi^* \mathcal{H}_e \phi d\mathbf{r} = 0.^7 \qquad (2.21)$$

Setting $\nabla_{\mathbf{R}} S = (\nabla_{\mathbf{R}_1} S, \ldots, \nabla_{\mathbf{R}_N} S)$, this is isomorphic to the Hamilton-Jacobi form

$$\frac{\partial S}{\partial t} + H(\mathbf{R}, \nabla_{\mathbf{R}} S) = 0 \qquad (2.22)$$

of the equations of motion of classical mechanics with the classical Hamilton function[8]

$$H(\mathbf{R}, \mathbf{P}) = T(\mathbf{P}) + V(\mathbf{R}) \qquad (2.23)$$

with $\mathbf{P} = (\mathbf{P}_1, \ldots, \mathbf{P}_N)$, where one puts

$$\mathbf{P}_k(t) \equiv \nabla_{\mathbf{R}_k} S(\mathbf{R}(t), t).$$

Here, $\mathbf{R}$ corresponds to generalized coordinates and $\mathbf{P}$ to their conjugated moments. Newton's equations of motion $\dot{\mathbf{P}}_k = -\nabla_{\mathbf{R}_k} V(\mathbf{R})$ associated to equation (2.22) are then

$$\frac{d\mathbf{P}_k}{dt} = -\nabla_{\mathbf{R}_k} \int \phi^* \mathcal{H}_e \phi d\mathbf{r} \quad \text{or} \qquad (2.24)$$

$$M_k \ddot{\mathbf{R}}_k(t) = -\nabla_{\mathbf{R}_k} \int \phi^* \mathcal{H}_e \phi d\mathbf{r} \qquad (2.25)$$

$$=: -\nabla_{\mathbf{R}_k} V_e^{Ehr}(\mathbf{R}(t)). \qquad (2.26)$$

The nuclei move now according to the laws of classical mechanics in an effective potential given by the electrons. This so-called *Ehrenfest potential* V_e^{Ehr} is a function of the nuclear coordinates $\mathbf{R}$ at time t. It results from an averaging over the degrees of freedom of the electrons, weighted by $\mathcal{H}_e$, where the nuclear coordinates are kept constant at their current positions $\mathbf{R}(t)$.

There is still the wave function χ of the nuclei in the equation for the electrons in the system for the TDSCF approach (2.16). Consistency requires it to be replaced by the position of the nuclei. Thus, if one replaces the probability density of the nuclei $|\chi(\mathbf{R}, t)|^2$ by the product of delta functions

---

[6] Because of this approximation step, the function ϕ is only an approximation of the original wave function ϕ in (2.16) and (2.17). To keep the notation simple we denote this approximation again by the symbol ϕ.

[7] An expansion of the right hand side of equation (2.19) with respect to $\hbar$ leads to a hierarchy of semi-classical methods [427].

[8] In the literature often the notation $\mathbf{Q}$ is found instead of $\mathbf{R}$ for the generalized classical coordinates. For the sake of simplicity, we will continue using the notation $\mathbf{R}$ in this chapter.

$\Pi_k \delta(\mathbf{R}_k - \mathbf{R}_k(t))$ in the limit $\hbar \to 0$ in (2.16), then one obtains for example for the position operator $\mathbf{R}_k$ with

$$\int \chi^*(\mathbf{R}, t)\mathbf{R}_k\chi(\mathbf{R}, t)d\mathbf{R} \quad \xrightarrow{\hbar\to 0} \quad \mathbf{R}_k(t) \tag{2.27}$$

the classical position $\mathbf{R}_k(t)$ as limit of the quantum mechanical expectations. Here, the delta functions are centered in the instantaneous positions $\mathbf{R}(t)$ of the nuclei given by (2.25). For (2.16), this classical limit process[9] leads to a time-dependent wave equation for the electrons

$$i\,\hbar\frac{\partial\phi_{\mathbf{R}(t)}(\mathbf{r}, t)}{\partial t} = -\sum_k \frac{\hbar^2}{2m_e}\Delta_{\mathbf{r}_k}\phi_{\mathbf{R}(t)}(\mathbf{r}, t) + V_e(\mathbf{R}(t), \mathbf{r})\phi_{\mathbf{R}(t)}(\mathbf{r}, t) \tag{2.28}$$

$$= \mathcal{H}_e(\mathbf{R}(t), \mathbf{r})\phi_{\mathbf{R}(t)}(\mathbf{r}, t), \tag{2.29}$$

that move in a self-consistent way with the nuclei, if the classical nuclei are propagated by (2.25). Note that now $\mathcal{H}_e$ and therefore the wave function ϕ of the electrons depend parametrically via V_e on the positions $\mathbf{R}(t)$ of the nuclei. The nuclei are thus treated as classical particles, whereas the electrons are still treated using quantum mechanics. In honor of Ehrenfest, who first posed the question how Newton's classical dynamics could be derived from Schrödinger's equation, one often calls approaches that are based on the equations

$$M_k\ddot{\mathbf{R}}_k(t) = -\nabla_{\mathbf{R}_k}V_e^{Ehr}(\mathbf{R}(t)), \tag{2.30}$$

$$i\,\hbar\frac{\partial\phi_{\mathbf{R}(t)}(\mathbf{r}, t)}{\partial t} = \mathcal{H}_e(\mathbf{R}(t), \mathbf{r})\phi_{\mathbf{R}(t)}(\mathbf{r}, t) \tag{2.31}$$

Ehrenfest molecular dynamics. Alternatively, one finds such approaches in the literature under the name QCMD (quantum-classical molecular dynamics model) [104, 204, 447]. Note again that the wave function $\phi_{\mathbf{R}(t)}$ of the electrons is here not equal to the wave function ϕ in (2.16), since an approximation was introduced by the limit process for the positions of the nuclei. The wave function of the electrons depends implicitly on $\mathbf{R}$ via the coupling in the system, which we expressed by the parametric notation $\phi_{\mathbf{R}(t)}$. In the following we will omit this parametrization for the sake of simplicity and will denote, if clear from the context, the electronic wave function just by ϕ.

2.2.2 Expansion in the Adiabatic Basis

The TDSCF approach leads to a mean field theory. One should keep in mind that transitions between different electronic states are still possible in this

---

[9] A justification of the transition from the Schrödinger equation to Newton's equation of motion of the nuclei is given by the theorem of Ehrenfest [381, 554] which describes the time evolution of averages of observables.

setting. This can be seen as follows: We expand the electronic wave function ϕ from (2.31) for fixed t in an appropriate basis $\{\phi_j\}$ of the electronic states

$$\phi_{\mathbf{R}(t)}(\mathbf{r}, t) = \sum_{j=0}^{\infty} c_j(t)\phi_j(\mathbf{R}(t), \mathbf{r}) \tag{2.32}$$

with complex coefficients $\{c_j(t)\}$ and $\sum_j |c_j(t)|^2 \equiv 1$. The $\{|c_j(t)|^2\}$ describe explicitly how the occupancy of the different states j evolves over time. A possible orthonormal basis, called *adiabatic* basis, results from the solution of the time-independent electronic Schrödinger equation

$$\mathcal{H}_e(\mathbf{R}, \mathbf{r})\phi_j(\mathbf{R}, \mathbf{r}) = E_j(\mathbf{R})\phi_j(\mathbf{R}, \mathbf{r}), \tag{2.33}$$

where $\mathbf{R}$ denotes the nuclear coordinates from equation (2.25) at the chosen time t. The values $\{E_j\}$ are here the energy eigenvalues of the electronic Hamilton operator $\mathcal{H}_e(\mathbf{R}, \mathbf{r})$, and the $\{\phi_j\}$ are the associated energy eigenfunctions.

For (2.30) and (2.31) one obtains with the expansion (2.32) the equations of motion in the adiabatic basis (2.33) as [447, 626, 627]

$$M_k\ddot{\mathbf{R}}_k(t) = -\sum_j |c_j(t)|^2 \nabla_{\mathbf{R}_k} E_j - \sum_{j,l} c_j^*(t)c_l(t)\,(E_j - E_l)\,d_k^{jl}, \tag{2.34}$$

$$i\,\hbar\dot{c}_j(t) = c_j(t)E_j - i\,\hbar\sum_{k,l} c_l(t)\dot{\mathbf{R}}_k(t)d_k^{jl}, \tag{2.35}$$

with the coupling terms given as

$$d_k^{jl} = \int \phi_j^* \nabla_{\mathbf{R}_k} \phi_l d\mathbf{r}, \tag{2.36}$$

$$d_k^{jj} \equiv 0. \tag{2.37}$$

Here, we used the properties

$$\int \phi_j^*(\mathbf{R}, \mathbf{r})\nabla_{\mathbf{R}_k}\mathcal{H}_e\phi_l(\mathbf{R}, \mathbf{r})d\mathbf{r} = (E_l(\mathbf{R}) - E_j(\mathbf{R}))\int \phi_j^*(\mathbf{R}, \mathbf{r})\nabla_{\mathbf{R}_k}\phi_l(\mathbf{R}, \mathbf{r})d\mathbf{r},$$

$$\int \phi_j^*(\mathbf{R}, \mathbf{r})\dot{\phi}_l(\mathbf{R}, \mathbf{r})d\mathbf{r} = \sum_{k=1}^{N} \dot{\mathbf{R}}_k(t)\int \phi_j^*(\mathbf{R}, \mathbf{r})\nabla_{\mathbf{R}_k}\phi_l(\mathbf{R}, \mathbf{r})d\mathbf{r}, \ \forall j \neq l,$$

of the adiabatic basis and used furthermore that ϕ and $\mathbf{R}$ in $V_e^{Ehr}(\mathbf{R}(t))$ can be treated as independent variables. This implies that the time-dependent wave function can be represented by a linear combination of adiabatic states and that its evolution in time is described by the Schrödinger equation (2.31). Here, $|c_j(t)|^2$ is the probability density that the system is in state ϕ_j at time point t.[10]

---

[10] This model can be modified by the assumption that the system remains in an adiabatic state until it jumps instantaneously to another adiabatic state. The coefficients $c_j(t)$ and the coupling terms d_k^{jl} serve as a criterion for such a jump. This assumption is made in the so-called *surface-hopping* method [300, 625].

2.2.3 Restriction to the Ground State

As a further simplification we will restrict the whole electronic wave function ϕ to a single state, typically the ground state ϕ_0 of $\mathcal{H}_e$ according to the stationary equation (2.33) with $|c_o(t)|^2 \equiv 1$ as in (2.32). We thus assume that the system remains in the state ϕ_0, and truncate the expansion (2.32) after the first term. This approximation is justified as long as the difference in energy between ϕ_0 and the first excited state ϕ_1 is everywhere large enough compared to the thermal energy $k_B T$ so that transitions to excited states[11] do not play a significant role.[12] The nuclei are then moved according to the equation of motion (2.25) on a single hypersurface of the potential energy

$$V_e^{Ehr}(\mathbf{R}) = \int \phi_0^*(\mathbf{R}, \mathbf{r}) \mathcal{H}_e(\mathbf{R}, \mathbf{r}) \phi_0(\mathbf{R}, \mathbf{r}) d\mathbf{r} \equiv E_0(\mathbf{R}). \qquad (2.38)$$

To compute this surface, the time-independent electronic Schrödinger equation (2.33)

$$\mathcal{H}_e(\mathbf{R}, \mathbf{r}) \phi_0(\mathbf{R}, \mathbf{r}) = E_0(\mathbf{R}) \phi_0(\mathbf{R}, \mathbf{r}) \qquad (2.39)$$

has to be solved for its ground state. Hence, we identified the Ehrenfest potential function V_e^{Ehr} just as the potential E_0 of the stationary electronic Schrödinger equation for the ground state. Note that E_0 is here a function of the nuclear coordinates $\mathbf{R}$.

2.2.4 Approximation of the Potential Energy Hypersurface and Classical Molecular Dynamics

As a consequence of (2.38), the computation of the dynamics of the nuclei can now be *separated* from the computation of the hypersurface for the potential energy. If we assume at first that we can solve the stationary electronic Schrödinger equation (2.33) for a given nuclear configuration, then we could derive an entirely classical approach by the following steps: First, the energy of the ground state $E_0(\mathbf{R})$ is determined for as many representative nuclear configurations $\mathbf{R}^j$ as possible from the stationary electronic Schrödinger equation (2.39). In this way, we evaluate the function $V_e^{Ehr}(\mathbf{R})$ at a number of points and gain a number of data points $(\mathbf{R}^j, V_e^{Ehr}(\mathbf{R}^j))$. From these discrete data points we then approximately reconstruct the global potential energy hypersurface for V_e^{Ehr}. For this, we compute an approximate potential surface by an expansion of many-body potentials in analytical form

---

[11] In the case of bound atoms the spectrum is discrete. The ground state is an eigenstate with the smallest energy level. The first excited state is an eigenstate with the second smallest energy level.

[12] So-called branching processes cannot be described this way in a satisfactory manner.

$$V_e^{Ehr} \approx V_e^{appr}(\mathbf{R}) = \sum_{k=1}^{N} V_1(\mathbf{R}_k) + \sum_{k<l}^{N} V_2(\mathbf{R}_k, \mathbf{R}_l) + \sum_{k<l<m}^{N} V_3(\mathbf{R}_k, \mathbf{R}_l, \mathbf{R}_m) + \dots,$$

$$(2.40)$$

which is appropriately truncated. With such an expansion the electronic degrees of freedom are replaced with interaction potentials V_n and are therefore no longer explicit degrees of freedom of the equations of motion. After the V_n are specified, the mixed quantum-mechanical and classical problem (2.30), (2.31) is reduced to a completely classical problem. We obtain Newton's equations of motion of classical molecular dynamics

$$M_k \ddot{\mathbf{R}}_k(t) = -\nabla_{\mathbf{R}_k} V_e^{appr}(\mathbf{R}(t)). \tag{2.41}$$

Here, the gradients can be computed analytically.

This method of classical molecular dynamics is feasible for many-body systems because the global potential energy gets decomposed according to (2.40). Here, in practice, the same form of the potential is used for the same kind of particles. For instance, if only a two-body potential function

$$V_e^{appr} \approx \sum_{k<l}^{N} V_2(\|\mathbf{R}_k - \mathbf{R}_l\|)$$

of the distance is used, only *one* one-dimensional function V_2 has to be determined.

This is certainly a drastic approximation that has to be justified in many respects and that brings a number of problems with it. It is not obvious how many and which typical nuclear configurations have to be considered to reconstruct the potential function from the potentials of these configurations with an error which is not too large. In addition, the error caused by the truncation of the expansion (2.40) plays certainly a substantial role. The precise form of the analytic potential functions V_n and the subsequent fitting of their parameters also have a decisive influence on the size of the approximation error. The assumption that the global potential function is represented well by a sum of simple potentials of a few generic forms and the transferability of a potential function to other nuclear configurations are further critical issues. Altogether, not all approximation errors can be controlled rigorously in this approach. Furthermore, quantum mechanical effects and therefore chemical reactions are excluded by construction. Nevertheless, the method has been proven successful, in particular in the computation of macroscopic properties.

The methods used in practice to determine the interactions in real systems are either based on the approximate solution of the stationary electronic Schrödinger equation (ab initio methods) and subsequent force-matching [208] or on the fitting (that is, parametrization) of given analytic potentials to experimental or quantum mechanical results. In the first approach, the potential is constructed implictly using ab initio methods. There, the electronic energy E_0 and the corresponding forces are computed approximately[13] for a number of chosen example configurations of the nuclei. By extrapolation/interpolation to other configurations an approximate potential energy hypersurface can be constructed that can in turn be approximated by simple analytic functions. In the second, more empirical approach, one directly chooses an analytic form of the potential which contains certain form functions that depend on geometric quantities such as distances, angles or coordinates of particles. Subsequently, this form is fitted by an appropriate determination of its parameters to available results from quantum mechanical computations or from actual experiments. In this way one can model interactions that incorporate different kinds of bond forces, possible constraints, conditions on angles, etc. If the results of the simulation are not satisfactory, the potentials have to be improved by the choice of better parameters or by the selection of better forms of the potential functions with other or even extended sets of parameters. The construction of good potentials is still a form of art and requires much skill, work, and intuition. Programs such as GULP [5, 244] or THBFIT [6] can help in the creation of new forms of potentials and in the fitting of parameters for solids and crystals.

Some Simple Potentials. The simplest interactions are those between two particles. Potentials that only depend on the distance $r_{ij} := \|\mathbf{R}_j - \mathbf{R}_i\|$ between any pair of particles are called pair potentials. Here, we use $(\mathbf{R}_1, \ldots, \mathbf{R}_N)$ as a notation for the classical coordinates $\mathbf{R}(t)$. The associated potential energy V has the form

$$V(\mathbf{R}_1, \ldots, \mathbf{R}_N) = \sum_{i=1}^{N} \sum_{j=i+1}^{N} U_{ij}(r_{ij}),$$

where U_{ij} denotes the potential acting between the particles i and j. Examples for such pair potentials U_{ij} between two particles are:

– **The Gravitational Potential**

$$U(r_{ij}) = -G_{Grav} \frac{m_i m_j}{r_{ij}}. \tag{2.42}$$

---

[13] The wave function in the electronic Schrödinger equation is still defined in a high-dimensional space. The coordinates of the electrons are in $\mathbb{R}^{3K}$. An analytic solution or an approximation by a conventional numerical discretization method is impossible in general. Therefore, approximation methods have to be used that substantially reduce the dimension of the problem. Over the years, many variants of such approximation methods have been proposed and used, such as the Hartree-Fock method, the density functional theory, configuration interaction methods, coupled-cluster methods, generalized valence bond techniques, the tight-binding approach, or the Harris functional method. An overview of the different approaches can be found for example in [526, 528].

– **The Coulomb Potential**

$$U(r_{ij}) = \frac{1}{4\pi\varepsilon_0} \frac{q_i q_j}{r_{ij}}. \tag{2.43}$$

– **The van der Waals Potential**

$$U(r_{ij}) = -a \left(\frac{1}{r_{ij}}\right)^6.$$

– **The Lennard-Jones Potential**

$$U(r_{ij}) = \alpha\varepsilon \left[\left(\frac{\sigma}{r_{ij}}\right)^n - \left(\frac{\sigma}{r_{ij}}\right)^m \right], \quad m < n. \tag{2.44}$$

Here, α is given as $\alpha = \frac{1}{n-m} \left(\frac{n^n}{m^m}\right)^{\frac{1}{n-m}}$. This potential is parametrized by σ and ε. The value ε describes the depth of the potential and thereby the strength of the repulsive and attractive forces. Materials of different strength can be simulated in this way. Increasing ε leads to stronger bonds and therefore harder materials. The value σ parametrizes the zero crossing of the potential. With $m = 6$ (as in the van der Waals force) and $n = 12$ the Lennard-Jones potential – as well as the resulting force – decreases very rapidly with increasing distance. Here, the choice $n = 12$ does not stem from physical considerations but merely from mathematical simplicity. For $(m, n) = (10, 12)$ we obtain the related potential function

$$U(r_{ij}) = A/r_{ij}^{12} - B/r_{ij}^{10},$$

which allows the empirical modeling of hydrogen bonds. The parameters A and B depend on the kind of the particular hydrogen bond and are in general fitted to experimental data.

– **The Morse Potential**

$$U(r_{ij}) = D(1 - e^{-a(r_{ij} - r_0)})^2. \tag{2.45}$$

D is the dissociation energy of the bond, a is an appropriately chosen parameter which depends on the frequency of the bond vibrations, and r_0 is a reference length.

– **Hooke's Law** (Harmonic Potential)

$$U(r_{ij}) = \frac{k}{2}(r_{ij} - r_0)^2.$$

Note that we omitted the indices i, j in the notation for the potentials U.

These simple potentials are certainly limited in their applications. However, noble gases can be represented well this way since their atoms are only attracted to each other by the van der Waals force. These simple potentials are also used outside of molecular dynamics, as for instance in the simulation of fluids on the microscale. However, more complex kinds of interactions, such as the ones that occur in metals or molecules, can not be simulated with such potentials in a realistic manner [209]. For this, other kinds of potential functions are needed that include interactions between several atoms of a molecule.

Since the eighties such many-body interactions have been introduced as potential functions. The various approaches involve density and coordination number, respectively, and exploit the idea that bonds are the weaker the higher the local density of the particles is. This led to the development of potentials with additional terms that most often consist of two components, a two-body part and a part which takes the coordination number (that is, the local density of particles) into account. Examples of such potentials are the glue model [209], the embedded atom method [174], the Finnis-Sinclair potential [232] and also the so-called effective-medium theory [336]. All these approaches differ strongly in the way how the coordination number is used in the construction of the potential. Sometimes different parametrizations are obtained even for the same material because of the different constructions. Special many-body potentials have been developed specifically for the study of crack propagation in materials [593].

Still more complex potentials are needed for instance for the modeling of semiconductors such as silicon. The potentials developed for these materials also use the concept of coordination number and bond order, that means that the strength of the bond depends on the local neighborhood. These potentials share a strong connection with the glue models. Stillinger and Weber [584] use a two-body and an additional three-body term in their potential. The family of potentials developed by Tersoff [603] was modified slightly by Brenner [122] and used in a similar form also in the modeling of hydrocarbons.

2.3 An Outlook on the Methods of Ab Initio Molecular Dynamics

Until now we have employed approximation methods for the approximate solution of the electronic Schrödinger equation only to obtain data for the specification and fitting of analytical potential function for the methods of classical molecular dynamics. But they can also be used in each time step of Newton's equation to directly compute the potential energy hypersurface for the actual nuclear coordinates. This is the basic idea of the so-called ab initio molecular dynamics. One solves the electronic Schrödinger equation approximately to determine the effective potential energy of the nuclei. From it one can compute the forces on the nuclei and move the nuclei according to Newton's equation of motion given these forces. This principle in its different variants forms the basis of the Ehrenfest molecular dynamics, the Born-Oppenheimer molecular dynamics and the Car-Parinello method.

Ehrenfest Molecular Dynamics. We consider again equations (2.30), (2.31) and assume that the system remains in a single adiabatic state, typically the ground state ϕ_0. Then, one obtains

$$M_k\ddot{\mathbf{R}}_k(t) = -\nabla_{\mathbf{R}_k} \int \phi_0^*(\mathbf{R}(t),\mathbf{r})\mathcal{H}_e(\mathbf{R}(t),\mathbf{r})\phi_0(\mathbf{R}(t),\mathbf{r})d\mathbf{r} \quad (2.46)$$

$$= -\nabla_{\mathbf{R}_k} V_e^{Ehr}(\mathbf{R}(t)),$$

$$i\,\hbar\frac{\partial\phi_0(\mathbf{R}(t),\mathbf{r})}{\partial t} = \mathcal{H}_e\phi_0(\mathbf{R}(t),\mathbf{r}), \qquad\qquad (2.47)$$

where $\phi_{\mathbf{R}(t)}(\mathbf{r},t) = c_0(t)\phi_0(\mathbf{R}(t),\mathbf{r})$ was assumed with $|c_0(t)|^2 \equiv 1$, compare (2.32).

Born-Oppenheimer Molecular Dynamics. In the derivation of the so-called Born-Oppenheimer molecular dynamics one uses the large difference in masses between electrons and atomic nuclei. The ratio[14] of the velocity v_K of a nucleus to the velocity of an electron v_e is in general smaller than 10^{-2}. Therefore, one assumes that the electrons adapt instantaneously to the changed nuclear configuration and so are always in the quantum mechanical ground state associated to the actual position of the nuclei. The movement of the nuclei during the adaptation of the electron movement is negligibly small in the sense of classical dynamics. This justifies to set

$$\Psi(\mathbf{R},\mathbf{r},t) \approx \Psi^{BO}(\mathbf{R},\mathbf{r},t) := \sum_{j=0}^{\infty} \chi_j(\mathbf{R},t)\phi_j(\mathbf{R},\mathbf{r}), \qquad (2.48)$$

which allows to separate the fast from the slow variables. In contrast to (2.14) the electronic wave functions $\phi_j(\mathbf{R},\mathbf{r})$ depend no longer on time but depend on the nuclear coordinates $\mathbf{R}$. Using a Taylor expansion of the stationary Schrödinger equation and several approximations that rely on the difference in masses between electrons and nuclei, see for example Chapter 8.4 in [546], the stationary Schrödinger equation can be separated into two equations, the electronic Schrödinger equation and an equation for the nuclei. The first equation describes how the electrons behave when the position of the nuclei is fixed. Its solution leads to an effective potential that appears in the equation for the nuclei and describes the effect of the electrons on the interaction between the nuclei. After restriction to the ground state and further approximations, the Born-Oppenheimer molecular dynamics results which is given by the equations

$$M_k\ddot{\mathbf{R}}_k(t) = -\nabla_{\mathbf{R}_k} \min_{\phi_0}\left\{\int \phi_0^*(\mathbf{R}(t),\mathbf{r})\mathcal{H}_e(\mathbf{R}(t),\mathbf{r})\phi_0(\mathbf{R}(t),\mathbf{r})d\mathbf{r}\right\}$$

$$=: -\nabla_{\mathbf{R}_k} V_e^{BO}(\mathbf{R}(t)), \qquad\qquad (2.49)$$

---

[14] The ratio of the mass m_e of an electron and the mass M_K of a nucleus is – except for hydrogen and helium – smaller than 10^{-4}. Furthermore, according to classical kinetic gas theory, the energy per degree of freedom of non-interacting particles is the same, thus it holds $m_e v_e^2 = M_K v_K^2$.

$$\mathcal{H}_e(\mathbf{R}(t), \mathbf{r})\phi_0(\mathbf{R}(t), \mathbf{r}) = E_0(\mathbf{R}(t))\phi_0(\mathbf{R}(t), \mathbf{r}).$$

With the forces $\mathbf{F}_k(t) = M_k\ddot{\mathbf{R}}_k(t)$ acting on the nuclei, their positions can be moved according to the laws of classical mechanics.[15]

In our case, in which we consider the ground state and neglect all coupling terms, the Ehrenfest potential V_e^{Ehr} agrees, according to equation (2.38), with the Born-Oppenheimer potential V_e^{BO}. However, the dynamics is fundamentally different. In the Born-Oppenheimer method, the computation of the electron structure is reduced to the solution of the stationary Schrödinger equation, which then is used to compute the forces acting at that time on the nuclei so that the nuclei can be moved according to the laws of classical molecular dynamics. The time-dependency of the state of the electrons is here exclusively a consequence of the classical motion of the nuclei and not, as in the case of the Ehrenfest molecular dynamics, determined from the time-dependent Schrödinger equation in the coupled system of equations (2.46). In particular the time evolution of the state of the electrons in the Ehrenfest method corresponds to a unitary propagation [360, 361, 375]. If the initial state is minimal, its norm and minimality are maintained [218, 605]. This is not true for the Born-Oppenheimer dynamics in which a minimization is needed in every time step.

A further difference of the two methods is the following: Let us assume that particle functions ψ_{α_i} are given from which, as for instance in the Hartree-Fock method,[16] with $\mathbf{r} = (\mathbf{r}_1, \ldots, \mathbf{r}_K)$ product functions $\psi_{\alpha_1 \ldots \alpha_K}^{SD}$, the so-called Slater determinants,[17] are formed by

$$\psi_{\alpha_1 \ldots \alpha_K}^{SD}(\mathbf{r}, t) = \frac{1}{\sqrt{K!}} \det \begin{vmatrix} \psi_{\alpha_1}(\mathbf{r}_1, t) & \psi_{\alpha_1}(\mathbf{r}_2, t) & \ldots & \psi_{\alpha_1}(\mathbf{r}_K, t) \\ \psi_{\alpha_2}(\mathbf{r}_1, t) & \psi_{\alpha_2}(\mathbf{r}_2, t) & \ldots & \psi_{\alpha_2}(\mathbf{r}_K, t) \\ \cdot & \cdot & \cdot & \cdot \\ \cdot & \cdot & \cdot & \cdot \\ \cdot & \cdot & \cdot & \cdot \\ \psi_{\alpha_K}(\mathbf{r}_1, t) & \psi_{\alpha_K}(\mathbf{r}_2, t) & \ldots & \psi_{\alpha_K}(\mathbf{r}_K, t) \end{vmatrix}. \quad (2.50)$$

For an approximate solution of the electronic Schrödinger equation one now expands the ground state $\phi_0(\mathbf{R}(t), \mathbf{r})$ with help of these products of particle functions as

$$\phi_0(\mathbf{R}(t), \mathbf{r}) = \sum_{\alpha_1, \ldots, \alpha_K} \gamma_{\alpha_1, \ldots, \alpha_K}(t)\psi_{\alpha_1 \ldots \alpha_K}^{SD}(\mathbf{r}, t) \quad (2.51)$$

with the coefficients

---

[15] There is also the approach to apply this method to every excited state ϕ_j without taking interferences into account, i.e., to proceed analogously to (2.33-2.36) and to neglect all or only certain coupling terms [304, 359].

[16] In density functional theory one uses a different kind of function for the particles, but the principle is the same.

[17] This means that the spin of the particles is neglected here.

$$\gamma_{\alpha_1,\ldots,\alpha_K}(t) := \int \psi^{*SD}_{\alpha_1\ldots\alpha_K}(\mathbf{r}, t)\phi_0(\mathbf{R}(t), \mathbf{r})d\mathbf{r}. \tag{2.52}$$

Then, one has to minimize in equation (2.49) under the constraint that the particle functions are orthonormal, $\int \psi^*_{\alpha_i}\psi_{\alpha_j}dr = \delta_{\alpha_i\alpha_j}$, since this is a necessary requirement for the expansion (2.51). Since the time evolution of the electrons under the Ehrenfest dynamics is a unitary propagation, the particle functions remain orthonormal if they were orthonormal at the initial time.

Car-Parrinello Molecular Dynamics. The advantage of the Ehrenfest dynamics is that the wave function stays minimal with respect to the current position of the nuclei. The disadvantage is that the size of the time step is determined by the motion of the electrons and is therefore "small". The size of the time step in the Born-Oppenheimer dynamics is determined by the motion of the nuclei, on the other hand, and is therefore certainly "larger". The disadvantage however is that a minimization is required in each time step. The Car-Parrinello molecular dynamics [137, 469] attempts to combine the advantages of both methods and to avoid their disadvantages. The fundamental idea is to transform the quantum mechanical separation of the time scales of the "fast" electrons and the "slow" nuclei into a classical adiabatic separation of energy scales within the theory of dynamical systems and to neglect the explicit time-dependency of the motion of the electrons [106, 465, 466, 515].

To understand the idea, we consider at first again the Ehrenfest and Born-Oppenheimer dynamics. If restricted to the ground state $\phi_0(\mathbf{R}, \mathbf{r})$, the central quantity

$$V_{El}(\mathbf{R}) := \int \phi^*_0(\mathbf{R}, \mathbf{r})\mathcal{H}_e(\mathbf{R}, \mathbf{r})\phi_0(\mathbf{R}, \mathbf{r})d\mathbf{r} = E_0(\mathbf{R})$$

is a function of the position of the nuclei $\mathbf{R}$. From the Lagrange function of classical mechanics for the motion of the nuclei

$$L(\mathbf{R}, \dot{\mathbf{R}}) = \sum_k^N \frac{1}{2}M_k\dot{\mathbf{R}}_k^2 - V_{El}(\mathbf{R}), \tag{2.53}$$

we obtain, using the appropriate Euler-Lagrange equations $\frac{d}{dt}\frac{\partial L}{\partial \dot{\mathbf{R}}_k} = \frac{\partial L}{\partial \mathbf{R}_k}$, the equation of motion (2.49)

$$M_k\ddot{\mathbf{R}}_k(t) = -\nabla_{\mathbf{R}_k}E_0(\mathbf{R}(t)). \tag{2.54}$$

One can regard the energy of the ground state $E_0 = V_{El}$ also as a functional of the wave function ϕ_0. If the wave function ϕ_0 has an expansion with now time-dependent particle functions $\{\psi_i(\mathbf{r}, t)\}$, analog to the expansion (2.51) in (one or several) Slater determinants (2.50), V_{El} can also be seen as a functional of the orbitals $\{\psi_i(\mathbf{r}, t)\}$. The force acting on the nuclei is obtained in classical mechanics as the derivative of a Lagrange function with respect to the positions of the nuclei. If one now also views the orbitals as "classical particles",[18] one can determine the forces acting on the orbitals as the functional derivative of an appropriate Lagrange function with respect to the orbitals. Then, a purely classical approach results in a Lagrange function of the form [137]

$$L_{CP}(\mathbf{R}, \dot{\mathbf{R}}, \{\psi_i\}, \{\dot{\psi}_i\}) = \tag{2.55}$$

$$\sum_k \frac{1}{2} M_k \dot{\mathbf{R}}_k^2 + \sum_i \frac{1}{2} \mu_i \int \dot{\psi}_i^* \dot{\psi}_i d\mathbf{r} - V_{El}(\mathbf{R}, \{\psi_i\}) + \varphi(\mathbf{R}, \{\psi_i\})$$

with the "fictitious masses" μ_i of the orbitals $\{\psi_i\}$ and a general, appropriately chosen constraint φ. A simple example for such a constraint is the orthonormality of the orbitals. This yields

$$\varphi(\mathbf{R}, \{\psi_i\}) = \sum_{i,j} \lambda_{ij} \left(\int \psi_i^* \psi_j d\mathbf{r} - \delta_{ij} \right)$$

with the Lagrange multipliers λ_{ij}. In this simple case φ does not depend (plane wave basis) or does only implicitly depend (Gaussian basis) on $\mathbf{R}(\mathbf{t})$. The respective Euler-Lagrange equations

$$\frac{d}{dt} \frac{\partial L}{\partial \dot{\mathbf{R}}_k} = \frac{\partial L}{\partial \mathbf{R}_k}, \quad \frac{d}{dt} \frac{\delta L}{\delta \dot{\psi}_i^*} = \frac{\delta L}{\delta \psi_i^*} \tag{2.56}$$

give Newton's equations of motion[19]

$$M_k \ddot{\mathbf{R}}_k(t) = -\nabla_{\mathbf{R}_k} \int \phi_0^* \mathcal{H}_e \phi_0 d\mathbf{r} + \nabla_{\mathbf{R}_k} \varphi(\mathbf{R}, \{\psi_i\}), \tag{2.57}$$

$$\mu_i \ddot{\psi}_i(\mathbf{r}, t) = -\frac{\delta}{\delta \psi_i^*} \int \phi_0^* \mathcal{H}_e \phi_0 d\mathbf{r} + \frac{\delta}{\delta \psi_i^*} \varphi(\mathbf{R}, \{\psi_i\}). \tag{2.58}$$

The nuclei move according to a physical temperature proportional to the kinetic energy $\sum_k M_k \dot{\mathbf{R}}_k^2$ of the nuclei. In contrast, the electrons move according to a "fictitious temperature" proportional to the fictitious kinetic energy $\sum_i \mu_i \int \dot{\psi}_i^* \dot{\psi}_i d\mathbf{r}$ of the orbitals.[20]

Let the initial state ϕ_0 at time t_0 be exactly the ground state. For a "low temperature of the electrons" the electrons move almost exactly on the Born-Oppenheimer surface. But the "temperature of the electrons" has to be "high" enough so that the electrons can adjust to the motion of the nuclei. The problem in practice is the "right temperature control". The subsystem of the physical motion of the nuclei described by equation (2.57) and the

---

[18] For this, one treats the orbitals in the context of a classical field theory.
[19] $\psi_i^*(\mathbf{r}, t)$ and $\psi_i(\mathbf{r}, t)$ are linearly independent for complex variations.
[20] The physical kinetic energy of the electrons is included in E_0.

subsystem of the fictitious orbital motions described by equation (2.58) have to be separated in such a way that the fast electronic subsystem stays "cold" for a long time and nevertheless immediately adjusts to the slow motion of the nuclei, while keeping the nuclei at the same time at their physical temperature (which is much higher). In particular, there is no transfer of energy allowed between the physical subsystem of the ("hot") nuclei and the fictitious subsystem of the ("cold") electrons. It is possible to satisfy these requirements if the force spectrum of the degrees of freedom of the electrons $f(\omega) = \int_0^\infty \cos(\omega t) \left(\sum_i \int \dot{\psi}_i^*(\mathbf{r}, t)\psi_i(\mathbf{r}, 0)dr \right) dt$ and that of the nuclei do not overlap in any range of frequencies [515]. In [106] it could be shown that the absolute error of the Car-Parrinello trajectory can be controlled relative to the trajectory determined by the exact Born-Oppenheimer surface by using the parameters μ_i.

The Hellmann-Feynman Theorem. In the molecular dynamics methods described above the force acting on a nucleus has to be determined according to the equations (2.46), (2.49) and (2.57). A direct numerical evaluation of the derivative

$$\mathbf{F}_k(\mathbf{R}) = -\nabla_{\mathbf{R}_k} \int \phi_0^* \mathcal{H}_e \phi_0 dr,$$

for instance using a finite difference approximation, is too expensive on the one hand and too inaccurate for dynamical simulations on the other hand. It is therefore desirable to evaluate the derivative analytically and to apply it directly to the different parts of $\mathcal{H}_e$. This is made possible by the following approach: Let q be any coordinate $(\mathbf{R}_k)_i, i \in \{1, 2, 3\}$ of any component $\mathbf{R}_k$ of $\mathbf{R}$. Keep now all other components of $\mathbf{R}$ and the other two coordinates of $\mathbf{R}_k$ fixed and only allow q to vary. Then, the electronic Hamilton operator $\mathcal{H}_e(\mathbf{R}, \mathbf{r}) = \mathcal{H}(q)$ depends on q (besides $\mathbf{r}$) according to equation (2.13) via the operators $V_{eK}(\mathbf{R}, \mathbf{r})$ and $V_{KK}(\mathbf{R})$. By the stationary electronic Schrödinger equation

$$\mathcal{H}(q)\phi_0(q) = E_0(q)\phi_0(q) \tag{2.59}$$

therefore also the state of the electrons ϕ_0 (beside $\mathbf{r}$) and the energy E_0 depend on q. If the electronic state is assumed to be normalized, that is, it satisfies $\int \phi_0^* \phi_0 dr = 1$, then a translation by q results in a force $F(q)$ of[21]

$$-F(q) = \frac{dE_0(q)}{dq} = \int \phi_0^*(q) \frac{d\mathcal{H}(q)}{dq} \phi_0(q) dr. \tag{2.60}$$

The justification for this result is provided by the **Hellmann-Feynman Theorem**:[22] Let $\phi_j(q)$ be the normalized eigenfunction of a self-adjoint op-

---

[21] An analogous results holds for the excited states ϕ_j with the associated eigenvalues $E_j(q)$ and the associated Schrödinger equation $\mathcal{H}(q)\phi_j(q) = E_j(q)\phi_j(q)$.

[22] The so-called Hellmann-Feynman theorem for quantum mechanical forces was proven originally in 1927 by Ehrenfest [204], it was discussed later by Hellman [311] and was rediscovered independently by Feynmann [225] in 1939.

erator $\mathcal{H}(q)$ associated to the eigenvalue $E_j(q)$ and q a real parameter, then it holds that

$$\frac{dE_j(q)}{dq} = \int \phi_j^*(q) \frac{d\mathcal{H}(q)}{dq} \phi_j(q) d\mathbf{r}. \tag{2.61}$$

This can be shown as follows: Using the product rule one obtains

$$\frac{dE_j(q)}{dq} = \int \phi_j^*(q) \frac{d\mathcal{H}(q)}{dq} \phi_j(q) d\mathbf{r} +$$
$$\int \frac{d\phi_j^*(q)}{dq} \mathcal{H}(q) \phi_j(q) d\mathbf{r} + \int \phi_j^*(q) \mathcal{H}(q) \frac{d\phi_j(q)}{dq} d\mathbf{r}.$$

The $\phi_j(q)$ are eigenfunctions associated to the eigenvalue $E_j(q)$, therefore it holds that

$$\frac{dE_j(q)}{dq} = \int \phi_j^*(q) \frac{d\mathcal{H}(q)}{dq} \phi_j(q) d\mathbf{r} +$$
$$E_j(q) \int \frac{d\phi_j^*(q)}{dq} \phi_j(q) d\mathbf{r} + E_j(q) \int \phi_j^*(q) \frac{d\phi_j(q)}{dq} d\mathbf{r}$$
$$= \int \phi_j^*(q) \frac{d\mathcal{H}(q)}{dq} \phi_j(q) d\mathbf{r} + E_j(q) \frac{d}{dq} \int \phi_j^*(q) \phi_j(q) d\mathbf{r}$$

and now the normalization condition for ϕ_j implies the theorem.

This allows a simple numerical computation of the forces between different bound atoms. Because of

$$\mathbf{F}_k(\mathbf{R}) = -\nabla_{\mathbf{R}_k} \int \phi_0^* \mathcal{H}_e \phi_0 d\mathbf{r} = -\int \phi_0^* \nabla_{\mathbf{R}_k} \mathcal{H}_e \phi_0 d\mathbf{r} \tag{2.62}$$

and

$$\nabla_{\mathbf{R}_k} \mathcal{H}_e = \nabla_{\mathbf{R}_k}(V_{ee} + V_{eK} + V_{KK}) = \nabla_{\mathbf{R}_k}(V_{eK} + V_{KK})$$

one obtains the force on the kth nucleus as

$$\mathbf{F}_k(\mathbf{R}) = -\int \phi_0^* \nabla_{\mathbf{R}_k}(V_{eK} + V_{KK}) \phi_0 d\mathbf{r}$$
$$= -\int \phi_0^* \nabla_{\mathbf{R}_k} V_{eK} \phi_0 d\mathbf{r} - \nabla_{\mathbf{R}_k} V_{KK} \tag{2.63}$$
$$= \frac{e^2}{4\pi\epsilon_0} \left(\int \phi_0^* \phi_0 \sum_{i=1}^{K} \sum_{j=1}^{N} \nabla_{\mathbf{R}_k} \frac{Z_j}{\| \mathbf{R}_j - \mathbf{r}_i \|} d\mathbf{r} - \nabla_{\mathbf{R}_k} \sum_{i<j}^{N} \frac{Z_i Z_j}{\| \mathbf{R}_i - \mathbf{R}_j \|} \right).$$

The derivatives now act directly on the potential functions V_{KK} and V_{eK} and can be computed analytically. The force $\mathbf{F}_k = \mathbf{F}_k(\mathbf{R})$ on the kth nucleus therefore results from the Coulomb forces (from the potential V_{KK}) acting between the nuclei and an additional effective force caused by the electrons. This effective force has the form of a Coulomb force induced by a hypothetical electron cloud with a density given by the solution of the electronic Schrödinger equation. In this way, the influence of the electrons on the nuclei is taken into account.

3 The Linked Cell Method for Short-Range Potentials

In Chapter 1 we introduced the particle model, first potential functions and the basic algorithm. Further potentials were presented in Section 2.2.4. So far, we left open how to evaluate the potentials or forces efficiently and how to choose a suitable time integration method. The following chapters will cover these issues. Note that the different methods and algorithms for the evaluation of the forces depend strongly on the kind of the potential used in the model. We will start in this chapter with the derivation of an algorithm for short-range interactions. This approach exploits the fast decay of a short-range potential function and the associated forces. Thus, short-range interactions can be approximated well if only the geometrically closest neighbors of each particle are considered. Note that the algorithm presented here also forms the basis for the methods discussed in the subsequent Chapters 7 and 8 for problems with long-range interactions.

We now consider a system which consists of N particles with masses $\{m_1, \cdots, m_N\}$ characterized by the positions $\{\mathbf{x}_1, \ldots, \mathbf{x}_N\}$ and the associated velocities $\{\mathbf{v}_1, \cdots, \mathbf{v}_N\}$ (respective momenta $\mathbf{p}_i = m_i \mathbf{v}_i$). $\mathbf{x}_i$ and $\mathbf{v}_i$ are here two-dimensional or three-dimensional vectors (one dimension for each direction) and are functions of time t. The space spanned by the degrees of freedom for the positions and velocities is called *phase space*. Each point in the $4N$-dimensional or $6N$-dimensional phase space represents a particular configuration of the system.

We assume that the domain of the simulation is rectangular, that is, $\Omega = [0, L_1] \times [0, L_2]$ in two dimensions, and $\Omega = [0, L_1] \times [0, L_2] \times [0, L_3]$ in three dimensions, respectively, with sides of the lengths L_1, L_2, and L_3. Depending on the specific problem certain conditions are imposed on the boundary that are introduced in the following without giving too many details at first. A more substantial description of these boundary conditions can be found in the subsequent application sections.

In periodic systems, as for example in crystals, it is natural to impose periodicity conditions on the boundaries. Periodic conditions are also used in non-periodic problems to compensate for the limited size of a numerical simulation domain Ω. In that case, the system is extended artificially by periodic continuation to the entire $\mathbb{R}^2$ or $\mathbb{R}^3$, respectively, compare Figure 3.1. Particles that leave the domain at one side reenter the domain at the opposite side. Also, particles located close to opposite sides of the domain interact with each other.

Reflecting boundary conditions arise in the case of a closed simulation box. A particle that gets closer than a certain distance to the wall is subject to a repulsive force which bounces the particle back off the wall.

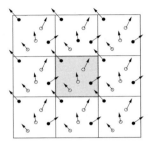

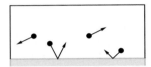

Fig. 3.1. Simulation domain with periodic boundary conditions in two dimensions. The simulation domain shown in grey is replicated in all directions. Particles leaving the simulation domain on one side reenter it at the opposite side.

Fig. 3.2. Reflecting boundary conditions in two dimensions. Particles hitting the boundary of the domain are reflected.

Outflow conditions are used for boundaries where particles can leave the simulation domain, whereas inflow boundary conditions allow new particles to enter the simulation domain at certain times across that boundary.

In addition there are a number of further boundary conditions that are tailored to specific problems. One example is a wall with a fixed given temperature. Particles hitting that wall are reflected but their velocities are changed depending on the temperature of the wall.

We assume now that the evolution in time of the considered system in the domain Ω is described by Hamilton's equations of motion

$$\dot{\mathbf{x}}_i = \nabla_{\mathbf{p}_i}\mathcal{H}, \quad \dot{\mathbf{p}}_i = -\nabla_{\mathbf{x}_i}\mathcal{H}, \ i = 1,\dots,N, \tag{3.1}$$

with the Hamiltonian $\mathcal{H}$. The dot $\dot{}$ denotes, as usually, the partial derivative with respect to time.

If the interactions between the particles are described by a conservative potential[1]

---

[1] In Chapter 2 we used the notation $(\mathbf{R}_1,\dots,\mathbf{R}_N)$ for the classical coordinates. Here and in the following, we will denote the positions of the particles, and thereby the independent variables in the potentials, by $(\mathbf{x}_1,\dots,\mathbf{x}_N)$. Specific examples for possible potentials can be found at the end of Section 2.2.4 as well as in the application Sections 3.6, 3.7.3 and 5.

$$V = V(\mathbf{x}_1, \ldots, \mathbf{x}_N) \tag{3.2}$$

which does not explicitly depend on time, and if Cartesian coordinates and velocities are used, the Hamiltonian reads

$$\mathcal{H}(\mathbf{x}_1, \ldots, \mathbf{x}_N, \mathbf{p}_1, \ldots, \mathbf{p}_N) = \sum_{i=1}^{N} \frac{\mathbf{p}_i^2}{2m_i} + V(\mathbf{x}_1, \ldots, \mathbf{x}_N). \tag{3.3}$$

From Hamilton's equations of motion (3.1) one obtains with $\mathbf{p}_i = m_i \mathbf{v}_i$ directly Newton's equations of motion

$$\begin{aligned} \dot{\mathbf{x}}_i &= \mathbf{v}_i, \\ m_i \dot{\mathbf{v}}_i &= \mathbf{F}_i, \end{aligned} \quad i = 1, \ldots, N, \tag{3.4}$$

or

$$m_i \ddot{\mathbf{x}}_i = \mathbf{F}_i, \quad i = 1, \ldots, N, \tag{3.5}$$

respectively, where the forces $\mathbf{F}_i$ only depend on the coordinates and are given by

$$\mathbf{F}_i = -\nabla_{\mathbf{x}_i} V(\mathbf{x}_1, \ldots, \mathbf{x}_N). \tag{3.6}$$

The expression $\nabla_{\mathbf{x}_i} V(\mathbf{x}_1, \ldots, \mathbf{x}_N)$ is here again an abbreviation for the term $\nabla_{\mathbf{y}} V(\mathbf{x}_1, \ldots, \mathbf{x}_{i-1}, \mathbf{y}, \mathbf{x}_{i+1}, \ldots, \mathbf{x}_N)$ which is evaluated at the point $\mathbf{y} = \mathbf{x}_i$, compare also the comment on page 19. If the initial positions and velocities of the particles are given, the evolution of the system in time depends only on the potential governing the interactions between the particles.[2]

The Hamiltonian (3.3) consists of the potential energy, given by the evaluation of the potential V at the positions of the particles, and the kinetic energy

$$E_{kin} = \sum_{i=1}^{N} \frac{\mathbf{p}_i^2}{2m_i} = \sum_{i=1}^{N} \frac{1}{2} m_i \mathbf{v}_i^2. \tag{3.8}$$

The total energy of the system is then given by $E = E_{kin} + V$ and the total derivative with respect to time reads

---

[2] For greater generality, one can consider the system of ordinary differential equations

$$M\ddot{\mathbf{x}} = -A\mathbf{x} + \mathbf{F}(\mathbf{x}). \tag{3.7}$$

This book's methods for the fast numerical solution of (3.5) can also be used to solve (3.7) efficiently. This generalization makes it possible to treat for example the smoothed particle hydrodynamic method [194, 254, 369, 394, 437] for the Euler and the Navier–Stokes equations or vortex methods [144, 156, 380, 408] for flow problems in the above framework.

$$\frac{dE}{dt} = \frac{dE_{kin}}{dt} + \frac{dV}{dt} = \sum_{i=1}^{N} m_i \mathbf{v}_i \dot{\mathbf{v}}_i + \frac{\partial V}{\partial t} + \sum_{i=1}^{N} \nabla_{\mathbf{x}_i} V \cdot \frac{\partial \mathbf{x}_i}{\partial t}.$$

Systems with potentials of the form (3.2) satisfy $\partial V/\partial t = 0$. Substituting the definition of the force (3.6) and taking Newton's equations of motion into account one obtains

$$\frac{dE}{dt} = \sum_{i=1}^{N} m_i \dot{\mathbf{v}}_i + \sum_{i=1}^{N} \nabla_{\mathbf{x}_i} V \mathbf{v}_i \tag{3.9}$$

$$= \sum_{i=1}^{N} m_i \dot{\mathbf{v}}_i - \sum_{i=1}^{N} \mathbf{F}_i \mathbf{v}_i = 0. \tag{3.10}$$

The energy E is therefore a constant of motion, meaning that it is conserved over time. For this reason, energy is a conserved quantity of the system [371].

Before we address the fast evaluation of the forces $\mathbf{F}_i$ and the computation of the energy, we first introduce in the next section a standard method for the time integration of the system.

3.1 Time Discretization – the Integration Method of Störmer-Verlet

The concept of discretization in numerical mathematics describes the transition from a problem that is posed on a continuous interval to a problem that is only posed at a finite number of points. Discretizations are primarily used in the solution of differential equations to transform the differential equation into a system of equations with a solution that approximates the solution of the differential equations only at the chosen points. In our context, this boils down to the computation of the new positions and velocities of the particles from the old positions, old velocities, and the corresponding forces.

Basic Discretization Formulae. We now decompose the time interval $[0, t_{end}] \subset \mathbb{R}$, on which the system of differential equations (3.4) is to be solved, into l subintervals of the same size, $\delta t := t_{end}/l$. In this way we obtain a grid that contains the points $t_n := n \cdot \delta t$, $n = 0, \ldots, l$, located at the ends of the subintervals. The differential equation is then only considered at those points in time. According to the definition of the derivative

$$\frac{dx}{dt} := \lim_{\delta t \to 0} \frac{x(t + \delta t) - x(t)}{\delta t}$$

of a differentiable function $x : \mathbb{R} \longrightarrow \mathbb{R}$, the differential operator dx/dt for continuous time is approximated at the grid point t_n by the discrete, one-sided difference operator

$$\left[\frac{dx}{dt}\right]_n^r := \frac{x(t_{n+1}) - x(t_n)}{\delta t} \tag{3.11}$$

by omitting the limit. Here, $t_{n+1} = t_n + \delta t$ is the next grid point to the right of t_n. A Taylor expansion of the function x at the point t_{n+1} according to

$$x(t_n + \delta t) = x(t_n) + \delta t \frac{dx}{dt}(t_n) + \mathcal{O}(\delta t^2) \tag{3.12}$$

implies a discretization error of the order $\mathcal{O}(\delta t)$ for the approximation of the first derivative. Thus, if the time step size is halved, one expects that the error caused by the time discretization is also approximately halved.

The differential operator dx/dt at the grid point t_n can be approximated alternatively by the central difference operator

$$\left[\frac{dx}{dt}\right]_n^c := \frac{x(t_{n+1}) - x(t_{n-1})}{2\delta t}. \tag{3.13}$$

A Taylor expansion yields in this case a discretization error of the order $\mathcal{O}(\delta t^2)$ for the approximation of the first derivative.

The second derivative d^2x/dt^2 can be approximated at the grid point t_n by the difference operator

$$\left[\frac{d^2x}{dt^2}\right]_n := \frac{1}{\delta t^2}\left(x(t_n + \delta t) - 2x(t_n) + x(t_n - \delta t)\right). \tag{3.14}$$

By Taylor expansion around both points $t_n + \delta t$ and $t_n - \delta t$ up to third order one obtains

$$x(t_n + \delta t) = x(t_n) + \delta t \frac{dx(t_n)}{dt} + \frac{1}{2}\delta t^2 \frac{d^2x(t_n)}{dt^2} + \frac{1}{6}\delta t^3 \frac{d^3x(t_n)}{dt^3} + \mathcal{O}(\delta t^4) \tag{3.15}$$

and

$$x(t_n - \delta t) = x(t_n) - \delta t \frac{dx(t_n)}{dt} + \frac{1}{2}\delta t^2 \frac{d^2x(t_n)}{dt^2} - \frac{1}{6}\delta t^3 \frac{d^3x(t_n)}{dt^3} + \mathcal{O}(\delta t^4). \tag{3.16}$$

Substituting these expansions into (3.14) yields directly

$$\left[\frac{d^2x}{dt^2}\right]_n = \frac{d^2x(t_n)}{dt^2} + \mathcal{O}(\delta t^2).$$

The discretization error for the approximation of the second derivative by (3.14) is therefore of the order $\mathcal{O}(\delta t^2)$.

The Discretization of Newton's Equations of Motion. An efficient and at the same time stable approach for the time discretization of Newton's equations (3.4) is the Verlet algorithm [323, 596, 645] which builds on the integration method of Störmer [588]. The algorithm is based on the difference operators introduced above. In the following different variants of the Störmer-Verlet method are derived.

Given a system of ordinary differential equations of second order in the form (3.5), the differential quotient can be replaced by a difference quotient at each time t_n, $n = 1, \ldots, l - 1$. Applying (3.14) one can determine the position at time t_{n+1} from the positions at time t_n and t_{n-1} and the force at time t_n. With the abbreviations $\mathbf{x}_i^n := \mathbf{x}_i(t_n)$ and analogous abbreviations for $\mathbf{v}_i$ and $\mathbf{F}_i$ one obtains at first

$$m_i \frac{1}{\delta t^2} \left(\mathbf{x}_i^{n+1} - 2\mathbf{x}_i^n + \mathbf{x}_i^{n-1} \right) = \mathbf{F}_i^n \tag{3.17}$$

and then

$$\mathbf{x}_i^{n+1} = 2\mathbf{x}_i^n - \mathbf{x}_i^{n-1} + \delta t^2 \cdot \mathbf{F}_i^n / m_i, \tag{3.18}$$

which involves the evaluation of the right hand side $\mathbf{F}_i$ at time t_n.[3] Given the initial positions $\mathbf{x}_i^0$ and the positions $\mathbf{x}_i^1$ in the first time step, all the later positions can be uniquely determined using this scheme. (3.18) is the standard form [645] of the Störmer-Verlet method for the integration of Newton's equations. For this method the positions at times t_n and t_{n-1} and the force at time t_n have to be stored. A disadvantage of the method in this form is the possibility of large rounding errors in the addition of values of very different size. The force term $\delta t^2 \cdot \mathbf{F}_i^n / m_i$ in (3.18), which is small because of the factor δt^2, is added to two much larger terms $2\mathbf{x}_i^n$ and $\mathbf{x}_i^{n-1}$ that do not depend on the time step δt. Also, (3.18) does not contain the velocities which are needed for instance in the computation of the kinetic energy. To this end, the velocity as the derivative of the position can be approximated using the central difference (3.13) according to

$$\mathbf{v}_i^n = \frac{\mathbf{x}_i^{n+1} - \mathbf{x}_i^{n-1}}{2\delta t}. \tag{3.19}$$

There are two other variants of the Störmer-Verlet method (equivalent in exact arithmetic) that are less susceptible to rounding errors than the variant (3.18). One is the so-called leapfrog scheme [323] in which the velocities are computed at $t + \delta t/2$. There, one first computes the velocities $\mathbf{v}_i^{n+1/2}$ from the velocities at time $t_{n-1/2}$ and the forces at time t_n according to

$$\mathbf{v}_i^{n+1/2} = \mathbf{v}_i^{n-1/2} + \frac{\delta t}{m_i} \mathbf{F}_i^n. \tag{3.20}$$

---

[3] To discretize these equations completely in time one has to choose an unique point in time at which the right hand side will be evaluated. If the right hand side is evaluated at time t_n, the resulting method is called an explicit time-stepping method. There, the values of the function at time t_{n+1} can be computed directly from those at previous times. Larger time steps are permitted by implicit time-stepping methods that evaluate the right hand side at time t_{n+1}. However, in these methods it is necessary to solve a linear or nonlinear system of equations in each time step. For a further discussion see Chapter 6.

The positions $\mathbf{x}_i^{n+1}$ are then determined as

$$\mathbf{x}_i^{n+1} = \mathbf{x}_i^n + \delta t \mathbf{v}_i^{n+1/2}, \qquad (3.21)$$

which involves the positions at time t_n and the velocities $\mathbf{v}_i^{n+1/2}$ that were just computed. Compared to the standard form (3.18) the effect of rounding errors is reduced. In addition the velocities are computed explicitly. However, the positions and velocities are given at different times. Thus, the velocity at time t_n must be computed for example as an average $\mathbf{v}_i^n = (\mathbf{v}_i^{n+1/2} + \mathbf{v}_i^{n-1/2})/2$. Only then, one can evaluate the kinetic and potential energy at the same time[4] t^{n+1}.

A different variant is the so-called Velocity-Störmer-Verlet method [596]. If one solves (3.19) for $\mathbf{x}_i^{n-1}$, substitutes the result into (3.18), and then solves for $\mathbf{x}_i^{n+1}$, one obtains

$$\mathbf{x}_i^{n+1} = \mathbf{x}_i^n + \delta t \mathbf{v}_i^n + \frac{\mathbf{F}_i^n \cdot \delta t^2}{2m_i}. \qquad (3.22)$$

Furthermore, (3.18) and (3.19) yields

$$\mathbf{v}_i^n = \frac{\mathbf{x}_i^{n+1} - \mathbf{x}_i^{n-1}}{2\delta t} = \frac{\mathbf{x}_i^n}{\delta t} - \frac{\mathbf{x}_i^{n-1}}{\delta t} + \frac{\mathbf{F}_i^n}{2m_i}\delta t.$$

Adding the corresponding expression for $\mathbf{v}_i^{n+1}$ one obtains

$$\mathbf{v}_i^{n+1} + \mathbf{v}_i^n = \frac{\mathbf{x}_i^{n+1} - \mathbf{x}_i^{n-1}}{\delta t} + \frac{(\mathbf{F}_i^{n+1} + \mathbf{F}_i^n)\delta t}{2m_i}. \qquad (3.23)$$

Using equation (3.19) it follows finally that

$$\mathbf{v}_i^{n+1} = \mathbf{v}_i^n + \frac{(\mathbf{F}_i^n + \mathbf{F}_i^{n+1})\delta t}{2m_i}. \qquad (3.24)$$

Equation (3.22) together with equation (3.24) yields the so-called Velocity-Störmer-Verlet method.

Figure 3.3 shows schematically the sequence of the steps in the computation for the three variants described in this section. The first row shows the procedure for the standard form (3.18). The second row illustrates the sequence of operations for the leapfrog scheme (3.20) and (3.21). The third row shows the order of computations in the Velocity-Störmer-Verlet method (3.22) and (3.24).

All three formulations need approximately the same amount of memory, the velocity variant of the Störmer-Verlet needs one auxiliary array to store intermediate results. Furthermore, the accuracy of all three variants is of

---

[4] Here, as starting values, the positions and velocities at time t_0 must be given. The velocity at time $t_{1/2}$ can then be computed by $\mathbf{v}_i^{1/2} = \mathbf{v}_i^0 + \frac{\delta t}{2m_i}\mathbf{F}_i^0$.

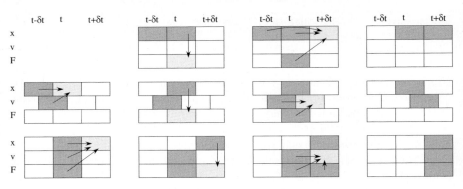

Fig. 3.3. Integration method of Störmer-Verlet: Three variants of the method. Standard form (3.18) (top), leapfrog scheme (3.20) and (3.21) (middle), Velocity-Störmer-Verlet method (3.22) and (3.24) (bottom).

second order, i.e. $\mathcal{O}(\delta t^2)$. The representation (3.22) and (3.24) is especially recommended for implementation since it is stable with respect to rounding errors and positions and velocities are available at the same time without additional computations.

The algorithm for the integration of the equations of motion (3.4) then proceeds as follows: With the given initial values for $\mathbf{x}_i^0$ and $\mathbf{v}_i^0$, $i = 1, \dots, N$, an outer loop starts at time $t = 0$ and increases the time by δt until the final time t_{end} is reached. If the values for time t_n are already known, the values for all the particles at time t_{n+1} can be computed according to (3.22) and (3.24). At first, we determine in a loop over all particles their new positions. Then, the new forces are determined. Finally, we compute the new velocities in a loop over all particles.[5] In this way we derive the complete Störmer-Verlet method in the velocity form. It is summarized in Algorithm 3.1.[6]

Thermodynamic quantities – as for example the kinetic or the potential energy – can be computed together at the same point in the program because the positions as well as the velocities are available for the same point in time, compare Figure 3.3 (lower left). In this way the kinetic energy

---

[5] Algorithms for the other variants can be implemented in an analogous manner.

[6] In Algorithm 3.1, three loops are needed inside the outside loop for the time step to compute the positions, forces, and velocities. One can reduce this to two loops over the particles by first computing the forces $\mathbf{F}_i$ in one loop, and then computing the velocities $\mathbf{v}_i$ and the positions $\mathbf{x}_i$ in another loop. As a result one obtains the position and the velocity at different points in time. If needed, the output and the computation of derived quantities such as temperature, kinetic or potential energy can be done in addition to the other computations within that second loop.

Algorithm 3.1 Velocity-Störmer-Verlet Method

```
//  start with initial data x, v, t
//  auxiliary vector F^old;
compute forces F;
while (t < t_end) {
  t = t + delta_t;
  loop over all i {                                        // update x
    x_i = x_i + delta_t * (v_i + .5 / m_i * F_i* delta_t); // using (3.22)
    F_i^old = F_i;
  }
  compute forces F;
  loop over all i                                          // update v
    v_i = v_i + delta_t * .5 / m_i * (F_i + F_i^old);      // using (3.24)
  compute derived quantities as for example kinetic or potential energy;
  print values of t, x, v as well as derived quantities;
}
```

$$E_{kin}^n = \frac{1}{2} \sum_{i=1}^{N} m_i (\mathbf{v}_i^n)^2 \tag{3.25}$$

at time t_n can be computed in the algorithm directly after the computation of the velocities $\mathbf{v}_i^n$. In the same way one obtains the potential energy $V^n = V(\mathbf{x}_1^n, \ldots, \mathbf{x}_N^n)$ at time t_n from the positions $\mathbf{x}_i^n$ of the particles at time t_n.

Critical properties of integration methods in molecular dynamics are efficiency, accuracy, and energy conservation. Accuracy specifies how much the numerically computed trajectory deviates from the exact trajectory after one time step. The error is usually given in powers of the time step δt. The energy is conserved along the trajectory of the particles for Hamiltonians that do not explicitly depend on time, compare (3.10). The numerical trajectory can deviate from the exact trajectory and thereby causes a small drift in the energy. Here, it is important to distinguish between errors caused by the finite accuracy of computer arithmetic and errors caused by the integration method itself even if infinitely accurate arithmetic is assumed.

Closely connected with these issues is the question whether the integration method has the properties of time reversibility and symplecticity. Time reversibility guarantees that, if the sign of the velocity is changed in the differential equation, the computed trajectory is followed exactly in inverse direction and the initial configuration is finally reached in the absence of numerical rounding errors. An integration method can be interpreted as a mapping in phase space, see Chapter 6. If the integration method is applied to a measurable set of points in the phase space, this set is mapped to another measurable set in the phase space. The integration method is called symplectic if the measure of both of those sets is equal. The mapping is then measure-preserving in the phase space and satisfies, as does the exact

Hamiltonian, the so-called Liouville theorem [554]. Symplectic methods exhibit excellent behavior with respect to energy conservation, see Section 6.1. Numerical approximations computed by symplectic methods can be viewed as exact solutions of slightly perturbed Hamiltonian systems. The perturbation can be analyzed by an asymptotic expansion in powers of the time step. A more detailed discussion of these aspects of integration methods as well as further remarks can be found in Chapter 6. Compared to many other integration methods, the Störmer-Verlet method has the advantage to be time-reversible and symplectic, as discussed in Section 6.2. Due to these properties and its simplicity it is the most commonly used method for the integration of Newton's equations of motion.

3.2 Implementation of the Basic Algorithm

With the techniques introduced up to now, we can already implement a prototype (see Algorithm 1.1) of the method of molecular dynamics. For this, we use the Velocity-Störmer-Verlet method for the time integration from Algorithm 3.1 and consider at first the gravitational potential (2.42) as an example for the interactions between the particles. The resulting method will be accelerated in the further course of the book by new methods for the evaluation of the forces and it will be adapted to other forms of the potential.

We first declare the following constants, data types and macros:

```
#define DIM 2
typedef double real;
#define sqr(x) ((x)*(x))
```

Later, in the three-dimensional case, DIM has to be set equal to 3. For a data structure of the particle we can directly use the variables for mass, position, velocity and force of the time integration.

Data structure 3.1 Particle

```
typedef struct {
  real m;              //  mass
  real x[DIM];         //  position
  real v[DIM];         //  velocity
  real F[DIM];         //  force
} Particle;
```

We adjust now the velocity variant of the Störmer-Verlet method from Algorithm 3.1 to the particle data structure 3.1. For this purpose we need a routine `compF_basis` that computes the force at time `t` at the current position of the particles, see Algorithm 3.6. In addition we need routines `compX_basis` and `compV_basis` that compute the positions and the velocities for the current time step, compare Algorithm 3.4. There, the routines `updateX` and `updateV` are called in a loop over all particles. These two routines are implemented in Algorithm 3.5. In the routine `compoutStatistic_basis` in Algorithm 3.2 we compute derived quantities as for instance the kinetic energy[7] and write their values in an appropriate form to a file. In the routine `outputResults_basis` we finally write the current time as well as the values for the positions and velocities of the particles into a file for postprocessing. The precise realization of these two routines is left to the reader. The overall resulting method is listed in Algorithm 3.2.

Algorithm 3.2 Velocity-Störmer-Verlet Method

```
void timeIntegration_basis(real t, real delta_t, real t_end,
                           Particle *p, int N) {
  compF_basis(p, N);
  while (t < t_end) {
    t += delta_t;
    compX_basis(p, N, delta_t);
    compF_basis(p, N);
    compV_basis(p, N, delta_t);
    compoutStatistic_basis(p, N, t);
    outputResults_basis(p, N, t);
  }
}
```

Algorithm 3.3 Computation of the Kinetic Energy

```
void compoutStatistic_basis(Particle *p, int N, real t) {
  real e = 0;
  for (int i=0; i<N; i++) {
    real v = 0;
    for (int d=0; d<DIM; d++)
      v += sqr(p[i].v[d]);
    e += .5 * p[i].m * v;
  }
  // print kinetic energy e at time t
}
```

---

[7] Also the computation of the potential energy, the temperature, the diffusion as well as of further quantities of statistical mechanics can be carried out here, compare Section 3.7.2.

Algorithm 3.4 Routines for the Velocity-Störmer-Verlet Time Step for a Vector of Particles

```
void compX_basis(Particle *p, int N, real delta_t) {
  for (int i=0; i<N; i++)
    updateX(&p[i], delta_t);
}
void compV_basis(Particle *p, int N, real delta_t) {
  for (int i=0; i<N; i++)
    updateV(&p[i], delta_t);
}
```

For the Velocity-Störmer-Verlet method we need an additional array `real F_old[DIM]` in the data structure for the particle 3.1 which stores the force from the previous time step.[8]

Algorithm 3.5 Routines for the Velocity-Störmer-Verlet Time Step for One Particle

```
void updateX(Particle *p, real delta_t) {
  real a = delta_t * .5 / p->m;
  for (int d=0; d<DIM; d++) {
    p->x[d] += delta_t * (p->v[d] + a * p->F[d]); // according to (3.22)
    p->F_old[d] = p->F[d];
  }
}
void updateV(Particle *p, real delta_t) {
  real a = delta_t * .5 / p->m;
  for (int d=0; d<DIM; d++)
    p->v[d] += a * (p->F[d] + p->F_old[d]);        // according to (3.24)
}
```

We will first carry out the computation of the force in a naive way by determining for each particle i its interaction with every other particle j. With

$$V(\mathbf{x}_1, \ldots, \mathbf{x}_N) = \sum_{i=1}^{N} \sum_{j=1, j>i}^{N} U(r_{ij})$$

we obtain the force on the particle i with

$$\mathbf{F}_i = -\nabla_{\mathbf{x}_i} V(\mathbf{x}_1, \ldots, \mathbf{x}_N) = \sum_{\substack{j=1 \\ j \neq i}}^{N} -\nabla_{\mathbf{x}_i} U(r_{ij}) = \sum_{\substack{j=1 \\ j \neq i}}^{N} \mathbf{F}_{ij}$$

---

[8] If we later want to use a different time integration scheme, we possibly need to add other auxiliary variables here.

just as the sum of the pairwise forces $\mathbf{F}_{ij} := -\nabla_{\mathbf{x}_i} U(r_{ij})$, where $r_{ij} := \|\mathbf{x}_j - \mathbf{x}_i\|$ denotes the distance between the particles i and j. This algorithm is given in Algorithm 3.6. The computation of the forces can be written as a double loop in which the function `force` is called with the addresses of the particles i and j.[9] It is clear that one needs $\mathcal{O}(N^2)$ operations to compute the forces in this way.

Algorithm 3.6 Computation of the Force with $\mathcal{O}(N^2)$ Operations

```
void compF_basis(Particle *p, int N) {
  for (int i=0; i<N; i++)
    for (int d=0; d<DIM; d++)
      p[i].F[d] = 0;                  // set F for all particles to zero
  for (int i=0; i<N; i++)
    for (int j=0; j<N; j++)
      if (i != j) force(&p[i], &p[j]);    // add the forces F_ij to F_i
}
```

In the following example, we use the gravitational potential (2.42). From the (scaled) potential $U(r_{ij}) = -m_i m_j / r_{ij}$ we obtain the force

$$\mathbf{F}_{ij} = -\nabla_{\mathbf{x}_i} U(r_{ij}) = \frac{m_i m_j}{r_{ij}^3} \mathbf{r}_{ij},$$

that particle j exerts on particle i. Here,

$$\mathbf{r}_{ij} := \mathbf{x}_j - \mathbf{x}_i$$

denotes the direction vector between the particle i and j at the positions $\mathbf{x}_i$, and $\mathbf{x}_j$ and r_{ij} denotes its length.[10] In Algorithm 3.7 the contribution $\mathbf{F}_{ij}$ is added to the previously initialized force $\mathbf{F}_i$.

To store a set of N particles we can use a vector of particles. Since we do not know the number N of particles at the compilation time of the program, we cannot use the declaration `Particle[N]`. Therefore, we have to dynamically allocate the necessary memory for that vector. Memory can be allocated and freed in C as shown in the code fragment 3.1.

We need a procedure in which the parameters of the simulation, as for instance the time step or the number of particles, are initialized. Furthermore,

---

[9] Pointers and the dereferencing of pointers are designated in C by the operator `*`. The inverse operation, the determination of the address of a variable, is written as `&`. In pointer arithmetic the pairs of expressions `p[0]` and `*p`, `p[i]` and `*(p+i)`, as well as `&p[i]` and `p+i` are equivalent.

[10] One should bear in mind that in some molecular dynamics books the distance vector is defined as $\mathbf{r}_{ij} := \mathbf{x}_i - \mathbf{x}_j$ and has the opposite direction. In the computation of the force one therefore obtains the opposite sign from the inner derivative of $\mathbf{r}_{ij}$ with respect to $\mathbf{x}_i$.

Algorithm 3.7 Gravitational Force between two Particles

```
void force(Particle *i, Particle *j) {
  real r = 0;
  for (int d=0; d<DIM; d++)
    r += sqr(j->x[d] - i->x[d]);              // squared distance r=r_{ij}^2
  real f = i->m * j->m /(sqrt(r) * r);
  for (int d=0; d<DIM; d++)
    i->F[d] += f * (j->x[d] - i->x[d]);
}
```

Code fragment 3.1 Allocate and Free Memory Dynamically

```
Particle *p = (Particle*)malloc(N * sizeof(*p));     // reserve
free(p);                                             // and release memory
```

a routine is needed to initialize the particles (mass, position, and velocity) at
the beginning of the simulation. The data for the particles could be given in
a file or could be created appropriately. We need to implement appropriate
functions `inputParameters_basis` and `initData_basis` for these initialization procedures. The output of the currently computed results takes place
in each time step in Algorithm 3.2 in the routines `compoutStatistic_basis`
and `outputResults_basis`. These routines have also to be implemented appropriately. The main program for the particle simulation can then look as
in Algorithm 3.8.

Algorithm 3.8 Main Program

```
int main() {
  int N;
  real delta_t, t_end;
  inputParameters_basis(&delta_t, &t_end, &N);
  Particle *p = (Particle*)malloc(N * sizeof(*p));
  initData_basis(p, N);
  timeIntegration_basis(0, delta_t, t_end, p, N);
  free(p);
  return 0;
}
```

With this program first numerical experiments can be carried out. The
implemented time integration method originated in astronomy. Delambre already used it in 1790 [566], and later Störmer [588] and others used it to
calculate the motion of electrically charged particles and the orbits of planets and comets. Following these applications we consider an easy problem

from astronomy. In a simplified model of our solar system we simulate the motion of the Sun, the Earth, the planet Jupiter, and a comet similar to Halley's Comet. We restrict the orbits to a two-dimensional plane and put the sun in the plane's origin. Each astronomical body is represented by one particle. Between any two particles acts the gravitational potential (2.42). A set of initial values is given in Table 3.1. The resulting orbits of the celestial bodies are shown in Figure 3.4.

$$
\begin{array}{lll}
m_{Sun} = 1, & \mathbf{x}^0_{Sun} = (0,0), & \mathbf{v}^0_{Sun} = (0,0), \\
m_{Earth} = 3.0 \cdot 10^{-6}, & \mathbf{x}^0_{Earth} = (0,1), & \mathbf{v}^0_{Earth} = (-1,0), \\
m_{Jupiter} = 9.55 \cdot 10^{-4}, & \mathbf{x}^0_{Jupiter} = (0,5.36), & \mathbf{v}^0_{Jupiter} = (-0.425,0), \\
m_{Halley} = 1 \cdot 10^{-14}, & \mathbf{x}^0_{Halley} = (34.75,0), & \mathbf{v}^0_{Halley} = (0,0.0296), \\
\delta t = 0.015, & t_{end} = 468.5 &
\end{array}
$$

Table 3.1. Parameter values for a simplified simulation of the orbit of Halley's Comet.

Fig. 3.4. Trajectories of Halley's Comet, the Sun, the Earth, and Jupiter in the simplified model.

For the visualization we have written the coordinates of the positions of the planets in each time step to a file using the routine `outputResults_basis`. This data can then be used by a visualization or graphics program to display the trajectories. For references to visualization tools see also Appendix A.2. We see that all the celestial bodies move approximately on elliptic Kepler orbits. The ellipses for the orbit of the Earth and of Jupiter do not precisely coincide which is due to mutual gravitational effects.[11] Since the Sun does not lie exactly in the center of gravity of the system we observe a slight movement of the Sun. The simulation time is chosen so that it covers a complete revolution of the comet in about 76 earth years while the other planets revolve appropriately many times around the Sun. Further numerical simulations of the dynamics of planetary systems can be found in [166].

---

[11] The ellipsoidal Kepler orbits of the planets around a central star are stable if one only considers the system which consists of the central star and a single planet. If one includes in addition the impact of the forces between the different planets, one obtains a system that depends very sensitively on its initial conditions. This was shown by Kolmogorov and later by Arnold and Moser using KAM theory [49]. Furthermore, relativistic effects and numerical rounding errors can contribute to physical or numerical instabilities of the system.

Modifications of the Basic Algorithm. For pairwise interactions Newton's third law states that $\mathbf{F}_{ji} + \mathbf{F}_{ij} = 0$. Thus, the force that particle j exerts on particle i is the same force, up to the sign, that particle i exerts on particle j, and the outer force on both particles vanishes.[12] Up to now the two forces $\mathbf{F}_i$ and $\mathbf{F}_j$ on two particles i and j were computed separately by evaluating both $\mathbf{F}_{ij}$ and $\mathbf{F}_{ji}$ separately and summing them up. But if $\mathbf{F}_{ij}$ has been evaluated, $\mathbf{F}_{ji}$ is known at the same time and does not have to be evaluated again. Thus, when $\mathbf{F}_{ij}$ is added to $\mathbf{F}_i$ in Algorithm 3.6, one can also add $\mathbf{F}_{ji} = -\mathbf{F}_{ij}$ to $\mathbf{F}_j$. This modification of the force computation is shown in Algorithm 3.9. In the loops over all particles only half of all index pairs (i,j) have to be considered, for instance $i < j$, see Algorithm 3.10.

Algorithm 3.9 Gravitational Force between two Particles

```
void force2(Particle *i, Particle *j) {
  ...                                      // squared distance r=r²ᵢⱼ
  real f = i->m * j->m /(sqrt(r) * r);
  for (int d=0; d<DIM; d++) {
    i->F[d] += f * (j->x[d] - i->x[d]);
    j->F[d] -= f * (j->x[d] - i->x[d]);    // modify both particles
  }
}
```

Algorithm 3.10 Computation of the Force with $\mathcal{O}(N^2/2)$ Operations

```
void compF_basis(Particle *p, int N) {
  ... // set F for all particles to zero
  for (int i=0; i<N; i++)
    for (int j=i+1; j<N; j++)
      force2(&p[i], &p[j]);                // add the forces Fᵢⱼ and Fⱼᵢ
}
```

In such a way almost half the operations for the force evaluation can be saved. Nevertheless, the method is still of order $\mathcal{O}(N^2)$. Simulations using the method in its original form are therefore very time-consuming and not practical for a larger number of particles. Further modifications are needed that lead to an improvement in the order of complexity. Such modifications will be described later in this chapter.

---

[12] The potentials that we have introduced in Chapter 1 all describe pairwise interactions. Later on we will also consider many-body potentials. For a general k-body potential one can save only a fraction $1/k$ of the operations using Newton's third law.

3.3 The Cutoff Radius

In every time step of the integration method, the computation of the forces on the particles requires the determination of the interactions of the particles and their appropriate summation. For interactions that are limited only to next neighbors of a particle it obviously does not make sense to sum over all particles to compute the forces. It is enough to sum over only these particles which contribute to the potential and to the force, respectively. One can proceed similarly with potentials and forces that decay rapidly with distance. In this context we say a function decays rapidly with the distance if it decays faster in r than $1/r^d$, where d is the dimension of the problem.

We consider as an example the Lennard-Jones potential (2.44) with $m = 12$ and $n = 6$. The potential between two particles with a distance of r_{ij} is given by

$$U(r_{ij}) = 4 \cdot \varepsilon \left(\left(\frac{\sigma}{r_{ij}} \right)^{12} - \left(\frac{\sigma}{r_{ij}} \right)^{6} \right) = 4 \cdot \varepsilon \left(\frac{\sigma}{r_{ij}} \right)^{6} \cdot \left(\left(\frac{\sigma}{r_{ij}} \right)^{6} - 1 \right). \quad (3.26)$$

The potential is parameterized by σ and ε. The value of ε determines the depth of the potential. A larger ε leads to more stable bonds. The value of σ determines the zero crossing of the potential.[13] Figure 3.5 shows the Lennard-Jones potential with the parameters $\sigma = 1$ and $\varepsilon = 1$.

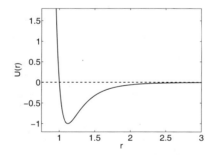

Fig. 3.5. Lennard-Jones potential with the parameters $\varepsilon = 1$ and $\sigma = 1$.

The potential function for N particles is obtained as the double sum

$$V(\mathbf{x}_1, \ldots, \mathbf{x}_N) = \sum_{i=1}^{N} \sum_{j=i+1}^{N} U(r_{ij})$$

$$= 4 \cdot \varepsilon \sum_{i=1}^{N} \sum_{j=i+1}^{N} \left(\frac{\sigma}{r_{ij}} \right)^{6} \cdot \left(\left(\frac{\sigma}{r_{ij}} \right)^{6} - 1 \right). \quad (3.27)$$

---

[13] The zero crossing of the associated force is given by $2^{\frac{1}{6}} \sigma$, see also (3.28).

The corresponding force $\mathbf{F}_i$ on the particle i is given by the gradient with respect to $\mathbf{x}_i$ as

$$\mathbf{F}_i = -\nabla_{\mathbf{x}_i} V(\mathbf{x}_1, \ldots, \mathbf{x}_N)$$

$$= 24 \cdot \varepsilon \sum_{\substack{j=1 \\ j \neq i}}^{N} \frac{1}{r_{ij}^2} \cdot \left(\frac{\sigma}{r_{ij}}\right)^6 \cdot \left(1 - 2 \cdot \left(\frac{\sigma}{r_{ij}}\right)^6\right) \mathbf{r}_{ij}. \qquad (3.28)$$

Here, $\mathbf{r}_{ij} = \mathbf{x}_j - \mathbf{x}_i$ is the direction vector between particles i and j at the positions $\mathbf{x}_i$ and $\mathbf{x}_j$. The force on particle i consists therefore of a sum over the forces $\mathbf{F}_{ij} := -\nabla_{\mathbf{x}_i} U(r_{ij})$ between the particles i and j,

$$\mathbf{F}_i = \sum_{\substack{j=1 \\ j \neq i}}^{N} \mathbf{F}_{ij}. \qquad (3.29)$$

The routine for the computation of the Lennard-Jones force on particle i is given in Algorithm 3.11.[14] Initially, we use the same values of the parameters σ and ε for all particles in the simulation and, for the sake of simplicity, we define them in globally declared variables.[15]

Algorithm 3.11 Lennard-Jones Force between two Particles

```
void force(Particle *i, Particle *j) {
  real r = 0;
  for (int d=0; d<DIM; d++)
    r += sqr(j->x[d] - i->x[d]);          // squared distance r=r²ᵢⱼ
  real s = sqr(sigma) / r;
  s = sqr(s) * s;                          // s=(σ/rᵢⱼ)⁶
  real f = 24 * epsilon * s / r * (1 - 2 * s);
  for (int d=0; d<DIM; d++)
    i->F[d] += f * (j->x[d] - i->x[d]);
}
```

The potential (as well as the corresponding force) decays very fast with the distance r_{ij} between the particles, compare Figure 3.5. The idea is now

---

[14] As in Algorithm 3.9 and 3.10, Newton's third law can be used to save about half of the computations.

[15] If not all particles which occur in (3.27) are the same, the parameters ε and σ depend on the types of particles involved in the interaction. One then writes ε_{ij} and σ_{ij} to explicitly indicate this dependence of the parameters on the pair of particles i and j. To ensure the symmetry $\mathbf{F}_{ij} + \mathbf{F}_{ji} = 0$ of the forces, so-called mixing rules may be used for the parameters associated to the particles, see for instance the Lorentz-Berthelot mixing rule in Algorithm 3.19 of Section 3.6.4. We then extend the data structure 3.1 by the values of σ and ε for each particle.

to neglect all contributions in the sums in (3.27) and (3.28) that are smaller than a certain threshold.

The Lennard-Jones potential (3.26) is then approximated by

$$U(r_{ij}) \approx \begin{cases} 4 \cdot \varepsilon \left(\left(\frac{\sigma}{r_{ij}} \right)^{12} - \left(\frac{\sigma}{r_{ij}} \right)^{6} \right) & r_{ij} \leq r_{\text{cut}}, \\ 0 & r_{ij} > r_{\text{cut}}, \end{cases}$$

i.e., it is cut off at a distance $r = r_{\text{cut}}$. The newly introduced parameter r_{cut}, which characterizes the range of the potential, is typically chosen as about $2.5 \cdot \sigma$. The potential function is thus approximated by

$$V(\mathbf{x}_1, \ldots, \mathbf{x}_N) \approx 4 \cdot \varepsilon \sum_{i=1}^{N} \sum_{\substack{j=i+1, \\ 0 < r_{ij} \leq r_{\text{cut}}}}^{N} \left(\frac{\sigma}{r_{ij}} \right)^{6} \cdot \left(\left(\frac{\sigma}{r_{ij}} \right)^{6} - 1 \right). \quad (3.30)$$

The force $\mathbf{F}_i$ in (3.28) on particle i is approximated similarly by

$$\mathbf{F}_i \approx 24 \cdot \varepsilon \sum_{\substack{j=1, j \neq i \\ 0 < r_{ij} \leq r_{\text{cut}}}}^{N} \frac{1}{r_{ij}^2} \cdot \left(\frac{\sigma}{r_{ij}} \right)^{6} \cdot \left(1 - 2 \cdot \left(\frac{\sigma}{r_{ij}} \right)^{6} \right) \mathbf{r}_{ij}. \quad (3.31)$$

Contributions to the force on particle i that stem from particles j with $r_{ij} \geq r_{\text{cut}}$ are neglected.[16] This introduces an error in the computation of the forces that slightly changes the total energy of the system. Furthermore, the corresponding potential and the force are no longer continuous and thus the total energy of the system is no longer exactly conserved. However, if the cutoff radius r_{cut} is chosen large enough, the effects resulting from these discontinuities are very small.[17]

We now assume that the particles are more or less uniformly distributed throughout the simulation domain. Then, r_{cut} can be chosen such that the number of remaining terms in the truncated sum in (3.30) and (3.31), respectively, is bounded independent of the number of particles N. The complexity of the evaluation of the potential and the forces is then proportional to N, i.e., it is of the order $\mathcal{O}(N)$ only. This results in a substantial reduction of the computational cost compared to the complexity of the order $\mathcal{O}(N^2)$ for the approach described in the last section. To develop and implement efficient algorithms for problems with short-range potentials, the only issue left is how

---

[16] In particular, in case of periodic boundary conditions and $r_{\text{cut}} < \min(L_1, L_2, L_3)$, particles no longer interact with their periodic copies.

[17] It is possible to introduce corrections to compensate for these effects. For instance the truncated potential can be shifted in such a way that it gets continuous again. However, this introduces an additional error in the energy. In other variants the potential is smoothed using an additional function so that the resulting potential is again continuous or even differentiable. A corresponding example is described in Section 3.7.3.

to manage the data so that for a given particle the neighboring particles it interacts with, can be found efficiently. In our context this can be done purely geometrically by a subdivision of the simulation domain into cells. Such an approach will be described in the following section.

3.4 The Linked Cell Method for the Computation of Short-range Forces

In this section we describe a method to approximately evaluate the forces and energies for rapidly decaying potentials. It is relatively easy to implement and it is at the same time very efficient.

The idea of the linked cell method is to split the physical simulation domain Ω into uniform subdomains (cells). If the length of the sides of the cells is chosen larger or equal to the cutoff radius r_{cut}, then interactions in the truncated potentials are limited to particles within a cell and from adjacent cells. Figure 3.6 shows an example in two dimensions in which the simulation domain is divided into cells of size $r_{\text{cut}} \times r_{\text{cut}}$.[18] The particle in the center of the circle in Figure 3.6 only interacts with particles inside the dark-grey shaded area and therefore only with particles from its cell or directly adjacent cells.[19]

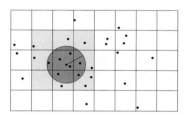

Fig. 3.6. Linked cell method: The simulation domain is decomposed into square cells of size $r_{\text{cut}} \times r_{\text{cut}}$. The dark-shaded area shows the domain of influence of a particle for the cutoff radius r_{cut}. This area is contained in the light-shaded area which consists of 3×3 cells.

The sums in (3.30) or (3.31) are now split into partial sums corresponding to the decomposition of the simulation domain into cells. For the force on particle i in cell ic one obtains a sum of the form

$$\mathbf{F}_i \approx \sum_{\substack{\text{cell } kc \\ kc \in \mathcal{N}(ic)}} \sum_{\substack{j \in \{\text{particles in cell } kc\} \\ j \neq i}} \mathbf{F}_{ij}, \tag{3.32}$$

---

[18] Non-square cells are certainly possible as well.

[19] Ultimately, the cells could be chosen to be smaller than r_{cut}, but then more cells are inside the cutoff radius.

where $\mathcal{N}(ic)$ denotes ic itself together with all cells that are direct neighbors of cell ic.

The question is now how to efficiently access the neighboring cells and particles inside these cells in an algorithm. For this purpose, appropriate data structures are needed to store particles and to iterate over all neighboring cells.

In two dimensions the position of a cell in the grid of all cells can be described by two indices (ic_1, ic_2). Each cell in the interior of the grid possesses eight neighboring cells with the corresponding indices $ic_1 - 1, ic_1, ic_1 + 1$ and $ic_2 - 1, ic_2, ic_2 + 1$. Cells at the boundary of the domain have correspondingly fewer neighboring cells except for the case of periodic boundary conditions where the grid of cells is also extended periodically, see Figure 3.7. In three dimensions a cell with indices (ic_1, ic_2, ic_3) has analogously 26 neighboring cells. The basic procedure is given in Algorithm 3.12.

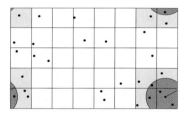

Fig. 3.7. Domain with periodic boundary conditions: The area of interaction of a particle within the cutoff radius r_{cut} is dark-shaded and the neighboring cells are light-shaded. The particle now interacts also with particles at the opposite sides of the domain.

Algorithm 3.12 Computation of the Force in the Linked Cell Method

```
loop over all cells ic
  loop over all particles i in cell ic {
    i->F[d] = 0 for all d;                    // set F_i to zero
    loop over all cells kc in N(ic)
      loop over all particles j in cell kc
        if (i != j)
          if (r_ij <= r_cut)
            force(&i, &j);                     // add F_ij to F_i
  }
```

The complexity of the computation of the forces on the particles amounts to $C \cdot N$ operations if the particles are distributed almost uniformly and thus the number of particles per cell is bounded from above. The constant C

depends quadratically on this bound. Compared to the naive summation of the force over all pairs of particles, the complexity is reduced from originally $\mathcal{O}(N^2)$ to now $\mathcal{O}(N)$.

The cells, as already mentioned, have to be at least as large as the cutoff radius to be able to apply the linked cell method in this form. We denote the dimension of the rectangular simulation domain along the dth coordinate by L_d and the number of grid cells along the dth coordinate by nc_d. Then, the relation $r_{\text{cut}} \leq L_d/nc_d$ has to hold. The largest number of cells nc_d per direction for the method in this form can therefore be computed by

$$nc_d = \left\lfloor \frac{L_d}{r_{\text{cut}}} \right\rfloor . \tag{3.33}$$

Here, $\lfloor x \rfloor$ denotes the largest integer $\leq x$.

3.5 Implementation of the Linked Cell Method

After the explanation of the principles of the numerical method, we will now discuss its actual implementation on a computer. We use as an example the short-range-truncated variant of the Lennard–Jones potential (3.30) with the corresponding force (3.31). Particles that are located in one grid cell will be stored in a linked list associated with that cell.

The List of Particles. So far, we have used one long vector to store all particles. In the linked cell method we are interested in the particles in each cell. Here, the number of particles within each cell can change from time step to time step, if particles move into or out of a cell. Therefore, we need a dynamic data structure for the particles of each cell. We use linked lists for this purpose, see also [559, 560]. These lists will be constructed from elements which consist of the actual particle and a pointer **next** to the next element of the respective list. These list elements are defined in data structure 3.2.

Data structure 3.2 Linked List

```
typedef struct ParticleList {
  Particle p;
  struct ParticleList *next;
} ParticleList;
```

Figure 3.8 shows an example for such a linked list. The end of the list is marked by an "invalid" pointer. For this, a memory address is used where under no circumstances valid data may reside. The predefined constant pointer NULL to this address has on many computer systems the value 0.

Fig. 3.8. A linked list.

A linked list is represented by a pointer to the first element. If we denote this root pointer by `root_list`, we can use it to find the first element of the list and then traverse the list one element after the other using the `next` pointer associated to each element. A complete loop over all elements of the list is given in the code fragment 3.2.

Code fragment 3.2 Loop over the Elements of a List

```
ParticleList *l = root_list;
while (NULL != l) {
  process element l->p;
  l = l->next;
}
```

or written in a different, but equivalent form

```
for (ParticleList *l = root_list; NULL != l; l = l->next)
  process element l->p;
```

We now have to implement the creation and modification of linked lists. For this purpose we write a routine that inserts one element into a list. Repeatedly inserting elements into an initially empty list, we can fill that list with the appropriate particles. The empty list is obtained by setting the root pointer `root_list` pointer to the value NULL, `root_list=NULL`. The insertion of an element into the list is easiest at the beginning of the list as is shown in Figure 3.9. To implement it we only have to change the values of two pointers. The details are shown in Algorithm 3.13.

Fig. 3.9. Inserting elements into a linked list.

For the sake of completeness we also have to discuss how an element can be removed from the list. If a particle leaves a cell and enters another cell, it must be removed from the list of the first cell and inserted into the list of the other cell. Figure 3.10 shows how the pointers must be changed to achieve that. In this figure the list entry marked grey in list 2 is inserted at the beginning of list 1 by insertion and removal operations.

Algorithm 3.13 Inserting Elements into a Linked List.

```
void insertList(ParticleList **root_list, ParticleList *i) {
  i->next = *root_list;
  *root_list = i;
}
```

The problem one encounters in the implementation of a removal operation is that the `next` pointer of the previous element in the list has to be changed. However, in a simply linked list there is no direct way to find the previous element.[20] Therefore, we must determine the pointer `(*q)->next` of the predecessor `**q` of `*i` already during the traversal of the list. The actual removal procedure looks then as in Algorithm 3.14.

Algorithm 3.14 Removing an Element in a Singly Linked List

```
void deleteList(ParticleList **q) {
  *q = (*q)->next;   // (*q)->next points to element to be removed
}
```

Here, the dynamically allocated memory for the element is not freed since the particle will be moved to another cell as in Figure 3.10.

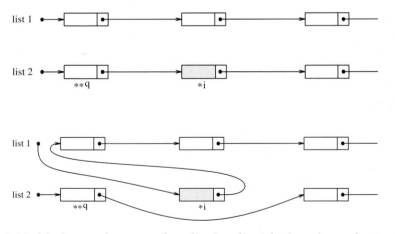

Fig. 3.10. Moving an element `*i` from list 2 to list 1 by inserting and removing. Initial state (top) and final state (bottom).

---

[20] This would be very easy to implement for doubly linked lists in which there are pointers for both directions.

It might be that some particles leave the simulation domain, for instance as a consequence of outflow boundary conditions. Then, the data structure associated to that particle has to be removed from its list and its dynamically allocated memory should possibly be released. If new particles can enter the simulation domain during the simulation, for instance because of inflow boundary conditions, data structures for these particles have to be created and inserted into the appropriate list. In the case of periodic boundary conditions, particles leaving the simulation domain on one side enter it at the opposite side of the domain, and they therefore just move from one cell to another cell which can be realized just as discussed above.

The Data Structure for the Particles. The grid cells into which the simulation domain has been decomposed are stored in a vector. A linked list of particles is associated to each cell, and its root pointer will be stored in that cell, i.e., the cell is represented as the root pointer of the list. All other parameters of the cell, such as the size and the position of the cell, can be determined from global parameters. The entire grid is then stored as a vector `grid` of cells of the type `Cell` that can be dynamically allocated as in code fragment 3.1. In the same way we allocate the memory for the particles which we then sort into the lists.

```
typedef ParticleList* Cell;
```

We now have to decide which cell will be stored at which position in the vector `grid`. The grid consists of $\prod_{d=0}^{DIM-1} nc[d]$ cells and the vector `grid` is therefore of just that length. We can enumerate the cells in the direction of the dth coordinate. For this, we use a multi-index `ic`. Now, a mapping is needed that maps this geometric index `ic` to the appropriate component $index(ic, nc)$ of the vector `grid`. This can be implemented in two dimensions for instance as

$$index(ic, nc) = ic[0] + nc[0] * ic[1].$$

In the general case one can use the macro from code fragment 3.3.

Code fragment 3.3 Macro for the Index Mapping

```
#if 1==DIM
#define index(ic,nc) ((ic)[0])
#elif 2==DIM
#define index(ic,nc) ((ic)[0] + (nc)[0]*(ic)[1])
#elif 3==DIM
#define index(ic,nc) ((ic)[0] + (nc)[0]*((ic)[1] + (nc)[1]*(ic)[2]))
#endif
```

With these data structures and declarations for the grid and its cells the evaluation of the forces by the linked cell method can be implemented as shown in Algorithm 3.15.

Algorithm 3.15 Computation of the Force in the Linked Cell Method

```
void compF_LC(Cell *grid, int *nc, real r_cut) {
  int ic[DIM], kc[DIM];
  for (ic[0]=0; ic[0]<nc[0]; ic[0]++)
    for (ic[1]=0; ic[1]<nc[1]; ic[1]++)
#if 3==DIM
      for (ic[2]=0; ic[2]<nc[2]; ic[2]++)
#endif
      for (ParticleList *i=grid[index(ic,nc)]; NULL!=i; i=i->next) {
        for (int d=0; d<DIM; d++)
          i->p.F[d] = 0;
        for (kc[0]=ic[0]-1; kc[0]<=ic[0]+1; kc[0]++)
          for (kc[1]=ic[1]-1; kc[1]<=ic[1]+1; kc[1]++)
#if 3==DIM
            for (kc[2]=ic[2]-1; kc[2]<=ic[2]+1; kc[2]++)
#endif
            {   treat kc[d]<0 and kc[d]>=nc[d]  according to boundary conditions;
              if (distance of i->p to cell kc <= r_cut)
                for (ParticleList *j=grid[index(kc,nc)];
                     NULL!=j; j=j->next)
                  if (i!=j) {
                    real r = 0;
                    for (int d=0; d<DIM; d++)
                      r += sqr(j->p.x[d] - i->p.x[d]);
                    if (r<=sqr(r_cut))
                      force(&i->p, &j->p);
                  }
            }
      }
}
```

In this implementation an extra check of the distance of the particle `i->p` to the cell `kc` has been inserted that was missing in the previous version of the linked cell method. Thus, instead of testing that the distance from the particle to each particle within the cell is beyond the cutoff radius, an entire neighboring cell with all its particles can be directly excluded from the force evaluation by such a check. This helps to improve the efficiency of the program.

The iteration over the grid cells, which in Algorithm 3.15 explicitly depends on the dimension, can be implemented in a more elegant fashion. For instance the macro from code fragment 3.4 could be used.

Code fragment 3.4 Macro for Dimension-Dependent Iteration

```
#if 1==DIM
#define iterate(ic,minnc,maxnc) \
for ((ic)[0]=(minnc)[0]; (ic)[0]<(maxnc)[0]; (ic)[0]++)
#elif 2==DIM
#define iterate(ic,minnc,maxnc) \
for ((ic)[0]=(minnc)[0]; (ic)[0]<(maxnc)[0]; (ic)[0]++) \
for ((ic)[1]=(minnc)[1]; (ic)[1]<(maxnc)[1]; (ic)[1]++)
#elif 3==DIM
#define iterate(ic,minnc,maxnc) \
for ((ic)[0]=(minnc)[0]; (ic)[0]<(maxnc)[0]; (ic)[0]++) \
for ((ic)[1]=(minnc)[1]; (ic)[1]<(maxnc)[1]; (ic)[1]++) \
for ((ic)[2]=(minnc)[2]; (ic)[2]<(maxnc)[2]; (ic)[2]++)
#endif
```

Using this macro, the loop over all grid cells can be written as

```
iterate(ic,nullnc,nc) {
...,
}
```

where `nullc` denotes a multi-index that has to be declared at an appropriate place and that is initialized to zero.

The time integration could be carried out as in Algorithms 3.2 and 3.4 if the particles would all be compactly stored in a contiguous vector. But the memory for the particles was allocated separately for each cell and each particle to be able to treat the possible change in the number of particles of each cell during the simulation. Therefore, the particles are traversed by an outer loop over all cells and an inner loop over all particles in the particle list of that cell, see Algorithm 3.16.

After the positions of the particles have been updated in the time step, not all particles will be necessarily located in the appropriate cell since some might have moved out of the cell. Therefore, we have to traverse the particles again in `compX_LC` to check if they left the cell and to move them to the new cell if necessary. This can be implemented using the techniques introduced above for the insertion and removal of particles. We obtain Algorithm 3.17.

The main program for the linked cell method needs just some small changes, see Algorithm 3.18. Essentially only the new data structures have to be initialized and added to the parameter lists of the appropriate routines. Furthermore, a new initialization routine `initData_LC` has to be provided in which the particles (their masses, positions, and velocities) are initialized at the start of the simulation. The particles have to be created in a suitable way (or they could be given in a file). In addition, the output routines `compoutStatistic_LC` and `outputResults_LC` in `timeIntegration_LC`

Algorithm 3.16 Routines for the Velocity-Störmer-Verlet Time Stepping with the Linked Cell Data Structure

```
void compX_LC(Cell *grid, int *nc, real *l, real delta_t) {
  int ic[DIM];
  for (ic[0]=0; ic[0]<nc[0]; ic[0]++)
    for (ic[1]=0; ic[1]<nc[1]; ic[1]++)
#if 3==DIM
      for (ic[2]=0; ic[2]<nc[2]; ic[2]++)
#endif
      for (ParticleList *i=grid[index(ic,nc)]; NULL!=i; i=i->next)
        updateX(&i->p, delta_t);
  moveParticles_LC(grid, nc, l);
}
void compV_LC(Cell *grid, int *nc, real *l, real delta_t) {
  int ic[DIM];
  for (ic[0]=0; ic[0]<nc[0]; ic[0]++)
    for (ic[1]=0; ic[1]<nc[1]; ic[1]++)
#if 3==DIM
      for (ic[2]=0; ic[2]<nc[2]; ic[2]++)
#endif
      for (ParticleList *i=grid[index(ic,nc)]; NULL!=i; i=i->next)
        updateV(&i->p, delta_t);
}
```

as well as the routine `timeIntegration_LC` have to be properly adapted to the linked cell data structure.

As seen already in the $\mathcal{O}(N^2)$-algorithm, about half of the operations can be saved if one exploits the symmetry of the forces. Then, one no longer has to iterate through all $3^{DIM} - 1$ neighboring cells but only through half of them. For all interactions within a cell also only half of them have to computed. However, one has to take the boundary conditions properly into account. In the case of periodic boundary conditions as in Figure 3.7, the summation over all neighboring cells also has to be adapted appropriately.

3.6 First Application Examples and Extensions

On the one hand, the method of molecular dynamics can be used to study the behavior of particle systems over time. On the other hand it can also be used to compute appropriate averages of microscopic quantities as approximations to relevant macroscopic quantities.

In this section we show some results of simulations that illustrate the dynamics of different particle systems. For the simulations we use the program described in the last section. We will first present the many different possibilities for applications for the program as described until now, and we

Algorithm 3.17 Sorting the Particles into the Appropriate Cells

```
void moveParticles_LC(Cell *grid, int *nc, real *l) {
  int ic[DIM], kc[DIM];
  for (ic[0]=0; ic[0]<nc[0]; ic[0]++)
    for (ic[1]=0; ic[1]<nc[1]; ic[1]++)
#if 3==DIM
      for (ic[2]=0; ic[2]<nc[2]; ic[2]++)
#endif
  { ParticleList **q = &grid[index(ic,nc)];    //  pointer to predecessor
    ParticleList *i = *q;
    while (NULL != i) {
       treat boundary conditions for i->x;
for (int d=0; d<DIM; d++)
         kc[d] = (int)floor(i->p.x[d] * nc[d] / l[d]);
       if ((ic[0]!=kc[0])||(ic[1]!=kc[1])
#if 3==DIM
         || (ic[2]!=kc[2])
#endif
                                    ) {
       deleteList(q);
       insertList(&grid[index(kc,nc)], i);
       } else q = &i->next;
      i = *q;
    }
  }
}
```

Algorithm 3.18 Main Program of the Linked Cell Method

```
int main() {
  int nc[DIM];
  int N, pnc;
  real l[DIM], r_cut;
  real delta_t, t_end;
  inputParameters_LC(&delta_t, &t_end, &N, nc, l, &r_cut);
  pnc=1;
  for (int d=0; d<DIM; d++)
    pnc *= nc[d];
  Cell *grid = (Cell*)malloc(pnc*sizeof(*grid));
  initData_LC(N, grid, nc, l);
  timeIntegration_LC(0, delta_t, t_end, grid, nc, l, r_cut);
  freeLists_LC(grid, nc);
  free(grid);
  return 0;
}
```

will discuss a few simple extensions that arise from the treated applications. These extensions include the implementation of various boundary conditions (periodic, reflecting, moving boundaries, heated boundaries) as well as mixing rules for the Lennard-Jones potential.

We start with examples in two dimensions since smaller numbers of particles are needed in that case. It is also easier to discover and correct possible mistakes in the program code. First, we consider two examples for the collision of two objects. Here, the respective values of the parameters of the potential function model solid and fluid bodies on a phenomenological level. Then, we simulate the dynamics of two fluids with different densities. Furthermore, we study fluids which exhibit the Rayleigh-Taylor instability and the Rayleigh-Bénard convection on the micro-scale. Finally, we consider surface waves in granular media. We always use the truncated Lennard-Jones interaction potential (3.30) and the resulting force (3.31) in this section.

3.6.1 Collision of Two Bodies I

We simulate the collision of two bodies made from the same material as a first easy example, see also [70, 71, 199, 200, 390]. A sketch of the initial configuration for the simulation can be found in Figure 3.11.

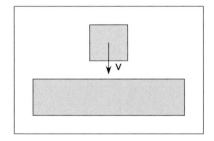

Fig. 3.11. Diagram for the collision of two bodies.

Here, the smaller body strikes the resting, larger body with high velocity. Outflow boundary conditions are used at the sides of the simulation domain. Particles leaving the domain are deleted.

At the beginning of the simulation the two bodies are put together from 40×40 and 160×40 particles of equal mass. These particles are arranged on a lattice of mesh size $2^{1/6}\sigma$ (corresponding to the minimum of the potential) according to Figure 3.11. The velocity of the particles in the moving body is initially set to the given velocity $\mathbf{v}$. In addition, the initial velocities of the particles in both bodies are superimposed with a small thermal motion which is chosen according to a Maxwell–Boltzmann distribution with mean velocity 0.1 per component, compare Appendix A.4.

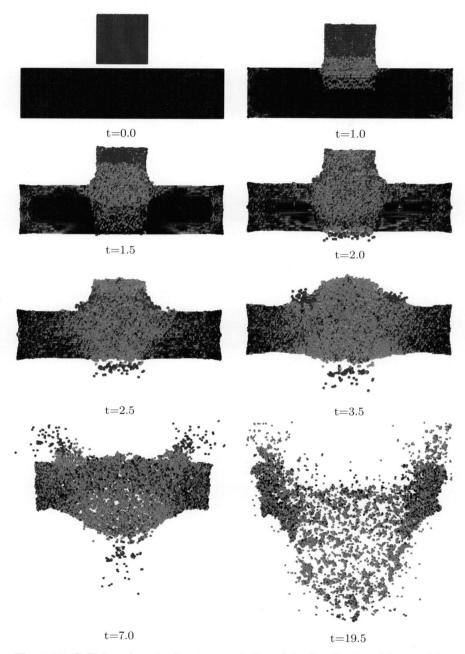

t=0.0

t=1.0

t=1.5

t=2.0

t=2.5

t=3.5

t=7.0

t=19.5

Fig. 3.12. Collision of two bodies, time evolution of the distribution of the particles.

Figure 3.12 shows the result of a simulation with the parameter values from Table 3.2.

$$L_1 = 250, \quad L_2 = 40,$$
$$\varepsilon = 5, \quad \sigma = 1,$$
$$m = 1, \quad \mathbf{v} = (0, -10),$$
$$N_1 = 1600, \quad N_2 = 6400,$$
$$r_{\text{cut}} = 2.5\sigma, \quad \delta t = 0.00005$$

Table 3.2. Parameter values for the simulation of a collision.

The color of each particle encodes its velocity. The parameter ε and σ are chosen here in such a way that the two solid bodies are relatively soft. Immediately after impact shock waves start to spread through the larger body, first along the surface, and then propagating into the interior of the body. Both objects are completely destroyed by the collision.

Because of the simple Lennard-Jones potential used in the simulation, this is only a phenomenological description of the collision of solid bodies. More realistic and quantitative simulations can be accomplished by the use of more sophisticated potentials, compare Chapter 5.

3.6.2 Collision of Two Bodies II

In this simulation we consider a drop which falls into a basin filled with fluid. The initial configuration is depicted in Figure 3.13.

Fig. 3.13. Falling drop, initial configuration.

Two extensions of the program are necessary for this simulation. One is the introduction of reflecting boundary conditions and the other is the treatment of an external gravitational field that acts on all particles.

Implementation of the Gravitational Field. An important element of this and following applications is the gravitational field $\mathbf{G}$, an extrinsic acceleration which affects all particles. In our example this acceleration acts in the $\mathbf{x}_2$ direction only. The resulting force in the two-dimensional case then has the form $\mathbf{F}_i^G = (0, m_i \cdot \mathbf{G}_2)$ for all particles i, with a given value for $\mathbf{G}_2$. For the implementation of such external forces we can introduce the global quantity G as a vector `real[DIM]` that has to be properly added to the forces on the particles. A possible way to implement this is to add the external forces $\mathbf{F}_i^G$ at the beginning of the function `compF_LC`. So far, the new forces F were set to zero, see Algorithm 3.15. Now, the force `i->p.F` on particle `i` will instead be initialized to the external force by code fragment 3.5.

Code fragment 3.5 Implementation of the Gravitational Field

```
for (int d=0; d<DIM; d++)
  i->p.F[d] = i->p.m * G[d];
```

Reflecting Boundaries. Reflecting boundary conditions are used on all sides of the simulation domain to realize fixed solid walls. Such conditions can be implemented by a repulsive force which acts on the particles that come close to the walls. Here, the magnitude of the force corresponds to the force from a virtual particle of the same mass that sits mirror-inverted on the other side of the boundary outside of the simulation domain, see Figure 3.14 for an example with two particles.

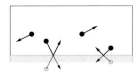

Fig. 3.14. Reflecting boundary conditions in two dimensions; particles hitting the boundary are reflected as if a particle would be located at the mirror image of the particle at the other side of the boundary.

With the Lennard-Jones potential (3.26), the additional repulsive force on particle i is

$$(\mathbf{F}_i)_2 = -24 \cdot \varepsilon \cdot \frac{1}{2r} \cdot \left(\frac{\sigma}{2r}\right)^6 \cdot \left(1 - 2 \cdot \left(\frac{\sigma}{2r}\right)^6\right) \qquad (3.34)$$

for the example of the lower wall. Here, r denotes the distance of the particle from the wall. This force is cut off at $2^{1/6}\sigma$ and thus has only a repulsive component. Note that the distance vector between particle i and its mirror image i' is $\mathbf{r}_{ii'} = \mathbf{x}_{i'} - \mathbf{x}_i = (0, -2r)^T$, which introduces a negative sign. Corresponding formulae can be derived for the force close to the other walls.[21]

---

[21] Alternatively, one can place the reflecting particle on the boundary. Instead of $2r$, (3.34) then contains the distance r.

Note furthermore that the time step has to be chosen small enough to guarantee the stability of the method.

The type of a wall can be described using a flag (reflecting=1, non-reflecting=0). For the four side edges in the two-dimensional case and the six side faces in the three-dimensional case we then need two parameters box_lower and box_upper as vectors int[DIM]. With their help the program can easily determine if a particle is in a cell with a reflecting boundary. If so, the additional force (3.34) has to be added to the force acting on that particle in the routine compF_LC.[22] The iteration over all neighboring cells in routine compF_LC in Algorithm 3.15 has to be properly modified for cells at the boundary. The non-existing neighboring cells are just skipped in our case.

Initial Conditions. At the beginning of the simulation the fluid completely fills the lower part of the simulation domain. The drop is located in $\mathbf{x}_1$ direction in the center of the simulation domain above the filled basin. The particles are placed on a regular mesh, in the drop as well as in the basin, and their velocity is perturbed with a slight thermal motion given by a Maxwell–Boltzmann distribution with a mean velocity 0.07 per component, compare Appendix A.4.

Velocity Scaling. Until time $t = 15$ we let gravity act only on the particles in the basin to bring them to rest. To this end, the velocity of the particles is scaled every 1000th time step to prevent too large values. To be more precise, the scaling proceeds as follows: The kinetic energy of the system at time t_n is given according to (3.25) by

$$E_{kin}^n = \frac{1}{2} \sum_{i=1}^{N} m_i (\mathbf{v}_i^n)^2.$$

If the system is to be transformed into one with a desired kinetic energy E_{kin}^D, we can achieve this by multiplying the velocities $\mathbf{v}_i^n$ of the particles in the routine compV_LC with the factor

$$\beta := \sqrt{E_{kin}^D / E_{kin}^n} \tag{3.35}$$

according to

$$\mathbf{v}_i^n := \beta \cdot \mathbf{v}_i^n.$$

---

[22] Another possibility to create reflecting boundary conditions is to move particles that would travel outside of the simulation domain across a reflecting boundary back into the simulation domain by mirroring their positions across the boundary and also to change the velocity of the particle in an appropriate way. For the case of the lower boundary this would consist in giving the $\mathbf{x}_2$ component of the new velocity a negative sign.

With such a scaling it holds that $\sum_{i=1}^{N} \frac{m_i}{2}(\beta \mathbf{v}_i^n)^2 = \beta^2 E_{kin}^n = E_{kin}^D$. In our simulation we use for the target kinetic energy the value $E_{kin}^D = 0.005 \cdot N$. In every time step the current value of E_{kin}^n is computed first. With that value the scaling factor β can be computed according to (3.35), and the velocities can then be scaled by this factor. After this scaling phase for the particles in the basin, the actual simulation begins: At $t = 15$ the velocity scaling is turned off and the particles in the drop are subjected to gravity. The drop then begins to fall into the basin.

Figure 3.15 shows the results of a simulation with 17227 particles in the basin and 395 particles in the drop with the parameter values from Table 3.3. The drop enters the fluid, displaces it and slowly dissolves. A wave is created that is subsequently reflected at the walls and leads to a swashing motion of the fluid in the basin. The slight asymmetry visible in the pictures results from the random (thermal) initial conditions for the particles.

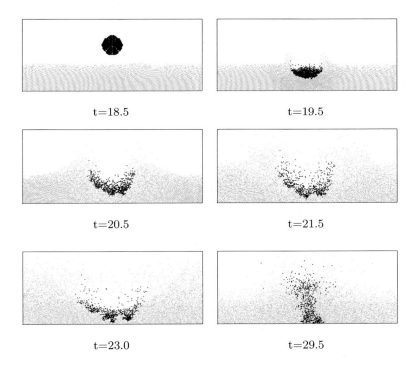

Fig. 3.15. Falling drop, time evolution of the distribution of the particles.

$$L_1 = 250, \qquad\qquad L_2 = 180,$$
$$\varepsilon = 1, \qquad\qquad \sigma = 1,$$
$$m = 1, \qquad\qquad \mathbf{G} = (0, -12),$$
$$N = 395 \text{ and } 17227, \qquad \delta t = 0.0005,$$
$$r_{\text{cut}} = 2.5\sigma$$

Table 3.3. Parameter values for the simulation of a falling drop.

3.6.3 Density Gradient

Now, we consider a flow driven by a density gradient. We start with a rectangular box which is divided into two chambers that contain the same amount of particles, compare Figure 3.16 (left). Here, the partition wall splits the simulation domain with the ratio 1:4. If we remove a part of the partition

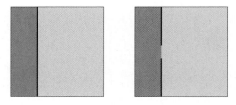

Fig. 3.16. Initial configuration for the density gradient simulation; closed partition wall (left), after opening a hole in the partition wall (right).

wall between the two chambers (Figure 3.16 (right)), the density difference between the two chambers results in a flow of particles from the domain with higher density to the domain with lower density until the densities in the two domains are approximately equal.

Reflecting boundary conditions are to be set at the sides of the simulation domain as well as at the partition wall separating the two chambers. No forces are acting through the partition wall. The partition wall can therefore be treated analogously to the reflecting external walls. To implement this, the iteration over the neighboring cells in the force computation routine compF_LC has to be appropriately changed and a repulsive force has to be added to the force of the particles close to the reflecting boundaries according to (3.34).

At the start of the simulation, the particles are positioned on a regular mesh in both chambers (different mesh sizes in the two chambers) and are subjected to a small thermal motion according to a Maxwell-Boltzmann distribution with $E_{kin}^D = 66.5 \cdot N$, compare Appendix A.4.

Figure 3.17 shows the evolution in time of a simulation of a total of 10920 particles with the parameter values given in Table 3.4. Particles flow from the chamber with higher density into the chamber with lower density. This

$$L_1 = 160, \qquad L_2 = 120,$$
$$\varepsilon = 1, \qquad \sigma = 1,$$
$$N = 10920, \qquad \delta t = 0.0005,$$
$$r_{\text{cut}} = 2.5\sigma, \qquad m = 1$$

Table 3.4. Parameter values for the simulation of the density gradient flow.

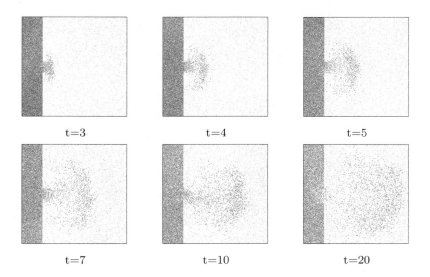

Fig. 3.17. Density gradient examples, evolution in time of the particle distribution.

results in a mushroom-like structure that grows into the chamber with lower density and slowly dissolves by mixing with other particles.

3.6.4 Rayleigh-Taylor Instability

The Rayleigh-Taylor instability is a well-known physical phenomenon in fluid dynamics. The mixing processes related to it arise on various length scales. They can be found inside exploding stars (supernovae) in astrophysics as well as in flow problems in microtechnology. This instability occurs if a fluid of higher density resides on top of a fluid with lower density while subjected to gravity. The instable situation resolves with the heavier fluid sinking down and displacing the lighter fluid. Here, characteristic structures emerge (see Figures 3.20 and 3.21) which depend on the differences in density and mass and on the strength of the gravitational (or other external) force field. On the macroscopic level, classical methods from fluid dynamics can be used to simulate the Rayleigh-Taylor instability. These methods allow a study of the phenomenon from the continuum point of view.

The Rayleigh-Taylor instability is self-similar. Therefore, it can also be observed on the mesoscopic level. The study of the related phenomena on the mesoscopic scale is of interest since this allows to better model and represent the physical processes in the boundary layers between different fluids. To this end, particle methods can be employed. It has already been shown that known hydrodynamic instabilities can develop in mesoscopic particle systems, compare [502, 505] and [32, 198, 494, 504].

In the following, we consider two examples for the Rayleigh-Taylor instability in two-dimensional systems. Figure 3.18 shows the initial configuration.

Fig. 3.18. Initial configuration for the Rayleigh-Taylor instability; the heavy fluid sits on top of the light fluid.

At the beginning of the simulation, the domain is filled completely with particles. The particles in the lower half of the simulation domain are chosen to be lighter than the particles in the upper half. Thus, the two types of particles possess different masses. Also the values of the parameter σ for the Lennard-Jones potential are different.

A Mixing Rule for the Lennard-Jones Potential. Since the model in this application consists of two different kinds of particles we now have to use parameters ε_{ij} and σ_{ij} in the interaction potential that depend on the particles i and j involved,

$$V = 4 \sum_{i=1}^{N} \sum_{\substack{j=1, j>i \\ 0<r_{ij}\leq r_{\mathrm{cut}}}}^{N} \varepsilon_{ij} \cdot \left(\frac{\sigma_{ij}}{r_{ij}}\right)^6 \cdot \left(\left(\frac{\sigma_{ij}}{r_{ij}}\right)^6 - 1\right). \tag{3.36}$$

Analogously, the resulting force reads

$$\mathbf{F}_i = 24 \sum_{\substack{j=1, j\neq i \\ 0<r_{ij}\leq r_{\mathrm{cut}}}}^{N} \varepsilon_{ij} \cdot \frac{1}{r_{ij}^2} \cdot \left(\frac{\sigma_{ij}}{r_{ij}}\right)^6 \cdot \left(1 - 2 \cdot \left(\frac{\sigma_{ij}}{r_{ij}}\right)^6\right) \mathbf{r}_{ij}.$$

If one only knows the parameters for the potentials between particles of the same type, one can try to deduce interaction parameters for the potential between particles of different types. This leads to so-called mixing rules for the potential parameters.

The procedure for the Lennard-Jones potential is the following: Let $(\sigma_{ii}, \varepsilon_{ii})$, $i = 1, 2$, be the parameters of the Lennard-Jones potential for particles of the first and the second type, respectively. One assumes that particles of different types also interact by a Lennard-Jones potential whose parameters $(\sigma_{ij}, \varepsilon_{ij})$ can be determined from $(\sigma_{ii}, \varepsilon_{ii}), i = 1, 2$. Here, the symmetry of the forces $\mathbf{F}_{ij} = -\mathbf{F}_{ji}$ and Newton's third law imply $\sigma_{ij} = \sigma_{ji}$ and $\varepsilon_{ij} = \varepsilon_{ji}$. A popular approach is the Lorentz-Berthelot mixing rule [34]. There, the parameters for the interactions between particles of different kinds are computed by

$$\sigma_{12} = \sigma_{21} = \frac{\sigma_{11} + \sigma_{22}}{2} \quad \text{(arithmetic mean) and}$$

$$\varepsilon_{12} = \varepsilon_{21} = \sqrt{\varepsilon_{11}\varepsilon_{22}} \quad \text{(geometric mean)}.$$

$$(3.37)$$

Although these mixing rules have been derived by empirical arguments only they nevertheless deliver satisfactory results in many cases. The implementation is given in the data structure 3.3, where now the parameters for each particle are stored, and in Algorithm 3.19, where the mixing rule is programmed. The previous global declarations of `sigma` and `epsilon` are omitted.

Data structure 3.3 Additional Particle Data for the Lennard-Jones Potential

```
typedef struct {
   ...                      // particle data structure 3.1
   real sigma, epsilon;  // parameters σ, ε
} Particle;
```

Algorithm 3.19 Lennard-Jones Force with Mixing Rule

```
void force(Particle *i, Particle *j) {
   real sigma = 0.5 * (i->sigma + j->sigma);    // Lorentz-Berthelot (3.37)
   real epsilon = sqrt(i->epsilon * j->epsilon);
   ...                              // force computation from Algorithm 3.11
}
```

The cutoff radius r_{cut} in these interactions should be chosen as the maximum of the two cutoff radii for the interactions of the particles of the same type.

Periodic Boundary Conditions. In the present simulation periodic boundary conditions are to be used on the vertical boundaries of the domain, while reflecting boundary conditions are imposed on the horizontal boundaries. Some changes in the code are necessary to implement periodic boundary conditions. For a given cell, the set of neighboring cells can now also include cells from the opposite side of the simulation domain, compare Figure 3.7. Furthermore, the distance to the closest periodic image is to be used in the computation of the forces and potentials between particles, see Figure 3.19.

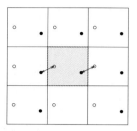

Fig. 3.19. Force computation for periodic boundary conditions: The simulation domain shown in grey is replicated in all spatial directions. Particles close to the boundaries interact with the closest periodic images of other particles.

In the program code this can be accounted for in the routine compF_LC in Algorithm 3.15. Conceptionally, kc[d] is just substituted by kc[d] modulo nc[d]. The computation of the distances of the particles also has to be changed appropriately. For instance, if we treat the case of the periodic left boundary, i.e. kc[0] == -1, we set kc[0] = nc[0]-1, before we iterate over the particles j of that cell. Inside the loop over the particles j, in the computation of the distance, the line r += sqr(j->p.x[0] - i->p.x[0]) has to be replaced by the expression r += sqr((j->p.x[0] - l[0]) - i->p.x[0]). Similar changes are necessary for the other coordinate directions. In the same way, the routine moveParticles_LC has to be modified to allow for the treatment of periodic boundary conditions. The positions x[d] of the particles that left the simulation box must be set to the positions of their periodic images inside the simulation box by properly adding or subtracting l[d].

Again, the initial velocities are chosen according to a Maxwell-Boltzmann distribution with $E_{kin}^{D} = 60 \cdot N$, compare Appendix A.4. In the simulation, a velocity scaling is carried out every 1000 time steps. The kinetic energy is set to $E_{kin}^{D} = 60 \cdot N$. Thus, in every time step in which a velocity scaling is to be performed, the current value of E_{kin}^{n} is computed, the scaling factor β is computing according to (3.35), and finally the velocities are scaled by that factor.

Figure 3.20 shows the result of a simulation with the parameters from Table 3.5. Here, a total number of 6384 particles is employed, where half of the particles belong to one type.

One can see the formation of the typical mushroom-shaped structures over the time of the simulation. The heavier particles (grey) sink down and displace

$$L_1 = 140, \qquad L_2 = 37.5,$$
$$\varepsilon_{up} = \varepsilon_{down} = \varepsilon = 1, \qquad \sigma_{up} = 0.9412, \quad \sigma_{down} = 1,$$
$$\mathbf{G} = (0,\text{-}12.44), \qquad m_{up} = 2, \qquad m_{down} = 1,$$
$$N = 6384, \qquad E_{kin}^{D} = 60 \cdot N,$$
$$r_{cut} = 2.5\sigma, \qquad \delta t = 0.0005$$

Table 3.5. Parameter values for the simulation of the Rayleigh-Taylor instability.

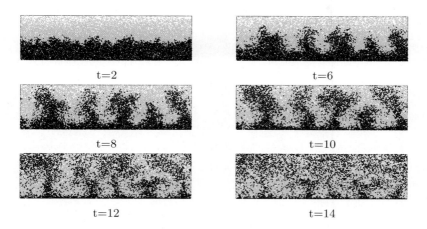

$$t=2 \qquad\qquad\qquad t=6$$

$$t=8 \qquad\qquad\qquad t=10$$

$$t=12 \qquad\qquad\qquad t=14$$

Fig. 3.20. Rayleigh-Taylor instability, time evolution of the particle distribution.

the lighter particles (black) which consequently rise up. In this simulation five mushroom-like structures of different sizes can be observed.

Figure 3.21 shows the result of a simulation with a different set of parameters. There, as before, the positions of the particles at different times are shown. The parameters associated to this simulation can be found in Table 3.6. The total number of particles is now 47704 and, in turn, half of

$$L_1 = 600, \qquad L_2 = 100,$$
$$\varepsilon_{up} = \varepsilon_{down} = \varepsilon = 1, \qquad \sigma_{up} = 1.1, \quad \sigma_{down} = 1.2,$$
$$\mathbf{G} = (0,\text{-}12.44), \qquad m_{up} = 2, \qquad m_{down} = 1,$$
$$N = 47704, \qquad E_{kin}^{D} = 60 \cdot N,$$
$$r_{cut} = 2.5\sigma, \qquad \delta t = 0.0005$$

Table 3.6. Parameter values for the simulation of the Rayleigh-Taylor instability.

the particles belong to one type. Again, the lighter particles are displaced by the heavier, sinking particles. The larger simulation domain and the larger number of particles result in a larger number of mushroom-shaped structures that are distinctly different in size and form from the ones in Figure 3.20.

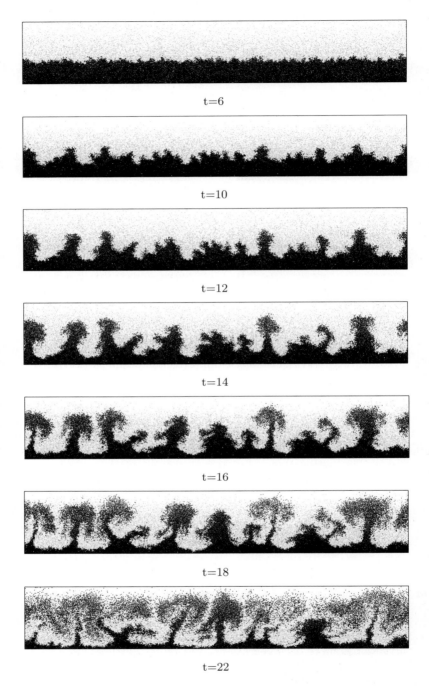

t=6

t=10

t=12

t=14

t=16

t=18

t=22

Fig. 3.21. Rayleigh-Taylor instability, time evolution of the particle distribution.

3.6.5 Rayleigh-Bénard Convection

In many technical applications, as for example in the growth of crystals, in cooling towers, power plants, heat reservoirs, but also in meteorology and oceanography, there is interest in the impact of differences in temperature on the flow field. Related phenomena also arise on microscopic scales. As an example for temperature-driven currents in microfluids we study here the so-called Rayleigh-Bénard convection. To this end, we consider a box with temperature-dependent horizontal walls and with periodic boundary conditions for the vertical walls, compare Figure 3.22.

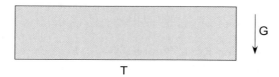

Fig. 3.22. Rayleigh-Bénard convection, configuration with heated lower wall.

If the lower wall is only slightly warmer than the upper wall, the heat is transported in the fluid by diffusion (conduction). If however the temperature difference between upper and lower wall exceeds a critical value, convection becomes a more efficient way to exchange heat between the walls. Then, convection cells form. Their number and shape depend on the temperature difference, the boundary conditions and also the initial conditions [73]. Thus, the flow is driven by the temperature difference, in which the processes

$$\text{heating} \rightarrow \text{expanding} \rightarrow \text{rising}$$
$$\text{cooling} \rightarrow \text{contracting} \rightarrow \text{sinking}$$

occur consecutively.

The analytic study of the behavior of a fluid on the macroscopic level is based on the Navier-Stokes equations which are the standard equations for the conservation of mass, momentum, and energy for fluids. In addition, the so-called Boussinesq approximation [108, 455] is often used. There, it is assumed that the density is the only material property of the fluid which depends on the temperature. Furthermore, it is assumed that the density depends linearly on the temperature. Numerical studies of the Rayleigh-Bénard problem that use the Navier-Stokes equations together with the Boussinesq approximation can be found in [355, 379, 457]. The results of physical experiments can be found in [667].

The simulation of such phenomena with molecular dynamics methods is also of interest. This surely can not replace the known techniques of CFD (computational fluid dynamics), but may help to study the relation between different scales of observation (microscopic and macroscopic scale, discrete and continuous description). Molecular dynamics simulations of a two-dimensional Rayleigh-Bénard problem can be found in [410, 504, 505]. [494] presents a quantitative comparison of molecular dynamics simulations with CFD simulations. A problem in the comparison of the results of molecular dynamics simulations with physical experiments is that the external forces (here gravity) and the density differences now have to be much larger to trigger convection. Moreover, the validity of the Boussinesq approximation is much in doubt [259]. Furthermore, in the three-dimensional case the resulting flow is in general three-dimensional, but, for a very large ratio of width to depth of the simulation box (Hele-Shaw flow) the problem can be replaced by a two-dimensional one.

We study here the formation of Rayleigh-Bénard cells in two simulations that differ in the ratio of length to width of the simulation box.

Boundary Conditions for a Heated Wall. Periodic conditions are imposed for the vertical boundaries. At the horizontal boundaries we use reflecting boundary conditions. Furthermore, the lower boundary is heated,[23] i.e., the particles will be accelerated when colliding with the wall. To this end, we simply multiply the x_2 component of the velocity of the particles which are reflected by the lower wall with the fixed factor 1.4. Thus, for those particles, we set

$$(\mathbf{v}_i^n)_2 := 1.4 \cdot (\mathbf{v}_i^n)_2$$

in the routine compF_LC.

Initial Conditions and Velocity Scaling. The heating of the lower wall adds energy to the system which causes the kinetic energy to increase during the simulation and leads to an increase in the velocities of the particles. To counteract this effect, the energy of the system is reduced by a scaling of the velocity every 1000 time steps using the factor β from (3.35) with a target kinetic energy $E_{kin}^D = 60 \cdot N$.

Figure 3.23 shows the time evolution of a simulation with the parameters from Table 3.7. Here, the length of the domain is four times larger than its width. The total number of particles is 9600. They are distributed on a regular mesh at the beginning of the simulation. The initial velocity of the particles is chosen according to a Maxwell–Boltzmann distribution with a kinetic energy of $E_{kin}^D = 90 \cdot N$, compare Appendix A.4.

To visualize the convection cells, at $t = 90$, the particles in the lower half of the simulation box are colored black and the other particles are colored grey, compare Figure 3.23 (upper left). At this point in time, the heat-driven convection has started to stabilize. The following pictures then show the movement of the colored particles. One clearly sees two Rayleigh-Bénard cells in which the particles are transported up and down.

---

[23] In Section 3.7, we will present a possibility to control the temperature of the system by a so-called thermostat.

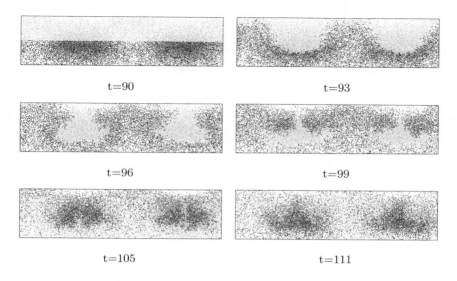

Fig. 3.23. Rayleigh-Bénard convection, time evolution of the particle distribution.

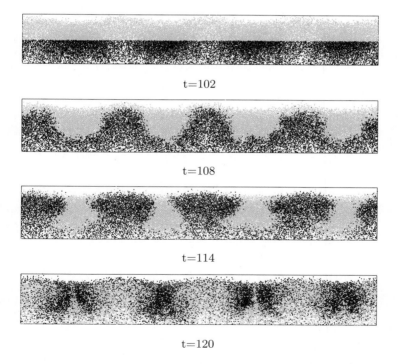

Fig. 3.24. Rayleigh-Bénard convection, time evolution of the particle distribution.

$$L_1 = 240, \qquad L_2 = 60,$$
$$\varepsilon = 1, \qquad \sigma = 1,$$
$$m = 1, \qquad G = (0, -12.44),$$
$$N = 9600, \qquad E_{kin}^D = 60 \cdot N,$$
$$r_{\text{cut}} = 2.5\sigma, \qquad \delta t = 0.0005$$

Table 3.7. Parameter values for the simulation of Rayleigh-Bénard convection.

Figure 3.24 shows the time evolution for a larger simulation box (ratio length to width of 6 : 1) with an appropriately increased number of particles. The length of the domain is $L_1 = 360$ and 14400 particles are used. All other parameters are kept at their previous values.

The color of the particles (black or grey) is now fixed at time $t = 102$, compare Figure 3.24 (top). The following pictures show again the movement of the particles. In this simulation four Rayleigh-Bénard cells are formed.

3.6.6 Surface Waves in Granular Materials

Granular materials [135, 154, 317, 337] occur in various forms in nature and technology. Examples are sand, grain, washing powder, sugar, or dust. They possess peculiar properties because they behave neither like solids nor like liquids. For example, shaking experiments with granular material that consists of grains of different sizes show that demixing and grain segregation processes may occur. Here, grains of different sizes are separated by vibrations (Brazil-nut effect) and clusters, patterns, and convection cells are spontaneously formed [92, 337, 425, 489].

In the following, we study the phenomenon of surface waves that develop when a thin granular layer is exposed to vibrations. There, different waveforms can be observed [425, 426, 629]. Numerical simulation can help to better understand the mechanism behind this behavior. The relation between the excitation frequency, i.e. the frequency of the external vibration, and the observed form of the wave has been investigated in [46, 506]. Comparisons between the results of simulations and experiments are found for example in [92]. To numerically simulate granular materials several methods are used: Direct Simulation Monte-Carlo methods (DSMC) [91, 440, 598], so-called "Event Driven" simulation methods (ED) [395, 396, 418, 440], Hybrid Simulation Monte-Carlo (HSMC) [439], and the molecular dynamics method.

We study surface waves in a granular medium in the two-dimensional case [506] with molecular dynamics simulations. To this end, we consider a system of particles subjected to gravity. Here, the bottom wall of the simulation box moves periodically up and down with a fixed frequency, compare Figure 3.25.

The oscillation of the lower wall causes the particles to vibrate. After some time, waves form at the free surface. The shape, amplitude, and frequency of these waves depend strongly on the amplitude and frequency of the oscillation

Fig. 3.25. Configuration for the simulation of surface waves in granular media: The lower wall oscillates with a given frequency f.

of the wall, but also on the strength of gravity, the mass of the particles, and the parameters of the potential function.

Moving Boundaries. In this application we impose periodic conditions on the vertical boundaries and reflecting boundary conditions on the horizontal boundaries. In addition, the lower boundary oscillates in vertical direction with a given frequency f and amplitude A. The vertical position of the lower boundary is then given by

$$(\mathbf{x}_{\text{wall}})_2(t) = A(1 + \sin(2\pi ft)).$$

To implement such a moving and reflecting boundary, we proceed in principle as for the resting reflecting boundary in Section 3.6.2. However, we now position the virtual particle in the implementation of the boundary condition directly on the boundary, compare footnote 21 on page 69. In addition, we have to take into account that the $\mathbf{x}_2$ position of the boundary depends on time. In a similar way as in (3.34) we obtain the $\mathbf{x}_2$ component of the additional repulsive force on particle i as

$$(\mathbf{F}_i)_2(t) = 24 \cdot \varepsilon \frac{1}{r_i^2(t)} \cdot \left(\frac{\sigma}{r_i(t)}\right)^6 \cdot \left(1 - 2 \cdot \left(\frac{\sigma}{r_i(t)}\right)^6\right) (\mathbf{r}_i)_2(t)$$

with $(\mathbf{r}_i)_2(t) := (\mathbf{x}_{\text{wall}})_2(t) - (\mathbf{x}_i)_2$ and $r_i(t) = ||(\mathbf{x}_{\text{wall}})_2(t) - (\mathbf{x}_i)_2||$. The strength of this force corresponds to the force from a virtual particle located at the moving lower boundary.

Additional Friction Terms in the Equations of Motion. The moving lower wall continuously supplies energy to the system. In this way the total energy of the system steadily increases, and, on average, the velocities of the particles grow larger and larger during the simulation. To counteract this effect, energy is removed from the system by an additional friction term in the equations of motion.

The force on a particle has now three parts: A Lennard-Jones term

$$\mathbf{F}_i = 24 \cdot \varepsilon \sum_{\substack{j=1 \\ j \neq i}}^{N} \frac{1}{r_{ij}^2} \cdot \left(\frac{\sigma_{ij}}{r_{ij}}\right)^6 \cdot \left(1 - 2 \cdot \left(\frac{\sigma_{ij}}{r_{ij}}\right)^6\right), \tag{3.38}$$

the gravitational force $\mathbf{G}$, and the additional frictional force

$$\mathbf{R}_i = \gamma \sum_{\substack{j=1 \\ j \neq i}}^{N} (\mathbf{v}_{ij} \cdot \mathbf{r}_{ij}) \frac{\mathbf{r}_{ij}}{r_{ij}^2}. \tag{3.39}$$

Here, $\mathbf{v}_{ij} := \mathbf{v}_j - \mathbf{v}_i$, and γ denotes an appropriately chosen constant.[24] The friction term is thus dependent on the velocity of the particles. The equation of motion for a particle i then reads

$$m_i \dot{\mathbf{v}}_i = \mathbf{F}_i + \mathbf{R}_i + \mathbf{G}.$$

Both formulae (3.38) and (3.39) are now again cut off at a distance of r_{cut}. In this way, the friction term is approximated by

$$\mathbf{R}_i \approx \gamma \sum_{\substack{j=1,j \neq i \\ r_{ij} < r_{\text{cut}}}}^{N} (\mathbf{v}_{ij} \cdot \mathbf{r}_{ij}) \frac{\mathbf{r}_{ij}}{r_{ij}^2}$$

and the linked cell method can directly be employed.

For the time integration we use the Velocity-Störmer-Verlet Algorithm 3.1. To this end, we discretize the frictional force at time t_{n+1} by

$$\mathbf{R}_i^{n+1} = \gamma \sum_{\substack{j=1,j \neq i \\ r_{ij}^n < r_{\text{cut}}}}^{N} (\mathbf{v}_{ij}^n \cdot \mathbf{r}_{ij}^n) \frac{\mathbf{r}_{ij}^n}{(r_{ij}^n)^2}.$$

This additional force has to be taken into account at an appropriate point in the computation of the velocities.

The parameter σ_{ii} of the particles i, i.e. the sigma in data structure 3.3, are randomly chosen from the interval $[0.9 \cdot \sigma, 1.1 \cdot \sigma]$. The interaction parameters σ_{ij} in (3.38) are determined according to the Lorentz-Berthelot mixing rule as the arithmetic mean of the σ_{ii} of the interacting particles.

Figure 3.26 shows the evolution in time of a simulation with the parameter values from Table 3.8. Here, the parameters f, A, and γ are the frequency and amplitude of the oscillation of the lower wall and the friction constant, respectively. The total number of particles is 1200. They are distributed at the beginning of the simulation on a regular mesh of the size 100×12, compare Figure 3.26 (upper left). The velocity of all particles at the start of the simulation vanishes, i.e. $\mathbf{v}_i = \mathbf{0}$, $i = 1, \ldots, N$.

After some time waves start to repeatedly appear and disappear with a certain frequency. The plots show about half of a period of this oscillatory behavior. The observed waves are entirely driven by the external excitation and break down immediately when energy is no longer fed into the system.

---

[24] Here, $(\mathbf{v}_{ij} \cdot \mathbf{r}_{ij})$ stands for the standard scalar product of the vectors $\mathbf{v}_{ij}$ and $\mathbf{r}_{ij}$.

$$L_1 = 180, \quad L_2 = 40,$$
$$\varepsilon = 0.1, \quad \sigma = 2.1,$$
$$m = 1, \quad \mathbf{G} = (0, -22.0),$$
$$N = 1200, \quad \delta t = 0.002,$$
$$f = 0.83, \quad A = 1.5,$$
$$r_{\text{cut}} = 3.0, \quad \gamma = 1$$

Table 3.8. Parameter values for the simulation of surface waves in a granular medium.

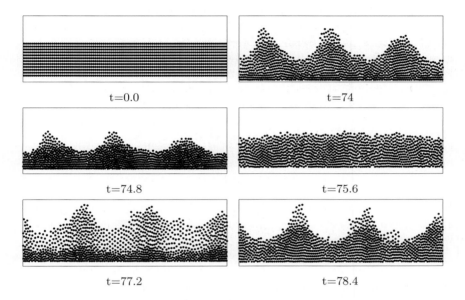

t=0.0   t=74

t=74.8   t=75.6

t=77.2   t=78.4

Fig. 3.26. Surface waves in a granular medium, time evolution of the particle distribution.

This is an important difference to wave phenomena in viscous fluids, in which a slow oscillation persists and the waves slowly die away.

Depending on the parameters used, different wave forms develop during several cycles of the oscillation of the lower wall. The number of wave crests is affected by the size of the box and the type of the boundary conditions at the walls. Furthermore, certain general qualitative observations can be made, such as ranges of parameters with stable patterns, instable transition regimes, and bifurcations in the parameter space. Here, oscillations with a frequency of half of the excitation frequency f and sometimes even deeper subharmonic frequencies were found [189, 426, 506]. One observes superpositions of standing or traveling waves or waves that repeat every few cycles or appear in every cycle but at different positions. Results of three-dimensional experiments can be found for instance in [92].

3.7 Thermostats, Ensembles, and Applications

So far, we changed the kinetic energy by a simple scaling of the velocity. This affects the temperature on the mesoscopic level. In this section we will consider methods to adjust and change the temperature of a simulated system in more detail. First, the number N of particles, the volume V of the simulation domain, and the temperature T (connected with a so-called heat bath) are fixed. In statistical mechanics this is called an NVT ensemble with thermostat. There are also approaches in which the number N of particles, the pressure P and the temperature T are fixed but the volume $V := |\Omega|$ and therefore the size of the simulation box can change. Such ensembles are called NPT ensembles. To this end, Newton's equations of motion have to be extended appropriately, which leads to the Parrinello-Rahman method [462]. Finally, we use these methods to simulate the cooling of argon. There, depending on the speed of cooling, the transition from fluid to solid results in either a crystalline phase or an amorphous glass phase.

Simulation with Thermostat. If a system is thermally and mechanically isolated, the total energy is constant in time according to (3.10). However, in some simulations the energy or the temperature should be changed over time for e.g. the following reasons: On the one hand, one may need the ability to control the temperature of the system to study physical or chemical phenomena such as phase transitions. On the other hand, for a simulation of an isolated system, the temperature may have to be adjusted to the desired value at the beginning.

In physical experiments, the temperature is kept constant by letting the considered system exchange heat with a significantly larger system, the so-called heat bath or thermostat. The influence of the small system on the temperature of the heat bath is negligible. The temperature of the heat bath is therefore assumed to be constant, i.e., it is equal to a given value. In the course of time the smaller system adopts the temperature of the heat bath. On the microscopic level the exchange of heat takes place by collisions of particles with the walls that separate the heat bath and the considered system. On average, the kinetic energy of the particles which hit these separation walls changes in dependence of the temperature of the heat bath. The resulting loss or increase in kinetic energy cools or heats the system until its temperature has reached that of the heat bath. To obtain the same effect in a simulation the system has to gain or loose energy in an appropriate way until the desired temperature is reached. This happens in a so-called equilibration phase. The overall procedure is shown in Figure 3.27. At the beginning the particles are assigned initial positions and initial velocities. Then, the temperature of the system is adjusted by the thermostat. Once the desired temperature is reached, the trajectories of the particles are computed and the relevant quantities are measured (production phase). The temperature adjustment and the production phase is repeated if measurements and simulations for other temperatures are desired.

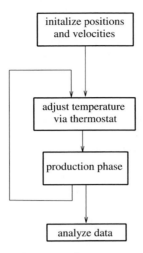

Fig. 3.27. Diagram of the simulation process with thermostat.

3.7.1 Thermostats and Equilibration

In this section we describe how to enforce a certain temperature on a three-dimensional system of particles.

The temperature T of a system and its kinetic energy E_{kin} are related by the equipartition theorem of thermodynamics [141] as

$$E_{kin} = \frac{3N}{2} k_B T. \qquad (3.40)$$

Here, N denotes the total number of particles in the system and $3N$ is the number of degrees of freedom of the system (one degree of freedom for each spatial direction).[25] Three degrees of freedom are to be subtracted if the center of gravity of the system is assumed to be at rest, three more degrees of

---

[25] This corresponds to the so-called atomic scaling. Furthermore, there is the non-periodic scaling with $3N - 6$ degrees of freedom in which the translation and rotation of the entire system are ignored. Then, there is the periodic scaling with $3N - 3$ degrees of freedom in which only the translation of the entire system is ignored, see also page 107. In particle systems with a reduced number of degrees of freedom caused for instance by rigid bonds between particles (as in molecules which will be discussed later), the kinetic energy is correspondingly reduced.

freedom are to be subtracted if also the rotations of the system do not play a role. The proportionality constant k_B is called the Boltzmann constant. The temperature is thus given as

$$T = \frac{2}{3Nk_B} E_{kin} = \frac{2}{3Nk_B} \sum_{i=1}^{N} \frac{m_i}{2} v_i^2. \tag{3.41}$$

Common thermostats have been derived by Andersen [42], Berendsen [80], and Nosé and Hoover [327, 451]. These thermostats all rely on a modification of the velocities which is either carried out explicitly by a scaling of the velocities of all particles or which is achieved implicitly by the introduction of an additional friction term in the equations of motion. We describe the method of velocity scaling and the method of Nosé-Hoover in more detail.

Velocity Scaling. In some of the preceding examples, for instance in the Sections 3.6.4 and 3.6.5, we already changed the kinetic energy of the system by a scaling of the velocities of the particles. In the following, we will reinterpret this procedure with respect to the temperature. Because of (3.40), a multiplication of the velocity by the factor

$$\beta := \sqrt{E_{kin}^D / E_{kin}} = \sqrt{T^D / T} \tag{3.42}$$

transforms the system from the temperature T to the temperature T^D, see also (3.35). A simple possibility to control the temperature is therefore to multiply the velocities of all particles at chosen times with the factor β (which depends on time), i.e., to set

$$\mathbf{v}_i^n := \beta \mathbf{v}_i^n.$$

To implement this, the current temperature $T(t)$ has to be computed according to (3.41). Then, from this value and the target temperature T^D, the appropriate value of β is determined by (3.42).

Depending on the current and the target temperature, the factor β can be relatively large or small, so that the scaling of the velocities could affect the distribution of energy in the system quite strongly. Therefore, instead of the above β, one often uses a modified version with damping parameter $\gamma \in [0, 1]$

$$\beta_\gamma = \left(1 + \gamma \left(\frac{T^D}{T(t)} - 1\right)\right)^{1/2} \tag{3.43}$$

to scale the velocities [80]. The choice $\gamma = 1$ leads to (3.42). In the case of $\gamma = 0$ the velocities are not scaled at all. If one uses a scaling factor proportional to the time step δt of the integration method, $\gamma \sim \delta t$, then the scaling of velocities in each time step according to (3.43) results in a rate of change of the temperature which is proportional to the difference in temperatures T and T^D, i.e.

$$\frac{dT(t)}{dt} \sim (T^D - T(t)).$$

The advantage of this procedure is its simplicity. However, this approach is not able to remove undesired or local correlations in the motion of the particles.

In our program the temperature is adjusted by velocity scaling as follows: After the computation of the new velocities in routine `updateV` the velocities are multiplied every kth time step by the factor β. In the intermediate steps the system is integrated without scaling. This allows the system to restore the "equilibrium" between potential and kinetic energy. This procedure is repeated until the target temperature is reached. To this end, the target temperature T^D and the parameter k have to be introduced as new variables in the code. In routine `updateV` one has to check if the velocities are to be scaled at the present time step and to execute the scaling when required. Here, the value of the current temperature $T(t)$ is determined by the Algorithm 3.3 (appropriately adapted to the linked cell data structure) and the relation (3.41).

Additional Friction Term in the Equations of Motion. A coupling to a heat bath can also be achieved by an additional friction term in the equations of motion, compare also Section 3.6.6. Newton's equations then have the form

$$\begin{aligned} \dot{\mathbf{x}}_i &= \mathbf{v}_i, \\ m_i \dot{\mathbf{v}}_i &= \mathbf{F}_i - \xi m_i \mathbf{v}_i, \end{aligned} \qquad i = 1, \ldots, N. \qquad (3.44)$$

The additional force $-\xi m_i \mathbf{v}_i$ on particle i is proportional to the velocity of the particle. Here, the function $\xi = \xi(t)$ can depend on time. It is positive if energy has to be removed from the system and negative if energy has to be injected into the system. The form of ξ determines how fast the temperature changes. In the literature one can find different proposals for the choice of ξ, see [291, 452]. Before we examine two examples in more detail, we discuss the impact of the friction term on the integration method.

Incorporation of the Friction Term into the Integration Method. If one substitutes for the derivative of the velocity $\dot{\mathbf{v}}_i$ at time t_n a one-sided difference operator according to (3.11) in the equations of motion (3.44), one obtains the discretization

$$m_i \frac{\mathbf{v}_i^{n+1} - \mathbf{v}_i^n}{\delta t} = \mathbf{F}_i^n - \xi^n m_i \mathbf{v}_i^n.$$

Here, we again use the notation $\xi^n = \xi(t_n)$. Solving for $\mathbf{v}_i^{n+1}$ gives

$$\mathbf{v}_i^{n+1} = \mathbf{v}_i^n + \delta t\, \mathbf{F}_i^n / m_i - \delta t\, \xi^n \mathbf{v}_i^n = (1 - \delta t\, \xi^n)\mathbf{v}_i^n + \delta t\, \mathbf{F}_i^n / m_i. \quad (3.45)$$

We see that the velocity $\mathbf{v}_i^n$ in the nth time step is now multiplied by the factor $1 - \delta t\, \xi^n$. For $\xi^n > 0$ this corresponds to a decrease and for $\xi^n < 0$

this corresponds to an increase in the velocity. Thus, the kinetic energy in the system – and also the temperature of the system – decreases or increases over time.

In the examples of the Sections 3.6.2, 3.6.4, and 3.6.5, we already scaled the velocities of the particles at certain times. Now, we see that this just corresponds to an additional friction term in the equations of motion. The relation to the factor β from (3.35), which we used up to now, is given by

$$\beta = 1 - \delta t\ \xi^n,$$

or

$$\xi^n = (1 - \beta)/\delta t.$$

In this way, also the scaling of velocities according to (3.42) or (3.43) can be interpreted as the introduction of an additional friction term in the equations of motion.

Consider now the implications of the friction term on the Velocity-Störmer-Verlet algorithm as described in Section 3.1. There, the force $\mathbf{F}_i^n$ in (3.22) and (3.24) has to be replaced by $\mathbf{F}_i^n - \xi^n m_i \mathbf{v}_i^n$, and $\mathbf{F}_i^{n+1}$ has to be replaced by $\mathbf{F}_i^{n+1} - \xi^{n+1} m_i \mathbf{v}_i^{n+1}$. One obtains from (3.22) the equation

$$\mathbf{x}_i^{n+1} = \mathbf{x}_i^n + \delta t(1 - \frac{\delta t}{2}\xi^n)\mathbf{v}_i^n + \frac{\mathbf{F}_i^n \cdot \delta t^2}{2m_i}. \tag{3.46}$$

Equation (3.24) yields after solving for the velocity $\mathbf{v}_i^{n+1}$ the equation

$$\mathbf{v}_i^{n+1} = \frac{1 - \frac{\delta t}{2}\xi^n}{1 + \frac{\delta t}{2}\xi^{n+1}}\mathbf{v}_i^n + \frac{1}{1 + \frac{\delta t}{2}\xi^{n+1}}\frac{(\mathbf{F}_i^n + \mathbf{F}_i^{n+1})\delta t}{2m_i}. \tag{3.47}$$

The friction term has therefore a double effect. For one, the velocity $\mathbf{v}_i^n$ is multiplied by $1 - \frac{\delta t}{2}\xi^n$. This factor enters into the computation of the new positions and velocities. In addition, in the computation of the new velocities, the result is scaled by the factor $1/(1 + \frac{\delta t}{2}\xi^{n+1})$. The extension (3.46) and (3.47) of the Velocity-Störmer-Verlet algorithm to additional friction terms can therefore be used to calibrate the temperature for a simulation or to change the temperature of a system during a simulation and thereby to supply energy to or to withdraw energy from the system.

About the Choice of $\boldsymbol{\xi}$. We will now examine two examples for the choice of ξ in some more detail. According to (3.40), a fixed constant temperature is equivalent to a constant kinetic energy and therefore equivalent to $dE_{kin}/dt = 0$. Using the equations of motion (3.44), one obtains

$$\frac{dE_{kin}}{dt} = \sum_{i=1}^{N} m_i \mathbf{v}_i \cdot \dot{\mathbf{v}}_i = \sum_{i=1}^{N} \mathbf{v}_i \cdot (\mathbf{F}_i - \xi m_i \mathbf{v}_i)$$

$$= -\sum_{i=1}^{N} \mathbf{v}_i \cdot (\nabla_{\mathbf{x}_i} V + \xi m_i \mathbf{v}_i) = -\left(\frac{dV}{dt} + \xi \sum_{i=1}^{N} m_i \mathbf{v}_i^2\right).$$

With the choice

$$\xi = -\frac{\frac{dV}{dt}}{\sum_{i=1}^{N} m_i \mathbf{v}_i^2} = -\frac{\frac{dV}{dt}}{2E_{kin}(t)},$$

one achieves $dE_{kin}/dt = 0$ and thereby a constant temperature [607]. The function ξ here corresponds to the negative ratio of the change in the potential energy to the current kinetic energy.

In the so-called Nosé-Hoover thermostat [327, 416, 451], the heat bath is considered as an integral part of the simulated system and directly enters the computations. The heat bath is represented by an additional degree of freedom that also determines the degree of coupling of the particle system to the heat bath. The evolution of the function ξ over time, which determines the strength of the friction, is described in this approach by the ordinary differential equation

$$\frac{d\xi}{dt} = \left(\sum_{i=1}^{N} m_i \mathbf{v}_i^2 - 3Nk_BT^D \right) /M, \qquad (3.48)$$

where $M \in \mathbb{R}^+$ determines the coupling to the heat bath and has to be chosen appropriately. A large value of M leads to a weak coupling.[26]

In addition to the integration of the equations of motion (3.44) one now also has to integrate the differential equation (3.48) for the friction term ξ.

Implementation. We now describe one possible implementation of the Nosé-Hoover thermostat in the framework of the Velocity-Störmer-Verlet method (3.22) and (3.24). A symplectic variant can be found in [98], see also Chapter 6. Depending on when the right hand side of (3.48) is used in the discretization, several different methods can be derived. One example is

---

[26] Hamiltonian system have the advantage that associated stable integrators can be constructed. But in general, the equations of motion (3.44) can not be derived from a Hamiltonian. Nevertheless, there is an equivalent formulation of the Nosé-Hoover thermostat that arises from the Hamiltonian

$$\mathcal{H}(\bar{\mathbf{x}}_1, \ldots, \bar{\mathbf{x}}_N, \bar{\mathbf{p}}_1, \ldots, \bar{\mathbf{p}}_N, \gamma, \bar{\mathbf{p}}_\gamma) = \sum_{i=1}^{N} \frac{\bar{\mathbf{p}}_i^2}{2m_i\gamma^2} + V(\bar{\mathbf{x}}_1, \ldots, \bar{\mathbf{x}}_N)$$
$$+ \frac{\bar{\mathbf{p}}_\gamma^2}{2M} + 3Nk_BT^D \ln(\gamma).$$

The variables with bars are so-called virtual variables that are related to the "physical" variables by $\bar{\mathbf{x}}_i = \mathbf{x}_i, \bar{\mathbf{p}}_i/\gamma = \mathbf{p}_i, \bar{\mathbf{p}}_\gamma/\gamma = \mathbf{p}_\gamma$ and which depend on a transformed time $\bar{t}$ where $d\bar{t}/\gamma = dt$ [607], see also Section 3.7.4. Thus, time depends implicitly on the variable γ. Now, certain equations of motion can be derived from this Hamiltonian, which, after some further reformulations with help of the definition $\xi := \gamma\mathbf{p}_\gamma/M$, lead to the equations of motion (3.44) and (3.48) in the "physical" variables [327].

the discretization of the right hand side by an average of the values at the old and the new time

$$\xi^{n+1} = \xi^n + \delta t \Big(\sum_{i=1}^{N} m_i (\mathbf{v}_i^n)^2 - 3Nk_BT^D +$$

$$\sum_{i=1}^{N} m_i (\mathbf{v}_i^{n+1})^2 - 3Nk_BT^D \Big) / 2M. \qquad (3.49)$$

The first step (3.46) of the method, i.e. the computation of the new positions $\mathbf{x}_i^{n+1}$ of the particles, can be implemented without a problem, since the necessary $\mathbf{x}_i^n, \mathbf{v}_i^n, \mathbf{F}_i^n$, and ξ^n are known from the previous time step. In the computation of the new velocities $\mathbf{v}_i^{n+1}$ at time t_{n+1} due to (3.47), one needs to know the friction coefficient ξ^{n+1}. But according to (3.49), this coefficient depends on $\mathbf{v}_i^{n+1}$. The relations (3.47) and (3.49) constitute a nonlinear system of equations for the velocities $\mathbf{v}_i^{n+1}, i = 1, \ldots, N$, and the friction term ξ^{n+1}. This system can be solved iteratively using Newton's method. To this end, one needs to compute and invert a Jacobian in each iteration step. In this special case the Jacobian is a sparse matrix and can be inverted easily. A few iteration steps of Newton's method are thus sufficient to solve the nonlinear system of equations to a satisfactory accuracy [239]. Alternatively, one can solve the nonlinear system with a simple fixed point iteration. Such an approach is possible for time integrators for general Hamiltonians as presented in Section 6.2.

Another possibility is a simpler approximation for ξ^{n+1}. For example, if one uses the approximation

$$\xi^{n+1} \approx \xi^n + \delta t \Big(\sum_{i=1}^{N} m_i (\mathbf{v}_i^n)^2 - 3Nk_BT^D \Big) / M$$

instead of (3.49), then ξ^{n+1} can be computed independently of $\mathbf{v}_i^{n+1}$. A better approximation is obtained by some kind of predictor-corrector method. To this end, one can proceed as follows [247]: With the abbreviation

$$\mathbf{v}_i^{n+1/2} = \mathbf{v}_i^n + \frac{\delta t}{2} (\mathbf{F}_i^n / m_i - \xi^n \mathbf{v}_i^n),$$

it holds with (3.47) that

$$\mathbf{v}_i^{n+1} = \frac{1}{1 + \delta t \cdot \xi^{n+1}/2} \Big(\mathbf{v}_i^{n+1/2} + \frac{\delta t}{2m_i} \mathbf{F}_i^{n+1} \Big) \qquad (3.50)$$

with an unknown value ξ^{n+1}. An approximation of ξ^{n+1} can be obtained from the predictor $\mathbf{v}_i^{n+1/2}$ as

$$\xi^{n+1} \approx \xi^n + \delta t \Big(\sum_{i=1}^{N} m_i (\mathbf{v}_i^{n+1/2})^2 - 3Nk_BT^D \Big) / M.$$

Afterwards, the new velocity $\mathbf{v}_i^{n+1}$ can be computed according to (3.50).

3.7.2 Statistical Mechanics and Thermodynamic Quantities

In this section we describe the necessary fundamentals of statistical mechanics in a nutshell. A general introduction to thermodynamics and statistical mechanics can be found in one of the many textbooks, see for example [141, 366, 512, 671].

In physics one distinguishes between so-called phenomenological thermodynamics (also called thermostatics) and statistical thermodynamics (also called statistical mechanics). In phenomenological thermodynamics one makes the following assumptions:

– Macroscopic systems in equilibrium possess reproducible properties.
– Macroscopic systems in equilibrium can be described by a finite number of state variables, as for example pressure, volume, temperature, etc.

Macroscopic here means that the physical system studied consists of so many microscopic degrees of freedom that the behavior of individual degrees of freedom is not relevant to the behavior of the complete system. Physical state variables, i.e. observables, rather correspond to averages over all microscopic degrees of freedom. Equilibrium here means that the macroscopic variables such as pressure or total energy are constant over time.

In statistical thermodynamics one derives the physical behavior of macroscopic systems from statistical assumptions about the behavior of the microscopic components (i.e. for instance single atoms or molecules). The objective is to deduce the parameters of phenomenological thermodynamics from the laws governing the forces between the microscopic components.

The Phase Space. We consider a system which consists of N particles. To study the dazzling array of possible states of this system one introduces the so-called *phase space* Γ. For the three-dimensional case this is the space spanned by the $6 \cdot N$ generalized positions and momenta. An element of the phase space then corresponds to a particular physical system with N particles. If the system at time t_0 occupies the point $(\mathbf{q}_0, \mathbf{p}_0)$ of the phase space, the evolution of the system over time is described by a trajectory

$$\Phi_{\mathbf{q}_0,\mathbf{p}_0,t_0} : \quad \mathbb{R}_0^+ \longrightarrow \mathbb{R}^{3N} \times \mathbb{R}^{3N},$$
$$\Phi_{\mathbf{q}_0,\mathbf{p}_0,t_0}(t) := (\mathbf{q}_1(t), \ldots, \mathbf{q}_N(t), \mathbf{p}_1(t), \ldots, \mathbf{p}_N(t))$$

in the phase space, compare figure 3.28.

If one considers a variety of different systems that consist of the same number of particles but cannot be distinguished on the macroscopic level, i.e., they possess the same total energy and the same volume, one obtains a "cloud" in the phase space. The set of all such physically similar systems that only differ in the particular positions and velocities of the individual particles but lead to the same values for the macroscopic state variables, is called *ensemble*, compare Figure 3.29.

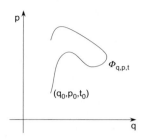

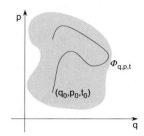

Fig. 3.28. Trajectory in phase space.   **Fig. 3.29.** An ensemble in phase space.

The trajectory of a system with a time-independent Hamiltonian cannot leave its ensemble. Thus the trajectory of the system is contained in the ensemble which is described by the quantities energy E, volume $V = |\Omega|$, and number of particles N. The set of all systems with the same energy, same volume, and same number of particles is called NVE ensemble. There are also other ensembles characterized by other state variables; an example is the NPT ensemble (P pressure, T temperature).

Statistical Averages. In statistical mechanics one studies how the systems in such an ensemble behave on average. To this end, one considers a collection of many similar systems with states that are statistically distributed in a suitable way, and introduces the concept of the phase density or N-particle distribution function

$$f_N : \mathbb{R}^{3N} \times \mathbb{R}^{3N} \times \mathbb{R}^+ \longrightarrow \mathbb{R}, \quad f_N : (\mathbf{q}, \mathbf{p}, t) \longmapsto f_N(\mathbf{q}, \mathbf{p}, t).$$

It is defined as the probability density of the system to be found in the subdomain $I := [\mathbf{q}, \mathbf{q} + d\mathbf{q}] \times [\mathbf{p}, \mathbf{p} + d\mathbf{p}]$ of the phase space. Thus, if $Prob(I)$ denotes the probability for the system to be inside the domain I, it holds that

$$Prob(I) = \int_I f_N(\mathbf{q}, \mathbf{p}, t) d\mathbf{q} d\mathbf{p}.$$

The determination of the probability density function for thermodynamic systems is one of the main tasks of statistical mechanics.[27]

To obtain macroscopic variables from the microscopic variables, i.e. from the positions and velocities of the particles, one is interested in the the so-called *ensemble average*

---

[27] The equation of motion for the N-particle distribution function is the Liouville equation [48]. It has the form of a conservation law and is derived analogously to the continuity equation in fluid dynamics using the transport theorem. So-called reduced distribution functions result for example in the Vlasov equation, the Boltzmann equation, or also the Navier-Stokes equations [56].

$$\langle A \rangle_\Gamma(t) := \frac{\int_\Gamma A(\mathbf{q}, \mathbf{p}) f_N(\mathbf{q}, \mathbf{p}, t) d\mathbf{q} d\mathbf{p}}{\int_\Gamma f_N(\mathbf{q}, \mathbf{p}, t) d\mathbf{q} d\mathbf{p}}, \tag{3.51}$$

which is just an average of a function A weighted by the N-particle distribution function f_N. Here, A is an arbitrary integrable function that depends on the coordinates and velocities of the particles and the time. The integration $\int_\Gamma$ denotes the integration over the ensemble Γ.

The macroscopic state variables of a thermodynamic system are averages over time of functions of the position and momentum coordinates of the particles of the system. The *ergodic hypothesis* [48] states that the limit of the time average

$$\langle A \rangle_\tau(\mathbf{q}_0, \mathbf{p}_0, t_0) := \frac{1}{\tau} \int_{t_0}^{t_0+\tau} A(\Phi_{\mathbf{q}_0, \mathbf{p}_0, t_0}(t)) dt \tag{3.52}$$

for $\tau \to \infty$ and the ensemble average (3.51) are equal. In particular, this implies that the limit of (3.52) for $\tau \to \infty$ does not depend on the initial coordinates $(\mathbf{q}_0, \mathbf{p}_0)$ nor on the initial time t_0 (except for a negligible set of exceptions). The equality with the ensemble average furthermore implies that the trajectory in (3.52) reaches every part of the ensemble over time. It also implies that the probability of the system being in a certain point in phase space correlates with the phase density f_N.

In a molecular dynamics simulation a particular trajectory $\Phi_{\mathbf{q}_0, \mathbf{p}_0, t_0}(t)$ of a system in phase space is computed approximately. This involves an approximation of $\Phi_{\mathbf{q}_0, \mathbf{p}_0, t_0}(t)$ at certain times $t_n = t_0 + n \cdot \delta t$, $n = 0, 1, \ldots$ For this reason, one can approximate the time average (3.52) by a sum[28]

$$\frac{\delta t}{\tau} \sum_{n=0}^{\lceil \tau/\delta t \rceil} A(\Phi_{\mathbf{q}_0, \mathbf{p}_0, t_0}(t_n)). \tag{3.53}$$

Here, the following questions have to be considered: Is the time interval τ large enough so that the limit of the time average is sufficiently well approximated?[29] Does one possibly measure in an unimportant part of phase space only? Does the ergodic hypothesis apply at all?

In the following, we employ thermostats and averaging in specific examples.

---

[28] In contrast, the Monte-Carlo method [428] relies on averaging according to (3.51). To this end, one selects points from the ensemble, evaluates the integrands in (3.51) in these points, and computes an average to approximate (3.51). These points are chosen according to a transition probability which depends on f_N.

[29] Symplectic integrators here guarantee that the approximated trajectory of a system does essentially not leave the ensemble, see also Section 6.1.

3.7.3 Liquid-Solid Phase Transition of Argon in the NVT Ensemble

We study a three-dimensional system which consists of 512 ($8 \times 8 \times 8$) particles in a cube with periodic boundary conditions. An initial configuration of the particles is given together with velocities corresponding to a temperature of 360 K (Kelvin). Initially, the particle system is relaxed with a constant temperature for 20 ps (picoseconds) and then it is subsequently cooled. First, a transition from the gas state to the liquid phase occurs, and then, a phase transition to the solid phase takes place. Depending on the cooling rate we obtain either a crystal or an amorphous, glass-like substance.

The material properties of the noble gas argon are now imposed onto the Lennard-Jones potential by the choice of the physical potential parameters as given in Table 3.9.

length	σ	$3.4 \cdot 10^{-10}$ m $= 3.4$ Å,
energy	ε	$1.65 \cdot 10^{-21}$ J $= 120$ K k_B,
mass	m	$6.69 \cdot 10^{-26}$ kg $= 39.948$ u,
time	$\sqrt{\frac{\sigma^2 m}{\varepsilon}}$	$2.17 \cdot 10^{-12}$ s $= 2.17$ ps,
velocity	$\sqrt{\frac{\varepsilon}{m}}$	$1.57 \cdot 10^2$ $\frac{m}{s}$,
force	$\frac{\varepsilon}{\sigma}$	$4.85 \cdot 10^{-12}$ N,
pressure	$\frac{\varepsilon}{\sigma^3}$	$4.22 \cdot 10^7$ $\frac{N}{m^2}$,
temperature	$\frac{\varepsilon}{k_B}$	120 K

Table 3.9. Parameters for argon and derived quantities.

Before we describe the simulation in more detail we introduce dimensionless equations at this point.

Dimensionless Equations – Reduced Variables. The idea is now to transform the variables in the equations of motions in such a way that their physical dimensions are reduced. This involves appropriately chosen reference quantities in the following way:

$$\text{dimensionless variable} = \frac{\text{variable with dimension}}{\text{reference quantity with the same dimension}}.$$

The reference quantities should be characteristic for the problem and they have to be constant. One goal of this approach is to obtain the relevant quantities and coefficients for the evolution of the system. In addition, the use of reduced variables avoids problems in the computation with unfavorably chosen physical units that could lead to large rounding errors. Furthermore, computations for one set of parameters can often directly be transformed to give results for another set of parameters.

As an example we consider the Lennard-Jones potential (3.26). Its parameters are the depth ε of the potential and the position σ where the potential crosses zero. Newton's equations of motion with this potential read

$$m\frac{\partial^2}{\partial t^2}\mathbf{x}_i = -24\varepsilon\sum_{\substack{j=1\\j\neq i}}^{N}\left(2\left(\frac{\sigma}{r_{ij}}\right)^{12} - \left(\frac{\sigma}{r_{ij}}\right)^{6}\right)\cdot\frac{\mathbf{r}_{ij}}{r_{ij}^2}. \qquad (3.54)$$

Now, characteristic reference quantities are chosen and the other quantities are scaled by them. Here, we use the length $\tilde{\sigma}$, the energy $\tilde{\varepsilon}$, and the mass $\tilde{m}$ and scale the quantities as follows:

$$\begin{aligned} m' = m/\tilde{m}, \quad &\mathbf{x}_i' = \mathbf{x}_i/\tilde{\sigma}, \quad \mathbf{r}_{ij}' = \mathbf{r}_{ij}/\tilde{\sigma}, \quad E' = E/\tilde{\varepsilon}, \quad V' = V/\tilde{\varepsilon}, \\ \sigma' = \sigma/\tilde{\sigma}, \quad &\varepsilon' = \varepsilon/\tilde{\varepsilon}, \quad T' = Tk_B/\tilde{\varepsilon}, \quad t' = t/\tilde{\alpha}, \end{aligned} \qquad (3.55)$$

where $\tilde{\alpha} = \sqrt{\frac{\tilde{m}\tilde{\sigma}^2}{\tilde{\varepsilon}}}$. With the relations

$$\frac{\partial\mathbf{x}_i}{\partial t} = \frac{\partial(\tilde{\sigma}\mathbf{x}_i')}{\partial t} = \tilde{\sigma}\frac{\partial\mathbf{x}_i'}{\partial t'}\frac{\partial t'}{\partial t} = \frac{\tilde{\sigma}}{\tilde{\alpha}}\frac{\partial\mathbf{x}_i'}{\partial t'} \qquad (3.56)$$

and

$$\frac{\partial^2\mathbf{x}_i}{\partial t^2} = \frac{\partial}{\partial t}\frac{\tilde{\sigma}}{\tilde{\alpha}}\frac{\partial\mathbf{x}_i'}{\partial t'} = \frac{\tilde{\sigma}}{\tilde{\alpha}^2}\frac{\partial^2\mathbf{x}_i'}{\partial t'^2},$$

one obtains by substitution into (3.54)

$$\tilde{m}m'\frac{\partial^2(\mathbf{x}_i'\tilde{\sigma})}{(\partial(t'\tilde{\alpha}))^2} = -24\sum_{\substack{j=1\\j\neq i}}^{N}\varepsilon'\tilde{\varepsilon}\left(2\left(\frac{\sigma'}{r_{ij}'}\right)^{12} - \left(\frac{\sigma'}{r_{ij}'}\right)^{6}\right)\frac{\mathbf{r}_{ij}'}{(r_{ij}')^2\tilde{\sigma}}$$

and therefore the equations of motion

$$m'\frac{\partial^2\mathbf{x}_i'}{\partial t'^2} = -24\sum_{\substack{j=1\\j\neq i}}^{N}\varepsilon'\left(2\left(\frac{\sigma'}{r_{ij}'}\right)^{12} - \left(\frac{\sigma'}{r_{ij}'}\right)^{6}\right)\frac{\mathbf{r}_{ij}'}{(r_{ij}')^2}.$$

With this dimensionless formulation, problems with very large or very small values of the variables no longer occur. Furthermore, systems with the same values for σ' and ε' behave in the same way, i.e., for two different systems with different physical parameters but the same values for σ' and ε' and the same initial conditions in reduced form one obtains the same trajectories in the reduced system.

Quantities as for instance the kinetic or the potential energy can be computed directly from the reduced variables. Using (3.56) and the definition of $\tilde{\alpha}$ one obtains for particles with the same mass m

$$E_{kin} = \tilde{\varepsilon}E_{kin}' = \frac{1}{2}\sum_i m\left(\frac{\partial\mathbf{x}_i}{\partial t}\right)^2 = \frac{1}{2}m'\tilde{\varepsilon}\sum_i\left(\frac{\partial\mathbf{x}_i'}{\partial t'}\right)^2 \qquad (3.57)$$

and

$$E_{pot} = \tilde{\varepsilon} E'_{pot} = \tilde{\varepsilon} \frac{1}{\tilde{\varepsilon}} E_{LJ} = \sum_{i,j,i<j} 4\varepsilon' \tilde{\varepsilon} \left(\left(\frac{\sigma'}{r'_{ij}} \right)^{12} - \left(\frac{\sigma'}{r'_{ij}} \right)^{6} \right). \quad (3.58)$$

For the temperature it holds that

$$T = T' \tilde{\varepsilon} / k_B. \quad (3.59)$$

In the following, we choose for simplicity $\tilde{\sigma} := \sigma$, $\tilde{\varepsilon} := \varepsilon$, and $\tilde{m} := m$. This directly implies $\sigma' = 1$, $\varepsilon' = 1$ and $m' = 1$.

Crystallization of Argon. The simulation runs as follows: The initial positions of the particles are chosen on a regular grid as shown in Figure 3.30. The simulation domain is a periodically continued cube with an edge length of 31.96 Å which corresponds to a scaled, dimensionless value of 9.4, see Tables 3.10 and 3.11. For heating and cooling we use the simple scaling method from Section 3.7.1. First, the system is brought up to a temperature of $T' = 3.00$ (which corresponds to 360 K). To this end, 50 integration steps are performed between two successive scalings of the velocity. After this initial temperature is reached, the system is cooled in steps of $7.8 \cdot 10^{-4}$ to a temperature of $T' = 0.5$, where again 50 integration steps are performed between the respective scalings.

$\varepsilon = 1.65 \cdot 10^{-21}$ J,	$\sigma = 3.4$ Å,	$m = 39.948$ u,
$L_1 = 31.96$ Å,	$L_2 = 31.96$ Å,	$L_3 = 31.96$ Å,
N $= 8^3$,	$T = 360$ K,	
$r_{cut} = 2.3\,\sigma$,	$r_l = 1.9\,\sigma$,	$\delta t = 0.00217$ ps

Table 3.10. Parameter values with units for the simulation of argon.

$\varepsilon' = 1$,	$\sigma' = 1$,	$m' = 1$,
$L'_1 = 9.4$,	$L'_2 = 9.4$,	$L'_3 = 9.4$,
N $= 8^3$,	$T' = 3.00$,	
$r'_{cut} = 2.3$,	$r'_l = 1.9$,	$\delta t' = 0.001$

Table 3.11. Parameter values as scaled quantities for the simulation of argon.

Besides the kinetic and potential energy of the system (equations (3.57) and (3.58)), we measure some further statistical data of the simulation that will be introduced in the following.

The instantaneous pressure P_{int} of a particle system consists of a kinetic part and a force term, i.e.

$$P_{int} = \frac{1}{3|\Omega|} \left(\sum_i m_i \dot{\mathbf{x}}_i^2 + \sum_i \mathbf{F}_i \mathbf{x}_i \right). \quad (3.60)$$

A time average of this quantity results in the internal pressure of the system. For constant volume $|\Omega|$, one can then detect phase transitions by abrupt changes in the pressure.

The diffusion of the particle system is measured as the mean standard deviation of the particle positions. To this end, we determine the distance[30] of the position of each particle at time t to its position at a reference time t_0, and compute

$$\text{Var}(t) = \frac{1}{N} \sum_{i=1}^{N} \|\mathbf{x}_i(t) - \mathbf{x}_i(t_0)\|^2 . \tag{3.61}$$

The diffusion equals zero at time t_0 and increases strongly at first. Therefore, one restarts the measurement in regular intervals. Here, one sets $t_0 = t$ and measures the diffusion after a fixed time anew. A transition from the gas phase to the liquid state and from the liquid to the solid state can be detected from the value of the diffusion. At a phase transition, the value of the diffusion decreases abruptly.

The radial distribution function $g(r)$ describes the probability to find a pair of particles with distance r. For its computation one determines all $\frac{N(N-1)}{2}$ distances, sorts them, and produces a histogram such that $\rho([r, r + \delta r))$ gives the number of particle pairs with distance in the interval $[r, r+\delta r)$. If one divides this number ρ by the volume

$$\frac{4\pi}{3} \left((r + \delta r)^3 - r^3 \right)$$

of the domain spanned by the range of distances r to $r + \delta r$, one obtains a particle density [282]. The continuous analog of this density is

$$g(r) = \frac{\rho(r)}{4\pi r^2 \int_{r'=0}^{R} \rho(r')} , \tag{3.62}$$

where we scale the absolute numbers additionally with the number of all particles up to a maximal distance R. To this end, we choose R as the largest distance r found in the histogram.

An example for a radial distribution function is shown in Figure 3.30. Here, small values of r up to a few atom distances are interesting. Then, the particle distances can be computed efficiently by using the linked cell method. Single peaks in the distribution function indicate fixed distances as in crystals while more uniform distributions indicate disordered states. Furthermore, one can compute a more accurate statistic by averaging the distances over several time steps.

---

[30] In the case of periodic boundary conditions, we have to compute the distance to the real position of the particle. To this end, if a particle leaves the simulation domain at one side and reenters it on the opposite side, the value of its initial position $\mathbf{x}_i(t_0)$ is to be corrected appropriately.

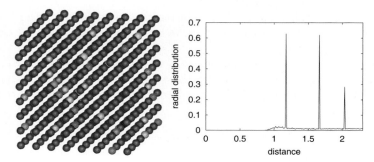

Fig. 3.30. Initial configuration and initial radial distribution.

In the following simulation, the particle interactions are governed by a modified Lennard-Jones potential

$$U_{ij}(r_{ij}) = 4 \cdot \varepsilon \cdot S(r_{ij}) \left(\left(\frac{\sigma}{r_{ij}} \right)^{12} - \left(\frac{\sigma}{r_{ij}} \right)^{6} \right). \tag{3.63}$$

Here, $S(r)$ is a smoothing function defined as

$$S(r) = \begin{cases} 1 & : \quad r \leq r_l, \\ 1 - (r - r_l)^2 (3r_{\text{cut}} - r_l - 2r)/(r_{\text{cut}} - r_l)^3 & : \quad r_l < r < r_{\text{cut}}, \\ 0 & : \quad r \geq r_{\text{cut}}. \end{cases} \tag{3.64}$$

This function guarantees that the potential as well as the forces continuously decrease to zero between r_l and r_{cut}. The parameters r_l and r_{cut} are chosen as $r_l = 1.9\sigma$ and $r_{\text{cut}} = 2.3\sigma$.

In Figure 3.31 the radial distribution function, the potential energy, the diffusion as standard deviation, and the pressure for the crystallization of argon are shown for the NVT ensemble. Temperature is controlled by a rescaling every 50 integration steps. We see that something drastic happens at $t' = 150$: We observe a sharp bend in the curve for the energy, a turning point in the pressure curve, and a jump in the time evolution of the diffusion (which is superimposed with a jump caused by the third rescaling step). The time $t' = 150$ corresponds to an actual temperature of 84 K. The temperature for the transition into the crystal state is approximately $T'_K = 0.71$ which translates to $T_K = 84$ K. This matches (up to an accuracy of 0.2%) with the physical melting point of argon at 83.8 K and (up to an accuracy of 4%) with the close-by boiling point at 87.3 K. Thus, in line with the accuracy of the simulation, we observe both transitions at the same time. The transition from the gas phase (over the liquid) to the solid state occurs and a crystal is formed in the process. The spatial arrangement of the argon atoms and the regularity of the crystal lattice can be seen from the graph of the radial distribution function in figure 3.31. In the gas phase, disorder is prevalent (we

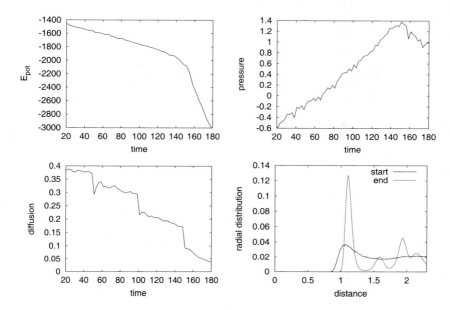

Fig. 3.31. Crystallization of argon in the NVT ensemble with rescaling. Potential energy, pressure, diffusion as standard deviation, and radial distribution function, in scaled units for the cooling phase.

have a smooth, almost constant function starting at $r' = 1$). In the crystal phase, characteristic peaks in the distribution function arise that correspond to the distances in the crystal lattice.

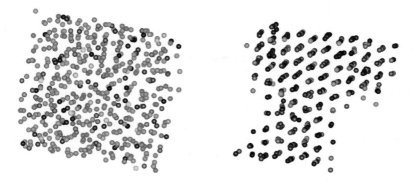

Fig. 3.32. Crystallization of argon in the NVT ensemble with rescaling, particle distribution for $t' = 140$ (left) and $t' = 250$ (right). The colors of the particles encode their velocities.

Figure 3.32 shows a snapshot of the argon particles at time $t' = 140$ (which is associated to the gas phase at 100 K) and at time $t' = 250$ (which is associated to the crystal phase at 60 K). In this simulation the system is kept at a constant temperature (60 K) from $t' = 150$ to $t' = 250$. We clearly see the resulting crystal structure. During the phase transition from liquid to solid most materials, including argon, release latent heat, and therefore the density and thus the volume change abruptly. In our simulation we work with the NVT ensemble in which the number of particles, the volume (i.e. the size of the simulation domain) and the temperature is kept constant. Therefore, the transition to the crystalline phase leads to an unphysical hole in the crystal as observed in Figure 3.32 (right).

Supercooling of Argon to a Glass State. In this example we consider the supercooling of argon. Many liquids do not change to a crystalline state but to a glass state if they are cooled fast enough. The theory of supercooling is based on the singular behavior of the solution of the so-called mode-coupling equations [517]. These equations are a simplified form of certain nonlinear equations of motion. Here, molecular dynamics methods can be applied with good success [358]. Simulations of the supercooling of argon and related substances can be found for instance in [210, 237, 269, 438].[31]

In our simulations, the interaction between particles is again realized by the modified Lennard-Jones potential (3.63) using the set of parameters from Table 3.11. The initial positions of the particles are chosen as the regular lattice from Figure 3.30.

The simulation proceeds similarly to the last experiment, only the cooling now happens significantly faster. First, the system is heated to the temperature $T' = 3.00$, with 25 integration steps between each rescaling of the velocities. The system is then subsequently cooled by a linear reduction of the temperature in steps of $2.5 \cdot 10^{-3}$ down to a temperature $T' = 0.02$. After each scaling an equilibration phase of 25 time steps is employed.

The graphs of the computed potential energy, the pressure, the diffusion, and the radial distribution function are shown in Figure 3.33. One can clearly recognize a phase transition in the time range $t' = 46$ to $t' = 48$. The time $t' = 47$ for the glass transition corresponds to a temperature of 38 K or approximately $T'_G = 0.3$. Thus, the phase transition occurs significantly later than in the previous experiment where a slower cooling rate was employed. It also takes place at a significantly lower critical temperature than the physical melting point of argon which is 83.8 K. The supercooled liquid is in a metastable state, the phase transition then occurs like a shock and results in a

---

[31] However, a problem is caused by the limited physical time for which the simulations can be run. This enforces cooling rates that are several times larger than the cooling rates in laboratory experiments. As a consequence, the measured critical temperature for the transition into the glass state is higher in simulations than in laboratory experiments.

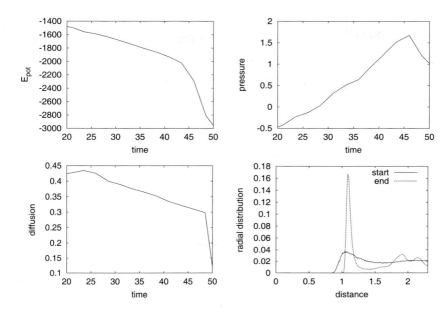

Fig. 3.33. Supercooled argon in the NVT ensemble with rescaling. Potential energy, pressure, diffusion, and radial distribution function in scaled units for the cooling phase.

characteristic amorphous glass state. The associated radial distribution function also supports such a finding. It differs significantly from the distribution function for a crystal structure and exhibits a more disordered, amorphous state.

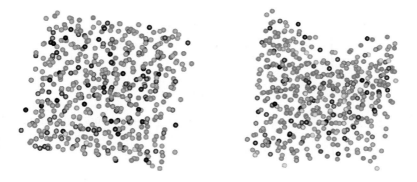

Fig. 3.34. Supercooled argon in the NVT ensemble with rescaling, particles for $t' = 42.5$ (left) and $t' = 47.5$ (right).

Figure 3.34 displays two configurations, one shortly before, and the other one shortly after the phase transition. Both pictures show disordered particle systems. The system is transformed during cooling into an amorphous solid state and stays in that state for the remaining simulation time. Thus, in contrast to the previous simulation, no crystal structure develops.

Again, one can observe an unphysical hole in the structure which stems from the abrupt increase in density at the phase transition. This would be avoided if the volume of the simulation domain could adapt to the new state. To allow such an adaptation one has to use a different ensemble that employs a variable volume instead of the NVT ensemble which has been used up to now. Such a technique and ensemble are discussed in the next section.

3.7.4 The Parrinello-Rahman Method for the NPT Ensemble

It is known that a system can be described by its Lagrangian. If momenta are introduced, the Lagrangian can be transformed into the corresponding Hamiltonian, and the equations of motion of the system can be derived, see Appendix A.1 for details. We will now follow such a procedure to formulate a method for a system in the NPT ensemble. The idea is to introduce additional degrees of freedom for the whole coordinate system in time and space which can be used to control the volume and the shape as well as the pressure and the temperature of the system [453]. We obtain these additional degrees of freedom by a transformation of the spatial coordinates $\mathbf{x}_i$ to scaled positional coordinates $\bar{\mathbf{x}}_i \in [0, 1)^3$ according to

$$\mathbf{x}_i = A\bar{\mathbf{x}}_i. \tag{3.65}$$

Here, $A = [\mathbf{a_0}, \mathbf{a_1}, \mathbf{a_2}]$ is a 3×3 matrix that depends on time. It is formed from the basis vectors $\mathbf{a_0}, \mathbf{a_1}, \mathbf{a_2}$ of the periodic simulation box. We furthermore scale the time t to $\bar{t}$ by virtue of

$$t = \int_0^{\bar{t}} \frac{d\tau}{\gamma(\tau)}, \quad \text{thus} \quad d\bar{t} = \gamma(\bar{t})dt.$$

Then, for the velocities, it holds

$$\dot{\mathbf{x}}_i(t) = \gamma(\bar{t})A(\bar{t})\dot{\bar{\mathbf{x}}}_i(\bar{t}).$$

The matrix A and the variable γ control the pressure and the temperature of the extended system. To this end, we also define the fictitious potentials of the thermodynamic variables P and T

$$V_P = P_{\text{ext}} \det A, \quad V_T = N_f k_B T^D \ln \gamma,$$

where P_{ext} denotes the external pressure of the system, T^D the target temperature, N_f the number of degrees of freedom, and $\det A$ the volume of the simulation box spanned by $\mathbf{a_0}, \mathbf{a_1}, \mathbf{a_2}$.

The extended Lagrangian for the NPT ensemble is now defined as[32]

$$\mathcal{L} = \frac{1}{2}\sum_{i=1}^{N} m_i\gamma^2\dot{\bar{\mathbf{x}}}_i^T A^T A\dot{\bar{\mathbf{x}}}_i + \frac{1}{2}W\gamma^2\text{tr}(\dot{A}^T\dot{A}) + \frac{1}{2}M\dot{\gamma}^2 \qquad (3.66)$$
$$-V(A\bar{\mathbf{x}}, A) - P_{\text{ext}}\det A - N_f k_B T^D \ln\gamma$$

where m_i denotes the mass of th ith particle, M denotes the (fictitious) mass of the Nosé-Hoover thermostat, and W denotes the (fictitious) mass of the so-called barostat (pressure controller). With the notation $G := A^T A$ we obtain as conjugated moments

$$p_{\bar{\mathbf{x}}_i} = \mathcal{L}_{\dot{\bar{\mathbf{x}}}_i} = m_i\gamma^2 G\dot{\bar{\mathbf{x}}}_i, \quad p_A = \mathcal{L}_{\dot{A}} = \gamma^2 W\dot{A}, \quad p_\gamma = \mathcal{L}_{\dot{\gamma}} = M\dot{\gamma}.$$

The Hamiltonian can therefore be written as

$$\mathcal{H} = \frac{1}{2}\sum_{i=1}^{N}\frac{p_{\bar{\mathbf{x}}_i}^T G^{-1} p_{\bar{\mathbf{x}}_i}}{m_i\gamma^2} + \frac{1}{2}\frac{\text{tr}(p_A^T p_A)}{\gamma^2 W} + \frac{p_\gamma^2}{2M} \qquad (3.67)$$
$$+V(A\bar{\mathbf{x}}, A) + P_{\text{ext}}\det A + N_f k_B T^D \ln\gamma.$$

The use of a constant time step in rescaled time would lead to variable time steps in physical time which complicates the implementation of an integration method. Therefore, we transform the time back to the original time. Furthermore, the equations of motion are simplified by a multiplication of the momenta with G^{-1} and by taking the logarithm of γ. Thus, we transform the variables a second time according to

$$\hat{\mathbf{x}}_i(t) := \bar{\mathbf{x}}_i(\bar{t}), \quad \hat{A}(t) := A(\bar{t}), \quad \hat{G}(t) := G(\bar{t}),$$
$$\eta(t) := \ln\gamma(\bar{t}), \quad p_{\hat{\mathbf{x}}_i} := G^{-1} p_{\bar{\mathbf{x}}_i}/\gamma, \quad p_{\hat{A}} := p_A/\gamma.$$

With this transformation, the Hamiltonian (3.68) becomes[33]

$$\mathcal{H} = \frac{1}{2}\sum_{i=1}^{N}\frac{p_{\hat{\mathbf{x}}_i}^T G p_{\hat{\mathbf{x}}_i}}{m_i} + \frac{1}{2}\frac{\text{tr}(p_{\hat{A}}^T p_{\hat{A}})}{W} + \frac{1}{2}\frac{p_\gamma^2}{M} \qquad (3.68)$$
$$+V(A\bar{\mathbf{x}}, A) + P_{\text{ext}}\det\hat{A} + N_f k_B T^D \eta,$$

---

[32] In (3.66), $A\bar{\mathbf{x}}$ is an abbreviation for $(A\bar{\mathbf{x}}_1, ..., A\bar{\mathbf{x}}_N)$. In the case of periodic boundary conditions, the potential V explicitly depends on A. For instance, for a pair potential it holds that

$$V(\mathbf{x}, A) = \frac{1}{2}\sum_{\mathbf{z}\in\mathbb{Z}^3}\sum_{\substack{i,j=1\\i\neq j \text{ if } \mathbf{z}=0}}^{N} U(\mathbf{x}_j - \mathbf{x}_i + A\mathbf{z}).$$

This additional dependence on A in the periodic case must be taken into account in any differentiation with respect to A.

[33] After this transformation, (3.68) is no longer a Hamiltonian of the system since Hamilton's equations can no longer directly be gained from it. Instead, they have to be derived from (3.68) while considering the transformations accordingly.

and the associated equations of motion read

$$\dot{\mathbf{x}}_i = \frac{p_{\hat{\mathbf{x}}_i}}{m_i}, \quad \dot{A} = \frac{p_{\hat{A}}}{W}, \quad \dot{\eta} = \frac{p_\gamma}{M}, \tag{3.69}$$

$$\dot{p}_{\hat{\mathbf{x}}_i} = -\hat{A}^{-1}\nabla_{\mathbf{x}_i}V - \hat{G}^{-1}\dot{\hat{G}}p_{\hat{\mathbf{x}}_i} - \frac{p_\gamma}{M}p_{\hat{\mathbf{x}}_i}, \tag{3.70}$$

$$\dot{p}_{\hat{A}} = -\sum_{i=1}^N \nabla_{\mathbf{x}_i}V\hat{\mathbf{x}}_i^T - \nabla_A V + \sum_{i=1}^N m_i\hat{A}\dot{\hat{\mathbf{x}}}_i\dot{\hat{\mathbf{x}}}_i^T - \hat{A}^{-T}P_{\text{ext}}\det\hat{A} - \frac{p_\gamma}{M}p_{\hat{A}}, \tag{3.71}$$

$$\dot{p}_\gamma = \sum_{i=1}^N \frac{p_{\hat{\mathbf{x}}_i}^T \hat{A}^T \hat{A}p_{\hat{\mathbf{x}}_i}}{m_i} + \frac{\text{tr}(p_{\hat{A}}^T p_{\hat{A}})}{W} - N_f k_B T^D. \tag{3.72}$$

For the stress tensor and thus the pressure one now obtains

$$\Pi_{\text{int}} = \frac{1}{\det\hat{A}} \sum_{i=1}^N \left(m_i\hat{A}\dot{\hat{\mathbf{x}}}_i\dot{\hat{\mathbf{x}}}_i^T \hat{A}^T - \nabla_{\mathbf{x}_i}V\hat{\mathbf{x}}_i^T \hat{A}^T \right), \quad P_{\text{int}} = \frac{1}{3}\text{tr}(\Pi_{\text{int}}).$$

By this approach we introduced with A nine new degrees of freedom into equation (3.65). Now, the question arises if these degrees of freedom are physically meaningful. Without a further restriction this is certainly not the case, since at least rotations of the system should be excluded. To this end, there exist the following three methods:

– The entries of the force $F_{\hat{A}} := \dot{p}_{\hat{A}}$ acting on $\hat{A}$ are set to zero below the diagonal, i.e.
$$F_{\hat{A}\alpha,\beta} = 0, \quad \alpha > \beta.$$

Thus, a reactive force that avoids rotations is directly applied.
– Symmetry is enforced by using

$$F_{\hat{A}}^S = \frac{1}{2}(F_{\hat{A}} + F_{\hat{A}}^T). \tag{3.73}$$

This eliminates the redundant degrees of freedom, but in general leads to distorted computational domains.
– Five constraints

$$\frac{\hat{A}_{\alpha,\beta}}{\hat{A}_{11}} = \frac{\hat{A}_{\alpha,\beta}^0}{\hat{A}_{11}^0}, \quad \alpha \le \beta, \tag{3.74}$$

are introduced, where the reference matrix $\hat{A}^0$ is chosen as the initial matrix A. Thus, due to these constraints and the symmetry of the stress tensor, only one degree of freedom $\hat{A}_{11}$ remains which can then be used for isotropic volume control. For a pressure control with isotropic volume, the entire trace of $F_{\hat{A}}$ can be used similarly to the so-called Andersen thermostat [42].

Implementation. In principle, the equations of motion (3.69)–(3.72) can be implemented as before. However, one has to pay attention to the integration method since after differentiation of $\dot{\mathbf{x}}_i = p_{\hat{\mathbf{x}}_i}/m_i$ and substitution of (3.70) the relation

$$\ddot{\mathbf{x}}_i = \frac{-\hat{A}^{-1}\nabla_{\mathbf{x}_i} V}{m_i} - \hat{G}^{-1}\dot{\hat{G}}\dot{\mathbf{x}}_i - \dot{\eta}\dot{\mathbf{x}}_i$$

holds. Thus, $\ddot{\mathbf{x}}_i(t)$ depends on $\dot{\mathbf{x}}_i(t)$. An analogous statement is valid for $\ddot{\hat{A}}$ and $\ddot{\eta}$. This problem can be solved for instance with a variant of the Störmer-Verlet method for general Hamiltonians or with a symplectic Runge-Kutta method, see Section 6.2.

To compute the temperature we have to determine the number of effective degrees of freedom. For systems of atoms without bonds this is just the number of atoms times three.[34] We here distinguish (compare the footnote on page 87)

- the *atomic scaling* $N_f = 3N$, where N is the number of atoms,
- the *non-periodic scaling* $N_f = 3N - 6$, since the translation and rotation of the center of gravity can be ignored,
- the *periodic scaling* $N_f = 3N - 3$, since in the periodic case only the translation of the center of gravity can be ignored.

Thus, we can define the temperature and the kinetic energy as

$$T_{\text{ins}} = \frac{2E_{\text{kin}}}{N_f k_B} \quad \text{and} \quad E_{\text{kin}} = \frac{1}{2}\sum_{i=1}^{N} \frac{p_{\hat{\mathbf{x}}_i}^T \hat{A}^T \hat{A} p_{\hat{\mathbf{x}}_i}}{m_i} .$$

This temperature can be used as target temperature T^D in the thermostat in (3.68). For the Parrinello–Rahman method we also need the fictitious masses W and M of the barostat and thermostat which have to be chosen appropriately [462].

3.7.5 Liquid-Solid Phase Transition of Argon in the NPT Ensemble

Again, we study the system of 512 argon particles from Section 3.7.3. The atoms are initially arranged in a periodic cube at a temperature of 360 K.

---

[34] So far we have used the atomic scaling of the particles for temperature control. If molecules are to be simulated, it makes sense to separate the center of gravity from the coordinates of the atoms. Then, the atoms are parameterized in a local coordinate system with respect to the common center of gravity such that a scaling does not tear the molecule apart. From this parameterization one obtains the corresponding Lagrangian and Hamiltonian and finally the equations of motion in the associated relative coordinates. Furthermore, in the case of molecules, the number of inner bonds must be subtracted from the number of degrees of freedom.

They are cooled with different cooling rates. In contrast to the previous experiments we now consider an NPT ensemble. The phase transition from the gas phase to the solid phase or to the supercooled liquid and glass state similarly occurs as before, but we now observe changes in the volume instead of the pressure.

For the implementation of the NPT ensemble we use the Parrinello–Rahman method from the previous Section 3.7.4 with a fictitious mass of $W = 100$. The temperature is controlled by a Nosé-Hoover thermostat with a fictitious mass of $M = 10$. The resulting system is integrated by a variant of the Störmer-Verlet method for general Hamiltonians, see also Section 6.2.

Crystallization of Argon. First, we slowly cool the system from 360 K to 60 K (i.e. from $T' = 3$ to $T' = 0.5$) within the physical time interval from 20 ps to 180 ps. The volume of the simulation domain is allowed to change.

We discussed several possibilities which make changes in the volume unique by imposing certain constraints. We now apply two of these approaches, the symmetric constraint (3.73) which enforces a symmetric stress tensor and the isotropic constraint (3.74) which results in an undistorted domain.

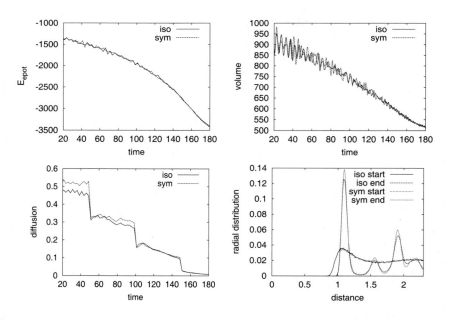

Fig. 3.35. Crystallization of argon in the NPT ensemble: Isotropic (3.74) and symmetric (3.73) constraints. Potential energy, volume, diffusion, and radial distribution function in scaled units for the cooling phase.

In Figure 3.35 the potential energy, the diffusion, the volume, and the radial distribution function are presented. In contrast to the experiments with the NVT ensemble, the phase transition can not be identified from the graphs anymore. The associated jump in the density is just compensated in the NPT ensemble by a change in the volume.[35] At the end of the simulation, however, the radial distribution function shows distinctive signs of a crystal structure with its specific peaks at characteristic lattice distances. These peaks are even more developed than in the simulation with constant volume in the NVT ensemble.

The characteristic differences of a simulation in the NPT ensemble compared to a simulation in the NVT ensemble can be seen in Figures 3.36 and 3.37.

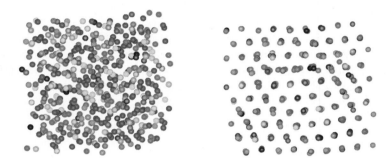

Fig. 3.36. Crystallization of argon in the NPT ensemble: Isotropic constraints (3.74), particles at $t' = 140$ (left) and $t' = 250$ (right).

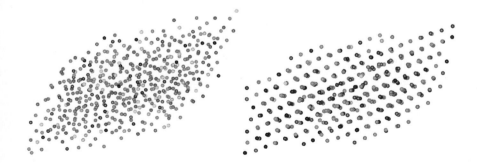

Fig. 3.37. Crystallization of argon in the NPT ensemble: Symmetric constraints (3.73), particles at $t' = 140$ (left) and $t' = 250$ (right).

---

[35] The jumps in the diffusion are caused by the restarts in the computation of the standard deviation.

The volume of the domain decreases during the cooling period. Thus, lattice imperfections and holes in the domain are avoided. Figure 3.36 shows how the simulation cube uniformly contracts (the atoms therefore appear somewhat larger than in Figure 3.32) and the atoms adopt positions in a crystal lattice. For the symmetric constraint we see in Figure 3.37 that the domain is highly distorted. The lattice structure of the annealed crystal is clearly visible.

Supercooling of Argon to a Glass State. Finally we carry out an experiment with a significantly larger cooling rate. To this end, we linearly reduce the temperature from 360 K to 2.4 K within 30 ps. Then, instead of a crystal, an amorphous substance is formed.

Figure 3.38 displays the measured values for symmetric constraints (3.73) and for isotropic constraints (3.74). The difference in the diffusion stems from the fact that the mean free path in the distorted domain is somewhat longer than in the isotropically contracted domain. Again, in contrast to the NVT simulation, the exact time for the phase transition can not be recognized from the graphs. But the radial distribution function clearly signals an amorphous, glass-like substance at the end of the simulation. There are no peaks in the distribution function that would be characteristic for crystals.

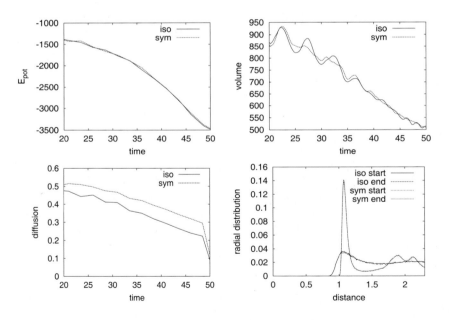

Fig. 3.38. Supercooled argon in the NPT ensemble: Isotropic (3.74) and symmetric (3.73) constraints. Potential energy, volume, diffusion, and radial distribution function in scaled units for the cooling phase.

This can also be seen in the Figures 3.39 and 3.40. The resulting structure differs significantly from the ordered structure of a crystal for both the
isotropic constraints and the symmetric constraints.

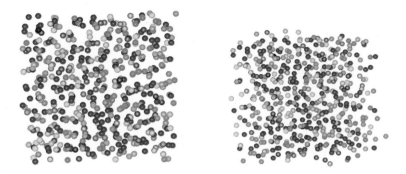

Fig. 3.39. Supercooled argon in the NPT ensemble: Isotropic constraints (3.74),
particles at $t' = 42.5$ (left) and $t' = 47.5$ (right).

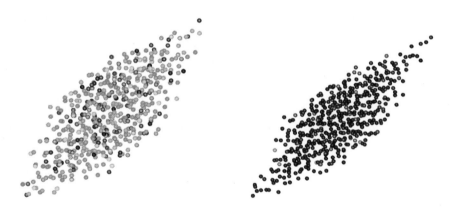

Fig. 3.40. Supercooled argon in the NPT ensemble: Symmetric constraints (3.73),
particles at $t' = 42.5$ (left) and $t' = 250$ (right).

4 Parallelization

In the following, we discuss the parallelization of the linked cell method from Chapter 3. We will use domain decomposition [568] as parallelization technique and MPI (message passing interface) [7] as a communication library. Parallelization is used to reduce the time needed to execute the necessary computations. This is done by distributing the computations to several processors, which can then execute these computations simultaneously, at least to some extent. In addition, parallelization also has the advantage that on a parallel computer there is often more memory available than on a single processor machine, and hence, larger problems can be tackled.

In the last years, the development of modern computers has led to more and more powerful scalable parallel computer systems, see Figure 1.2. By now, such systems allow molecular dynamics simulations with many hundreds or thousands of millions of particles. The proper usage of parallel computers used to be an art since programming systems were very machine specific and programs developed on those machines were hard to test and difficult to port to other machines. Nowadays, however, there are (almost) fully developed programming environments that allow debugging of parallel codes and also ensure portability between different parallel computers.

We start with an overview of parallel computers and different parallelization strategies. Then, in Section 4.2, we present domain decomposition as parallelization strategy for the linked cell method. In Section 4.3 we discuss in detail its implementation with MPI. Finally, in Section 4.5, we present some application examples for our parallelized algorithm. We extend examples from Section 3.6 from the two- to the three-dimensional case.

4.1 Parallel Computers and Parallelization Strategies

Taxonomy of Parallel Computers. Since 1966 (Flynn [234]) parallel computer systems are categorized depending on whether the data stream and/or the instruction stream are processed in parallel. In this way, the fundamental types SISD (single instruction/single data stream – the classical microprocessor), SIMD (single instruction/multiple data stream) and MIMD (multiple instruction/multiple data stream) can be distinguished.[1]

Older parallel computers by MasPar and the Connection Machine Series by Thinking Machines or some current designs such as the "array processor experiment" (APE) fall for instance into the class of SIMD computers. On these computers, programs are executed on an array of very many, but simple processors. However, this particular architecture nowadays plays a minor role and is used only for some specific applications. Vector computers, such as Cray T90 and SV1/2, NEC SX-5 or Fujitsu VPP, also fall into the class of SIMD computers. In such computers, the same instructions are executed in a quasi-parallel manner using the assembly line principle. In a certain sense, RISC (reduced instruction set computer) microprocessors also belong to this class. A RISC processor usually only executes simple instructions, but with very high speed. Every such instruction is again split into smaller subinstructions that are processed in a pipeline on the instruction level. In this way, the processor always works on several instructions at the same time (i.e. in parallel).[2]

Most of the current parallel computers are of the MIMD type. In MIMD computers, every processor executes its own sequence of instructions in the form of its own program. Here, one has to distinguish between MIMD multi-processor systems with shared or with distributed memory.

Systems with *shared* memory have a large global memory to which different processors have read and write access. The shared memory may be realized as one large bank of memory, as several distributed smaller memory banks, or even as memory distributed to all processors, compare Figure 4.1. However, such systems, regardless of realization, allow the programmer to access the entire shared memory (at least virtually) within a global address space.

Systems with shared memory permit a relatively easy parallelization of sequential programs since memory is globally addressable and therefore no significant changes have to be made to the data structures. In addition, sections of the program that can be executed independently can easily be determined and executed in parallel. Control directives have to be inserted into the

---

[1] For a detailed review of current parallel computers see [8].

[2] Further developments led to very long such pipelines (super-pipelining). Nowadays, the increased integration density of transistors on chips allows several arithmetic units and pipelines to be placed on one processor, leading to superscalar processors (post-RISC). In very long instruction word (VLIW) processors such as Intel's Itanium, several instructions that are explicitly independent of each other are specified in one long instruction word. In both post-RISC and VLIW architectures, several instructions are processed in parallel. This parallelization on the lowest level is controlled either by a dispatcher unit in the processor or by the compiler, depending on the processor type. It does not have a direct impact on programming. However, a clever memory layout and ordering of instructions in programs can lead to further speedups even on such RISC processors.

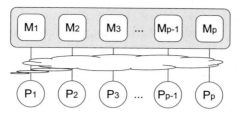

Fig. 4.1. System with (virtual) shared memory. The network between the memory banks and the processors can be realized in different ways and is not drawn in detail.

sequential program that identify these parallelizable sections.[3] It is important that each processor performs approximately the same amount of work, and that different operations do not interfere with each other. For example, the latter can be the case if two processors try to write to the same memory location at the same time. The exact timing of the two different processors will determine which processor writes last and thereby determines the value of the memory location after the write operations.[4]

Widely accepted programming models are "multi-threading", which operates on a low level of abstraction, and designs based on it that work with compiler directives, such as "OpenMP" [9]. OpenMP has now been standardized, and therefore parallel programs using it can be ported to other hardware platforms to a large extent. Essentially, only a few additional directives and possibly some restructuring of the code are needed to adapt a sequential program to parallel computers with shared memory.

However, for large numbers of processors, memory bandwidth becomes the bottleneck of such systems and limits their performance. One observes a decline in performance starting at a certain number of processors which depends on the hardware used for the parallel system. Therefore, large parallel computers are designed with a multilevel memory hierarchy that only gives the impression of a global address space, but exhibits significant differences in performance depending on the distance between processor and memory module. Examples for systems with shared memory are the Fire E25k by SUN,

---

[3] There are also special auto-parallelizing compilers which automatically recognize such sections in the code. However, these compilers are of limited efficiency and they are only suitable for certain easy standard situations.

[4] There are several ways to synchronize processors to prevent such situations. One possibility consists in protecting locations in memory with semaphores that signal whether a different processor wants to access the same memory locations. Often it is more advantageous to distribute the operations to the processors in an appropriate way, together with barriers at which all processors are explicitly synchronized with each other. Barriers prevent processors from executing operations after the barrier as long as there is at least one processor still executing operations before the barrier.

the HP 9000 SuperDome series and the HP AlphaServer series, the multiprocessor servers from the IBM pSeries 690 or the Cray XMT. Typically, these computers consist of 16 to 64 processors of the same type, equipped with a shared common global memory. The SGI Altix 4000 series can be cited as a further example. It provides a virtual global address space, but explicitly introduces the notion of nonuniform memory access (NUMA).

It should be mentioned that writing efficient parallel programs for parallel computers with shared memory and large numbers of processors is difficult. When optimizing such programs, one always has to take into account that different parts of memory are accessible at different speeds, starting from caches over local memory to memory modules located further away.

In contrast, in systems with *distributed* memory, every processor has its own local memory which it can access and work with, see Figure 4.2. Memory is addressed locally and references to other memory are meaningless since the processor cannot access that memory directly. Hence, to make parallel programming possible, data have to be exchanged explicitly between processors, or more precisely, between their local memories. This requires appropriate hard- and software for fast data exchange.

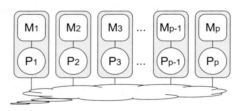

Fig. 4.2. System with distributed memory. The network between different processors can be realized in different ways and is not drawn in detail.

On the hardware side, parallel computers with distributed memory need fast connections between processors to allow the fast exchange of data between the distributed memory. These connections can be realized in a single stage over a bus (shared medium) or a crossbar switch. There are variants of multi-stage processor networks which directly use the communication channels of the processors to build rings, grids, or tori, and there are switched networks in which (crossbar) switches are used to also form rings, tori, trees, or more general networks. One condition for any powerful processor network is that the network hardware has to allow multiple or competing paths between processors. It is also possible to construct efficient networks for very large numbers of processors using small switches which are combined into so-called "fat" trees or two- or three-dimensional tori.

Examples for systems with distributed memory are for instance the IBM pSeries 690 (RS/6000-SP) which uses a fat tree of crossbar switches as connection between processors, the Cray XT4, XT3 and T3E series which employs a three-dimensional torus made from directly coupled processor-memory modules, modern architectures such as the IBM Blue Gene which employs a mixture of three-dimensional torus and tree, the Hitachi SR series which uses a large central crossbar switch, or Beowulf type cluster computers which use PCs and Ethernet type networks. In most cases, cluster computers are low cost parallel computers built from mass market components. Ethernet is often used as network technology in such systems [581]. The performance of these computers is, however, limited by the relatively high latency, the low bandwidth, and the poor scalability of the network. The use of other standardized high performance network technologies such as GigabitEthernet, Myrinet or Infiniband can significantly improve the overall performance of cluster computers for a number of applications [557]. Some examples are the HP BladeSystem c-Class, the SGI Altix XE Series or the IBM System Cluster 1600.

The computing nodes are connected in certain topologies, as shown in Figure 4.3, depending on the network technology and the network protocol. The networks differ in the number of connections per processor and per switch, in the network distance between two computing nodes, and in the overall performance of the network, measured for instance by the bisection bandwidth. Attempts to adapt the communication pattern of a parallel program to the existing processor network and thereby to develop programs for hypercubes or a specific kind of torus have not paid off. This is due to the fact that switching and routing techniques are constantly updated and improved. Therefore, we will not consider the actual structure of the parallel computer in the following. We will just assume an abstract computer in which every processor can communicate efficiently with every other processor.

The distinction between parallel computers with shared memory and with distributed memory is becoming blurred by recent hybrid designs. In such designs, several multiprocessor systems with shared memory are connected to form a larger parallel computer, see Figure 4.4. Overall, one obtains a com-

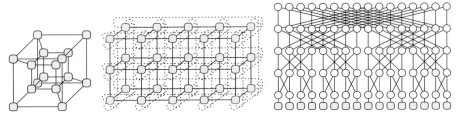

Fig. 4.3. Topologies of parallel computers: $d = 4$-dimensional hypercube with 2^d processors in which each processors has d connections (left); a $d = 3$-dimensional torus, in which each processor has $2d$ connections (center); a $d = 4$-stage fat tree built from switches connecting 2^d processors (right).

puter with distributed global memory, but programs can use programming techniques for shared memory for the several processors located together on one node. In this way, the data and the problem do not have to be partitioned into that many subproblems despite the often large number of processors. Examples for such computers are, besides some of the parallel computers already mentioned above, the IBM ASCI White system consisting of 512 RS/6000 SP nodes with 16 CPUs per node, the SGI ASCI Blue Mountain system consisting of 48 nodes with 128 CPUs per node, the Compaq ASCI-Q with 375 nodes with 32 CPUs per node, the Hitachi SR-8000 with, for instance, 144 nodes with eight processors per node, or the "earth simulator" by NEC with 640 nodes and eight vector processors per node. Finally, the computer with the highest performance in the world at the writing of this book also falls into this category. It is an IBM BlueGene/L computer system installed at the Lawrence Livermore National Laboratory which consists of 65,536 dual processor nodes connected by a $32 \times 32 \times 64$ 3D-torus.

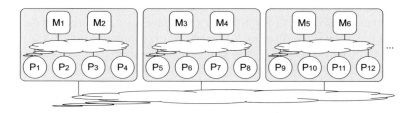

Fig. 4.4. System with distributed memory constructed from smaller multiprocessor systems with shared memory. There are local networks between the processors and the shared memory, and a global network between the multiprocessor systems, both only symbolically represented and not drawn in detail.

On computers with distributed memory, the memory of each processor has to contain all necessary data for all of its operations. The processors can communicate between each other by messages, where one processor sends data to another processor. Thus, a program for a parallel computer with distributed memory does not only consist of a sequence of computations, but data must also be sent and received at appropriate points in the program. These operations for sending and receiving data have to be specified explicitly by the programmer.

In the development of parallel computers, several different approaches have been pursued for the specification of those communication operations. Earlier "message passing" libraries such as PVM (parallel virtual machine) [10], Parmacs (parallel macros), NX/2, Picl, or Chimp have finally led to a uniform standard called "MPI" (Message Passing Interface) [7, 271, 272, 273]. There exists at least one (optimized) implementation of MPI for every current parallel computer. For tests and experiments one can use the free

MPI implementations "MPICH" (developed at Argonne National Lab) [11] or "LAM" (now developed and supported at Notre Dame University) [12], which simulate a parallel computer on a single computer or on a cluster of workstations. A short introduction to MPI can be found in Appendix A.3.

In this book we deal exclusively with the parallelization for computers with distributed memory.

Parallelization Strategies. The parallelization of a sequential molecular dynamics code depends strongly on the type of parallel computer on which the program will be executed. To this end, different techniques have been developed for the parallelization of algorithms for problems with short-range potentials, see for instance [69, 227, 280, 484]. For systems with shared memory the parallelization is relatively straightforward. As already mentioned, it is sufficient to insert compiler directives (in C as `#pragma`) that indicate the sections and loops which can be processed in parallel. In this way, no substantial change of the program is necessary. Compilers on sequential machines ignore these compiler directives.

However, explicit communication of data is needed for systems with distributed memory. A naive approach to parallelization is called the *replicated data* approach, see [126, 150, 178, 309, 339, 384, 544, 571]. Each processor keeps a copy of *all* data in its memory, but it works only on the part of the data associated to it. Thus, after a change of the data associated to a processor, a global exchange of data is necessary to ensure that all data are consistent among all processors. One disadvantage of this approach is the relatively large communication complexity, since every change of data will be communicated even to processors that do not need that piece of data in their computation. This seriously impacts parallel efficiency. Also, memory is wasted by storing all data on all processors and the problem size is limited by the size of the local memory on a single processor.

For the simulation of a system of N particles on a parallel computer with P processors, the replicated data method operates as follows. The N particles are partitioned into P subsets. Every processor works only on the N/P particles assigned to it. For example, for the basic force computation this means that each processor computes all the sums for the subset of particles i associated to it

$$\mathbf{F}_i = \sum_{j=1, j \neq i}^{N} \mathbf{F}_{ij}, \qquad (4.1)$$

where $\mathbf{F}_{ij}$ again denotes the force from particle i on particle j. We restrict ourselves to pair potentials here. To be able to execute these computations, each processor needs the positions, and possibly parameters for the potentials of all N particles as copies (replicated data) [484]. Therefore, a processor needs in each time step a current copy of all particles. Each processor then has to receive N pieces of data. In total this algorithm achieves a parallel complexity of the order $\mathcal{O}(N^2/P)$ for the CPU time, but its communication and memory complexities are of the order $\mathcal{O}(N)$. Thus, for an increasing number of processors, the communication and memory complexities dominate the total wall-clock time. If we now consider only potentials with finite range or potentials that can be truncated and are still well approximated by their truncated versions (as in the Linked-Cell-Method), the situation changes. Again each processor processes only N/P particles out of the total N particles. However, for the computation of the forces on a particle according to (4.1), one no longer has to take into account interactions with all particles, but only interactions with particles that are sufficiently close-by. The complexity of the computation on each processor then reduces from $\mathcal{O}(N^2/P)$ to $\mathcal{O}(N/P)$ but the communication complexity remains $\mathcal{O}(N)$, see also Table 4.1.[5] Thus, for increasing numbers of processors the computation complexity decreases correspondingly, but the communication complexity does not. The entire method does not scale with P. Therefore, the replicated data approach is not ideally suited for the parallelization of our linked cell algorithm.

A more suitable approach is parallelization by *data partitioning*. Here, each processor stores only the data of the particles that it needs during the computation. These are on the one hand the N/P particles that are assigned to the processor in the parallel computation, and on the other hand the particles that interact with those particles. The assignment of particles to processors could be done for example according to the particle number, or according to other, not necessarily geometric criteria.[6] After a subset of particles is assigned to a processor, the particles interacting with these particles can be determined easily by the linked cell approach. To this end, only particles in adjacent cells have to be examined. The number of particles to be stored on each processor is therefore of the order $\mathcal{O}(N/P)$. Furthermore, each processor has to receive at most $\mathcal{O}(N/P)$ particle data in each communication step, and it has to send some part of its own $\mathcal{O}(N/P)$ particle data. Since the number of communicated particles is of the same order as the number of computations, all complexities – computation complexity, communication complexity, memory complexity – decrease with increasing number of processors, in contrast to the replicated data approach. The entire method thus scales as $\mathcal{O}(N/P)$. However, the amount of data communicated is also of the order $\mathcal{O}(N/P)$.

---

[5] This argument is valid for all short-range potentials, including many-body potentials which we will consider in Chapter 5. It is valid in particular for the angle potentials for polymer chains [349], as long as there are only a bounded number of interactions per particle.

[6] This approach to parallelization is quite successful for the simulation of polymer chains if the particles are distributed among processors according to the linear order induced by the polymer chain. In this way only neighboring particles and particles in regions where different parts of the polymer chain come close are needed in the force computation [349].

A related approach is the parallelization by (static) *domain decomposition*. In this approach the data are partitioned and assigned to processors in such a way that as little communication as possible is needed. To this end, the simulation domain is decomposed into subdomains and each processor is associated to a subdomain, see for example [149, 212, 662]. Each processor then computes the trajectories of the particles that are located in its subdomain. If a particle moves from one subdomain to another subdomain, then the particle also changes its "owner", i.e. the processor it is associated with. Assuming equidistribution of the particles in the domain of our problem, $\mathcal{O}(N/P)$ particles are assigned to each processor which then computes the forces on the particles assigned to it. Since the particles have been distributed to the processors according to their locations and the geometry of the domain, most of the particles that are needed to compute the short-range interactions are already in the subdomain and therefore on the same processor. For short-range potentials, the particles for which the processor misses data are located in adjacent subdomains close to their subdomain boundaries. In each communication step only the data of these particles from adjacent subdomains have to be communicated. In this way, the number of particles for which data have to be received or sent decreases to $\mathcal{O}(\sqrt{N/P})$ in two dimensions and to $\mathcal{O}\left((N/P)^{2/3}\right)$ in three dimensions. Furthermore, particle data only have to be exchanged with processors associated to adjacent subdomains. The number of adjacent subdomains is independent of the total number of processors. The resulting parallel program therefore scales with increasing number of processors and with increasing number of particles; the complexity of the entire computation is of the order $\mathcal{O}(N/P)$.

We assume in the remainder of this chapter that the particles are uniformly distributed within the simulation domain. Then, a subdivision of the simulation domain into subdomains of equal size implies an approximately equal load for all processors.[7]

Table 4.1 summarizes the properties of the different parallelization strategies, compare also [484]. Altogether, domain decomposition proves to be the most appropriate strategy for the parallelization of our sequential linked cell algorithm because of its relatively low communication requirements. In addition, domain decomposition techniques are a good match with the cell concept of the linked cell method. Therefore, we assume in the following a parallel

---

[7] If we consider short-range forces and use a geometric decomposition of the domain it might happen that the particles are not uniformly distributed. However, in this case *dynamic* domain decomposition can ensure an almost uniform distribution of the N particles to the P processors. Dynamic domain decomposition means that the decomposition of the domain into subdomains changes over time and the particles are redistributed according to the new decomposition. Whether the communication complexity is of the same order as for static domain decomposition in the uniform case depends on further properties of the distribution, see [277, 433].

computer with distributed memory and use static domain decomposition as parallelization strategy for the linked cell method.

	computation	communication	memory
replicated data	$\mathcal{O}(N/P)$	$\mathcal{O}(N)$	$\mathcal{O}(N)$
data partitioning	$\mathcal{O}(N/P)$	$\mathcal{O}(N/P)$	$\mathcal{O}(N/P)$
domain decomposition	$\mathcal{O}(N/P)$	$\mathcal{O}(\sqrt{N/P})$ or $\mathcal{O}\left((N/P)^{2/3}\right)$	$\mathcal{O}(N/P)$

Table 4.1. Comparison of parallelization strategies for the force computation; given are the complexities for the computation, the communication complexities, and the memory requirements per processor for N particles on P processors for short-range forces.

4.2 Domain Decomposition as Parallelization Strategy for the Linked Cell Method

We now turn to the parallelization of the sequential program described in Chapter 3.

Domain Decomposition and Parallel Computing. According to Chapter 3, the main idea of the sequential linked cell method for the simulation of problems with short-range potentials is to decompose the simulation domain Ω into cells with edges that are at least as long as the cutoff radius r_{cut} of the potential. Because the range of the potential is bounded by the cutoff radius, the interactions with particles inside one of the cells can be computed in one loop over the particles in the cell and the particles in directly adjacent cells, compare Figure 3.6. Given suitable data structures for the particles in each cell, particles within the cutoff radius can be found and accessed fast. Altogether, this leads to an efficient force computation.

The decomposition into cells fits well with the decomposition of the simulation domain into subdomains for the parallelization. The domain is decomposed in such a way that the subdomain boundaries coincide with cell boundaries in the linked cell decomposition. To this end, we decompose the simulation domain Ω into np[d] parts in the direction of the dth coordinate. For simplicity we assume that the number of linked cells nc[d] along that direction is a multiple of the number of subdomains np[d] along that direction. In this way, we obtain a total of $\prod_{d=0}^{\mathrm{DIM}-1}$ np[d] subdomains Ω_{ip} with multi-indices ip $= (\mathrm{ip}[0], \ldots, \mathrm{ip}[\mathrm{DIM}-1])$ in the range from $(0, \ldots, 0)$ to np $= (\mathrm{np}[0] - 1, \ldots, \mathrm{np}[\mathrm{DIM}-1] - 1)$ where

$$\Omega_{\mathrm{ip}} = \bigotimes_{d=0}^{\mathrm{DIM}-1} \left[\mathrm{ip}[d] \frac{\mathrm{l}[d]}{\mathrm{np}[d]}, \ (\mathrm{ip}[d]+1) \frac{\mathrm{l}[d]}{\mathrm{np}[d]} \right[. \tag{4.2}$$

Figure 4.5 shows such a geometric subdivision of a rectangular domain Ω into six subdomains $\Omega_{(0,0)}$ to $\Omega_{(1,2)}$ for the two-dimensional case. The subdomains $\Omega_{\mathtt{ip}}$ are all of the same size and they are subdivided into $\prod \frac{\mathtt{nc[d]}}{\mathtt{np[d]}}$ cells.

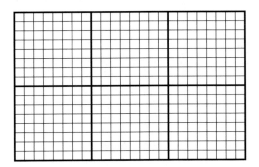

Fig. 4.5. Decomposition of the simulation domain Ω into six subdomains. The global domain Ω is divided along the boundaries of the linked cells into subdomains such that each processor owns the same number of cells.

Now, each processor is associated with one such subdomain and processes the particles inside its subdomain $\Omega_{\mathtt{ip}}$. Thus, the linked cell method runs on each processor and computes the forces, energies, new positions, and velocities for the particles inside its subdomain. But to compute the forces on some of the particles inside its subdomain, the processor needs the positions of particles within distance r_{cut} which can be situated in adjacent subdomains. Such particles are located in those cells from adjacent subdomains that are within a distance of r_{cut} from the processor's subdomain boundary. They are called its *border neighborhood*. To store the data for these particles from the adjacent subdomains, every subdomain is extended by one cell in each coordinate direction, as shown in Figure 4.6. The particle data for these cells are stored in particle lists as described in Section 3.5.

If the data for the particles in the border neighborhood have been copied from these adjacent subdomains to the local processor, the local processor can then compute – independently from all other processors – all the forces, the new velocities, and the new positions of particles of its subdomain. In this way, all subdomains are processed in parallel, and the forces, velocities, and positions of particles can be computed in parallel. After the new positions of the particles have been computed, particles that have left a subdomain must be assigned and transported to a new processor. Furthermore, the data of particles needed for the computation of the forces in other subdomains have to be exchanged between processors. These data will be exchanged in a communication phase among the processors and stored in the border neigh-

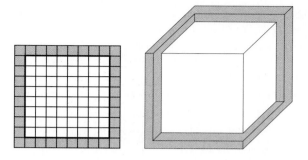

Fig. 4.6. Subdomain assigned to a processor (light) and border neighborhood made from cells belonging to adjacent subdomains (grey). Two-dimensional case (left) and part of the border neighborhood in the three-dimensional case (right). Given particle data for its subdomain and its border neighborhood, the interactions for all cells in the subdomain can be computed locally, i.e., no further data are needed.

borhoods prior to the computation of the forces. This communication phase is described in more detail in the following.

Communication. In the Velocity-Störmer-Verlet Algorithm 3.1 from Section 3.1, the forces on the particles are computed first and then the particles are moved from their old positions $\mathbf{x}_i^n$ to their new positions $\mathbf{x}_i^{n+1}$. As we already noted, every processor needs data from particles in cells adjacent to its boundary (from its border neighborhood) to be able to compute the forces on particles in its own subdomain. Therefore, the particle data from adjacent subdomains that are needed for the force computation must be exchanged before the computation of the forces can take place. It may also happen that particles leave their subdomains and enter the border neighborhood when they are moved. These particles are then assigned to different processors. After the new positions of the particles have been computed, particles that have left their subdomain must be "transported" to their new processors.

To exchange data among processors, every processor collects the appropriate particle data in buffers, and the contents of the buffers are then sent over the network to the appropriate neighboring processors. These processors receive the data in buffers and then insert them into the appropriate data structures. One has to ensure that data are exchanged in as few communication steps as possible since establishing a connection between processors takes a relatively long time. Furthermore, only as much data as necessary should be sent because the exchange of data also takes a relatively long time compared to the execution of instructions on the processors. If particles have left their subdomain, all of their data must be sent. In the communication for the force computation there is a further saving possible by just sending the new positions of the particles and not their velocities or other data.

Let us consider the communication of one processor with its neighbors. Figures 4.7 and 4.8 show a schematic sketch for the two- and three-dimensional case. The particles that lie in the border neighborhoods of other processors have to be sent to these processors. In Figure 4.7 (left), data are sent in the $\mathbf{x}_2$ direction. Let us focus on the processor assigned to the subdomain $\Omega_{(1,1)}$ in the center of the 3×3 array. The particles in the cells marked in light gray are sent along the arrows to the two neighboring processors, are received by these processors, and are sorted into the particle lists on these processors. In return, the processor for subdomain $\Omega_{(1,1)}$ receives data from these two neighbors. The processor thus sends two data packets and receives two data packets. In a second communication step, data are exchanged between processors in the $\mathbf{x}_1$ direction, see Figure 4.7 (right). To save communication steps with diagonally adjacent subdomains such as $\Omega_{(0,0)}$ or $\Omega_{(0,2)}$, also those cells from the *corners* of the border neighborhood are sent which were just received by the respective processor in the first communication step. The analogous three-dimensional case is sketched in Figure 4.8. To this end, the different parts of the border neighborhood are transported in three steps. In this way, a processor has to communicate in total with $2d$ neighbors in d dimensions, and does not have to communicate to all eight direct neighbors in two dimensions, or all 26 direct neighbors in three dimensions, respectively. Processors at the boundary of the simulation domain Ω send and receive fewer messages, or (in the case of periodic boundary conditions) exchange data with their neighbors at the other side of the domain. The data in the particle lists for the border neighborhoods are no longer needed after the forces have been computed and can then be deleted.

A further communication phase is needed to transport those particles with new positions outside their old subdomains to their new subdomains. We assume that the time step is small enough so that each particle travels across at most one cell per time step. Then, data have to be exchanged only with neighboring processors, and the border neighborhood of the old processor contains all particle data that have to be sent. In such a way, we obtain a communication pattern which corresponds to that of the force computation, only traversed in the *reverse* order (and with different particle data). This can be seen in Figures 4.9 and 4.10. Again, data are exchanged in d steps. In the first step, data are exchanged with the neighbors in $\mathbf{x}_1$ direction. In the second step, data (including the data from diagonal neighbor cells) are sent to the processors neighboring each processor in the $\mathbf{x}_2$ direction and, in the three-dimensional case, data are exchanged with the neighboring processors in $\mathbf{x}_3$ direction. Structuring the communication in this way avoids as before the direct communication with all neighbor processors that only have an edge (in the three-dimensional case) or a corner in common with the processor in question.

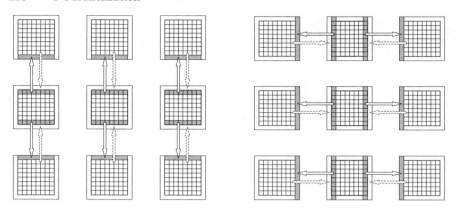

Fig. 4.7. In the force computation by the linked cell method, the particle data structures for the border neighborhoods must contain the correct values. Instead of exchanging data with each of the $3^d - 1$ neighboring processors, the communication proceeds in d steps with communication with two processors in each step. The figure shows the communication patterns for one subdomain and its neighbors. Data is exchanged in both directions as indicated by the arrows; the arrows drawn with solid lines show the communication from the subdomains drawn at the center, the arrows drawn with dashed lines show the communication to the subdomains drawn at the center. First, rows in x_2 direction (left) are exchanged and then columns in x_1 direction (right). In the second step, processors send not only data from their subdomains, but also data from parts of their border neighborhood (the corners) which were just received in the first communication step.

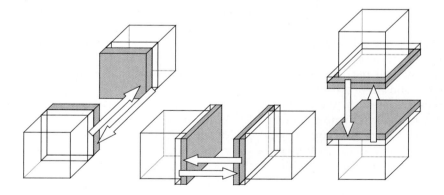

Fig. 4.8. Communication for the force computation by the linked cell method in three dimensions. In the first step, data are exchanged in x_3 direction (left), in the second step, data are exchanged in x_2 direction (center), and in the third step, data are exchanged in x_1 direction (right). As shown, data are always exchanged in both directions. Thus, with only six send and receive operations data are exchanged with all 26 neighbor processors.

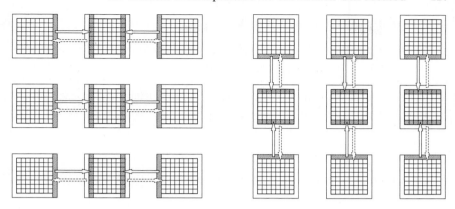

Fig. 4.9. Particles that have moved out of the subdomain of their processor are sent to the new owner. The transmission of the corresponding particle data from the border neighborhoods proceeds in d steps with communication between two processors per step. In the two-dimensional case shown, columns in x_1 direction are exchanged first (left) and then rows in x_2 direction (right). This is exactly adjoint to the communication pattern used in the communication before the force computation. If appropriate, data are also sent in the opposite direction, as indicated in the figure.

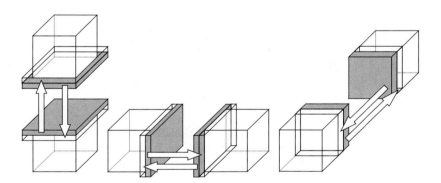

Fig. 4.10. Communication pattern in three dimensions for the transport of particles that have moved out of their subdomain. In the first step, data are exchanged in the x_1 direction (left), in the second step, data are exchanged in x_2 direction (center), and in the third step, data are exchanged in x_3 direction (right). If appropriate, data are also sent in the opposite direction.

4.3 Implementation with MPI

We consider parallel computers with distributed memory. The goal is to write a program that runs simultaneously on all of the processors and communicates explicitly by send and receive operations with the programs on the other processors. We will develop a *single* program that runs on each processor independently, but only works on the data it owns. The global execution of the program is synchronized by the receiving of necessary data from other processors. We use the *Message Passing Interface* (MPI) as programming environment [7]. MPI essentially provides a library

- that allows to simultaneously start a certain number of processes (on one[8] or several computers),
- that allows to uniquely identify processes by a process number, and
- provides functions through which the different processes can communicate with each other.[9]

MPI is a very powerful and complex library with more than 120 different functions. Fortunately, the parallelization of many methods – as for instance our linked cell method – can be implemented with only a few of these functions (see Appendix A.3 and [272, 273, 458] for details). The most important functions are:

- MPI_Init():
  Initalizes the MPI library environment.
- MPI_Finalize():
  Terminates the MPI library environment.
- MPI_Comm_size():
  Determines the number **numprocs** of processes started.
- MPI_Comm_rank():
  Determines the number of the local process **myrank** $\in \{0, \ldots, \text{numprocs} - 1\}$.
- MPI_Send() or MPI_Isend():
  Sends an MPI message.
- MPI_Recv():
  Receives an MPI message.

The local number of the process **myrank** and the total number **numprocs** of processes can be used to control the behavior of particular processes. In our case we use **myrank** and **numprocs** to determine which part of the data, i.e. which cells, are handled by a particular process.

---

[8] Note that codes parallelized with MPI can also run on sequential machines with an appropriate MPI implementation. In this way, it is possible to simulate a parallel computer on a sequential machine. This approach allows to recognize errors in the code before production runs are started on a cluster computer or a large parallel computer.

[9] Usually one process is started on each available processor. Therefore, we will not always distinguish in the following in detail between an MPI process and the corresponding processor of the parallel computer.

Subdomain. We now assume that a partition of the domain into subdomains is given. The corresponding data which describe that partition are collected in the data structure 4.1 SubDomain.

Data structure 4.1 Subdomain, Cells, and Neighbor Processes of Ω_{ip}

```
typedef struct {
  real l[DIM];              // size of simulation domain
  int nc[DIM];              // number of cells in simulation domain

  // additional parameters for the parallelization
  int myrank;               // process number of the local process
  int numprocs;             // number of processes started
  int ip[DIM];              // position of process in the process mesh
  int np[DIM];              // size of process mesh, also number of subdomains
  int ip_lower[DIM];        // process number of the neighbor processes
  int ip_upper[DIM];

  int ic_start[DIM];        // width of border neighborhood, corresponds to
                            // the first local index in the interior of the subdomain
  int ic_stop[DIM];         // first local index in the upper border neighborhood
  int ic_number[DIM];       // number of cells in subdomain, including border
                            // neighborhood
  real cellh[DIM];          // dimension of a cell
  int ic_lower_global[DIM];    // global index of the first cell of the subdomain
} SubDomain;
```

For this, we recall that the entire domain Ω has already been decomposed into nc[d] cells for the sequential linked cell method. We now assign the cells with the (global) indices ic_lower_global[d] to ic_lower_global[d] + (ic_stop[d] - ic_start[d]) to its associated process. To this end, analogously to the linked cell mesh, we organize the processes conceptionally into a mesh which contains all processes $0, \ldots, \text{numprocs} - 1$ (see Figure 4.5). We thus assign to each process a multi-index ip based on its process number myrank. This multi-index identifies the process and thereby its associated subdomain Ω_{ip} in the mesh of processes (see code fragment 4.1).
In this way we can represent the subdomain Ω_{ip} associated to the process with the multi-index ip as

$$\Omega_{\text{ip}} = \bigotimes_{d=0}^{\text{DIM}-1} \left[\text{ip}[d] \frac{l[d]}{\text{np}[d]}, (\text{ip}[d] + 1) \frac{l[d]}{\text{np}[d]} \right[$$

and assign it to the process myrank for local computations. In addition, during these local computations, each process needs data from a border

Code fragment 4.1 Initialization of the SubDomain Data Structure

```
#if 1==DIM
#define inverseindex(i,nc,ic) \
((ic)[0]=(i))
#elif 2==DIM
#define inverseindex(i,nc,ic) \
((ic)[0]=(i)%(nc)[0], (ic)[1]=(i)/(nc)[0])
#elif 3==DIM
#define inverseindex(i,nc,ic) \
((ic)[0]=(i)%(nc)[0], \
(ic)[1]=((i)/(nc)[0])%(nc)[1], \
(ic)[2]=((i)/(nc)[0])/(nc)[1])
#endif

void inputParameters_LCpar(real *delta_t, real *t_end, int* N,
                           SubDomain *s, real* r_cut) {
  // set parameters as in the sequential case
  inputParameters_LC(delta_t, t_end, N, s->nc, s->l, r_cut);
  // set additional parameters for the parallelization
  MPI_Comm_size(MPI_COMM_WORLD, &s->numprocs);
  MPI_Comm_rank(MPI_COMM_WORLD, &s->myrank);
```

// set np[d] so that $\prod_{d=0}^{DIM-1} np[d] = numprocs$

```
  // determine position of myrank in the process mesh np[d]
  int iptemp[DIM];
  inverseindex(s->myrank, s->np, s->ip);
  for (int d=0; d<DIM; d++)
    iptemp[d] = s->ip[d];
  for (int d=0; d<DIM; d++) { // determine neighbor processes
    iptemp[d] = (s->ip[d] - 1 + s->np[d]) % s->np[d];
    s->ip_lower[d] = index(iptemp, s->np);
    iptemp[d] = (s->ip[d] + 1 + s->np[d]) % s->np[d];
    s->ip_upper[d] = index(iptemp, s->np);
    iptemp[d] = s->ip[d];
  }
  for (int d=0; d<DIM; d++) { // set local parameters
    s->cellh[d]      = s->l[d] / s->nc[d];
    s->ic_start[d]   = (int) ceil(*r_cut / s->cellh[d]);
    s->ic_stop[d]    = s->ic_start[d] + (s->nc[d] / s->np[d]);
    s->ic_number[d]  = (s->ic_stop[d] - s->ic_start[d]) +
                       2 * (s->ic_start[d]);
    s->ic_lower_global[d] = s->ip[d] * (s->nc[d] / s->np[d]);
  }
}
```

neighborhood that is `ic_start[d]` cells wide.[10] A process therefore stores `ic_number[d]` cells, including the cells in the border neighborhood, for each of the coordinate directions `d=0,...,DIM-1`.

As already explained in the previous section, direct communication with neighboring processes in the diagonal directions can be avoided by the sequencing of communication steps for each coordinate direction. In each coordinate direction `d`, a process has to exchange data only with two other processes. The *process numbers* of these processes will be stored in the multi-indices `ip_lower[d]` and `ip_upper[d]`. For instance, `ip_lower`[0] will contain the process number of the left neighbor. The entire initialization of the data structure `SubDomain` can be implemented as shown in code fragment 4.1. There, we compute the appropriate values for the subdomain Ω_{ip} from the dimensions `l[d]` of the simulation domain Ω, the process number `myrank`, and the number `numprocs` of processes started.

In the next step of the parallelization, the sequential linked cell code has to be adapted to this new generalized description of the domain and the subdomains. The necessary changes are relatively small. Essentially, one only has to replace the parameters `l` and `nc` with an instance `s` of the new data structure `SubDomain`. All loops over cells in routines like `compX_LC` now only run over the local subdomain, meaning that instead of loops such as

```
for (ic[d]=0; ic[d]<nc[d]; ic[d]++)
```

now loops such as

```
for (ic[d]=s->ic_start[d]; ic[d]<s->ic_stop[d]; ic[d]++)
```

are used.[11] Also, the calling sequences for the macro `index` have to be changed from `index(ic,nc)` to `index(ic, s->ic_number)`.

Main Program. The changes in the main program are small, see Algorithm 4.1. The more significant changes are the initialization and termination of the MPI library and the changes necessary to use the data structure `SubDomain` describing the subdomain Ω_{ip} assigned to the local process.

In `inputParameters_LCpar`, a subdomain Ω_{ip} associated to a process is determined from the process number `myrank`, as described in the code fragment 4.1. Of course, the partition into subdomains Ω_{ip} has to be taken into account in the setup of the particles in `initData_LCpar`. For instance, each

---

[10] We implement here a general communication subroutine that allows an arbitrary width of the border neighborhood. If one can ensure that particles can move at most across one cell in each time step and that r_{cut} is chosen appropriately, the border neighborhood is only one cell wide and `ic_start[d]` can be set to one.

[11] If the macro `iterate` is used, calls of the form `iterate(ic, nullnc, nc)` have to be replaced by `iterate(ic, s->ic_start, s->ic_stop)`.

process could read or create only particles in its subdomain.[12] Alternatively, one process could read or create all particles, then sorts them into the appropriate domains and sends them to the appropriate processes, if there is enough memory available on that single processor to allow for such an approach. The routine `timeIntegration_LCpar` for the time integration can be implemented analogously to the sequential case. The parallel computation and output of further quantities is implemented in the routine `compoutStatistic_LCpar`. The parallel output of the positions and velocities of the particles is implemented in `outputResults_LCpar`.

Algorithm 4.1 Main Program: Parallel Linked Cell Method

```
#include <mpi.h>
int main(int argc, char *argv[]) {
  int N, pnc, ncnull[DIM];
  real r_cut, delta_t, t_end;
  SubDomain s;
  MPI_Init(&argc, &argv);
  inputParameters_LCpar(&delta_t, &t_end, &N, &s, &r_cut);
  pnc = 1;
  for (int d = 0; d < DIM; d++) {
    pnc *= s.ic_number[d];
    ncnull[d] = 0;
  }
  Cell *grid = (Cell*) malloc(pnc*sizeof(*grid));
  initData_LCpar(N, grid, &s);
  timeIntegration_LCpar(0, delta_t, t_end, grid, &s, r_cut);
  freeLists_LC(grid, ncnull, s.ic_number, s.ic_number);
  free(grid);
  MPI_Finalize();
  return 0;
}
```

Parallel Force Computation and Parallel Particle Moving. To be able to actually compute with multiple processes, we still have to insert the communication steps described above in our linked cell code. First, a process needs certain particle data from its neighboring processes *before* the computation of the forces. Thus, we have to appropriately generalize the function `compF_LC` to `compF_LCpar`, see Algorithm 4.2. After the forces have been computed, the particles are moved. Hence, all copies of particles from neighboring

---

[12] Note that the parallel generation of initial values with a Maxwell-Boltzmann distribution is in general not an easy task. It involves the parallel generation of random numbers.

Algorithm 4.2 Parallel Linked Cell Force Computation

```
void compF_LCpar(Cell *grid, SubDomain *s, real r_cut) {
  compF_comm(grid, s);
  compF_LC(grid, s, r_cut); // sequential version adapted to s
  delete particles in the border neighborhood;
}
```

processes no longer contain valid data and can therefore be deleted right after the force computation.

In addition, a data exchange between neighboring processes is necessary *after* the particles are moved since particles can leave the subdomain Ω_{ip} assigned to the local process. These particles (i.e. all their data) have to be sent to a neighboring process. We thus extend the function moveParticles_LC accordingly to moveParticles_LCpar, see Algorithm 4.3. If necessary, the appropriate boundary conditions have to be taken into account here as well.

Algorithm 4.3 Parallel Sorting of the Particles into their New Cells

```
void moveParticles_LCpar(Cell *grid, SubDomain *s) {
  moveParticles_LC(grid, s); // sequential version adapted to s
  moveParticles_comm(grid, s);
  delete particles in the border neighborhood;
}
```

Before we consider how to exactly implement the functions compF_comm and moveParticles_comm, we first implement a single communication step of the communication patterns discussed in the last section (see Figures 4.7 to 4.10).

One-dimensional Communication. We implement the communication with neighboring processes in *one* routine sendReceiveCell in Algorithm 4.4. With this routine we can then express the communication patterns from Figures 4.7 to 4.10. The idea is to carry out a one-dimensional data exchange between the process and its two neighboring processes ip_lower[d] and ip_upper[d] for each dimension separately. The process sends the contents of cells from two opposite parts of the subdomain and possibly from parts of the border neighborhood to the appropriate neighboring processes and receives in exchange particles that have to be sorted into its cells.

The communication pattern should be implemented in such a way that it uses as few send and receive operations as possible and that it transmits only data which are necessary for the computation. It would be inefficient if the content of each cell needed in the computation would be sent in a separate message. Instead, all data that have to be sent from the local process to a particular process should be sent together in one message. But the content of the cells is stored in linked lists, see data structure 3.2, and pointers are used to connect the elements of the list. Hence, those elements do not have to be contiguous in memory and it is not possible to just copy the area of the memory directly. Therefore, we have to convert the linked list data structure and the data contained in the list elements into a different (contiguous) format.[13] We will use a vector ip_particle of particle data for that. The routine sendReceiveCell is kept as general as possible so that it can be reused easily. Thus, it is necessary to describe the local cells from grid that have to be sent or received in a way that is sufficiently general to allow such a reuse. Since all cells are stored in a vector grid and since indices are computed from ranks, it is enough to store for both neighboring processes lowerproc and upperproc the range of indices of the cells that have to be sent (icstart to icstop) and the range of indices for the cells that have to be received (icstartreceive to icstopreceive).

In a first step we determine the vector ic_length of the numbers of particles from the cells that have to be transmitted. Next, we store all the corresponding particle data in the ordering of the cells ic in the long vector ip_particle. Each cell corresponds to a contiguous section in this vector. To determine to which cell ic a particle belongs, one could look at the coordinates of the particle. In our case it is easier, however, to (re-)compute the appropriate cell ic from the vector ic_length. This easy computation uses the data in icstartreceive, icstopreceive, and ic_lengthreceive, see Algorithm 4.4.

We use non-blocking communication in which the execution of send and receive operations can overlap if the parallel computer supports it.[14] In a

---

Algorithm 4.4 Sending and Receiving of Particle Lists

```
void sendReceiveCell(Cell *grid, int *ic_number,
                int lowerproc, int *lowericstart, int *lowericstop,
                int *lowericstartreceive, int *lowericstopreceive,
                int upperproc, int *uppericstart, int *uppericstop,
                int *uppericstartreceive, int *uppericstopreceive) {
   MPI_Status status; MPI_Request request;
   int sum_lengthsend = 0, sum_lengthreceive = 0;
   int k = 0, kreceive = 0, ncs = 1;
   int *ic_lengthsend = NULL, *ic_lengthreceive = NULL, ic[DIM];
   Particle *ip_particlesend = NULL, *ip_particlereceive = NULL;
```

---

[13] In principle it is possible to leave this conversion to the MPI library if one introduces appropriate data structures and their corresponding memory layout there.

[14] A parallel example program is given in Appendix A.3.

```
// sending to lowerproc, receiving from upperproc
sum_lengthsend = sum_lengthreceive = k = kreceive = 0; ncs = 1;
for (int d = 0; d < DIM; d++)
  ncs *= lowericstop[d] - lowericstart[d];
ic_lengthsend = (int*)malloc(ncs*sizeof(*ic_lengthsend));
ic_lengthreceive = (int*)malloc(ncs*sizeof(*ic_lengthreceive));
iterate (ic, lowericstart, lowericstop) {
  ic_lengthsend[k] = lengthList(grid[index(ic,ic_number)]);
  sum_lengthsend += ic_lengthsend[k++];
}
MPI_Isend(ic_lengthsend, ncs, MPI_INT, lowerproc, 1,
          MPI_COMM_WORLD, &request);
MPI_Recv(ic_lengthreceive, ncs, MPI_INT, upperproc, 1,
          MPI_COMM_WORLD, &status);
MPI_Wait(&request, &status);
free(ic_lengthsend);
for (k=0; k<ncs; k++)
  sum_lengthreceive += ic_lengthreceive[k];
sum_lengthsend *= sizeof(*ip_particlesend);
ip_particlesend = (Particle*)malloc(sum_lengthsend);
sum_lengthreceive *= sizeof(*ip_particlereceive);
ip_particlereceive = (Particle*)malloc(sum_lengthreceive);
k = 0;
iterate(ic, lowericstart, lowericstop)
  for (ParticleList *i = grid[index(ic,ic_number)]; NULL != i;
       i = i->next)
    ip_particlesend[k++] = i->p;
MPI_Isend(ip_particlesend, sum_lengthsend,
          MPI_CHAR, lowerproc, 2, MPI_COMM_WORLD, &request);
MPI_Recv(ip_particlereceive, sum_lengthreceive,
          MPI_CHAR, upperproc, 2, MPI_COMM_WORLD, &status);
MPI_Wait(&request, &status);
free(ip_particlesend);
kreceive = k = 0;
iterate(ic, uppericstartreceive, uppericstopreceive) {
  for (int icp=0; icp<ic_lengthreceive[kreceive]; icp++) {
    ParticleList *i = (ParticleList*)malloc(sizeof(*i));
    i->p = ip_particlereceive[k++];
    insertList(&grid[index(ic,ic_number)], i);
  }
  kreceive++;
}
free(ic_lengthreceive);
free(ip_particlereceive);
// sending to upperproc, receiving from lowerproc
...
}
```

first step, the vectors `ic_length` are exchanged. Afterwards, both the sender and the receiver know the number of particles to be transmitted and therefore the lengths of the messages. In a second step, the current particle data from `ic_particle` are transmitted. These two steps are implemented in Algorithm 4.4. In this algorithm, only half of the communication step is shown: Sending data to `lowerproc` and receiving data from `upperproc`. The other half of the communication step can be implemented in the same way, only `lower` and `upper` have to be exchanged.

Communication for Force Computation and Particle Moving. To be able to fully describe the communication steps for the force computation and the moving of the particles, we need an additional routine `setCommunication` (see Algorithm 4.5) which determines the subdomains for the communication pattern described in the previous section, compare Figure 4.7. In `setCommunication` we compute the appropriate ranges of the indices, `icstart` to `icstop`, which describe the portion of the local domain that has to be sent to other processes, as well as the indices `icstartreceive` to `icstopreceive` that describe the portion of the border neighborhood for which particles have to be received from other processes.

Using this routine, one can now easily and concisely implement the communication phase prior to the force computation according to Figure 4.7. The cells in the border neighborhoods are filled with the appropriate copies of particle data from neighboring processes, as written in Algorithm 4.6. For subdomains next to the boundary of the simulation domain, the communication routines may have to be changed to take different boundary conditions into account.

Particles are transported with the exactly reverse order of communication steps in Algorithm 4.7, where particles are moved to their new cells. Here, all particles that entered border neighborhoods have to be sent to their new owners, i.e., we send the associated particle information from cells in the border neighborhood to cells in the interior of their new subdomain. Since the routines `sendReceiveCell` and `setCommunication` have been written for a very general setting, this case can be implemented completely analogously to the previous one.

The computation of the energy can also be distributed to all processes. To compute the potential or kinetic energy of the system, which is needed for instance to compute the temperature, the energies of the particles in the subdomains are computed in parallel and the local results are then globally added together. This can be implemented in a communication step with `MPI_Allreduce` after the new velocities have been computed.

As in the $\mathcal{O}(N^2)$-method and in the sequential linked cell algorithm, about half of the computations can be saved if one uses the fact that the forces are symmetric. Then, not all $3^{\texttt{DIM}} - 1$ neighboring cells have to be traversed but it is sufficient to traverse just half of them. In the same way, interactions within one cell can be computed with half the number of operations. Here,

Algorithm 4.5 Communication Pattern According to Figure 4.7

```
void setCommunication(SubDomain *s, int d,
                      int *lowericstart, int *lowericstop,
                      int *lowericstartreceive, int *lowericstopreceive,
                      int *uppericstart, int *uppericstop,
                      int *uppericstartreceive, int *uppericstopreceive) {
  for (int dd = 0; dd < DIM; dd++) {
    if (d == dd) { // only border neighborhood
      lowericstart[dd] = s->ic_start[dd];
      lowericstop[dd]  = lowericstart[dd] + s->ic_start[dd];
      lowericstartreceive[dd] = 0;
      lowericstopreceive[dd]   = lowericstartreceive[dd] +
                                 s->ic_start[dd];
      uppericstop[dd]          = s->ic_stop[dd];
      uppericstart[dd]         = uppericstop[dd] - s->ic_start[dd];
      uppericstopreceive[dd]   = s->ic_stop[dd] + s->ic_start[dd];
      uppericstartreceive[dd]  = uppericstopreceive[dd] -
                                 s->ic_start[dd];
    }
    else if (dd > d) { // including border neighborhood
      int stop = s->ic_stop[dd] + s->ic_start[dd];
      lowericstartreceive[dd] = lowericstart[dd] = 0;
      lowericstopreceive[dd]   = lowericstop[dd]  = stop;
      uppericstartreceive[dd]  = uppericstart[dd] = 0;
      uppericstopreceive[dd]   = uppericstop[dd]  = stop;
    }
    else { // excluding border neighborhood
      lowericstartreceive[dd] = lowericstart[dd] = s->ic_start[dd];
      lowericstopreceive[dd]   = lowericstop[dd]  = s->ic_stop[dd];
      uppericstartreceive[dd]  = uppericstart[dd] = s->ic_start[dd];
      uppericstopreceive[dd]   = uppericstop[dd]  = s->ic_stop[dd];
    }
  }
}
```

one can either exchange the forces on the particles in the cells of the border neighborhood in an extra communication phase and add them up, or one can avoid communication but then incurs additional computation using the original, less efficient algorithm in and close to the border neighborhood. In both approaches, boundary conditions have to be taken into account properly.

If we merge the loop on **updateX** with the loop on the force computation **compF** in the implementation of the Velocity-Störmer-Verlet method, the two communication routines **moveParticles_comm** and **compF_comm** are called directly after each other in the resulting program. This suggests that it should be possible to further optimize the communication among the processes. Instead of first sending particles to their new processes and then sending copies

Algorithm 4.6 Communication before the Force Computation

```
void compF_comm(Cell *grid, SubDomain *s) {
  int lowericstart[DIM], lowericstop[DIM];
  int uppericstart[DIM], uppericstop[DIM];
  int lowericstartreceive[DIM], lowericstopreceive[DIM];
  int uppericstartreceive[DIM], uppericstopreceive[DIM];
  for (int d = DIM-1; d >= 0; d--) {
    setCommunication(s, d, lowericstart, lowericstop,
                     lowericstartreceive, lowericstopreceive,
                     uppericstart, uppericstop,
                     uppericstartreceive, uppericstopreceive);
    sendReceiveCell(grid, s->ic_number,
                    s->ip_lower[d], lowericstart, lowericstop,
                    lowericstartreceive, lowericstopreceive,
                    s->ip_upper[d], uppericstart, uppericstop,
                    uppericstartreceive, uppericstopreceive);
  }
}
```

Algorithm 4.7 Communication for Moving the Particles

```
void moveParticles_comm(Cell *grid, SubDomain *s) {
  int lowericstart[DIM], lowericstop[DIM];
  int uppericstart[DIM], uppericstop[DIM];
  int lowericstartreceive[DIM], lowericstopreceive[DIM];
  int uppericstartreceive[DIM], uppericstopreceive[DIM];
  for (int d = 0; d < DIM; d++) {
    setCommunication(s, d, lowericstartreceive, lowericstopreceive,
                     lowericstart, lowericstop,
                     uppericstartreceive, uppericstopreceive,
                     uppericstart, uppericstop);
    sendReceiveCell(grid, s->ic_number,
                    s->ip_lower[d], lowericstart, lowericstop,
                    lowericstartreceive, lowericstopreceive,
                    s->ip_upper[d], uppericstart, uppericstop,
                    uppericstartreceive, uppericstopreceive);
  }
}
```

of them to other processes, one can merge both communication steps into one.

In this new communication step the data from all the particles that have to be moved to new processes are transmitted to their new owners in only one message. Thus, in each time step, every process sends only one message instead of the original two messages to each of its $2 \cdot$DIM directional neighboring processes.

4.4 Performance Measurements and Benchmarks

For sufficiently large numbers of particles and enough processors the parallel program can be significantly faster than the sequential program. How can one evaluate this increase in performance? One fundamental measure is the so-called speedup

$$S(P) := \frac{T}{T(P)}, \tag{4.3}$$

where P denotes the number of processors used in the computation, $T(P)$ denotes the time needed by the parallel computation on these P processors, and T denotes the time needed for the same problem by the sequential program.[15] Instead of the time T needed to execute the (best) sequential program one often uses $T(1)$, i.e. the time needed to execute the parallel program on one processor. The speedup lies between 1 and P.[16] A speedup equal to the number of processors P, i.e. $S(P) = P$, is called linear. In general, this optimal value P for the speedup is not obtained because parts of the program may be hard to parallelize, because the loads of the different processors may be hard to balance, or because the communication may require a non-negligible time.

Another measure for the improvement of performance is the parallel efficiency

$$E(P) := \frac{T}{P \cdot T(P)} = \frac{S(P)}{P} \tag{4.4}$$

or $E(P) = T(1)/(P \cdot T(P))$, in which the speedup is divided by the number of processors. The optimal value here is $E(P) = 1$ or 100%, the minimum is $1/P$. The efficiency measures how well the processors are used in parallel. In practice one often observes parallel efficiencies significantly smaller than one, especially for large numbers of processors. The sequential program is accelerated by the parallelization, but some performance is lost because of communication and unbalanced load.

One estimate for the speedup and the efficiency that is often cited in this context was introduced by Amdahl [39]. It was originally introduced for vector computers but can similarly be applied to parallel computers. In the estimate, it is assumed that a certain portion of the program cannot be

---

[15] In MPI one can use the function `MPI_Wtime()` to measure the wall-clock time. The function is called once at the beginning and once at the end of the part of the program that is to be measured. Afterwards the time needed for that part of the program is obtained from the difference of the two results.

[16] If a speedup higher than P is obtained, one could execute the parallel algorithm also as P processes on one processor and one should then obtain an improved sequential execution time T barring cache effects. For the other extreme case of a speedup less than one, which possibly should be called a "slow down", one could improve the speedup by using less processors, keeping some processors idle. In this way a speedup can be obtained that always satisfies $1 \leq S(P) \leq P$.

parallelized (or vectorized). From this, it is relatively easy to derive an upper bound for the speedup. Let α be the portion of the execution time T that has to be executed sequentially. Assume that the remainder $\gamma = 1 - \alpha$ can be parallelized perfectly. Then, it holds that $T(P) = \alpha T + \gamma T/P$ and therefore

$$S(P) = \frac{T}{\alpha T + \gamma T/P} = \frac{1}{\alpha + (1 - \alpha)/P}.$$

Assuming that arbitrarily many processors are available, we obtain

$$S(P) \to 1/\alpha \quad \text{and} \quad E(P) \to 0 \quad \text{for} \quad P \to \infty.$$

Thus, the maximal speedup that can be realized, is bounded by the portion α of the algorithm that cannot be parallelized. Analogously, the efficiency becomes arbitrarily small. If only one percent of the algorithm cannot be parallelized then only a speedup of 100 can be obtained. This is a serious limitation that has long been used as an argument against massively parallel computation, i.e. against the use of a very large number of processors. In addition, the execution time $T(P)$ might increase for an increasing number of processors because of increased communication between processors and load unbalance.

Note that Amdahl's law assumes a fixed size of the problem. But an increase in the number of processors allows also to solve larger problems. Here, the sequential portion of the algorithm can even become smaller since a one-time fixed effort for program management or subproblems of lower complexity might have a decreasing impact. Consequently, one could indeed obtain good parallel efficiencies for sufficiently large problems. The problem is that in general one cannot determine the execution time T for the sequential program for such large problems. This is due to the fact that usually one processor alone does not have enough memory or that the time needed to run the sequential program on one processor is too long. Therefore, one often uses a properly scaled value, the so-called scaleup

$$S_C(P, N) := \frac{T(P, N)}{T(\kappa P, \kappa N)}, \quad \kappa > 1. \tag{4.5}$$

Here, $T(P, N)$ denotes the time the parallel program needs to execute $\mathcal{O}(N)$ operations on P processors. Assuming that the operations can be distributed optimally to the processors, one obtains for $T(\kappa P, \kappa N)$ always the same value, independent of the factor κ by which the number of operations and the number of processors have been multiplied.[17] The load per processor is then constant and the scaleup (4.5) is one. Loss of efficiency can be recognized by a smaller scaleup.[18]

---

[17] κ is often set to 2 or 2^{DIM} with DIM denoting the dimension of the problem.

[18] The scaleup should be larger than $1/\kappa$. If it would be smaller, one could use P processors instead of κP processors in the computation and would need less parallel computing time assuming that the P processors have enough memory.

Amdahl's estimate gives a different result for the scaleup than for the speedup. If we denote by γ the portion of the execution time of the program that is executed in parallel on P processors and by α the portion that has to be executed sequentially, one obtains

$$T(P, N) \;=\; \alpha T + \gamma T \frac{N}{P} \;=\; \alpha T + \gamma T \frac{\kappa N}{\kappa P} \;=\; T(\kappa P, \kappa N).$$

Therefore, the scaleup is one.

An example. We now test the performance increase which stems from the parallelization of the algorithm. We show results for the simulation of one time step for a system with N particles in a three-dimensional simulation box with periodic boundary conditions. The interaction of the particles is modeled by a Lennard-Jones potential. The particles are distributed uniformly over the simulation domain, with a constant particle density $\rho' := \frac{N}{|\Omega'|} = \frac{N}{L_1' \cdot L_2' \cdot L_3'} = 0.8442$. Hence, a static decomposition of the simulation domain into subdomains of equal size should lead to a reasonable load balance.

At the start of the simulation, the particles are arranged on a regular face-centered cubic lattice, see Section 5.1.1. In addition, the initial velocities of the particles are superimposed by a small thermal motion. The thermal motion is chosen according to a Maxwell-Boltzmann distribution with a reduced temperature of $T' = 1.44$, compare also Appendix A.4. The external forces are set to zero. Volume, number of particles, and energy are kept constant during the simulation, i.e., we consider an NVE ensemble.

$$
\begin{array}{ll}
\varepsilon = 1, & \sigma = 1, \\
m = 1, & \\
\rho' = 0.8442, & T' = 1.44, \\
r_{cut} = 2.5\sigma, & \delta t = 0.00462
\end{array}
$$

Table 4.2. Parameter values for the first parallel example.

The parameter values used in the simulation are shown in Table 4.2. The size of the simulation domain is computed from the number N of particles as $L_1' = L_2' = L_3' = (N/\rho')^{1/3}$. The domain is subdivided into approximately cubic subdomains so that the number of subdomains in each direction is always a power of two and the number of subdomains in different directions differs at most by a factor of two. The parallel program was run on a Cray T3E parallel computer[19] in single precision (32 bit, `float` in C, or `real*4` in Fortran). The local processors in this computer were 600 MHz DEC Alpha processors.

---

[19] The Cray T3E-1200 at the John von Neumann Institute for Computing (NIC) at the Research Centre Jülich.

Table 4.3 shows the execution time. We can observe in each row that a doubling of the number of processors approximately halves the execution time, given a large enough number of particles per processor. Also, the execution time approximately doubles when the number of particles is doubled, as seen in the columns of the table. In the diagonal we observe that doubling the number of processors and the number of particles results in an almost constant execution time. No results are given for large numbers of particles and small numbers of processors since the main memory of the processors is no longer large enough for the resulting problem.

					processors				
particles	1	2	4	8	16	32	64	128	256
16384	0.9681	0.4947	0.2821	0.1708	0.1076	0.0666	0.0403	0.0269	0.0202
32768	1.8055	1.0267	0.5482	0.3266	0.1685	0.1089	0.0683	0.0422	0.0294
65536	3.3762	1.9316	1.0786	0.6346	0.3220	0.1724	0.1112	0.0707	0.0433
131072	6.0828	3.4387	2.0797	1.1637	0.5902	0.3316	0.1892	0.1139	0.0713
262144	12.6010	6.2561	3.6770	2.0825	1.1610	0.6544	0.3570	0.1937	0.1221
524288	26.2210	14.2460	6.5593	3.7391	2.0298	1.1569	0.6521	0.3354	0.1960
1048576		28.7260	13.1900	6.9030	3.7369	2.1510	1.1642	0.5968	0.3482
2097152			26.8290	14.0750	6.9057	3.8103	2.0890	1.1748	0.6768
4194304				28.2560	15.0430	6.9920	3.8077	2.0546	1.1915
8388608					29.4340	14.9250	7.7331	4.0058	2.1944
16777216						28.5110	14.7590	7.7412	3.9246

Table 4.3. Parallel execution time (in seconds) for one time step on the Cray T3E.

					processors				
particles	1	2	4	8	16	32	64	128	256
16384	1	1.95	3.43	5.66	8.99	14.53	24.02	35.98	47.92
32768	1	1.75	3.29	5.52	10.71	16.57	26.43	42.78	61.41
65536	1	1.74	3.13	5.32	10.48	19.58	30.36	47.75	77.97
131072	1	1.76	2.92	5.22	10.30	18.34	32.15	53.40	85.31
262144	1	2.01	3.42	6.05	10.85	19.25	35.29	65.05	103.20
524288	1	1.84	3.99	7.01	12.91	22.66	40.21	78.17	133.78

Table 4.4. Speedup for one time step on the Cray T3E.

In Table 4.4 we show the speedup which corresponds to the execution times from the previous table. Here, $T(1)$ was used for T, i.e., the time needed by the sequential program is taken as the time needed by the parallel program on one processor. One can observe how the program is accelerated by the use of larger and larger numbers of processors. Ideally, the program would be accelerated by a speedup factor of P for P processors. We observe a value of 133 for 256 processors and 524288 particles. It is typical that the largest accelerations are observed for large problems and can therefore be found in the last row of the table. The reason is the relatively small loss due to sequential parts of the program and due to communication among processors (compare also the discussion of Amdahl's law above).

The parallel efficiency can be computed directly from the speedup. The efficiencies are given in Table 4.5. For one processor, the value is one, by definition. The efficiency decreases with increasing numbers of processors, but also increases with larger problem sizes. Furthermore, one can observe cache effects in the two Tables 4.4 and 4.5. For instance the efficiency is larger than one for the case of 262144 particles on two processors. On one processor, obviously a critical number of particles and thus memory size is exceeded which slows down the access to memory and thus leads to longer execution times.

particles	1	2	4	8	16	32	64	128	256
					processors				
16384	1	0.978	0.857	0.708	0.562	0.454	0.375	0.281	0.187
32768	1	0.879	0.823	0.691	0.669	0.518	0.413	0.334	0.239
65536	1	0.873	0.782	0.665	0.655	0.612	0.474	0.373	0.304
131072	1	0.884	0.731	0.653	0.644	0.573	0.502	0.417	0.333
262144	1	1.007	0.856	0.756	0.678	0.601	0.551	0.508	0.403
524288	1	0.920	0.999	0.876	0.807	0.708	0.628	0.610	0.522

Table 4.5. Parallel efficiency for one time step on the Cray T3E.

particles	1	2	4	8	16	32	64	128
					processors			
16384	0.942	0.902	0.863	1.013	0.988	0.975	0.955	0.915
32768	0.934	0.951	0.863	1.014	0.977	0.979	0.966	0.974
65536	0.981	0.928	0.926	1.075	0.971	0.911	0.976	0.991
131072	0.972	0.935	0.998	1.002	0.901	0.928	0.976	0.932
262144	0.884	0.953	0.983	1.026	1.003	1.003	1.064	0.988
524288	0.912	1.080	0.950	1.000	0.943	0.993	1.092	0.963
1048576		1.070	0.937	0.999	0.980	1.029	0.991	0.881
2097152			0.949	0.935	0.987	1.000	1.016	0.986
4194304				0.960	1.007	0.904	0.950	0.936
8388608					1.032	1.011	0.999	1.020

Table 4.6. Scaleup for $\kappa = 2$ for one time step on the Cray T3E.

We have used a factor of two in the problem size and the number of processors for the computation of the scaleup in Table 4.6, i.e., we used $\kappa = 2$. We can therefore also list values for problems that no longer fit into the main memory of one processor (for $N > 524288$). All in all, we obtain very good scaleup values close to one. This is because the parallel algorithm is of a type that only needs communication between local neighbors. Some scaleup values are even a little bit larger than one which is caused by the specific distribution and balance of the parallel data. We can only distribute entire cells which means that, for certain numbers of processors, a processor has to process one row of cells more than another processor. But a problem twice the size can be distributed somewhat better to twice as many processors.

In the following, we study our parallel implementation of the linked cell method on a different type of parallel computer. Besides conventional super-computers as the Cray T3E, an alternative has evolved, namely clusters of workstations and PCs. In such parallel systems, commodity computers are connected by Ethernet or faster networks and are run as parallel comput-ers using software packages such as PVM or MPI [248, 272]. These parallel computers are often called Beowulf class supercomputers [581]. These "do-it-yourself" machines differ in several aspects from commercial parallel com-puters, as for instance in the performance of the communication network but also in the software. As an example for this class of cluster computers we show execution times for the Linux cluster "Parnass2" of the Institute for Numerical Simulation at the University of Bonn in Table 4.7. It consists of PCs, with two 400 MHz Intel Pentium II processors each, that are coupled by a high speed network made by Myricom [557]. On this cluster computer we use the MPI implementation SCore [13].

particles	processors							
	1	2	4	8	16	32	64	128
16384	0.9595	0.4953	0.2778	0.1591	0.0997	0.0627	0.0407	0.0253
32768	1.9373	1.0068	0.5394	0.3162	0.1712	0.1055	0.0730	0.0443
65536	3.7941	1.9502	1.0478	0.5983	0.3127	0.1743	0.1239	0.0721
131072	7.2012	3.6966	2.0093	1.1917	0.5697	0.3132	0.2152	0.1088
262144	16.859	7.6617	4.1565	2.0424	1.0624	0.6156	0.4054	0.1971
524288	32.072	14.879	8.4316	4.4442	2.1176	1.1310	0.6817	0.4113
1048576	64.407	32.069	16.369	8.5576	4.2066	2.0963	1.1645	0.6901
2097152	121.92	60.422	32.737	17.273	8.0585	4.0968	2.3684	1.1889
4194304	248.37	118.07	61.678	32.746	16.285	8.2477	4.2860	2.2772
8388608			119.68	64.623	31.615	15.562	8.4831	4.4283
16777216						31.837	16.740	8.5574

Table 4.7. Parallel execution time (in seconds) for one time step on the PC cluster "Parnass2", connected with a Myrinet network.

A comparison of the execution times in Table 4.3 with those in Table 4.7 shows no significant differences. The Cray T3E is approximately 10% faster. The parallel program scales similarly on both machines. This demonstrates that the PC cluster is nearly as adequate as a Cray T3E for this kind of problems. Since such do-it-yourself clusters cost much less to buy, assemble, and build, they are serious competitors to commercial parallel computers.

In closing, we give results for the Cray T3E for another implementation of the short-range force evaluation according to Plimpton [483] that is based on the neighbor-list algorithm of Verlet [645] and is often cited as benchmark. Table 4.8 shows the execution times of this algorithm for the example from Table 4.3.

particles	processors								
	1	2	4	8	16	32	64	128	256
16384	0.1936	0.1030	0.0538	0.0288	0.0152	0.0080	0.0043	0.0025	0.0019
32768	0.3842	0.1985	0.1036	0.0587	0.0286	0.0150	0.0080	0.0043	0.0024
65536		0.4112	0.2059	0.1096	0.0563	0.0285	0.0151	0.0080	0.0043
131072			0.4265	0.2176	0.1120	0.0562	0.0285	0.0152	0.0080
262144				0.4290	0.2176	0.1091	0.0555	0.0286	0.0151
524288					0.4440	0.2177	0.1095	0.0562	0.0286
1048576						0.4424	0.2180	0.1123	0.0562
2097152							0.4294	0.2181	0.1092
4194304								0.4443	0.2179
8388608									0.4430

Table 4.8. Results for Plimpton's implementation with neighborhood lists. Parallel execution times (in seconds) for one time step on the Cray T3E.

A comparison of the results shows that the runtimes for Plimpton's implementation are consistently smaller than those for our implementation. Essentially, this can be traced back to a different implementation of the force computation for the short-range potentials. Unlike our linked cell algorithm, in which at every time step the neighborhoods are computed anew, Plimpton's implementation uses neighbor-lists that are reused for several time steps. For each particle a vector with pointers to particles within a distance of at most 2.8σ is stored, which is somewhat larger than r_{cut}. The implementation assumes that during these time steps (here 20) no other particle (but the ones in the list) will come closer than $r_{cut} = 2.5\sigma$ to the particle. In this way, one avoids the expensive testing of the distances of the particles for those particles that are in neighboring cells in the linked cell method but further away than r_{cut}. An easy estimate involving the volumes of the sphere and of the cells shows that a sizeable percentage of particles in the neighbor cells will lie outside of the cutoff radius. Also, vectors can be processed more efficiently than linked lists. However, the construction of the neighbor-lists can be significantly more expensive than one step of the linked cell method. To further optimize the implementation, Plimpton's method sets the values of σ, ε and m to one which saves multiplications, and works in single precision. In total this gives an acceleration by a factor of about five. Plimpton's program, however, is no longer flexible, and is – as are other typical benchmark codes – very much adapted to the specific situation of the benchmark example.

4.5 Application Examples

The parallelization of our linked cell code allows to treat problems with larger numbers of particles. In particular, problems in three dimensions can now properly be dealt with.

4.5.1 Collision of Two Bodies

We now extend the two-dimensional simulation of a collision between two bodies of the same material from Section 3.6.1 to the three-dimensional case, see also [199]. The setting of the simulation can be found in Figure 4.11 (upper left). The smaller body hits the larger, resting body with high velocity. Standard outflow boundary conditions are used at the sides of the simulation domain. We observe the evolution of the system for given parameters ε and σ of the Lennard-Jones potential.

At the beginning of the simulation, the particles within the two bodies are arranged on a regular mesh with mesh width $2^{1/6}\sigma$ (which corresponds to the minimum of the potential). The two bodies consist of $10 \times 10 \times 10$ and $10 \times 30 \times 30$ cubic cells with four particles each, giving a cubic face centered lattice, see Section 5.1.1. The velocities of the particles in the moving body are set to the given parameter $\mathbf{v}$. In addition, the initial velocity of the particles is superimposed by a small thermal motion chosen according to a Maxwell-Boltzmann distribution with an average velocity of 3.4 per component. There are no external forces in this simulation.

Figure 4.11 shows the result of a simulation with the parameter values from Table 4.9.

$$
\begin{array}{lll}
L_1 = 150\sigma, & L_2 = 150\sigma, & L_2 = 150\sigma, \\
\varepsilon = 120, & \sigma = 3.4, & \\
m = 39.95, & \mathbf{v} = (0, 0, -20.4), & \\
\text{distance between particles} = 2^{1/6}\sigma, & N_1 = 4000, & N_2 = 36000, \\
r_{cut} = 2.5\sigma, & \delta t = 0.001 &
\end{array}
$$

Table 4.9. Parameter values for the simulation of a collision in three dimensions.

The color of the particles indicates their velocity, as in the results of the two-dimensional simulation in Section 3.6.1. Immediately after the impact the smaller body disintegrates. It is completely destroyed by the collision. Furthermore, the impact punches a hole through the larger body and plastically deforms it.

Because of the simple Lennard-Jones potential used in this simulation, we obtain only a first, phenomenological description of the collision of solid bodies. More realistic simulations can be obtained by the use of more complicated potentials, compare Section 5.1. Further simulations of collisions can be found in [69, 71, 390].

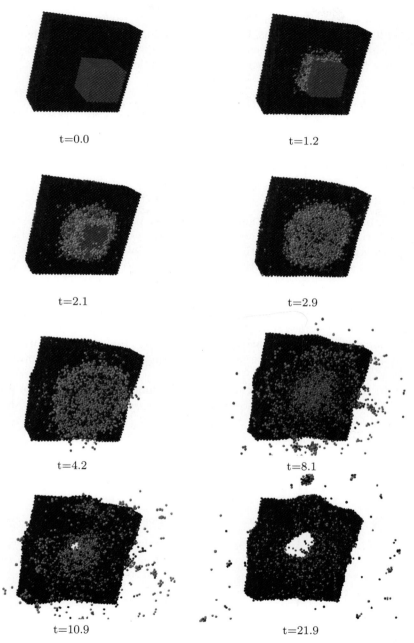

t=0.0

t=1.2

t=2.1

t=2.9

t=4.2

t=8.1

t=10.9

t=21.9

Fig. 4.11. Collision of two bodies, time evolution of the particle distribution.

4.5.2 Rayleigh-Taylor Instability

We now extend the simulation of the Rayleigh-Taylor instability from Section 3.6.4 to three dimensions. Again, a fluid is put on top of another fluid with lower density. The heavier fluid sinks down and thereby displaces the lighter fluid. We use the parameters from Table 4.10.

$$
\begin{aligned}
&L_1 = 90, && L_2 = 80, && L_3 = 22.5 \\
&\sigma = 1.15, && \varepsilon = 2.0, && N = 96 \times 84 \times 22 = 177408, \\
&m_{up} = 80, && m_{down} = 40, && T' = 100, \\
&G = (0, 0, -1.2435), && r_{cut} = 2.5\sigma, && \delta t = 0.001
\end{aligned}
$$

Table 4.10. Parameter values for the simulation of a Rayleigh-Taylor instability.

A total of 177408 particles is employed. At the start of the simulation, the particles are arranged on a regular grid. One half of the particles, the ones that are in the upper layer, have a mass of 80, and the other particles have a mass of 40. The initial velocities are chosen according to a Maxwell-Boltzmann distribution with a temperature parameter $T' = 100$. During the computation, the velocities are scaled every 10 time steps according to (3.35) and (3.40) using $T' = 100$. One can observe how typical mushroom-like structures are formed during the simulation. The heavier particles sink down and displace the lighter particles that therefore rise up. In this simulation we can observe the emergence of 4×4 mushroom structures of approximately equal size, see Figure 4.12.

In a further simulation with slightly changed parameters (as given in Table 4.11), one large mushroom-like structure emerges. This is shown in Figure 4.13.

$$
\begin{aligned}
&L_1 = 50, && L_2 = 40, && L_3 = 45 \\
&\sigma = 1.15, && \varepsilon = 2.0, && N = 54 \times 42 \times 46 = 104328, \\
&m_{up} = 80, && m_{down} = 40, && T' = 100, \\
&G = (0, 0, -0.6217), && r_{cut} = 2.5\sigma, && \delta t = 0.001
\end{aligned}
$$

Table 4.11. Parameter values for the simulation of a Rayleigh-Taylor instability.

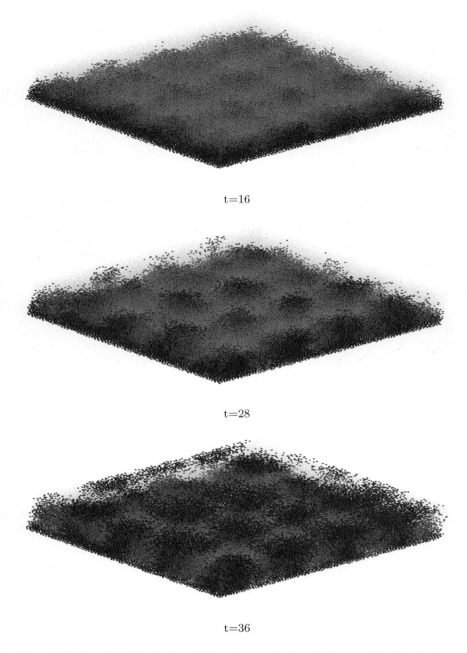

t=16

t=28

t=36

Fig. 4.12. Rayleigh-Taylor instability, time evolution of the particle distribution.

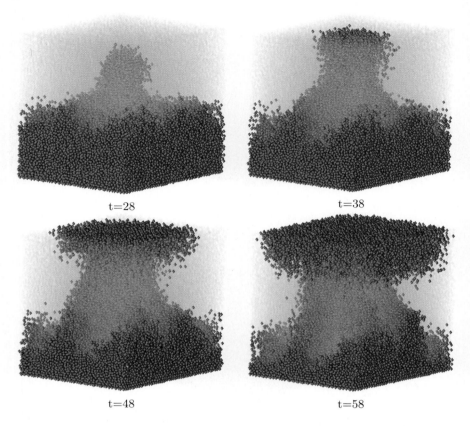

t=28   t=38

t=48   t=58

Fig. 4.13. Rayleigh-Taylor instability, time evolution of the particle distribution.

5 Extensions to More Complex Potentials and Molecules

So far, only pair potentials were used in the previous simulations. In this chapter, we now consider several applications which need more complex potentials, and we discuss the necessary changes in the algorithms. We start with three examples for many-body potentials, the potential of Finnis and Sinclair [232, 330, 593], the EAM potential [64, 173, 174], and the potential of Brenner [122]. The Finnis-Sinclair potential and the EAM potential describe metallic bonds. We use these potentials to simulate microcracks and structural changes in metallic materials. The Brenner potential describes hydrocarbon bonds. We employ it in the simulation of carbon nanotubes and carbon buckyballs. Furthermore, we extend our code to potentials with fixed bond structures. We then simulate simple networks of atoms and chain molecules such as alkanes and polymers. Finally, we give an outlook on the implementation of more complex biomolecules and proteins.

5.1 Many-Body Potentials

The application of appropriate potentials is crucial to obtain reliable results in particle simulations. Simple pair potentials such as the Lennard-Jones potential can not reproduce the specific properties of materials such as metals or hydrocarbons with sufficient accuracy [174]. Instead, so-called many-body potentials are necessary, in which the force between any two particles depends on the position of several neighboring particles.

These potentials are of the form

$$V(\mathbf{x}_1, \ldots, \mathbf{x}_N) = \sum_{i=1}^{N} \left(\sum_{\substack{j=1 \\ i \neq j}}^{N} U(r_{ij}) - S_i(r_{i1}, \ldots, r_{iN}) \right) \tag{5.1}$$

with a repulsive part U, realized by a pair potential, and an attractive, more complicated part S_i. There, interactions between multiple particles are taken into account, so that the force on a particle depends on the number and position of neighboring particles. Such potentials allow for the modeling of the specific bonds in metallic materials. They are superior to conventional pair potentials such as the Lennard-Jones potential and enable a physically more accurate description.

In recent years, many empirical and semi-empirical approaches have been developed for the simulation of metals and related covalent materials. In 1983, Daw and Baskes [173, 174] published the so-called embedded atom method (EAM) for transition metals with face-centered cubic (fcc) lattices, see Figure 5.1. This approach has been extended by Johnson and Adams to metals with body-centered cubic (bcc) lattices [29, 344]. It was further improved to the MEAM potential [63, 65] which also takes the bond neighborhood into account. Abell and Tersoff took this method up and developed a similar potential for silicon [28, 602]. Their potential in turn was used by Brenner [122] as the starting point for his hydrocarbon potential, which we will treat in more detail in Section 5.1.3.

An alternative approach was developed by Finnis and Sinclair in 1984 [232]. They derived a many-body potential from the tight-binding technique, which is a semi-empirical approximation of the density functional theory. The Finnis-Sinclair potential was originally developed for transition metals with body-centered cubic lattices. Rosato et al. [529] and Sutton and Chen [593] introduced variants for transition metals with face-centered cubic lattices (fcc) and with hexagonal closest packed lattices (hcp). A generalization to metallic alloys can be found in [497]. In the following, we present the Finnis-Sinclair potential in more detail.

5.1.1 Crack Propagation in Metals – the Finnis-Sinclair Potential

The analysis of the formation and propagation of cracks in metals on a microscopic level is an interesting application of the method of molecular dynamics to material sciences. Here, one is especially interested to reproduce the experimentally observed propagation of cracks [229, 230] and thus to find explanations for the specific behavior of cracks on the microscopic level. It has been observed in physical experiments that there is a maximum velocity of crack propagation. Up to a certain velocity, a straight crack with smooth edges forms, but when the velocity reaches this critical value, the dynamics of crack propagation changes dramatically. The velocity of the crack propagation starts to oscillate and rough crack surfaces are produced. Furthermore, bifurcations of the crack trajectory can be observed. The simulation of cracks on the microscopic level by a molecular dynamics approach allows for a detailed investigation and an accurate determination of the velocity of the crack propagation [330]. To this end, the propagation of a crack can be followed automatically via an analysis of the computed stresses. The results of such simulations can then be used to improve macroscopic continuum models of crack formation [496].

Metals can be classified into several distinct groups according to their structural type, i.e. the arrangement of their atoms to a crystal lattice [524]. For instance, in copper, silver, gold and nickel, the atoms are arranged in a cubic face-centered Bravais-lattice (fcc), in which atoms are positioned at every vertex and at the center of each face of a cubic unit cell (compare Figure 5.1, lower left). Then, in iron the atoms are located on a cubic body-centered Bravais-lattice (bcc). In such a lattice, an atom is placed at each vertex and at the center of a cubic unit cell (compare Figure 5.1, upper left). Furthermore, there is the arrangement in a hexagonal closest (spherical) packing (hcp), compare Figure 5.1, right, which can be found in magnesium or zinc. There are also more general cubic-like and rhombic lattices.[1] The preferred lattice types of the most important elements can be found in Figure 5.2.

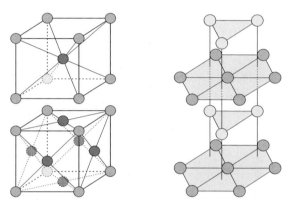

Fig. 5.1. Lattice types. Upper left: Body-centered cubic (bcc). Lower left: Face-centered cubic (fcc). Right: Hexagonal closest packing (hcp).

Potentials that describe metallic bonds have to be adapted to the respective lattice structure. Here, different forms and parameters of the potentials exist, depending on the arrangement of atoms in the metal. In the following, we use the Finnis-Sinclair potential

$$V = \varepsilon \sum_{i=1}^{N} \left(\sum_{\substack{j=i+1}}^{N} \left(\frac{\sigma}{r_{ij}} \right)^n - c \left[\sum_{\substack{j=1 \\ j \neq i}}^{N} \left(\frac{\sigma}{r_{ij}} \right)^m \right]^{1/2} \right) \tag{5.2}$$

for fcc metals according to Sutton and Chen [593]. Like the simple Lennard-Jones potential (3.27), this potential features both a repulsive and an attractive part. The repulsive part is realized by a pair potential, while the

---

[1] The valence electrons of the atoms are delocalized in metallic bonds. This decreases the kinetic energy and contributes to the strength of the bonds. In so-called transition metals, the d- and f-orbitals of different atoms also interact covalently with each other. This often results in closest packed structures.

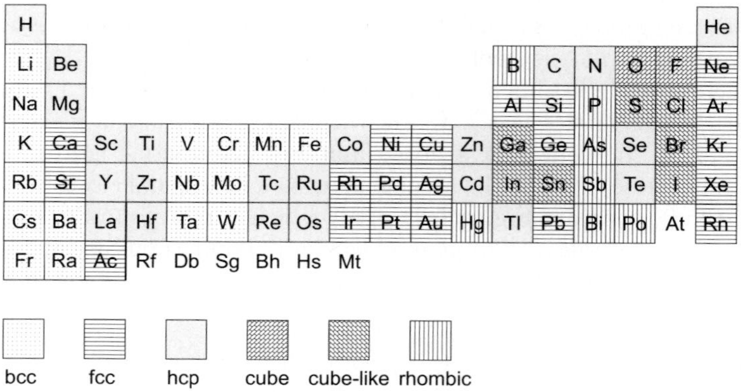

Fig. 5.2. Preferred lattice types under normal conditions for some important elements.

attractive part is realized by a many-body potential. The material dependent parameters n, m, ϵ, and σ play the same role as in the Lennard-Jones potential. The additional parameter c is related to the material and describes the specific type of lattice. Several sets of parameters are given in Table 5.1.

	m	n	ε	σ	c
copper	6	9	$1.2382 \cdot 10^{-2}$eV	3.61 Å	39.432
silver	6	12	$2.5415 \cdot 10^{-3}$eV	4.09 Å	144.41
gold	8	10	$1.2793 \cdot 10^{-2}$eV	4.08 Å	34.408

Table 5.1. Parameters for the Sutton-Chen potential for some fcc metals [593].

Using the abbreviation

$$S_i = \sum_{\substack{j=1 \\ j \neq i}}^{N} \left(\frac{\sigma}{r_{ij}} \right)^m \tag{5.3}$$

the potential can also be written as

$$V = \varepsilon \sum_{i=1}^{N} \left(\sum_{j=i+1}^{N} \left(\frac{\sigma}{r_{ij}} \right)^n - c\sqrt{S_i} \right). \tag{5.4}$$

The gradient of the potential gives the force on the individual particles. With $\mathbf{r}_{ij} = \mathbf{x}_j - \mathbf{x}_i$, this force is

$$\mathbf{F}_i = -\varepsilon \sum_{\substack{j=1 \\ i \neq j}}^{N} \left(n \left(\frac{\sigma}{r_{ij}} \right)^n - \frac{c\,m}{2} \left(\frac{1}{\sqrt{S_i}} + \frac{1}{\sqrt{S_j}} \right) \left(\frac{\sigma}{r_{ij}} \right)^m \right) \frac{\mathbf{r}_{ij}}{r_{ij}^2}. \tag{5.5}$$

Now, the more complicated term $(S_i^{-1/2} + S_j^{-1/2})$ appears in addition to the terms present in the force for the Lennard-Jones potential (3.28). Due to this term, the force from particle j on particle i depends on the positions of all other particles.

Since the terms in the sum decay rapidly with r_{ij}, the sums in (5.3), (5.4) and (5.5), respectively, do not have to be evaluated for all particles, but can be cut off as in the case of the Lennard-Jones potential. With a threshold parameter r_{cut} one then obtains the approximations

$$S_i \approx \bar{S}_i = \sum_{\substack{j=1, j\neq i \\ r_{ij} < r_{\text{cut}}}}^{N} \left(\frac{\sigma}{r_{ij}}\right)^m, \tag{5.6}$$

$$V \approx \varepsilon \sum_{i=1}^{N} \left(\sum_{\substack{j=i+1 \\ r_{ij} < r_{\text{cut}}}}^{N} \left(\frac{\sigma}{r_{ij}}\right)^n - c\sqrt{\bar{S}_i} \right) \tag{5.7}$$

and

$$\mathbf{F}_i \approx -\varepsilon \sum_{\substack{j=1, i\neq j \\ r_{ij} < r_{\text{cut}}}}^{N} \left(n\left(\frac{\sigma}{r_{ij}}\right)^n - \frac{cm}{2}\left(\frac{1}{\sqrt{\bar{S}_i}} + \frac{1}{\sqrt{\bar{S}_j}}\right)\left(\frac{\sigma}{r_{ij}}\right)^m \right)\frac{\mathbf{r}_{ij}}{r_{ij}^2}. \tag{5.8}$$

We abbreviate

$$\mathbf{F}_{ij} = -\varepsilon \left(\frac{\sigma}{r_{ij}}\right)^m \cdot \left(n\left(\frac{\sigma}{r_{ij}}\right)^{n-m} - \frac{cm}{2}\left(\frac{1}{\sqrt{\bar{S}_i}} + \frac{1}{\sqrt{\bar{S}_j}}\right)\right)\frac{\mathbf{r}_{ij}}{r_{ij}^2} \tag{5.9}$$

for the force from particle j on particle i. Then, we obtain

$$\mathbf{F}_i \approx \sum_{\substack{j=1, i\neq j \\ r_{ij} < r_{\text{cut}}}}^{N} \mathbf{F}_{ij}.$$

Now, we study the propagation of cracks in metals on the molecular level using the potential (5.2). At the beginning of the simulation, the particles are placed on a face-centered cubic lattice. At one side, a small dislocation is introduced: The first 10 or 20 particles in the middle of the sample are pulled apart by a small force which acts along a line (2D) or a plane (3D), respectively, compare Figure 5.3. This force decreases linearly to zero from the exterior of the metal to the interior. Furthermore, the bottom and the top sides of the metal cube are pulled apart by a constant force which acts on the particles at these boundaries, compare Figure 5.3 for a sketch in the two-dimensional case.

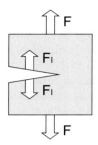

Fig. 5.3. Basic configuration for the simulation of crack formation in metals in the two-dimensional case; on the upper and lower side the metal is pulled with a given force F; on the left hand side, a defect has been introduced.

Implementation. Now, several changes and extensions of the code are necessary due to the many-body part of the potential. For the computation of the potential energy V and the forces $\mathbf{F}_i$, the terms $\bar{S}_i$, $i = 1, \ldots, N$, from (5.6) are needed. After they have been determined, the potential energy V and the forces $\mathbf{F}_i$ can be computed as previously with the linked cell method according to (5.7) and (5.8). Therefore, in each time step, the values for $\bar{S}_i$ are calculated and saved before the force computation. This can be done with one loop over all particles. To this end, for a fixed i all the particles in the same cell and in the neighboring cell are visited – as described in algorithm 3.15 – and the appropriate interaction terms are added up.

In the implementation, we therefore complement the data structure 3.1 `Particle` by an additional value `real s`, which contains the value of $\bar{S}_i$ for particle i. Furthermore, a new routine `compS_LC` has to be implemented,

```
void compS_LC(Cell *grid, int *n, real r_cut);
```

which computes the values $\bar{S}_i$ according to (5.6). This routine is called at the beginning of the routine `compF_LC` in algorithm 3.15 before any forces are computed. Additionally, we have to modify the computation of the potentials and forces inside of the routine `force` according to (5.4), (5.8) and (5.9).

Parallelization. The parallelization of the linked cell method was already described in Chapter 4. The parallel algorithm now has to be supplemented by the parallel computation of the values $\bar{S}_i$ according to (5.6). This can be done analogously to the explanations in Section 4.2. If the data of the particles in the cells of the border neighborhood of the respective subdomain are available, $\bar{S}_i$ can be computed inside each subdomain in parallel.

Additionally, the following problem occurs: All $\bar{S}_j$ with $\|\mathbf{x}_j - \mathbf{x}_i\| < r_{\text{cut}}$ are needed to compute the force $\mathbf{F}_i$ on particle i. Because of the cutoff in the summation for the $\bar{S}_j$, the corresponding particles lie either in the same cell as particle i or in a cell directly adjacent to it. Close to the boundary of the local simulation domain it may happen that the correct value has been computed by a neighboring process and is therefore not known to the process that computes the force F_i. Thus, an additional communication step is needed,[2] in which the $\bar{S}_j$ of the particles in the cells of the border neighborhood are exchanged between processes. In the algorithm this is implemented directly after the parallel computation of the $\bar{S}_j$. Again, the communication routines from Section 4.3 can be used to realize this communication step.

Examples. Here, we present the results of one simulation for the two-dimensional and the three-dimensional case, respectively. Figure 5.4 shows the results of a two-dimensional simulation with the parameter values from Table 5.2.[3] One can see how the sound waves induced by the crack spread through the material.[4] Figure 5.5 shows the result of a three-dimensional simulation with the parameter values from Table 5.3.

$$
\begin{aligned}
L_1 &= 200, & L_2 &= 200, \\
\varepsilon &= 1, & \sigma &= 1, \\
c &= 10.7, \\
m &= 6, & n &= 12, \\
\text{lattice spacing} &= 1.1875, & T &= 1 \text{ K}, \\
r_{\text{cut}} &= 5.046875,
\end{aligned}
$$

Table 5.2. Parameter values for the simulation of a microcrack in silver (2D).

The propagation and thus the velocity of a crack can be tracked automatically if one associates to each particle i the 3×3 matrix

$$
\boldsymbol{\sigma}_i = -\frac{1}{2} \frac{1}{|\Omega|} \sum_{\substack{j=1, i\neq j \\ r_{ij} < r_{\text{cut}}}}^{N} \mathbf{F}_{ij} \cdot \mathbf{r}_{ij}^T, \tag{5.10}
$$

---

[2] This second communication step can be completely avoided, if the data of the particles are saved and exchanged in a border neighborhood of double size. Then, the $\bar{S}_j$ needed for the force computation by any process can be computed from data of that process. This however causes a moderate increase in the memory requirements.

[3] The parameter c depends on the type of lattice and on the number of neighboring atoms. Therefore, its values in two and three dimensions are in general different.

[4] These waves emanate from the crack, propagate through the material, and are reflected at the boundary of the sample, which influences the propagation velocity and the trajectory of the crack. This can be avoided by damping the acoustic waves at the boundary of the sample. The implementation of such a damping consists of the introduction of an additional friction-like term (compare Section 3.7.1) that only affects particles close to the boundary of the sample.

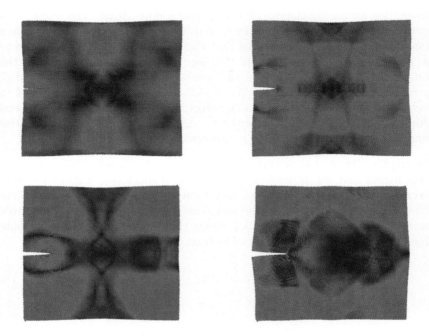

Fig. 5.4. Propagation of a microcrack in silver in two dimensions.

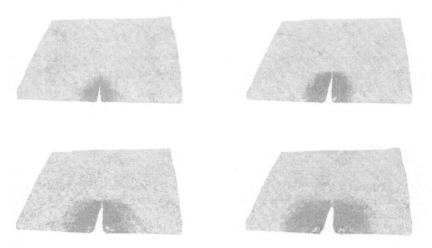

Fig. 5.5. Propagation of a microcrack in silver in three dimensions.

$$L_1 = 80, \qquad L_2 = 80, \quad L_3 = 80,$$
$$\varepsilon = 1, \qquad \sigma = 1, \qquad c = 144.41,$$
$$m = 6, \qquad n = 12,$$
$$\text{lattice spacing} = 1.21875, \quad T = 1 \text{ K},$$
$$r_{\text{cut}} = 4$$

Table 5.3. Parameter values for the simulation of a microcrack in silver (3D).

where $|\Omega|$ denotes the volume of the simulation domain.[5] This matrix can be interpreted as stress tensor at the position of the particle i (where the stress caused by the thermal movement of the particles is not taken into account, see also (5.27) and (5.28)). The propagation of the crack can be tracked with help of the position of the maximal $\|\boldsymbol{\sigma}_i\|$. Here, one assumes that the measured stress is maximal at the tip of the crack. An alternative formula for the stress tensor can be found in [330].

The Lennard-Jones potential has been used in [464] to simulate the formation of cracks[6]. More detailed studies of crack propagation with the Finnis-Sinclair potential can be found in [330, 685]. In [496] a microscopic molecular dynamics model in the immediate vicinity of the crack tip is coupled with a macroscopic finite element elasticity model and applied to the propagation of cracks in a two-dimensional silver plate. The obtained multiscale model successfully reproduces the crack velocity, the change of the roughness of the crack surface over time, and the macroscopic trajectory of the crack.

The formation of cracks plays an important role not only in metals, but also in semiconductors. For the latter case, most work in the literature uses a potential introduced by Stillinger and Weber [584] or refined variants of it. Results and references can be found for silicon in [409], and for gallium arsenide and silicon nitride in [444, 639, 641]. In these articles, also the fracture behavior of ceramics and the propagation of cracks in layers of graphite is studied using the Brenner potential (see Section 5.1.3). Numerical experiments for the crack dynamics in quasi-crystals can be found in [432]. Finally, [326, 691, 692] describe simulations of the fracture dynamics and the propagation of cracks in copper with more than 35 million particles. These simulations are based on a variant of the EAM potential which was developed in [648]. We explain the basic principles of the EAM in the following section.

---

[5] Here, $\mathbf{r}_{ij}^T$ stands for the transpose of the vector $\mathbf{r}_{ij} = \mathbf{x}_j - \mathbf{x}_i$. The product $\mathbf{F}_{ij} \cdot \mathbf{r}_{ij}^T$ is a 3×3 tensor with the entries $(\mathbf{F}_{ij})_l (\mathbf{r}_{ij})_m$ in the lth row and mth column.

[6] As we already noted, pair potentials such as the Lennard-Jones potential cannot model the specific properties of metals accurately enough.

5.1.2 Phase Transition in Metals – the EAM Potential of Daw and Baskes

The *embedded atom method* (EAM) is another approach to derive potentials that can be used in the simulation of metals and metal alloys by molecular dynamics. These potentials have been used successfully in the modelling of crack formation, surface reactions, epitaxial growth, the austenite-martensite transformation, and phase transitions in solids, nanoparticles, and thin films [350]. The embedded atom method was introduced by Daw and Baskes [173] and has been further developed in a number of articles [64, 65, 649]. In the meantime it has become a standard in the simulation of metals by the molecular dynamics method.

The basic idea is as follows: Each energy potential induces an electron density. The other way around, Hohenberg and Kohn [325] showed that also the electron density uniquely determines the potential.[7] This principle is used in the EAM to construct the potential by means of the electron density: Each atom i is surrounded by its host material, i.e. the bulk of all the other atoms. Its energy therefore depends on the electron density of the host material at position $\mathbf{x}_i$, in other words, the atom is *embedded* in the electron density of the host. Via an appropriately chosen embedding function $\mathcal{F}_i$ we can specify the energy V_i^{emb} of the ith atom which is contributed by the embedding into the electron density of the host material as

$$V_i^{\mathrm{emb}} = \mathcal{F}_i(\rho_i^{\mathrm{host}}). \tag{5.11}$$

Here, ρ_i^{host} describes the electron density of the host material without atom i at the point $\mathbf{x}_i$. Now, we make the simplifying assumption that the electron density ρ_i^{host} is just the sum of all the electron densities ρ_j^{atom} of all the atoms j with $j \neq i$. We furthermore assume that these densities only depend on the distance. Thus we set

$$\rho_i^{\mathrm{host}} = \sum_{j=1, j \neq i}^{N} \rho_j^{\mathrm{atom}}(\|r_{ij}\|). \tag{5.12}$$

The embedding potential (5.11) is now combined with a pair potential of the form[8]

$$V_i^{\mathrm{pair}} = \frac{1}{2} \sum_{j=1, j \neq i}^{N} \phi_{ij}(\|r_{ij}\|).$$

Here, the functions ϕ_{ij} only depend on the types of the atoms i and j. The interaction between different atom types is often well approximated by the

---

[7] In Hohenberg and Kohn's work the potential is generated by the energy density of all atoms whereas in the EAM it is generated by the energy density of the all atoms except of atom i.

[8] The use of V^{emb} alone results in unrealistic physical properties and conflicts with experimental measurements on solids.

geometric mean of the interaction between atoms of the same type, i.e. $\phi_{ij} = \sqrt{\phi_{ii}\phi_{jj}}$ [174]. Then, V_i^{pair} can be written in the form

$$V_i^{\text{pair}} = \frac{1}{2} \sum_{j=1, j \neq i}^{N} \frac{Z_i(\|r_{ij}\|) Z_j(\|r_{ij}\|)}{\|r_{ij}\|} \tag{5.13}$$

with appropriately chosen functions Z_i which only depend on the distance. The Z_i can be interpreted as "effective charges". Thus, one obtains an energy similar to the electrostatic charge. Here, the functions Z_i describe short-range effects, they decay very rapidly and are practically zero for distances of just a few Å.

Altogether, we obtain the potential

$$V = \sum_{i=1}^{N} \mathcal{F}_i \left(\sum_{j=1, j \neq i}^{N} \rho_j^{\text{atom}}(\|r_{ij}\|) \right) + \frac{1}{2} \sum_{i=1}^{N} \sum_{j=1, j \neq i}^{N} \frac{Z_i(\|r_{ij}\|) Z_j(\|r_{ij}\|)}{\|r_{ij}\|} \tag{5.14}$$

where the embedding functions $\mathcal{F}_i$, the effective charges Z_i, and the functions ρ_j^{atom} in the electron density have to be determined depending on the material under consideration. In the case of $\mathcal{F}_i$ and Z_i, this is done in a semi-empirical way. To this end, $\mathcal{F}_i$ and Z_i are modeled as cubic[9] splines, where the coefficients are calculated by a weighted least squares approach such that typical physical properties like lattice constants, elastic constants, vacancy formation energies, and others are reproduced as accurately as possible. The density functions ρ_j^{atom} are determined by Hartree-Fock approximations, see [151] for details. More recent and more accurate data can be found in [130].[10]

Implementation. The part of the potential (5.14) given by the embedding in the host (5.11) needs some extensions to the force computation in the linked cell algorithm in Section 3.5, analogous to the extensions necessary for the potential of Finnis-Sinclair which we presented in the previous Section 5.1.1.

For the computation of the energy V and the forces $\mathbf{F}_i$, we need the functions $\rho_j^{\text{atom}}(r)$, $Z_i(r)$ and $\mathcal{F}_i(\rho)$ and their derivatives. Usually, these functions are chosen to be B-splines. To this end, one approximates the function between any two nodes as cubic polynomials that match the given values at the nodes (interpolation) and furthermore give a globally twice differentiable function. To obtain a unique characterization, boundary conditions for the first derivative (for $Z_i(r)$) or for the second derivative (for $\mathcal{F}_i(\rho)$) are additionally prescribed, see also Table 5.4. From these values one first computes the second derivative of the spline approximation at all nodes by the solution of a tridiagonal system of linear equations. The subsequent evaluation of the

---

[9] A linear $\mathcal{F}_i$ would result in a pair potential.

[10] It does not pay off to invest too much effort in the electron density, because simple Hartree-Fock calculations produce in general a much higher accuracy of the electron densitiy than an empirically fitted embedding function.

spline approximation at a point r can then be obtained with the values and derivatives at the two neighboring nodes of r from a cubic polynomial or a B-spline. The details can be found in many standard books in numerical analysis such as [176, 180, 587]. The evaluation for arbitrary values of r and ρ has to be implemented in appropriate procedures. We will describe such procedures later in the algorithms 7.3 and 7.8 in Section 7.3.2 for general degrees of the polynomials. Special attention has to be given to the efficiency and speed of the evaluation of the spline approximation since it will be used in any computation involving $Z_i(r)$ and $\mathcal{F}_i(\rho)$.

The functions $\rho_j^{\mathrm{atom}}(r)$ and $Z_i(r)$ are constructed in such a way that they decay to zero within the threshold radius r_{cut}. Then, the linked cell method 3.15 with threshold radius r_{cut} can be applied to compute forces and energies. The pair potential V_i^{pair}, which contains the effective charge functions Z_i, can be computed directly according to (5.13). However, for the computation of the forces caused by the electron densities ρ_i^{host} from (5.12) we need an intermediary step, just as in the computation of the terms $\bar{S}_i$ in the Finnis-Sinclair potential. The electron density ρ_i^{host} for all particles i is computed in a first linked cell loop. In a second loop, the electron density is used together with the embedding function $\mathcal{F}_i$ as in (5.11) to compute energies and forces.

For this purpose, we add the variable `real rho_host` to the data structure 3.1 `Particle` which we use to store the value of ρ_i^{host} for the appropriate particle i. In a new subroutine `compRho_LC` we determine the value of that variable according to (5.12). The force computation has to be adapted accordingly. The parallelization of the linked cell algorithm proceeds as in Section 4.2, extended as in the previous Section 5.1.1. To compute ρ_i^{host} in parallel, an additional communication step is required, or, alternatively, a border neighborhood of double the size is used and the ρ_i^{host} are computed redundantly by respective processes for particles i close to the subdomain boundaries.

Example. In the following experiments, we simulate structural changes in metal alloys using the EAM potential.[11] We know from Section 5.1.1 that crystalline metals can have different structural types, i.e. different arrangements of the atoms in the crystal lattice, compare Figure 5.1. In some elements such as iron, but also in some alloys made from metals with different lattice types, a change in temperature can cause a change in structural type. One example is the transition from β- to γ-iron at 1185 K. Here, one speaks of a temperature-induced structure conversion or a phase transition. Such transitions in alloys were first observed by Adolf Martens around 1890 when he studied alloys under a microscope [450]. The phase transition can occur globally in the entire sample or only locally in regions of the size of several atom layers which then leads to microstructures.

---

[11] In our example of a iron nickel alloy, ferromagnetic effects are ignored and the electrostatic forces are modeled by the EAM potential.

We now study this dependence of the phase on the temperature and the conversion from one phase to another for the case of a nanoparticle. To this end, we consider an alloy which consists of 80% iron and 20% nickel atoms. For a complete description of the associated EAM potential (5.12) we need a concrete functional form for the electron density $\rho_j^{\text{atom}}(r)$. Here, we proceed as follows: The electrons from the outer orbitals are most important for the electrostatic interactions between atoms. Analogous to the wave functions of the stationary Schrödinger equation, their density is approximated by functions of the form

$$\rho^{*-\text{orbital}}(r) = \frac{1}{4\pi}\left|\sum_k C_k \frac{(2\zeta_k)^{n_k+1/2}}{\sqrt{(2n_k)!}}r^{n_k-1}e^{-\zeta_k r}\right|^2 , \tag{5.15}$$

where n_k is a positive integer which relates to the principal quantum number. The empirical parameters C_k and ζ_K depend on the material and the orbital, respectively. The electron density $\hat{\rho}_j^{\text{atom}}(r)$ is now assumed to be a linear combination of the electron densities of the orbitals involved. In the case of an iron nickel alloy, the outer electrons only occupy $4s$ and $3d$ orbitals. We therefore set

$$\hat{\rho}_j^{\text{atom}}(r) = N_s\rho^{4s-\text{orbital}}(r) + (N - N_s)\rho^{3d-\text{orbital}}(r) ,$$

where N denotes the total number of the outer electrons of the respective atom and N_s denotes the empirically determined number of electrons in the $4s$ orbital. Detailed values of the different parameters for iron and nickel atoms are given in Table 5.4, see also [151]. We now cut off the density at the radius r_{cut} and translate it properly to achieve a continuous transition. We then obtain

$$\rho_j^{\text{atom}}(r) \approx \begin{cases} \hat{\rho}_j^{\text{atom}}(r) - \hat{\rho}_j^{\text{atom}}(r_{\text{cut}}), & r \le r_{\text{cut}}, \\ 0, & r > r_{\text{cut}}. \end{cases} \tag{5.16}$$

The embedding functions $\mathcal{F}_i$ and the effective charges Z_i are approximated by cubic splines whose nodes and values are given in Table 5.4. The parameters for iron and nickel are taken from [429] and [350, 430], respectively.[12] The nodes and values describe a cubic spline for each $\mathcal{F}_i$ and Z_i. Here, the effective charge Z_i is set to zero outside of the interval which is spanned by the given nodes, whereas the embedding function $\mathcal{F}_i$ is linearly continued there.

We now arrange 2741 particles in form of a ball on a bcc lattice with the lattice constant a_0 for iron from Table 5.4. To this end, 80% of the particles represent iron atoms and 20% represent nickel atoms, with iron and nickel atoms randomly distributed. This system represents a metallic nanoparticle which we now study in the NVT ensemble for different temperatures. The

---

[12] Note that these improved parameters differ somewhat from the original EAM parameters for nickel from [173].

		nickel			iron	
k	n_k	ζ_k [Å^{-1}]	C_k	n_k	ζ_k [Å^{-1}]	C_k
4s 1	1	54.87049	-0.00389	1	51.08593	-0.00392
2	1	38.47144	-0.02991	1	35.92446	-0.03027
3	2	27.41786	-0.03189	2	25.54344	-0.02829
4	2	20.87506	0.15289	2	19.14388	0.15090
5	3	10.95341	-0.20048	3	9.85795	-0.21377
6	3	7.31714	-0.05423	3	6.56899	-0.05096
7	4	3.92519	0.49292	4	3.63805	0.50156
8	4	2.15217	0.61875	4	2.03603	0.60709
3d 1	3	12.67158	0.42120	3	11.46739	0.40379
2	3	5.43072	0.70658	3	4.94799	0.71984

	nickel			iron	
N	N_s	r_{cut} [Å]	N	N_s	r_{cut} [Å]
10	0.85	4.64453	8	0.57	4.40905

	nickel			iron	
ρ [Å^{-3}]	$\mathcal{F}(\rho)$ [J]	$\mathcal{F}''(\rho)$	ρ [Å^{-3}]	$\mathcal{F}(\rho)$ [J]	$\mathcal{F}''(\rho)$
0	0	0	0	0	0
0.01412	-5.87470e-19		0.00937	-6.15372e-19	
0.02824	-8.63439e-19		0.01873	-9.32104e-19	
0.05648	-5.78532e-19		0.03746	-6.60495e-19	
0.06495	0	0	0.04308	0	0

	nickel			iron	
r [Å]	$Z(r)$ [e]	$Z'(r)$	r [Å]	$Z(r)$ [e]	$Z'(r)$
0	28.0	0	0	26.0	0
2.112	0.9874		2.00921	1.4403	
2.4992	0.1596		2.49716	0.2452	
2.992	0.0	0	2.69808	0.1491	
			2.87030	0.0734	
			3.44436	0	0

	nickel			iron	
ρ_0 [Å^{-3}]	a_0 [Å]	lattice type	ρ_0 [Å^{-3}]	a_0 [Å]	lattice type
0.02824	3.52	fcc	0.01873	2.87	bcc

Table 5.4. Parameters of the EAM potential for nickel and iron. Values of the coefficients ζ_k in the electron densities $\rho^{4s-\text{orbital}}$ and $\rho^{3d-\text{orbital}}$ of the respective orbitals, the total number N of the outer electrons, and the number N_s of the outer electrons in the 4s orbital. Values of the cutoff radius r_{cut}, the nodes ρ_j and r_j, and the values of the embedding function $\mathcal{F}_i$ and the effective charge Z_i at those nodes. Number ρ_0 of electrons per Å^3 and lattice parameter a_0 (edge length of the unit cube in the crystal, compare Figure 5.1) at equilibrium.

temperature is adjusted after each time step by the scaling (3.42) of the velocities as described in Section 3.7.1. The parameters of the simulation are listed in Table 5.5.

$$L_1 = 60 \text{ Å}, \qquad L_2 = 60 \text{ Å}, \qquad L_3 = 60 \text{ Å},$$
$$N = 2741, \qquad N_{\text{nickel}} = 2196, \qquad N_{\text{iron}} = 545,$$
$$\text{Sphere with radius} = 20 \text{ Å}, \qquad \text{Lattice} = \text{bcc}, \qquad a_0 = 2.87 \text{ Å},$$
$$\delta t = 10^{-15} \text{ s}, \qquad t_{\text{end}} = 7.96 \cdot 10^{-11} \text{ s},$$
$$T(t = 0) = 100 \text{ K}, \qquad T(t = t_{\text{end}}) = 530 \text{ K},$$
$$m_{\text{nickel}} = 58.6934 \text{ u}, \qquad m_{\text{iron}} = 55.845 \text{ u}$$

Table 5.5. Parameter values for the simulation of an iron nickel alloy.

We start the simulation at a temperature of 100 K. Here, the bcc structure is preserved over time. We then increase the temperature of the system linearly. At about 480 K, a conversion of the arrangement of the atoms can be observed: First, local domains with fcc structure are formed, which subsequently grow with increasing temperature. Figure 5.6 shows the results of the simulation for temperatures of 100 K, 480 K and 530 K. One can clearly see two different areas in the cross-section, one with an fcc structure and one with an hcp structure. The lower part of the nanoparticle also exhibits an fcc structure, but this structure is not aligned with the plane of the cross-section and can therefore not be seen in the figure. This configuration is essentially preserved when the temperature is further increased to 800 K.

The associated radial distribution functions (3.62) are shown in Figure 5.7. The position and the relative height of the maxima clearly show a transition from a pure bcc lattice at 100 K to a closest packed lattice with fcc and hcp portions at 530 K which additionally contains some small perturbations and dislocations.

This phase transition of the iron nickel alloy can be reversed to some extent. To this end, the temperature must be reduced very slowly and carefully. The use of the NPT ensemble is also advantageous. Here, the transition to the original state is somewhat retarded. This delay can also be seen from macroscopic variables such as the size of the nanoparticle or the electric resistance [350]. Such a behavior is typical for many phase transition processes and is known as *hysteresis* [124]. If the cooling occurs too fast, the crystal structure of the high temperature phase is preserved.

More experiments with iron nickel alloys and iron aluminum alloys can be found in [350, 429, 430]. There, periodic crystals and thin layers are studied. Chemical vapor deposition processes with ionic clusters are simulated in [281]. The formation of cracks and dislocations in metals is studied in [278, 322, 432, 475, 531, 691, 692].

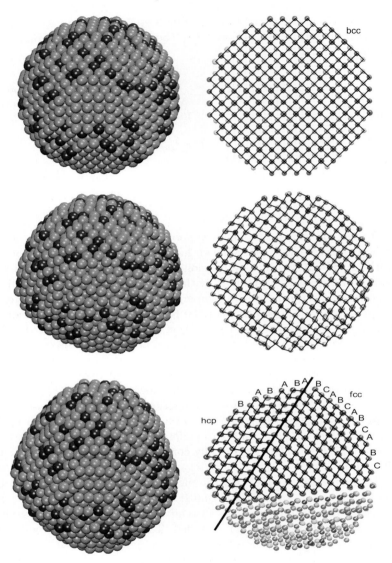

Fig. 5.6. Simulation of the heating of an iron nickel nanoparticle. Starting (top), intermediate (middle) and final state (bottom) of the simulation at 100 K, 480 K and 530 K, respectively. Left column: Three-dimensional pictures of the shape of the nanoparticle with iron atoms (green) and nickel atoms (blue). Right column: Two-dimensional cross-sections through the lattice structure in the x_2-x_3 plane, approximately three atom layers thick, atom colors represent their x_1 coordinate. Pure bcc crystal (top) and structure with both fcc and hcp parts and dislocations (bottom).

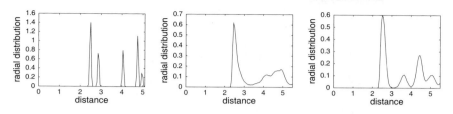

Fig. 5.7. Radial distribution functions of the configurations of the iron nickel nanoparticle at a temperature of 100 K (left), 480 K (middle) and 530 K (right).

5.1.3 Fullerenes and Nanotubes – the Brenner Potential

In the last years, new forms of carbon with remarkable structures have been discovered. Examples are fullerene balls [364] and single- or multi-wall carbon nanotubes [333]. Fullerene balls consist of 60 carbon atoms that are arranged in pentagons and hexagons, as on a soccer ball.[13] They promise a wide range of applications and are studied in a new subfield in chemistry, the so-called fullerene chemistry.

A single-wall (n, m) carbon nanotube is a rolled-up sheet of carbon atoms arranged in a honeycomb pattern, compare Figures 5.8 and 5.9. Here, the pair of integers (n, m) describes a vector $n\mathbf{a} + m\mathbf{b}$ along which the carbon sheet is rolled up, where $(\mathbf{a}, \mathbf{b})$ is a given pair of base vectors, compare Figure 5.8 (right). The angle θ between the vector $n\mathbf{a} + m\mathbf{b}$ and the base vector $\mathbf{a}$ is called the chiral angle. Depending on its value, nanotubes are classified into zig-zag nanotubes ($\theta = 0$), armchair nanotubes ($\theta = 30°$) and chiral nanotubes ($0 < \theta < 30°$).

The diameter and chiral angle of a (n, m) nanotube can be derived from geometric considerations.[14] We obtain a diameter of $0.078\sqrt{n^2 + nm + m^2}$ and a chiral angle $\theta = \arctan[\sqrt{3}m/(m + 2n)]$, see also [670]. Information about the exact geometry and structure of nanotubes can be found in [51]. There, also the structure of various caps is described which may close the ends of a nanotube.

The diameter of nanotubes is on the order of nanometers, as suggested by their name. Their length, on the other hand, can extend up to the order of micrometers [279]. Possible applications for nanotubes range from the storage of hydrogen for fuel cells to composite materials with improved mechanical properties. In the meantime, there are already first prototypes of products using nanotube technology, among them flat field emission displays. The special electrical properties of nanotubes – depending on their type, they behave

---

[13] These carbon molecules resemble the famous structures and buildings of the architect Buckminster Fuller. They are named buckminsterfullerenes, or in short buckyballs, in his honor.

[14] Here, the honeycomb structure and the bond length of 0.14 nanometers between two carbon atoms are taken into account.

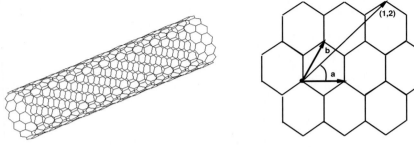

Fig. 5.8. Example of the structure of a carbon nanotube.

Fig. 5.9. A honeycomb structure made from carbon atoms. Shown are the base vectors $(\mathbf{a}, \mathbf{b})$ and an example of a vector $n\mathbf{a} + m\mathbf{b}$ with $(n, m) = (1, 2)$. The corresponding chiral angle θ is the angle between the vectors $n\mathbf{a} + m\mathbf{b}$ and $\mathbf{a}$.

either like metals or like semiconductors – may enable the production of extremely small microelectronic components in the future. An extensive discussion of the properties of nanotubes can be found in [670], further information is given in [26, 27].

The production of fullerenes and nanotubes and the experimental study of their material properties is difficult. Computer simulations are therefore an important tool to gain further insight. Ab initio calculations of the electron structure of C_{60} buckyballs by Hartree-Fock and density functional methods have been carried out in [520]. But these approaches are not practicable for larger molecules such as long nanotubes because of their computational complexity. Here, molecular dynamical simulations can be a possible alternative. To this end, the use of a many-body potential is again necessary for realistic results. In the following, we present the potential of Brenner [122] in more detail. It generalizes the potentials of Abell and Tersoff [28, 602] for carbons to hydrocarbons.

The Brenner Potential. In this section we describe the potential of Brenner [122] and discuss its implementation in a linked cell code. A newer version can be found in [123]. The potential is given by

$$V = \sum_{i=1}^{N} \sum_{j=1, j>i}^{N} f_{ij}(r_{ij}) \frac{c_{ij}}{s_{ij} - 1} \left[U_R(r_{ij}) - \bar{B}_{ij} U_A(r_{ij}) \right] \tag{5.17}$$

with a repulsive and an attractive part

$$U_R(r_{ij}) = e^{-\sqrt{2s_{ij}}\beta_{ij}(r_{ij} - r_{ij,0})} \text{ and } U_A(r_{ij}) = s_{ij} e^{-\sqrt{\frac{2}{s_{ij}}}\beta_{ij}(r_{ij} - r_{ij,0})}, \tag{5.18}$$

respectively. Here, the parameter c_{ij} relates to the minimum of the potential whereas the parameter $r_{ij,0}$ yields the distance at equilibrium between atoms

i and j. The parameters s_{ij}, β_{ij}, $r_{ij,0}$ and c_{ij} are constants that reflect the types of the atoms i and j. They have different values depending on whether carbon or hydrogen atoms are present. The possible combinations can be expressed as $s_{ij} \in \{s_{CC}, s_{HH}, s_{CH}, s_{HC}\}$, and analogously for β_{ij}, $r_{ij,0}$ and c_{ij}. These parameters are symmetric with respect to the atom types, for instance $s_{CH} = s_{HC}$. Their specific values are listed in Table 5.6. These parameters were determined by Brenner by an elaborate fitting to values from theoretical and experimental tests. In the special case $s_{ij} = 2$, the two terms U_R and U_A just reduce to the Morse potential (2.45).

The function f_{ij} is defined as

$$f_{ij}(r) = \begin{cases} 1 & \text{for } r < r_{ij,1}, \\ \frac{1}{2}\left[1 + \cos\left(\pi \frac{r - r_{ij,1}}{r_{ij,2} - r_{ij,1}}\right)\right] & \text{for } r_{ij,1} \leq r < r_{ij,2}, \\ 0 & \text{for } r_{ij,2} \leq r. \end{cases} \quad (5.19)$$

It is equal to one inside the sphere with radius $r_{ij,1}$, zero outside of the sphere with radius $r_{ij,2}$, and it decays continuously in-between the spheres from one to zero. The function f_{ij} ensures that the potential V has a short range. The values of $r_{ij,1}$ and $r_{ij,2}$ for the combinations CC, CH, HC, HH can be found in Table 5.6.

carbon	hydrogen	hydrocarbons
$r_{CC,0} = 1.39$ Å,	$r_{HH,0} = 0.74144$ Å,	$r_{CH,0} = 1.1199$ Å,
$c_{CC} = 6.0$ eV,	$c_{HH} = 4.7509$ eV,	$c_{CH} = 3.6422$ eV,
$\beta_{CC} = 2.1$ Å^{-1},	$\beta_{HH} = 1.9436$ Å^{-1},	$\beta_{CH} = 1.9583$ Å^{-1},
$s_{CC} = 1.22$,	$s_{HH} = 2.3432$,	$s_{CH} = 1.69077$,
$r_{CC,1} = 1.7$ Å,	$r_{HH,1} = 1.1$ Å,	$r_{CH,1} = 1.3$ Å,
$r_{CC,2} = 2.0$ Å,	$r_{HH,2} = 1.7$ Å,	$r_{CH,2} = 1.8$ Å,
$\delta_C = 0.5$,	$\delta_H = 0.5$,	
$\alpha_{CCC} = 0.0$,	$\alpha_{HHH} = 4.0$,	$\alpha_{HHC}, \alpha_{CHH},$
		$\alpha_{HCH}, \alpha_{HCC} = 4.0$ Å^{-1}

Table 5.6. Parameters for the Brenner potential.

Apart from the so-called bond order term $\bar{B}_{ij}$, this potential is a simple pair potential. However, the factor $\bar{B}_{ij}$ alters – similarly to the Finnis-Sinclair potential (5.2) in the last section – the attractive part of the potential. It reflects the kind of bond between the atoms i and j and takes the atom configuration in the local neighborhood of these two atoms into account. Here, the number of carbon and hydrogen atoms in the immediate neighborhood plays an important role and detailed knowledge about the bonding behavior of carbon and hydrogen (occupation numbers, coordination numbers, conjugated systems) enters into the potential. To this end, estimates for the number of carbon and hydrogen atoms in the neighborhood to each particle are needed to adapt the potential to the local bonding situation.

To be precise, let N_i^C and N_i^H denote the number of carbon and hydrogen atoms bonded with the carbon atom i, respectively.[15] These numbers can be approximated using the function f_{ij} by

$$N_i^C = \sum_{j \in C} f_{ij}(r_{ij}) \quad \text{and} \quad N_i^H = \sum_{j \in H} f_{ij}(r_{ij}). \tag{5.20}$$

Here, $\sum_{j \in C}$ and $\sum_{j \in H}$ denotes the summation over all carbon and hydrogen atoms, respectively. Because of the local support of f_{ij}, the summation extends only to the neighbors of atom i within a distance of $r_{ij,2}$, but not over all atoms. Furthermore, we denote by $N_i = N_i^C + N_i^H$ the number of all atoms that interact with the carbon atom i.

We now define a continuous function

$$N_{ij}^{conj} = 1 + \sum_{k \in C,\ k \neq i,j} f_{ik}(r_{ik}) F(N_k - f_{ik}(r_{ik})) +$$
$$\sum_{k \in C,\ k \neq i,j} f_{jk}(r_{jk}) F(N_k - f_{jk}(r_{jk})) \tag{5.21}$$

with

$$F(z) = \begin{cases} 1 & \text{for } z \leq 2, \\ \frac{1}{2}\left[1 + \cos\left(\pi(z-2)\right)\right] & \text{for } 2 < z < 3, \\ 0 & \text{for } z \geq 3, \end{cases} \tag{5.22}$$

similarly to the definition of f_{ij}. The value of N_{ij}^{conj} depends via N_k on the number of carbon and hydrogen atoms in the local neighborhood of carbon atom i and j, respectively. In this way the potential can be adapted to the known interaction between carbon atoms, between hydrogen atoms, and between carbon and hydrogen atoms while taking the configuration of neighboring atoms of the atoms i and j into account. The function N_{ij}^{conj} is continuous also for the case where bonds are broken or formed.

Furthermore, we define values B_{ij} by

$$B_{ij} = \Big(1 + H_{ij}(N_i^H, N_i^C) + \tag{5.23}$$
$$\sum_{\substack{k=1 \\ k \neq i,j}}^{N} G_i(\theta_{ijk}) f_{ik}(r_{ik}) \exp\left(\alpha_{ijk}(r_{ij} - R_{ij} - r_{ik} + R_{ik})\right)\Big)^{-\delta_i}.$$

The number B_{ij} describes the bonding state of atom i with respect to atom j. Here, the H_{ij} denote two-dimensional cubic splines that depend on the atom types of i and j. The H_{ij} smooth the transition from the bound state

---

[15] In the following, the values for N_i^C and N_i^H are only needed for carbon atoms i, not for hydrogen atoms i.

to the unbound state. For more details see Table A.1 in Appendix A.5. The δ_i and α_{ijk} are fitting parameters, their values are given in Table 5.6. The function G_i depends on the angle θ_{ijk} between the lines through the particles i and j and the particles i and k, respectively. The index i of G_i can either be C or H, depending on whether the particle i models a carbon or a hydrogen atom. In the case of a hydrogen atom, G_H is constant and equals to 12.33, in the case of a carbon atom, G is of the form

$$G_C(\theta_{ijk}) = a_0 \left(1 + c_0^2/d_0^2 - c_0^2/\left(d_0^2 + (1 + \cos\theta_{ijk})^2\right)\right), \qquad (5.24)$$

where a_0=0.00020813, c_0=330, d_0=3.5, see also [122]. Note that B_{ij} is not symmetric in its indices.

The empirical bond order function $\bar{B}_{ij}$ in (5.17) is now computed as the arithmetic mean of the two bond orders B_{ij} and B_{ji} and an additional three-dimensional spline K that interpolates between the values at the neighbors, compare Table A.1 in Appendix A.5,

$$\bar{B}_{ij} = (B_{ij} + B_{ji})/2 + K(N_i, N_j, N_{ij}^{conj}). \qquad (5.25)$$

The spline K also models the influence of radicals for pairs of atoms with different coordination numbers. More information and explanations can be found in [28, 122, 602].

The Brenner potential is able to reproduce intramolecular energies and bonds in carbon and hydrocarbon molecules. It also permits the breaking and forming of bonds. The basic structure of the bond order term can already be found in the approaches of Abell [28] and Tersoff [602], which in turn are based on the concept of bond order proposed by Pauling [468]. Brenner generalized this approach in such a way that non-local effects from conjugated bonds are taken into account as well. The computation of the force $\mathbf{F}_i$ on a particle i as the negative gradient of the Brenner potential is a tedious task which we leave as an exercise to the reader.

In addition to the Brenner potential, we employ a smoothed Lennard-Jones potential between the particles i and j to account for intermolecular van der Waals forces. It has the form

$$U_{ij}(r) = \begin{cases} 0, & r < r_{ij,2}, \\ S_1(r), & r_{ij,2} \leq r < r_{ij,3}, \\ 4\varepsilon_{ij}\left(\left(\frac{\sigma_{ij}}{r}\right)^{12} - \left(\frac{\sigma_{ij}}{r}\right)^6\right), & r_{ij,3} \leq r < r_{ij,4}, \\ S_2(r), & r_{ij,4} \leq r < r_{ij,5}, \\ 0, & r_{ij,5} \leq r. \end{cases}$$

Here, $S_1(r)$ and $S_2(r)$ denote cubic splines that are given by the values and derivatives at the respective boundary points. The values of the parameters of the potential U_{ij} are given in Table 5.7.

	ε_{ij}[eV]	σ_{ij}[Å]	$r_{ij,2}$[Å]	$r_{ij,3}$[Å]	$r_{ij,4}$[Å]	$r_{ij,5}$[Å]
C–C	4.2038×10^{-3}	3.37	2.0	3.2	9.875	10.0
H–H	5.8901×10^{-3}	2.91	1.7	2.76	9.875	10.0
C–H	4.9760×10^{-3}	3.14	1.8	2.98	9.875	10.0

Table 5.7. Parameter values for a Lennard-Jones-like potential from [407, 436] that is employed in addition to the Brenner potential to model van der Waals interactions. The values of the parameters ε_{CH} and σ_{CH} are given by the Lorentz-Berthelot mixing rules: $\varepsilon_{CH} = \sqrt{\varepsilon_{CC}\varepsilon_{HH}}$ and $\sigma_{CH} = (\sigma_{CC} + \sigma_{HH})/2$.

Comments on the Implementation. A naive, direct implementation of the Brenner potential would lead to a complexity of the order $\mathcal{O}(N^4)$. Here, a factor of N^2 is due to the double sum over i and j in (5.17), an additional factor of N is caused by the sum over the particles in the computation of $\bar{B}_{ij}$ in (5.23), and another factor of N stems from the dependence of F on N_k in (5.21). In total this results in four nested loops over all particles.

However, we do not have to compute the sum over all combinations of particles since $r_{ij,2}$ in (5.19) acts as a cutoff parameter in the sums in (5.17) and (5.20). An efficient implementation can therefore use the linked cell method as described in Chapter 3. Then, just sums over particles in neighboring cells must be formed. But note that the implementation is not as simple as in the case of pair potentials. In particular one needs for $\bar{B}_{ij}$ cutoff regions around both particles i and j in the double sum over the particles in (5.17), see Figure 5.10.

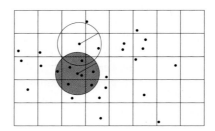

Fig. 5.10. Cutoff regions around particle i and particle j in two dimensions.

Then, similar to the evaluation of the Finnis-Sinclair potential in Section 5.1.1, we can precompute certain quantities, compare also [133]. In this way the complexity can be reduced to $C^3 \cdot N$ with a constant C that depends on the particle density.

In the following, we present the results of several numerical computations which involve the Brenner potential. We consider the collision reaction of a C_{60} fullerene with dehydrobenzene and the behavior of carbon nanotubes under tensile load. In addition, we study the reinforcement of a polyethylene matrix with nanotubes. Finally, we present the execution times for the parallelized code on a Cray T3E and a PC cluster.

Example: Collision of Dehydrobenzene with a Buckyball. First, we consider several cases of the collision of a dehydrobenzene molecule (C_6H_4) with a C_{60} fullerene, see also [301]. A dehydrobenzene molecule is moved with a given velocity towards a resting C_{60} molecule. The simulation uses the Brenner potential with the parameters from Table 5.6 and Table A.1 in Appendix A.5 and the modified Lennard-Jones potential (5.26) for the intermolecular interactions. Data for the different molecules can be found in Appendix A.5. All simulations were run for 1 ps with an initial temperature of 316 K and time steps of 0.035 fs. Depending on the velocity of the dehydrobenzene molecule, three possible interactions with the C_{60} molecule can be observed: Elastic collision, collision reaction, or destruction of both molecules. From our experiments we identified the following velocity ranges for these three cases:

– elastic collision: $14.17 - 28.34$ Å/ps
– collision reaction: $35.43 - 99.20$ Å/ps
– destruction of the molecules: ≥ 106.29 Å/ps

Figure 5.11 shows a few snapshots of the simulation of an elastic collision.

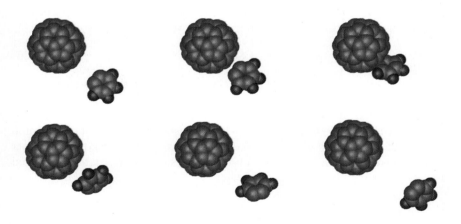

Fig. 5.11. Elastic collision of a dehydrobenzene molecule with a C_{60} fullerene.

Example: Nanotube under Tensile Load. We now carry out several tensile and bending experiments with $(7, 0)$ zigzag nanotubes. The initial configurations are taken from [51]. In the following, we consider in particular the tensile load case. In this experiment, a constant force pulls at both ends of the nanotube and, as a result, the nanotube is stretched.

On the macroscopic level, a tensile force on a rod-shaped body of length l causes an elongation Δl. The size of this elongation depends on the physical dimensions of the body, the strength of the material, and the acting tensile load. The strain ε is the elongation Δl of the tube, normalized by the length of the tube at rest without any acting force, i.e. $\varepsilon = \Delta l / l$. With the new length $\tilde{l} := l + \Delta l$ the following relation holds

$$\tilde{l} = (1 + \varepsilon)\, l,$$

see also Figure 5.12.

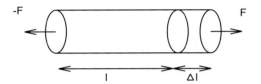

Fig. 5.12. A tensile load experiment.

If we generalize this one-dimensional concept of strain to three dimensions and thus also allow shear forces, we can describe the strain with help of a strain matrix $\boldsymbol{\epsilon} \in \mathbb{R}^{3 \times 3}$ where the off-diagonal elements of the matrix now describe shear strains. For our atomic system we obtain analogously

$$\tilde{\mathbf{x}}_i = (\mathbf{1} + \boldsymbol{\epsilon})\mathbf{x}_i, \tag{5.26}$$

where $\tilde{\mathbf{x}}_i$ denotes the new positions of the atoms and $\mathbf{1}$ denotes the identity matrix.

The deformation of the system with the strain matrix $\boldsymbol{\epsilon}$ results in stress. This stress is then described by the stress tensor $\boldsymbol{\sigma} \in \mathbb{R}^{3 \times 3}$. The diagonal elements of this tensor give the uniaxial stresses, the off-diagonal elements describe shear stresses. The stress tensor consists of a kinetic and a potential part and is defined by

$$\boldsymbol{\sigma}_{\alpha\beta} = \frac{1}{|\Omega|} \frac{d}{d\boldsymbol{\epsilon}_{\alpha\beta}} \Bigg(E_{\mathrm{kin}}((\mathbf{1} + \boldsymbol{\epsilon})\mathbf{v}_1, \ldots, (\mathbf{1} + \boldsymbol{\epsilon})\mathbf{v}_N)$$

$$- V((\mathbf{1} + \boldsymbol{\epsilon})\mathbf{x}_1, \ldots, (\mathbf{1} + \boldsymbol{\epsilon})\mathbf{x}_N) \Bigg) \Bigg|_{\epsilon = 0}$$

$$= \frac{1}{|\Omega|} \sum_{i=1}^{N} m_i (\mathbf{v}_i)_\alpha (\mathbf{v}_i)_\beta + (\mathbf{F}_i)_\alpha (\mathbf{x}_i)_\beta, \quad \alpha, \beta \in \{1, 2, 3\}, \tag{5.27}$$

where we denote the kinetic energy by E_{kin}, the potential energy by V, and the force on particle i by $\mathbf{F}_i = -\nabla_{\mathbf{x}_i} V$. The volume of the simulated region

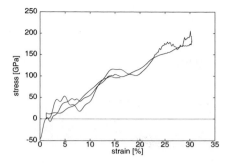

Fig. 5.13. Stress-strain diagram for a carbon nanotube under tensile load. The component in the direction of the tensile force is shown. Constant tensile force 2.8 nN in axial direction from both sides – the nanotube does not snap.

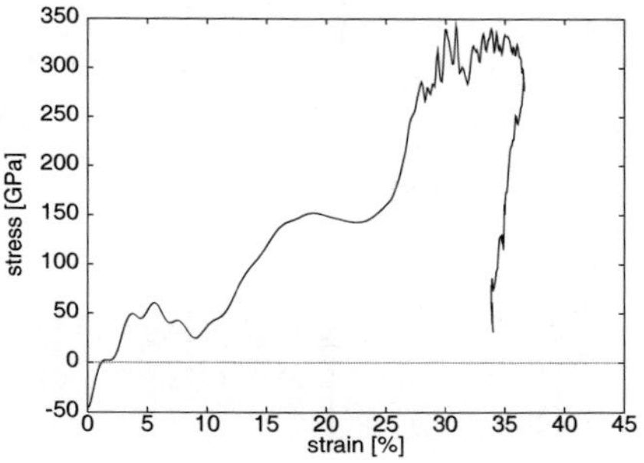

Fig. 5.14. Stress-strain diagram for a carbon nanotube under tensile load. The component in the direction of the tensile force is shown. Constant tensile force 3.2 nN in axial direction from both sides – the nanotube snaps.

– here the volume of the nanotube as hollow cylinder with a wall thickness of 3.4 Å (the thickness of one layer of graphite) – is denoted by $|\Omega|$. The stress of the system is thus distributed onto the individual atoms. We define the stress of a single atom as

$$\sigma_{\alpha\beta,i} = m_i(\mathbf{v}_i)_\alpha(\mathbf{v}_i)_\beta + (\mathbf{F}_i)_\alpha(\mathbf{x}_i)_\beta.$$

In the case of a two-body potential we can write

$$\sigma_{\alpha\beta} = \frac{1}{|\Omega|}\sum_{i=1}^{N}\left(m_i(\mathbf{v}_i)_\alpha(\mathbf{v}_i)_\beta - \frac{1}{2}\sum_{\substack{j=1 \\ j\neq i}}^{N}(\mathbf{F}_{ij})_\alpha(\mathbf{r}_{ij})_\beta\right), \qquad (5.28)$$

where $\mathbf{r}_{ij} = \mathbf{x}_j - \mathbf{x}_i$ denotes again the distance vector and $\mathbf{F}_{ij}$ denotes the force between particles j and i. This formula holds, unlike (5.27), also in the periodic case, compare Section 3.7.4.

Now, we simulate the behavior of a nanotube composed of 308 carbon atoms. Here and in the following, we use again the parameters from Table 5.6 and Table A.1 in Appendix A.5. The tube is stretched by a constant tensile force of 2.8 nN which acts on both ends. Figure 5.13 shows the corresponding stress-strain diagram. The maximum of the observed strain amounts to 30% of the relaxed initial state of the tube. If the tensile force is increased to 3.2 nN, the nanotube snaps, as shown in Figure 5.14. Tear-off occurs at a strain of approximately 35%. Our measurements yield a Young's modulus – that is the derivative of the stress with respect to the strain and therefore the slope in the stress-strain diagram – of about 1 TPa.[16] Figure 5.15 shows a few snapshots shortly before tear-off.

Corresponding experiments can be carried out in an analogous way also for more complex structures. Figure 5.16 presents several snapshots of a simulation of the tensile load case with a multi-walled nanotube.

Fig. 5.15. Stretching a nanotube with constant force 3.2 nN.

Fig. 5.16. Tensile load experiment with a multi-walled carbon nanotube.

Finally, we consider an example for a bending experiment with a carbon nanotube. Here, load is applied with an angle of 45 degrees at both ends of the tube. Figure 5.17 shows the results. When the tube is released, it flips back to its initial configuration. Similar experiments can be found in [579, 670].

---

[16] For comparison: The Young's modulus of diamond also amounts to approximately 1 TPa.

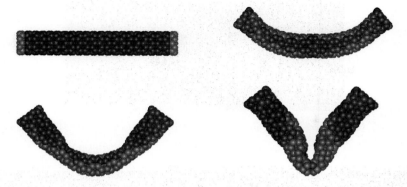

Fig. 5.17. Bending experiment with a nanotube.

Example: Polyethylene with an Embedded Nanotube. The reinforcement of materials by embedded nanotubes is an area of intensive research.
Here, a focus of interest are the properties of the composite material. Do the
nanotubes reinforce the composite material at all, and if so, how is the load
transfered from the host material to the nanotubes? Simulations can help to
find answers to these questions. As a result one obtains a prediction of the
functional relation between the Young's modulus of the single components
and the Young's modulus of the composite material.

As an example we consider polyethylene. It consists of chains of methylene
monomeres (CH_2). We incorporate a nanotube into the polyethylene matrix
to possibly reinforce it. For the numerical simulation of such composite materials, we now have to take the atoms of the nanotube and the polyethylene
molecules as well as their interactions into account. Since polyethylene is a
hydrocarbon, the potential of Brenner can be directly employed together with
the Lennard-Jones potential (5.26).

The Figures 5.18 and 5.19 show pictures of a tensile load experiment[17]
with a capped $(10, 10)$-nanotube which is embedded into a polyethylene matrix. Here, the nanotube consists of 1020 carbon atoms whereas the polyethylene matrix is made from eight chains of 1418 CH_2 molecules each (together
with CH_3 molecules at the beginning and the end of each chain). Note that
the nanotube is initially positioned with an angle of approximately 15 degrees
to the $\mathbf{x}_1$ axis to avoid symmetry effects. The bond lengths at the start of the
simulation are 1.53 Å for C-C bonds and 1.09 Å for C-H bonds. The size of
the simulation domain is 53.7 Å×53.7 Å×133.95 Å. The composite material
is now subjected to a tensile force at the left and right side of the sample.

---

[17] In this simulation an external stress – which increases linearly over time – is
applied to the system. Starting from the configuration of the old time step, the
configuration for the next time step is computed by a local minimization of the
potential energy.

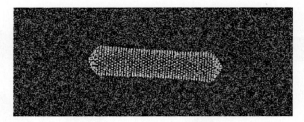

Fig. 5.18. Nanotube-polyethylene composite, relaxed state; the length of the nanotube is 63 Å.

Fig. 5.19. Nanotubes-polyethylene composite at 16% elongation of the nanotube; the length of the nanotube is 73 Å.

Figure 5.20 shows the resulting stress-strain diagram. One can see that the polyethylene is significantly reinforced by the embedded nanotube. The composite material has a Young's modulus of about 14 GPa. This value has to be compared to the Young's modulus of the single components: polyethylene has a modulus of 1.2 GPa, whereas the nanotube has a modulus of 526 GPa.

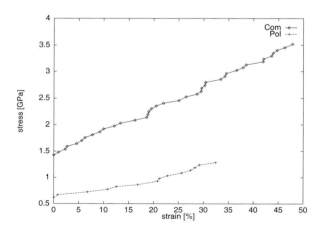

Fig. 5.20. Stress-strain diagram of pure polyethylene (Pol) and the nanotube-polyethylene composite (Com). Here, the stress component in the direction of the traction is shown.

Especially for long nanotubes and large amounts of polyethylene, such simulations can only be run on large parallel computers because of their huge computational requirements. Thus, in the following, we present a few results of the parallelization.

Parallelization. The parallelization of the code is based on the domain decomposition strategy [133] as described in Chapter 4. Now, the complicated many-body part of the Brenner potential necessitates several communication steps in each time step for an efficient parallel computation of the forces and the energies. We here consider a tensile load experiment with nanotubes of different lengths. As in Section 4.4, we use the Cray T3E-1200 and the cluster Parnass2 which consists of Intel Pentium II PCs (400 MHz). The simulation domain is subdivided into subdomains along the longitudinal axis of the nanotube. The domain decomposition is static. It has been chosen in such a way that a good load balancing is ensured during the entire run of the simulation. A fifth order generalization of the Störmer-Verlet method has been employed to integrate the equations of motion, compare also Section 6.2.

time	processors							
particles	1	2	4	8	16	32	64	128
10.000	**8.80**	4.55	2.31	1.17	0.62	0.46		
20.000	17.78	**9.12**	4.56	2.31	1.17	0.75	0.35	
40.000	35.86	18.56	**9.19**	4.60	2.32	1.19	0.61	0.44
80.000	**72.15**	36.83	18.46	**9.22**	4.61	2.56	1.28	0.71
160.000	146.16	**75.17**	37.80	18.63	**9.23**	4.73	2.38	1.40
320.000	292.37	150.86	**74.47**	37.20	18.64	**9.26**	4.67	2.59
640.000		298.32	151.28	**74.89**	37.78	21.02	**10.42**	5.09
1.280.000			301.37	151.08	**76.64**	37.55	21.46	**10.12**
2.560.000				306.77	153.68	**75.54**	40.25	20.78
5.120.000					309.03	152.47	**90.07**	39.39
10.240.000						307.83	157.71	**82.56**

Table 5.8. Parallel execution times (in seconds) for one time step on a PC cluster, simulation of a tensile load case.

	processors							
	1	2	4	8	16	32	64	128
speedup	1.00	1.94	3.93	7.86	15.69	31.57	62.61	112.89
efficiency	1.00	0.97	0.98	0.98	0.98	0.99	0.98	0.88

Table 5.9. Speedup and parallel efficiency for one time step on a PC cluster, simulation of a tensile load case with 320 000 atoms.

Table 5.8 shows the parallel execution times for one time step of the algorithm on the PC cluster. The associated speedups and parallel efficiencies can be found in Table 5.9. Table 5.10 gives the parallel execution times for the Cray T3E. The corresponding speedups and parallel efficiencies are given in Table 5.11.

time particles	processors								
	1	2	4	8	16	32	64	128	256
10.000	**26.04**	13.43	6.76	3.46	1.85	1.05			
20.000	52.13	**26.06**	13.45	6.77	3.43	1.87	1.06		
40.000	105.29	52.65	**26.04**	13.22	6.74	3.46	1.87	1.05	
80.000	211.68	105.43	52.23	**25.95**	13.07	6.79	3.47	1.88	1.07
160.000	419.90	210.32	103.76	52.88	**26.07**	13.12	6.66	3.46	1.89
320.000		420.78	213.01	105.81	52.40	**25.96**	13.14	6.70	3.47
640.000			421.11	207.02	103.74	52.39	**26.09**	13.14	6.77
1.280.000				422.67	209.53	107.35	52.37	**25.99**	13.29
2.560.000					427.65	215.25	103.83	52.64	**25.93**
5.120.000						427.10	215.20	104.69	52.67
10.240.000							425.90	212.74	105.72
20.480.000								429.83	216.98
40.960.000									426.85

Table 5.10. Parallel execution times (in seconds) for one time step on the Cray T3E-1200, simulation of a tensile load case.

	processors								
	1	2	4	8	16	32	64	128	256
speedup	1.00	1.99	4.04	7.94	16.10	32.00	63.04	121.35	222.17
efficiency	1.00	0.99	1.01	0.99	1.01	1.00	0.99	0.94	0.86

Table 5.11. Speedup and parallel efficiency for one time step on the Cray T3E-1200, simulation of a tensile load case with 160 000 atoms.

On both machines an almost linear speedup and a good scaling is observed. If the number of processors is doubled, the running time is halved. If the number of atoms is increased by a factor of two and if the number of processors is doubled as well, the running time stays approximately constant. Note that the PC cluster with Myricom network is even slightly faster than the Cray T3E for this application. It thus represents a cost-efficient and competitive alternative to larger high performance computers and supercomputers.

5.2 Potentials with Fixed Bond Structures

The potentials described up to now were particularly appropriate to describe materials composed of single atoms. In many applications – especially in the simulation of polymers, proteins, DNA, and other biomolecules – molecules are composed from atoms with fixed, given bond structure. In such cases, the potential used must embody that given inner structure of the molecule. To this end, we have to take into account the bond lengths, the bond angles and the torsion angles between different atoms of the molecule. Parameters and data for such potentials are made available for instance by the program packages CHARMM [125], Amber [471] or Gromos [634].

In the following, we describe the necessary extensions of the previous simulation code to potentials with given inner bond structure. First, we discuss the implementation of such bonding relations for harmonic potentials in the example of a two-dimensional grid structure. Then, we introduce a new data structure for linear chain molecules using polymers as an introductory example. Here, an additional vector with pointers to the atoms of the chain molecule is stored for each molecule. If several molecules are modeled, a two-dimensional array of pointers is used as a further data structure. We also study potentials for bond lengths, bond angles and torsion angles. Using these potentials, we investigate the properties of alkanes (butane, decane, eicosane). Finally, we discuss how more complex molecules can be modeled with pointer data structures, give a short outlook on how to employ data from the RCSB Protein Data Bank PDB to set up simulations of biomolecules, and discuss how such simulations could be carried out.

5.2.1 Membranes and Minimal Surfaces

In the following, we consider an elastic membrane (imagine for example a very thin sheet of rubber) that is subjected to certain external forces and thus changes its form accordingly. In a stable equilibrium, the membrane adopts a position in which the potential energy is at its minimum. If the membrane is deflected from its equilibrium, a counterforce results that opposes the deflection and tries to pull the membrane back to the starting position. In the continuous case, it makes sense to assume that the potential energy of the membrane is proportional to the change of area compared to the area in the stable equilibrium, i.e.[18]

$$V(q) \sim \int_\Omega \sqrt{1 + ||\nabla q||^2} dx,$$

---

[18] Because of $\sqrt{1 + ||\nabla q||^2} = 1 + \frac{1}{2}||\nabla q||^2 + \mathcal{O}(||\nabla q||^4)$, the integrand can be approximated well by $1 + \frac{1}{2}||\nabla q||^2$ for small $||\nabla q||$. We then obtain $V(q) \dot\sim \int 1 + \frac{1}{2}||\nabla q||^2 dx$. Since we are interested in the deflection with minimal energy, the term $\int 1 dx$ does not influence the position of the minima and can be omitted in the minimization.

where q describes the deflection from the resting position. If one limits the deflection to small amplitudes $||\nabla q|| \ll 1$, the Taylor series gives in first approximation

$$V(q) \sim \frac{1}{2} \int_{\Omega} ||\nabla q||^2 dx.$$

We now use a simple model of a system of many coupled harmonic oscillators to describe membranes in this section, compare Figure 5.21.

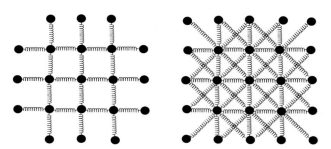

Fig. 5.21. Systems of coupled harmonic oscillators in two dimensions.

For this purpose we employ a harmonic potential

$$U(r_{ij}) = \frac{k}{2}(r_{ij} - r_0)^2 \quad \text{resp.} \quad U(r_{ij}) = \frac{k}{2}(r_{ij} - \sqrt{2}r_0)^2. \qquad (5.29)$$

Every particle therefore only interacts with its direct neighbors in the grid of harmonic oscillators. Contrary to the applications considered before, the particles now have fixed neighborhood relations that do not change during the simulation. In a later section we will give a general modification of the linked cell method for this case. But here, we proceed for the sake of simplicity as follows: We restrict ourselves to a uniform $m_1 \times m_2$ grid of particles which are pairwise bound as in Figure (5.21, right). The potential (5.29, right) only acts between diagonally coupled neighbors while the potential (5.29, left) is used for the other bonds. This choice ensures that a uniform initial configuration with a grid spacing of r_0 in both direction yields a minimizer of the energy, compare Figure 5.22.

After an appropriate initialization of the particles, the simple linked cell method can be used as follows: Since the particles form a two-dimensional grid, we directly use the cell structure `grid` from the two-dimensional case, compare also Section 3.5, and associate a fixed particle to each cell. The cells of the linked cell method thus degenerate to trivial cells which only contain a single particle, compare Figure 5.23. The two-dimensional grid `grid[index(ic,nc)]` now points directly to the single particle in the cell `ic:=` (i_1, i_2). The particle itself still has three-dimensional coordinates.

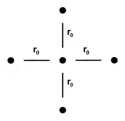

Fig. 5.22. Basic setup of an initial configuration for the membrane. Grid points are placed in both spatial directions at a distance of r_0. With a choice of the potential as in (5.29, right) for diagonal bonds and as in (5.29, left) for the other bonds, the simulation starts at a minimizer of the energy.

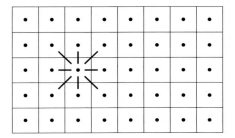

Fig. 5.23. The linked cell method in two dimensions. In this case every cell only contains a single particle. The computation of the forces is limited to the eight neighboring cells in two dimensions.

During the computation of the forces, one can access the interacting pairs simply using the grid `grid` and can thus evaluate the potential (5.29) easily. The particle associated to cell `ic=` (i_1, i_2) interacts with the particles from the four neighboring cells

$$(i_1 \pm 1, i_2), \quad (i_1, i_2 \pm 1),$$

by the potential (5.29, left), and by the potential (5.29, right) with the particles from the four neighboring cells

$$(i_1 \pm 1, i_2 \pm 1).$$

Particles at the border of the grid have an appropriately smaller number of neighbors, for instance, for a particle on the left boundary line only the neighboring cells

$$(i_1, i_2 \pm 1), \quad (i_1 + 1, i_2 \pm 1) \quad \text{and} \quad (i_1 + 1, i_2)$$

have to be taken into account.

Furthermore, the procedure compX_LC has to be modified slightly. Up to now it called the algorithm 3.17 moveParticles_LC which checks if particles have left the cell and inserts them into their new cell if necessary. This call is omitted in our case. We keep the particles in the same two-dimensional cell structure during the entire simulation. The cells no longer correspond to a geometric decomposition of the simulation box, but only serve as a data structure which allows direct access to every particle and its eight neighbors. The position of the particles are updated with the Störmer-Verlet method, as described in Sections 3.1 and 3.2.

Minimal Surfaces. We consider a system of harmonic oscillators which models a cylindrical membrane clamped between two annuli. The initial configuration is shown in Figure 5.24 (left). In Figure 5.24 (right) we give a smoothed graphical representation of the same configuration. There, individual particles are no longer shown, but a surface is fitted to the positions of the particles which is then rendered by a visualization software package.

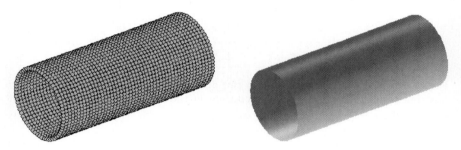

Fig. 5.24. Minimal surfaces: Particle representation and smoothed representation of the initial configuration.

The tube itself consists of 50×64 (axial $\times$ radial) particles. Such a tube can be constructed from a planar two-dimensional mesh with 64 mesh points in $\mathbf{x}_1$ direction and 50 mesh points in $\mathbf{x}_2$ direction which is closed to a torus and thus is extended periodically across the left and right edge. Then, the right neighbors of the particle on the right edge which is associated to cell (i_1, i_2) are contained in the cells with the indices $(i_1 + 1 - m_1, i_2)$ and $(i_1 + 1 - m_1, i_2 \pm 1)$, and analogously for the left edge. Figure 5.25 shows the result of two simulations with the parameters given in Table 5.12. Here, the potential (5.29) is used.

In the simulation which resulted in Figure 5.25 (left), the tube was pulled apart at both ends with a constant radial force. In the simulation which led to Figure 5.25 (right), the tube was pulled apart in the middle in circular direction with constant radial force, while the ends of the tube were held fixed. In both cases $||\mathbf{F}_{pull}|| = 0.1$ was used.

$$L_1 = 200, \qquad L_2 = 200, \qquad L_3 = 200,$$
$$r_0 = 2.2, \qquad k = 400, \qquad m = 0.1,$$
$$N = 50 \times 64, \qquad \delta t = 0.01, \qquad t_{end} = 7.5$$

Table 5.12. Values of the parameters for the simulation of minimal surfaces.

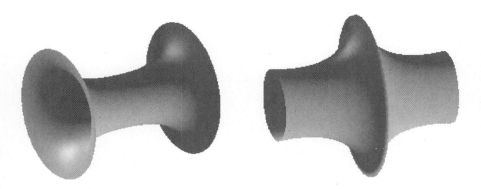

Fig. 5.25. Minimal surfaces: Stationary final configurations of the two simulations. The tube is pulled apart in circular direction at both ends (left). The tube is pulled apart in circular direction in the middle (right).

Cloth. We now consider a two-dimensional system which consists of 50×50 particles. It models a cloth. Initially, all particles are placed at the bottom of the three-dimensional simulation box. The five adjacent particles associated to the cell numbers $(37, 25)$, $(37, 24)$, $(37, 26)$, $(36, 25)$ and $(38, 25)$ are then pulled up with a constant force $\mathbf{F}_{pull} = (0, 1.5, 0)^T$, compare Figure 5.26 (upper left). Here, every inner particle interacts with its eight next neighbors. The particles at the edges interact only with their existing neighbors, compare Figure 5.21 (right). The potential between the particles is given by (5.29). To avoid self-penetration, every particle is additionally equipped with a three-dimensional repulsive potential $(1/r)^{12}$. For the computation of the resulting force, the standard three-dimensional linked cell method can be used. Figure 5.26 shows the cloth at different stages of the simulation with the parameter values given in Table 5.13. As before, the pictures show a surface that has been fitted to the positions of the particles.

$$L_1 = 200, \qquad L_2 = 200, \; L_3 = 200,$$
$$N = 50 \times 50, \; m = 1, \qquad k = 300,$$
$$r_0 = 2.2, \qquad \delta t = 0.01$$

Table 5.13. Values of the parameters for the simulation of a cloth.

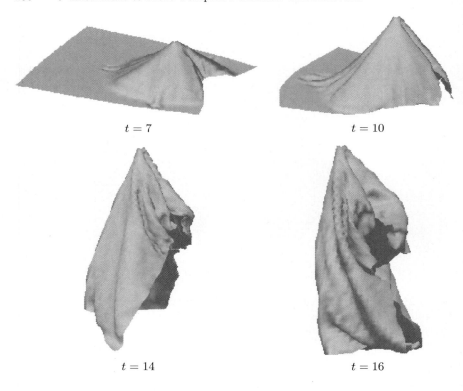

$t = 7$ $t = 10$

$t = 14$ $t = 16$

Fig. 5.26. Time evolution of the upwards pulled cloth.

5.2.2 Systems of Linear Molecules

Now, we study systems of molecules. To this end, we need to extend our
data structures properly to be able to describe molecules as sets of atoms or
particles, respectively, which are bonded in a certain way. We limit ourselves
at first to molecules with a linear internal structure, compare Figure 5.27.
We will treat molecules with a more complicated internal structure later in
Section 5.2.3.

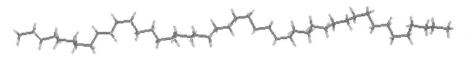

Fig. 5.27. Example of a linear molecule (polyethylene).

Intramolecular Interaction. In molecular systems we can distinguish between intermolecular and intramolecular interactions. Intermolecular interactions act between the atoms of *different* molecules. Intramolecular interactions act between the atoms within a *single* molecule. They are generally much stronger than intermolecular interactions. To model intramolecular interactions, essentially three different potential functions are commonly used, the so-called bond potential, the angle potential, and the torsion potential, compare also Figure 5.28.

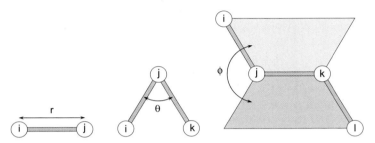

Fig. 5.28. Graphical representation of a valence bond potential (left), an angle potential (middle), and a torsion potential (right).

Valence or bond forces exist between any two bonded atoms, see Figure 5.28 (left). These interaction can often be described well in the form of a harmonic potential, i.e.

$$U_b(r) = \frac{1}{2} k_b (r_{ij} - r_0)^2. \tag{5.30}$$

In this potential, r_{ij} denotes the bond length between the atoms i and j, r_0 denotes the equilibrium distance and k_b denotes the force or spring constant.

Angular forces are three-body forces between three successive bonded atoms, compare Figure 5.28 (middle). The associated angle potential is characterized by an angle $\theta = \theta_{ijk}$ between the atoms i, j and k, which varies around an equilibrium value θ_0. For this potential one often uses the form

$$U_a(\theta) = -k_\theta (\cos(\theta - \theta_0) - 1) \tag{5.31}$$

or (for small deviations from the equilibrium value θ_0)

$$U_a(\theta) \approx \frac{1}{2} k_\theta (\theta - \theta_0)^2. \tag{5.32}$$

The previous potentials only depended on the distance r_{ij} between two atoms. Now, the potential depends on the angle determined by three successive atoms i, j, and k. To compute the forces on the atoms which result from such a potential, one has to choose an appropriate coordinate system. A natural choice would be the inner angle coordinates instead of cartesian coordinates. But for now, we keep the cartesian coordinate system. The reason is the following: During a simulation one has to compute the intramolecular interactions due to chemical bonds as well as intermolecular nonbonded interactions. Since the latter account for most of the work, we decided to also employ the cartesian coordinates for the angle (and torsion) bonds. This, however, forces us to perform a coordinate transformation from the cartesian coordinates $\mathbf{x}_i$ of particle i to the angles every time we compute the angular forces. For three successive bound particles i, j, and k, the associated angle θ can be computed from the cartesian coordinates $\mathbf{x}_i, \mathbf{x}_j$, and $\mathbf{x}_k$ using the relation

$$\theta = \theta_{ijk} = \arccos\left(\frac{\langle \mathbf{r}_{ij}, \mathbf{r}_{kj}\rangle}{\|\mathbf{r}_{ij}\|\|\mathbf{r}_{kj}\|}\right). \tag{5.33}$$

Here, we again write $\mathbf{r}_{ij} := \mathbf{x}_j - \mathbf{x}_i$. In the simulation we have to sum over all possible angles $\theta = \theta_{ijk}$. The resulting local forces are added to the forces associated to the respective particles i, j and k. In the case of the particle j, i.e. the middle particle, the force can simply be determined by the relation $\mathbf{F}_{j,\theta_{ijk}} = -\mathbf{F}_{i,\theta_{ijk}} - \mathbf{F}_{k,\theta_{ijk}}$. Moreover, it holds that

$$\begin{aligned}
\mathbf{F}_{i,\theta_{ijk}} &= -\nabla_{\mathbf{x}_i} U_a(\theta_{ijk}) \\
&= -k_\theta(\theta_{ijk} - \theta_0) \cdot \frac{\partial\theta_{ijk}}{\partial\mathbf{x}_i} \\
&= -k_\theta(\theta_{ijk} - \theta_0) \cdot \frac{\partial\theta_{ijk}}{\partial\mathbf{x}_i} \underbrace{\frac{\partial\cos\theta_{ijk}}{\partial\theta_{ijk}}\frac{\partial\theta_{ijk}}{\partial\cos\theta_{ijk}}}_{=1} \\
&= -k_\theta(\theta_{ijk} - \theta_0) \cdot \frac{\partial\cos\theta_{ijk}}{\partial\mathbf{x}_i}\frac{\partial\theta_{ijk}}{\partial\cos\theta_{ijk}}\underbrace{\frac{\partial\theta_{ijk}}{\partial\theta_{ijk}}}_{=1} \\
&= -k_\theta(\theta_{ijk} - \theta_0) \cdot \frac{\partial\cos\theta_{ijk}}{\partial\mathbf{x}_i} \cdot \left(-\frac{1}{\sin\theta_{ijk}}\right) \\
&= k_\theta\frac{(\theta_{ijk} - \theta_0)}{\sin\theta_{ijk}} \cdot \frac{\partial\cos\theta_{ijk}}{\partial\mathbf{x}_i}.
\end{aligned}$$

$$\tag{5.34}$$

We still have to compute the term

$$\frac{\partial\cos\theta_{ijk}}{\partial\mathbf{x}_i}.$$

To this end, we define

$$S := \langle\mathbf{r}_{ij},\mathbf{r}_{kj}\rangle \text{ and } D := \|\mathbf{r}_{ij}\|\|\mathbf{r}_{kj}\|,$$

and, using the identity $\cos\theta_{ijk} = S/D$ (compare (5.33)), we obtain

$$\frac{\partial \cos \theta_{ijk}}{\partial \mathbf{x}_i} = -\frac{1}{D}\left(\mathbf{r}_{kj} - \frac{S}{D^2}\mathbf{r}_{ij}\|\mathbf{r}_{kj}\|^2\right).$$

$\mathbf{F}_{k,\theta_{ijk}}$ can be computed in an analogous way. The total force on particle i from angle potentials is thus determined by summing over all triples k,l,m of successive bound particles according to

$$\mathbf{F}_i = \sum_{\substack{\theta_{klm} \\ k<m}} \mathbf{F}_{i,\theta_{klm}}.$$

Torsion forces are four-body forces between four successive bonded atoms. For a schematic representation of their effect, see Figure 5.28 (right). For the associated potential one often uses the form

$$U_t(\phi_{ijkl}) = \sum_{n=1}^{3} k_{\phi_n}(\cos(n\phi - \delta_n) + 1). \tag{5.35}$$

Here, $\phi = \phi_{ijkl}$ denotes the angle between the planes spanned by the atoms i,j,k and j,k,l. The parameters δ_n are appropriate phase shifts and the parameters k_{ϕ_n} denote appropriately chosen force constants. For small deviations from the equilibrium state ϕ_0 one may approximate this potential by a harmonic potential $U_t(\phi_{ijkl}) = \frac{1}{2}k_\phi(\phi - \phi_0)^2$ with an appropriately chosen constant k_ϕ. Alternatively, it is also customary to use a trigonometric polynomial of the form

$$U_t(\phi_{ijkl}) = \sum_{n} k_{\phi_n}\cos^n\phi. \tag{5.36}$$

Here, we again have to transform the cartesian coordinates into angular coordinates. This is somewhat more complicated for torsion angles. Here, we have the relation[19]

$$\phi = \phi_{ijkl} = \pi \pm \arccos\left(\left\langle \mathbf{r}_{ij} - \left\langle \mathbf{r}_{ij}, \frac{\mathbf{r}_{kj}}{r_{kj}}\right\rangle \mathbf{r}_{kj}, \mathbf{r}_{lk} - \left\langle \mathbf{r}_{lk}, \frac{\mathbf{r}_{kj}}{r_{kj}}\right\rangle \mathbf{r}_{kj}\right\rangle\right). \tag{5.37}$$

---

[19] This is the so-called scalar product definition for the torsion angle since it only involves the computation of scalar products $\langle .,.\rangle$. In the literature on can find an alternative definition

$$\phi = \phi_{ijkl} = \pi \pm \arccos\left(\frac{\langle \mathbf{r}_{ij} \times \mathbf{r}_{jk}, \mathbf{r}_{jk} \times \mathbf{r}_{kl}\rangle}{\langle \|\mathbf{r}_{ij} \times \mathbf{r}_{jk}\|, \|\mathbf{r}_{jk} \times \mathbf{r}_{kl}\|\rangle}\right),$$

which needs cross products. This definition is used in many articles, however, it leads to more complicated expressions in the computation of the forces.

The sign is given by the scalar quantity

$$\text{sign}(\det(\mathbf{r}_{ij}, \mathbf{r}_{jk}, \mathbf{r}_{kl})) = \text{sign}(\langle \mathbf{r}_{ij}, \mathbf{r}_{jk} \times \mathbf{r}_{kl} \rangle).$$

In the case of a torsion angle of zero or ± 180 degree, all atoms lie in the same plane. But the associated configurations differ substantially, compare Figure 5.29. The configuration with a torsion angle of zero degrees is called trans configuration, the configuration with an angle of ± 180 degrees is called cis configuration.[20]

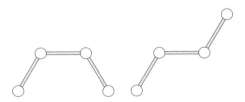

Fig. 5.29. Cis configuration and trans configuration.

The gradient of the torsion potential and hence the forces acting on the particles are derived in the same way as for the angle potential. Their detailed computation is left as an exercise to the reader. Here, one should bear in mind that the often used factorization of the chain rule

$$-\nabla_{\mathbf{x}_i} U_t = -(dU_t/d\phi)(d\phi/d\cos\phi)(\partial\cos\phi/\partial\mathbf{x}_i)$$

has a singularity at $\phi = 0$ or $\phi = \pi$ because of the factor $d\phi/d\cos\phi = -\sin^{-1}\phi$. This singularity can be avoided if one employs instead the chain rule in the form

$$-\nabla_{\mathbf{x}_i} U_t = -(dU_t/d\phi)(\partial\phi/\partial\mathbf{x}_i).$$

Then, with the definition (5.37) of the torsion angle, formulas for the torsion forces can be derived that are free of singularities. For more details, see [60, 74].

Altogether, the total potential includes the sum over all bonded and non-bonded interactions and can be summarized as follows:

$$V = \underbrace{\sum_{\substack{\text{bonds} \\ (i,j)}} \frac{1}{2}k_b(r_{ij} - r_0)^2 + \sum_{\substack{\text{angles} \\ (i,j,k)}} \frac{1}{2}k_\theta(\theta_{ijk} - \theta_0)^2}_{\text{bonded interactions}}$$

---

[20] We use in (5.37) the so-called polymer convention for the choice of origin of ϕ. One also encounters the IUPAC convention which differs by a factor π. Then, the cis configuration is associated to an angle of zero degrees and the trans configuration is associated to an angle of ± 180 degrees.

$$+ \sum_{\substack{\text{torsion} \\ (i,j,k,l)}} \sum_{n=1}^{3} k_{\phi_n}(\cos(n\phi - \delta_n) + 1)$$

$$\underbrace{\phantom{+ \sum_{\text{torsion}} \sum_{n=1}^{3} k_{\phi_n}}}_{\text{bonded interactions}}$$

$$+ \underbrace{\sum_{\substack{i,j \\ i<j}} 4\varepsilon \left[\left(\frac{\sigma_{ij}}{r_{ij}} \right)^{12} - \left(\frac{\sigma_{ij}}{r_{ij}} \right)^{6} \right]}_{\text{nonbonded interactions}}. \tag{5.38}$$

Here, the parameters k_b, r_0, k_θ, θ_0, k_{ϕ_n}, δ_n and ε are constants since we only consider polymers made from a single kind of monomer with each monomer considered as a particle.[21] In principle, the linked cell method of Chapter 3 can be directly used to compute the forces and potentials for molecules. In addition to the intermolecular Lennard-Jones potential, now also intramolecular potentials have to be taken into account. The intramolecular forces are obviously short-ranged, since they only resemble interactions between bonded, i.e. neighboring particles.

Implementation for Linear Molecules. In this section we describe a possible implementation for linear molecules which is based on the code for the linked cell method from Chapter 3. In addition to the forces considered there, now also *intramolecular* forces must be dealt with. The function compF_LC (algorithm 3.15), which computes the intermolecular forces, can be used almost without changes. The only point to be additionally considered is whether the intermolecular forces should also act within a molecule. Usually one employs *no* intermolecular Lennard-Jones (and electrostatic) interactions between four successive bonded atoms. Instead of removing these terms from the explicit summation, it may be advantageous to sum them separately, save them, and subtract them again in the routine for the intramolecular forces.

An extra routine compMol_LC is now introduced in which intramolecular forces are computed. Since the molecules have linear structure, the atoms can be numbered uniquely in the order in which they appear in the linear molecular chain. This linear arrangement simplifies the access to the other atoms within the molecule which is needed in the different force computations. To this end, we introduce the number of the associated molecule and the number of the atom within the molecule into the particle data structure 3.1.

```
int MolNo;  //  number of molecule
int AtomNo; //  number of atom
```

---

[21] In general, however, these parameters depend on the types of the involved atoms i, j, k, l. Then, k_b, r_0 and ε are functions of (i, j), while k_θ and θ_0 are functions of (i, j, k) and k_{ϕ_n} and δ_n are functions of (i, j, k, l).

The intramolecular forces can then be computed within an outer loop over all molecules and an inner loop over all the atoms (or particles, respectively,) in each molecule.

The ordering of the molecules and atoms is realized using an additional array M, see also Figure 5.30. Here, $M[i][j]$ is a pointer to the jth atom in the ith molecule. This additional array must be built and initialized at the beginning of the simulation. In this way, the particles involved in the computation of the intramolecular forces within a molecule can be found easily via the vector $M[i]$. The bond parameters k_b, r_0, k_θ, θ_0, k_{ϕ_n}, δ_n, σ, and ε can be stored globally in our example, since they are constant.

```
struct Particle *M[][];
```

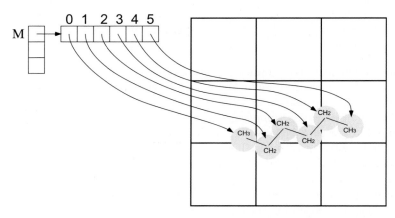

Fig. 5.30. Data structure for linear molecules, here an alkane molecule in the linked cell grid.

Parallelization. In the domain decomposition approach as in Chapter 4, it now can happen that a molecule does not lie entirely within the subdomain associated to a single process, but that different parts of the molecule lie in different subdomains which are associated to different processes. To save memory, the array M should not be stored globally. Instead, every process should store only the information of M which are necessary for the force computations for the part of the molecule by the respective process. Thus, each process ip keeps a *local* array M associated to Ω_{ip} in which the memory addresses of the atoms can be found with positions inside the subdomain Ω_{ip} of this process. To compute the forces within the molecules close to the edge of the subdomain, information from neighboring subdomains is needed. Then, data in M from particles in the border neighborhood of neighboring processes have to be exchanged and inserted and removed accordingly. This again can be implemented analogously to algorithm 4.6. Furthermore, if during the simulation a particle moves from one cell to another cell, this particle is inserted into the particle list of the new cell and is removed from the particle list of the old cell, compare Section 3.5. Now, the local data structures M have to be updated accordingly.

Example: Butane. The simplest organic compounds are composed of carbon and hydrogen. These hydrocarbons can be divided into three groups based on their chemical properties: Saturated hydrocarbons (alkanes or paraffins), unsaturated hydrocarbons (alkenes or olefins) and aromatic hydrocarbons. In the following, we study the family of alkanes in more detail.

Starting from the smallest possible saturated hydrocarbon, methane CH_4, other hydrocarbon molecules can be formed by successively adding methylene groups CH_2 to a chain. The stoichiometric formulas of the resulting hydrocarbons have the general form C_nH_{2n+2}, $n = 1, 2, 3, \ldots$ The chemical properties of the compound change only slightly with increasing number of carbon atoms, whereas the physical properties generally change more significantly. This sequence of hydrocarbon chains constitutes the family of alkanes. The fourth element C_4H_{10} of the series is butane, see Figure 5.31 and 5.32, the tenth, $C_{10}H_{22}$, is decane, and the twentieth, $C_{20}H_{42}$, is eicosane.

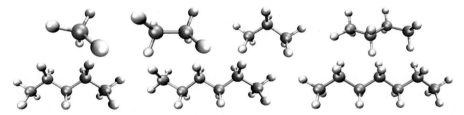

Fig. 5.31. The first elements of the alkane sequence: methane, ethane, propane, butane (upper row), pentane, hexane, heptane (lower row).

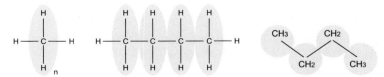

Fig. 5.32. Principle of construction of alkanes (left), butane as example (right).

In the following, we discuss a simple model for alkanes, the united atom model [125]. In this model, each carbon atom together with its bonded hydrogen atoms is modeled as one larger particle (monomer). Monomeres are then joined together into chains with the bond, angle, and torsion angle potentials that have already been introduced in (5.30), (5.31), and (5.35), compare also Figure 5.28. In addition, nonbonded interactions are modeled by means of the Lennard-Jones potential. This potential is usually applied between the monomers of different alkane molecules and between monomers inside a single alkane molecule if these are more than three bonds apart in the molecular chain. The various parameters of the potential are chosen in such a way that the simulation results fit well with real measurements.

For the butane molecule we have to deal with four bonded monomers. The two inner monomers represent a CH_2 group, and the first and last monomer represent a CH_3 group. The intramolecular forces are valence forces resulting from (5.30) that act between two neighboring (bonded) monomers. In addition, angular forces between two successive bonds are taken into account. Differing from the expressions (5.31) and (5.32), we use

$$U_a(\theta) = k_\theta(\cos\theta - \cos\theta_0)^2 \tag{5.39}$$

for the angle potential. We furthermore use a potential function given by [614] for the torsion angle adapted to the simulation of butane of the form (5.36)

$$\begin{aligned} U_t(\phi) = [&1.116 - 1.462\cos(\phi) - 1.578\cos^2(\phi) + 0.368\cos^3(\phi) + 3.156\cos^4(\phi) \\ & + 3.788\cos^5(\phi)]K_\phi\,. \end{aligned}$$

Force terms again result from the gradients of the potentials. The angle $\phi = \phi_{ijkl}$ is computed as shown in equation (5.37). These bond forces and their force constants account for the natural form of the butane polymer chain after appropriate equilibration, see Figure 5.31. The short-range intermolecular interactions are modeled by the Lennard-Jones potential. Here, the Lennard-Jones potential is only used for interactions of monomers from different butane molecules, but not for interactions between monomers from the same butane molecule. The actual parameters of the potential functions are given in SI units in Table 5.14, following [614].

$$
\begin{array}{lll}
k_b = 17.5\frac{\text{MJ}}{\text{mol·nm}^2}, & r_0 = 1.53 \text{ Å}, & \text{valence potential,} \\
k_\theta = 65\frac{\text{kJ}}{\text{mol}}, & \theta_0 = 109.47 \text{ degree}, & \text{angle potential,} \\
K_\phi = 8.31451\frac{\text{kJ}}{\text{mol}}, & & \text{torsion potential,} \\
\sigma = 3.923 \text{ Å}, & \varepsilon = 0.5986\frac{\text{kJ}}{\text{mol}}, & \text{Lennard-Jones potential.}
\end{array}
$$

Table 5.14. Parameter values for the potential functions for the simulation of alkanes.

Our program uses unscaled variables. For this reason a scaling analogous to (3.55) in Section 3.7.3 is necessary beforehand. We use for the scaling $\tilde{\sigma} = 10^{-9}$ m, $\tilde{\varepsilon} = 1$ kJ/mol, $\tilde{m} = 1$ u and $\tilde{\alpha} = \tilde{\sigma}\sqrt{\tilde{m}/\tilde{\varepsilon}} = 10^{-12}$ s $= 1$ ps.

In the following, we study 64 butane molecules in a cubic simulation box with periodic boundary conditions at a temperature of 296 K. This temperature is significantly higher than the boiling point of butane of approximately 274 K, i.e., we consider the gaseous phase of butane. An initial distribution for the 64 molecules is obtained either by random or equidistant translations of the configuration of one molecule. To this end, one can use the data set 5.1.

Code fragment 5.1 PDB entry for butane

```
ATOM      1   CH3   BUTL   1    2.142    1.395   -8.932   1.00   0.00
ATOM      2   CH2   BUTL   1    3.631    1.416   -8.537   1.00   0.00
ATOM      3   CH2   BUTL   1    4.203   -0.012   -8.612   1.00   0.00
ATOM      4   CH3   BUTL   1    5.691    0.009   -8.218   1.00   0.00
CONECT    1   2
CONECT    2   1    3
CONECT    3   2    4
CONECT    4   3
```

The PDB file format[22] is an international standard in widespread use for the storage of three-dimensional structure data of proteins and other biomolecules [85]. The relevant data for our computations are contained in columns 6–8. These columns give the x_1, x_2, and x_3 coordinates of the four carbon atoms of butan. They also serve as the centers of the four monomers. In our simulation, we use the parameters from Tables 5.14 and 5.15. Note here that the same average mass m is used for all monomers for reasons of simplicity, even though the monomers at the beginning and the end of the chain contain a different number of bonded hydrogen atoms than the two inner ones.

$$L_1 = 2.1964 \text{ nm}, \quad L_2 = 2.1964 \text{ nm}, \quad L_3 = 2.1964 \text{ nm},$$
$$m = 14.531 \text{ u}, \quad T = 296 \text{ K},$$
$$r_{\text{cut}} = 2.5\ \sigma, \quad \delta t = 0.0005 \text{ ps}$$

Table 5.15. Parameter values for the simulation of butane.

A butane molecule can assume different spatial configurations. Depending on the value of the torsion angle, one distinguishes between a trans configuration with a torsion angle $|\phi| < \frac{\pi}{3}$ and a cis configuration with all other torsion angles.

---

[22] We will present further information about this format in Section 5.2.3.

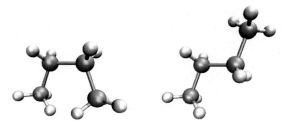

Fig. 5.33. Configurations of butane: Cis and trans configuration, distinguished by the torsion angle.

We report the percentages of the butane molecules in cis and trans configuration in Table 5.16. We averaged over all torsion degrees of freedom of all molecules in each given time interval. Furthermore, we measure the pressure of the complete system according to (3.60) and indicate the average energies for the four different parts of the overall potential. It can be observed from the values in Table 5.16 that butane molecules prefer the trans configuration. From the histogram of torsion angles in Figure 5.34 we see that the angle $\phi = 0$, i.e. the trans configuration, predominates. In the cis configuration, the angle $\phi = 2\pi/3$ and its symmetric counterpart $\phi = 4\pi/3$ occur most often. These two cases taken together are half as likely as the trans configuration.

| time | trans | cis | pressure | LJ | torsion | angle | bond |
ps	(%)	(%)	GPa	kJ/mol	kJ/mol	kJ/mol	kJ/mol
46-54	66.77	33.22	0.1664	-17.9573	2.2954	2.2653	3.3158
86-94	74.94	25.05	0.1602	-18.1049	2.1674	2.3692	3.3321
46-94	69.79	30.20	0.1622	-18.0180	2.2395	2.2476	3.3244

Table 5.16. Simulation of 64 butane molecules, statistical measurements. The energies given are per butane molecule.

Finally, the energies associated to the bond, angle, torsion, and Lennard-Jones potentials are shown in Figure 5.35. The intramolecular bond forces are relatively large due to the parameters of the model. But in reality not all of the degrees of freedom for all parts of the potential are active at a temperature of 296 K. This is especially the case for bonds and angles. In the Störmer-Verlet time integration scheme, these parts of the overall potential now serve to approximately fix the bond angles and the distances between atoms, since these degrees of freedom are virtually frozen for the simulated temperatures.

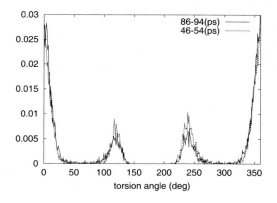

Fig. 5.34. Simulation of 64 butane molecules, histogram of the torsion angles.

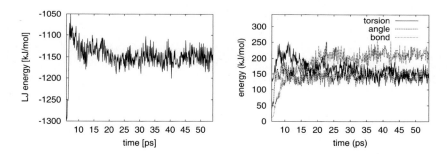

Fig. 5.35. Simulation of 64 butane molecules, time evolution of the energies.

Example: Decane. We study in the following the alkane of length ten, the so-called decane, in its gaseous phase. We consider 27 molecules of decane, in total 270 monomeres, in a cubic box with periodic boundary conditions at a temperature of 296 K. Again, each carbon atom together with its bonded hydrogen atoms is modeled as a monomer and serves as a particle[23] in the united atom model. These particles are then joined to chain molecules of length ten by means of the bond, angle, and torsion potentials (5.30), (5.39), and (5.36). Further nonbonded interactions are modeled by the Lennard-Jones potential. They act between particles from different decane molecules and also between particles from the same decane molecules if these particles are more than three bonds apart in the molecular chain. For our simulation, we use the values from Table 5.17 and the parameters from Table 5.14, see also [614].

---

[23] CH_2 for an inner particle and CH_3 for the first and last particle in a decane chain molecule. For reasons of simplicity, we again use the same average mass m for all particles in our simulation.

$$L_1 = 2.0598 \text{ nm}, \quad L_2 = 2.0598 \text{ nm}, \quad L_3 = 2.0598 \text{ nm},$$
$$m = 14.228 \text{ u}, \quad T = 296 \text{ K},$$
$$r_{\text{cut}} = 2.5 \, \sigma, \quad \delta t = 0.0005 \text{ ps}$$

Table 5.17. Parameter values, simulation of decane.

The initial configuration is constructed as follows: First, 270 single particles are randomly scattered in the simulation domain. Then, particles that are close together are combined to chains of length ten, initial velocities are assigned according to a Maxwell-Boltzmann distribution (Appendix A.4) and an *equilibration* is carried out. During this equilibration phase of approximately 10 ps, the natural geometry of the decane chain molecules slowly emerges. After the completion of this equilibration phase, the particle positions represent realistic initial data that can now be used to start the actual simulation.

Again, we measure the percentages of molecules in trans and cis configuration by averaging over all occurring torsion degrees of freedom of all molecules in the given time interval. In addition, we also measure the pressure in the complete system according to (3.60) as well as the averaged energies associated to the four different parts of the overall potential. The results are shown in Table 5.18.

time	trans	cis	pressure	LJ	torsion	angle	bond
ps	(%)	(%)	GPa	kJ/mol	kJ/mol	kJ/mol	kJ/mol
46-94	77.33	22.66	0.2849	-58.4045	15.4678	9.8089	10.8906

Table 5.18. Simulation of 27 decane molecules, statistical measurements. The energies given are per decane molecule.

Again, we see that the trans configuration is clearly more likely than the cis configuration. From the histogram of the torsion angles in Figure 5.36, we see that this is mainly due to the angle $\phi = 0$. In the cis configuration one basically observes the angle $\phi = 2\pi/3$ and its symmetric counterpart $\phi = 4\pi/3$. Compared to the simulation of butane, this effect is even stronger: The angles $\phi = 2\pi/3$ and $\phi = 4\pi/3$ occur more rarely, the largest part of the bonds oscillates around the angle $\phi = 0$.

In contrast to butane, there is also significantly more energy in the torsion angles, see Figure 5.37. One reason is that in molecules with longer chains, the ratio of torsion degrees of freedom to bond degrees of freedom is substantially larger. In total, the oscillations of the torsion angles are more realistic in a physical sense. The potential forms and their parameters which we used, have ultimately been developed for large molecules and therefore do not model short chain molecules such as butane as accurately.

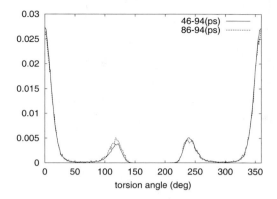

Fig. 5.36. Simulation of 27 decane molecules, histogram of torsion angles.

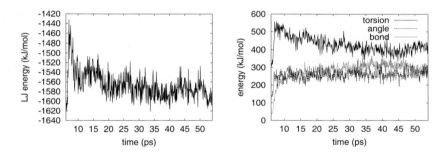

Fig. 5.37. Simulation of 27 decane molecules, time evolution of the energies.

Extension: Diffusion of Gas Molecules in Alkanes. Again, we study alkanes. Now, 30 chains of 20 monomers each (20-alkane or eicosane) are simulated in a cubic box of length 25 Å with periodic boundary conditions at a temperature of 360 K. This temperature lies significantly above the boiling point of eicosane, i.e., we consider eicosane gas. The parameters for the potentials and the simulation are taken from the Tables 5.14 and 5.17 of the previous section. Also, the initial configuration for the simulation is generated by a similar procedure as before. Furthermore, some small molecules are added – in our case oxygen (O_2). To this end, we distribute 20 O_2 molecules randomly in the domain, i.e., we dissolve O_2 in eicosane gas. Each oxygen molecule is modeled as one particle. The oxygen particles interact with the eicosane monomers by means of the Lennard-Jones potential. The appropriate parameters for the potential are determined by the Lorentz-Berthelot mixing rule (algorithm 3.19 in Chapter 3.6.4) with the parameters for oxygen as given in Table 5.19.

$$\varepsilon_O = 940 \ \text{J/mol}$$
$$\sigma_O = 3.43 \ \text{Å}$$

Table 5.19. Parameter values for oxygen.

We now simulate the diffusion of the oxygen particles in the eicosane gas. This diffusion process is strongly influenced by the eicosane chain molecules. After an initial equilibration phase of 10 ps at a time step of $\Delta t = 5 \cdot 10^{-4}$ ps, the molecular dynamics simulation is continued over a longer time period and the motion of the oxygen molecules is observed. We measure the diffusion of the oxygen and eicosane molecules. The diffusion of oxygen is computed from the mean standard deviation of the particle positions which is given by equation (3.61) in Section 3.7.3. Analogously we compute the diffusion of eicosane.

Figure 5.38 shows the time evolution of the resulting diffusion of the eicosane molecules and the oxygen molecules. As expected, the mobility of the smaller and lighter oxygen molecules is higher than the mobility of the eicosane molecules. However, at a temperature of 360 K, one would expect much more than just a difference by a factor of three. This can be explained with the geometrical structure of the eicosane molecules: The motion of the oxygen molecules in the simulation domain is hampered by the long eicosane molecules. The respective energies are given in Figure 5.39. Again, the energy of the Lennard-Jones potential dominates, which acts here also between the oxygen and the eicosane. The intramolecular energies are comparable to the ones observed in the simulation of the decane molecules, but they do not oscillate as strongly. Here, the torsion energy is larger than the energy of the other bonded potentials. This is due to the length of the molecular chains.

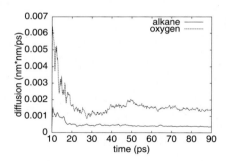

Fig. 5.38. Simulation of eicosane molecules (20-alkane) together with oxygen molecules, time evolution of the diffusion coefficients starting at time 10 ps.

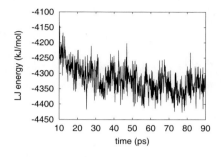 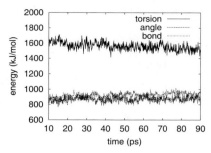

Fig. 5.39. Simulation of eicosane molecules together with oxygen molecules, diagrams of the energies.

Alkanes are a simple special case of polymers. In general, polymers are materials in which single molecules can be very large – up to 1000 atoms and more. Every polymer molecule is composed of monomers. The monomers are basic units that consist of very special atoms that predominately bind in form of a long chain. There are synthetic polymers and biopolymers. Biopolymers are polymers which occur in nature, such as for example DNA, RNA, proteins, polysaccharides, and others. In the case of proteins, the monomers are the amino acids. Synthetic polymers are the basis of all forms of the materials known as plastics (for example polyethylene, PVC, nylon, polyester). Besides the material properties of the polymer, the process of polymerization itself is also of particular interest. Here, chain molecules are formed from the monomers by chemical reactions, e.g. catalysis, such that networks of enlaced polymer molecules are built. Further properties of interest are the degree and spatial structure of the interlacing. Computer simulations of polymers are extremely expensive because of the large size of the molecules and the resulting runtime complexities. To this end, the simulation of polymers with the molecular dynamics method[24] is a relatively new area of research which certainly allows for improvements in many respects. Here, the use of long-range electrostatic forces (the Coulomb potential) is an important ingredient for the simulation of the polymerization process, which needs large amounts of computation time. In addition, the time scale of the phenomena of interest plays a decisive role. Unfortunately, it is often much larger than the simulation times which can be achieved by current molecular dynamics methods on computers presently available, see also the discussion in Chapter 9.

---

[24] Furthermore, Monte-Carlo methods are used successfully in such studies. There, synthetic polymers are often represented using the bead-spring model or the Rousse model. Further reading in this direction can be found for instance in [197, 363, 442].

5.2.3 Outlook to More Complex Molecules

Up to now we only dealt with simple linear chain molecules and their implementation. Now, we discuss in brief several modifications that allow the simulation of molecules with more complex structure. Furthermore, we present techniques that allow us to gain molecular data and appropriate potential parameters from databases and other program packages as input for our simulations.

Data Structure. A molecule has been represented in Section 5.2.2 as a logical vector of atoms. Using this data structure, we implemented alkane chain molecules via the united atom model. But there is an abundance of more complex molecular structures beyond such chain molecules. Already a simple isoalkane, in which the chain bifurcates, or a cyclic alkane, which contains a ring structure, can not be represented in such a way. There are also ring-like molecules, such as benzene, and molecules with more general graphs. Further examples are proteins and other biomolecules. Their structures can become arbitrarily complex with an increasing number of atoms.

To be able to represent the neighborhood/bonding relations of atoms in such complex molecules, we need a new data structure for the molecules. Instead of one vector we now use pointers that represent the graph of atoms and bonds within the molecule. We encountered such techniques in a simple form already in Section 3.5 where we considered linked lists. Now, we introduce pointers for each particle to all its neighboring particles. This way, more complex molecular structures can be represented and also different strands of a molecule can be tracked. In our case it is practical to assume an upper limit `MAXNEIGHBOR` on the number of neighboring atoms[25] which we define to be four, for example. For the particles, we then may use data structure 5.1, see also Figure 5.40, where additional memory can be allocated for bond parameters if necessary.[26]

For a linear molecule, this data structure reduces to a doubly-linked list. In principle, a singly-linked structure would be enough for most operations. However, for example when particles cross subdomain boundaries, i.e. in the

---

[25] To keep the number of neighboring atoms variable, one can alternatively use a list of pointers to the neighboring atoms. This can results in a net saving of storage space in particular cases compared to the use of preallocated vectors of fixed size.

[26] There are different possibilities to manage the bond parameters. First, the parameters could be stored individually for each bond, for instance together with the pointers to the neighboring atoms. One disadvantage is the variable number of possible bonds, so that a lot of memory may remain unused if vectors of fixed size are preallocated. Alternatively, the parameters could be stored in global (associative) tables (sorted by atoms types) in which they can be easily found and accessed. Since often there are significantly more atoms than different choices of bond parameters, storage space can be saved this way.

Data structure 5.1 General molecules as connectivity graphs of atoms.

```
#define MAXNEIGHBOR 4
typedef struct {
    ...                          // data structure for particles 3.1
    struct Particle *neighbor[MAXNEIGHBOR];
} Particle;
```

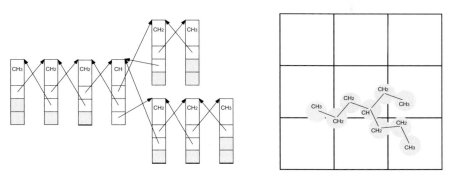

Fig. 5.40. Data structure for molecules as graph of particles and bonds for the example of an isoalkane molecule. Branched structure (left) and molecule in the linked cell grid (right).

parallel code, the insertion and removal of particles is easier to implement in a doubly-linked structure. The array M[i][j] from Section 5.2.2 is then no longer needed.

The doubly-linked structure of the molecule has to be allocated and initialized correctly at the start of the simulation. The computation of forces for the nonbonded interactions is carried out as usual with the linked cell method. The intramolecular forces can be computed in a loop over all particles and their pointers to neighboring particles. Here, the action of the bonding forces onto each single particle is computed. Alternatively, one could also use Newton's third law to compute forces on groups of particles (mostly four neighboring atoms) if one ensures (for example with additional flags) that each bond potential is only computed once in the loop. Usually, some particles from the same molecule (e.g. four consecutive ones) are not to interact through the Lennard-Jones potential. As already noted in Section 5.2.2 for linear molecules, instead of checking in the linked cell method for each interaction whether it has to be computed and added, it can be advantageous to compute and add such interactions in the linked cell method and then subtract them again in the computation of the intramolecular forces, if needed.

Parallelization. The parallelization of our linked data structures requires somewhat more effort than for the vector structure of linear molecules. It has to be taken into account that a long molecule could extend across subdomain (and therefore, process) boundaries and that parts of molecules could move across process boundaries.

If at first we assume a static data distribution, we only have to ensure that the necessary copies of the particles in the border neighborhoods are available. If now in addition the pointers of the particles in the border neighborhood point to the appropriate copy of the neighboring particle, we only have to prevent the possibility that a single force term is taken into account more than once.

However, the case of a dynamical data distribution poses a challenge, since parts of molecules can move from one process to another. Even though the particles in the border neighborhoods can be sent as before, the pointers to the neighboring particles have to be set correctly. Since the value of a pointer, namely the memory address, does not make sense for the other processes in a distributed memory system, a way has to be found to transmit the graph structure across process boundaries. To this end, a unique numbering of the particles as in Section 5.2.2 is an elegant possibility. Pointers are then translated into its corresponding particle numbers, these are sent to their new process and the new process translates the particle numbers back into (local) pointers. If a search of the particle numbers in the corresponding cells of the new process is too time-consuming in this back-translation, hash techniques can be used instead [357].

Potentials. The overall potential function is basically constructed from the potentials for linear molecules which we introduced in the previous section. They consist of bonded and nonbonded terms. The bonded terms are composed of harmonic bond potentials (5.30), angle potentials (5.31) for interactions of three bonded atoms, and torsion potentials for interactions between four bonded atoms. Here, we encounter a new form of torsion potential besides the form in (5.35) for four consecutively bonded atoms i-j-k-l, namely at branching points of the molecule, were several strands are connected. To this end, we apply the so-called improper torsion potential

$$U_{ut}(\omega) = \frac{1}{2} k_\omega (\omega - \omega_0)^2. \tag{5.40}$$

Now, in contrast to the conventional torsion potential, the atoms i, j, k, l which determine the torsion angle are not bonded in a chain but in a star-like fashion. Figure 5.41 shows a sample configuration for such an interaction.[27] Here, $\omega = \omega_{ijkl}$ denotes the angle between the two planes which are spanned by the atoms i, j, k and j, k, l. Analogously, several improper torsion potentials can be applied to atoms with more than three bonds.

---

[27] Occasionally, improper torsion potentials are also applied at atoms that are not branching points in the molecule. Thus, one can impose additional restrictions on the geometry of the molecule, as for example the planarity of sp^2-hybridized carbons in carboxyl groups.

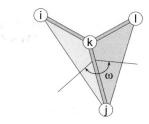

Fig. 5.41. A configuration of atoms i, j, k, l and the improper torsion angle ω_{ijkl} between the planes spanned by atoms i, j, k and j, k, l.

Geometry and Parameters for the Potentials. Besides their applications in the material sciences, molecular dynamics methods serve as an important tool in the study of biomolecules. The RCSB[28] Protein Data Bank (PDB) [85] provides three-dimensional structural data of proteins. It delivers the atoms of the molecules with their coordinates and the bonds between them.[29] An example can be found in code fragment 5.2, another example was already given in code fragment 5.1 for the simulation of butane.

Code fragment 5.2 The Structure of Ant Acid in PDB format.

ATOM	1	C	UNK	1	-0.014	1.204	0.009	1.00	0.00
ATOM	2	O	UNK	1	0.002	-0.004	0.002	1.00	0.00
ATOM	3	O	UNK	1	1.139	1.891	0.001	1.00	0.00
ATOM	4	H	UNK	1	-0.957	1.731	0.016	1.00	0.00
ATOM	5	H	UNK	1	1.126	2.858	0.007	1.00	0.00
CONECT	1	2	2	3	4				
CONECT	2	1							
CONECT	3	1	5						
CONECT	4	1							
CONECT	5	3							

The data contained in the PDB have mostly been found experimentally by x-ray crystallography [113, 516] or solution-NMR (nuclear magnetic resonance) [159, 113]. In general, x-ray crystallography cannot determine the position of hydrogen atoms, neither can it distinguish between nitrogen, oxygen,

---

[28] The PDB is currently maintained by the Research Collaboratory for Structural Bioinformatics (RCSB), which took over management from Brookhaven National Laboratory in 1999.

[29] Here, also atoms with different chemical types of bonds are distinguished, which is expressed by suffixes to the atom identifiers. Unfortunately, not all bonds between atoms are included. Furthermore, some particular groups of atoms such as amino acids may be abbreviated by one symbol instead of listing all the atoms in the group.

and carbon atoms. Thus, these details have to be manually added afterward using additional chemical knowledge. Furthermore, with x-ray crystallography, the proteins must usually be analyzed in their crystal (dehydrated) form. Consequently, only the structural data of a dehydrated and thus deformed state can be gained this way.

Thus, before any realistic simulation of bioproteins can start, missing hydrogen atoms and the surrounding water or other solvent molecules have to be added to the structural data from the PDB database. The missing hydrogen atoms can be produced using a "hydrogen generator" as available, for instance, in HyperChem [14], which uses additional chemical knowledge. The surrounding water can be generated as follows: First, a grid of water molecules is produced, where the single molecules are randomly rotated and slightly perturbed in their positions. Then, those water molecules are omitted which overlap with the protein or are too close to it. To this end, the distance of a water molecule to the protein molecule can be computed efficiently with the linked cell method. Alternatively, water molecules can also be deposited close to the protein molecule in a layer by layer fashion. Furthermore, salt and mineral atoms dissolved in the water have to be added for some simulations. The basic structure of the overall initialization process is given in Figure 5.42.

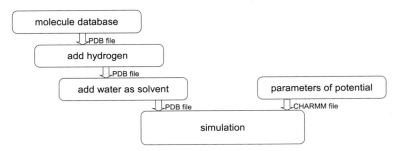

Fig. 5.42. Preparatory steps for protein simulations: After the file containing the coordinates of the basic atoms and the bonds of a molecule is read in from the PDB database, its structure is completed with the help of a hydrogen generator by filling in missing hydrogen atoms. Afterwards, the structure is surrounded by water molecules, and finally the parameters of the potentials are read from a parameter file.

Figure 5.43 shows the difference between a structure taken from the database, the structure with missing hydrogen atoms added, and the final structure with surrounding water molecules added, as used in the computation.

After the geometric configuration of the atoms is established, the bonds between atoms have to be described. For that purpose we use the potentials which were previously introduced. The force constants k_b, k_θ, k_{ϕ_n}, δ_n, k_ω,

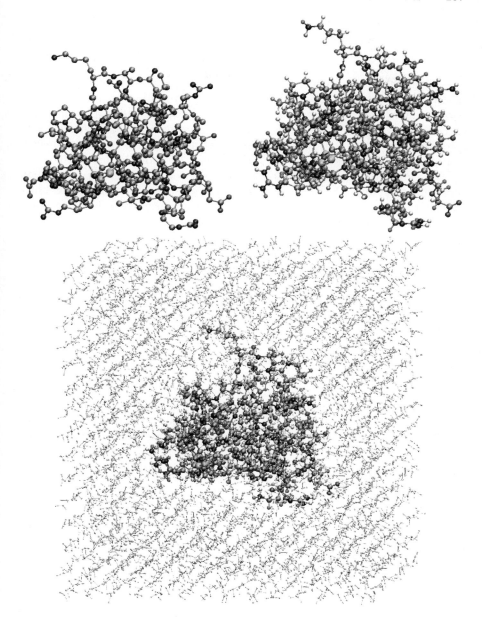

Fig. 5.43. The BPTI protein, as given in the PDB database (upper left), with added hydrogen atoms (upper right), and with surrounding water molecules (below).

the equilibrium parameters r_0, θ_0, ϕ_0, ω_0, and the Lennard-Jones constants σ and ε depend, however, on the atoms involved in the bond. Every combination of atoms therefore needs its own set of parameters. Here, equilibrium distances and angles can be determined geometrically, while force constants are identified using the energies and natural frequencies of the bonds. Typically, such parameters are derived by a fitting of the potential functions to experimental data. Many teams have intensively worked (and are still working) on this problem. Nowadays, extensive libraries of such parameter sets are available. Prominent representatives are the parameter sets contained in the program packages CHARMM [125], Amber [471] or Gromos [634].

A section of a CHARMM parameter file, for example, can then look like code fragment 5.3.[30] The potential functions, constants and their units are first described in comments, and then the parameter values for particular atom combinations are listed. In detail, the parameters for valence bonds are given first. For example, the bond between hydrogen and carbon atoms has an equilibrium distance $r_0 = 1.11$ Å and a force constant of $k_b = 330$ kcal/mol/Å^2. Then, the angle potentials follow. For instance, the triple of hydrogen, oxygen, and carbon atom here has an equilibrium angle of $\theta_0 = 108$ degrees and a force constant of $k_\theta = 65$ kcal/mol/degree^2. Furthermore, parameters for the torsion potentials and improper torsion potentials are analogously specified. Here, the wild card X stands for arbitrary atoms. Finally the parameters ε and $\sigma/2$ of the Lennard-Jones potential are given. The parameters for pairs of atoms are computed with the Lorentz-Berthelot mixing rule from the values for a single type of atom, as described in the comment.

With help of a proper interface to the PDB data base and the associated parameter files from Amber, CHARMM, or Gromos, potential functions for almost all known proteins can be generated. Examples for simulations with such complex molecules can be found in Chapter 9.

---

[30] Here, differently bonded atoms are denoted by identifiers which are derived from the atom name. The identifier X is used as a wild card which stands for an arbitrary atom. In addition, atom groups such as amino acids are described, and hydrogen bridge bonds, ions, and other irregularities are specified. Note that each program package uses slightly different potentials which require accordingly different parameter sets.

Code fragment 5.3 Parameters for the Potentials in CHARMM Parameter File Format.

```
BONDS
!V(bond) = Kb(b - b0)**2
!Kb: kcal/mole/A**2
!b0: A
!atom type  Kb      b0
H    C      330.0   1.11
O    C      620.0   1.23
O    H      545.0   0.96


ANGLES
!V(angle) = Ktheta(Theta - Theta0)**2
!Ktheta: kcal/mole/rad**2
!Theta0: degrees
!atom types    Ktheta  Theta0
H    O    C    65.0    108.0
O    C    O    100.0   124.0
O    C    H    50.0    121.7


DIHEDRALS
!V(dihedral) = Kchi(1 + cos(n(chi) - delta))
!Kchi: kcal/mole
!n: multiplicity
!delta: degrees
!atom types        Kchi  n    delta
X    C    O    X   2.05  2    180.0


IMPROPER
!V(improper) = Kpsi(psi - psi0)**2
!Kpsi: kcal/mole/rad**2
!psi0: degrees
!atom types        Kpsi   psi0
O    X    X    C   120.0  0.00


NONBONDED
!V(Lennard-Jones) = Eps,i,j[(Rmin,i,j/ri,j)**12 - 2(Rmin,i,j/ri,j)**6]
!epsilon: kcal/mole, Eps,i,j = sqrt(eps,i * eps,j)
!Rmin/2: A, Rmin,i,j = Rmin/2,i + Rmin/2,j
!atom  epsilon  Rmin/2
C      -0.110   2.0000
H      -0.046   0.2245
O      -0.120   1.7000
```

6 Time Integration Methods

With a discretization in time like the Störmer-Verlet method from Section 3.1, the solution of the continuous Newton's equations is computed only approximately at selected points along the time axis. There, approximations to the values at later points in time are computed from the values of the approximations at previous points in time in an incremental fashion. Now, we consider time integration methods in a more general framework. First, we discuss local and global error estimates. It will turn out that one has to be careful in the interpretation of results from integrations over long times with many time steps. We thus study the conservation properties of time integration schemes and introduce so-called symplectic integrators. Then, we give techniques to construct more general schemes. Finally, we discuss possibilities to speed up the time integration. To this end, we focus on three approaches:

- Higher order schemes. Their errors show a better order of convergence with respect to the time step. Thus, for a given (small enough) time step, the use of a higher order scheme decreases the error in the simulation. Alternatively, also a larger time step can be selected to obtain a solution of the same accuracy but now with a smaller number of time steps.
- Methods with multiple time steps. Especially for molecular problems, the different length scales for the bond, angle, torsion angle, Lennard-Jones, and Coulomb parts of the overall potential function imply different time scales in the problem. With an appropriate separation, these different time scales can be treated individually using time steps of different size. A typical representative of such methods is the impulse or r-Respa method [274, 275, 621, 623] presented in Section 6.3.
- Freezing high frequency modes by enforcing additional constraints. If the bond and angle forces are kept fixed, a larger time step can be used without compromising the stability of the time integration scheme. This approach is presented in more detail in Section 6.4, for which the SHAKE and RATTLE methods [43, 534] give an example.

In Section 3.1 we already introduced the Störmer-Verlet method as one possibility for the discretization of Newton's equations of motion

$$m\ddot{\mathbf{x}} = \mathbf{F}(\mathbf{x}). \tag{6.1}$$

In the following, we will consider other integration methods. To this end, we rewrite (6.1) in the Hamilton formalism. This is possible since the total energy of the mechanical system is conserved.[1] We obtain

$$\dot{\mathbf{q}} = \nabla_{\mathbf{p}}\mathcal{H}(\mathbf{q}, \mathbf{p}), \qquad \dot{\mathbf{p}} = -\nabla_{\mathbf{q}}\mathcal{H}(\mathbf{q}, \mathbf{p}) \tag{6.2}$$

with the positions $\mathbf{q}$ and momenta $\mathbf{p}$, which correspond to $\mathbf{x}$ and $m\dot{\mathbf{x}}$. In our case, the Hamiltonian is

$$\mathcal{H}(\mathbf{q}, \mathbf{p}) = T(\mathbf{p}) + V(\mathbf{q}) = \frac{1}{2}\mathbf{p}^T m^{-1}\mathbf{p} + V(\mathbf{q}), \tag{6.3}$$

where m denotes the masses of the particles and V denotes the potential. In this way, (6.2) turns into the system

$$\dot{\mathbf{q}} = m^{-1}\mathbf{p}, \qquad \dot{\mathbf{p}} = -\nabla_{\mathbf{q}}V(\mathbf{q}) \tag{6.4}$$

of differential equations of first order.

6.1 Errors of the Time Integration

Local Error. In general, a one-step method[2] for the time integration of Newton's equation of motion (6.2) can be written as

$$\begin{pmatrix} \mathbf{q}^{n+1} \\ \mathbf{p}^{n+1} \end{pmatrix} = \Psi(\mathbf{q}^n, \mathbf{p}^n, \delta t) := \begin{pmatrix} \Psi_1(\mathbf{q}^n, \mathbf{p}^n, \delta t) \\ \Psi_2(\mathbf{q}^n, \mathbf{p}^n, \delta t) \end{pmatrix} \tag{6.5}$$

with a function Ψ that specifies the method. If we denote by $(\mathbf{q}, \mathbf{p})$ the exact solution of (6.1) that intersects $(\mathbf{q}^n, \mathbf{p}^n, t_n)$ at time t_n, then the the corresponding propagation function Φ can be defined by

$$\Phi(\mathbf{q}^n, \mathbf{p}^n, \delta t) := \begin{pmatrix} \mathbf{q}(\delta t + t_n) \\ \mathbf{p}(\delta t + t_n) \end{pmatrix}.$$

Then, for an integration method of order p and a sufficiently smooth solution $(\mathbf{q}(t), \mathbf{p}(t))$, the estimate

$$\|\Psi(\mathbf{q}^n, \mathbf{p}^n, \delta t) - \Phi(\mathbf{q}^n, \mathbf{p}^n, \delta t)\| = \mathcal{O}(\delta t^{p+1}) \tag{6.6}$$

holds [297], and consequently, the local error in each time step satisfies

---

[1] Note however that the temperature is not conserved for a system with only a finite number of particles. In the case of an ergodic system, the instantaneous temperature fluctuates around a mean value which depends only on the energy of the initial data.

[2] One-step methods need only one previous, old value $\begin{pmatrix} \mathbf{q}^n \\ \mathbf{p}^n \end{pmatrix}$ to determine the new value $\begin{pmatrix} \mathbf{q}^{n+1} \\ \mathbf{p}^{n+1} \end{pmatrix}$. In contrast, multi-step methods refer to the values at several previous points in time.

$$\begin{aligned}\|\mathbf{q}^{n+1} - \mathbf{q}(\delta t + t_n)\| &= \mathcal{O}(\delta t^{p+1}), \\ \|\mathbf{p}^{n+1} - \mathbf{p}(\delta t + t_n)\| &= \mathcal{O}(\delta t^{p+1}).\end{aligned} \tag{6.7}$$

Thus, we can approximate Newton's equations of motion very well across one time step δt: The accuracy can be improved by using either a smaller time step δt or, alternatively, an integration scheme with higher order p. Also, values derived from $\mathbf{q}$ and $\mathbf{p}$, such as energy and momentum, can be computed very accurately across one time step.

Global Error. Consider now a fixed time interval from t_0 to $t_{\text{end}} = t_0 + n \cdot \delta t$. Hence, starting from t_0, n time steps of size δt are needed to reach t_{end}. Let us again consider an integration method of order p, and let Ψ have Lipschitz constant M.[3] Furthermore, let $(\mathbf{q}, \mathbf{p})$ denote the exact solution through $(\mathbf{q}^0, \mathbf{p}^0, t_0)$. Then, we obtain for the global error of the time integration from t_0 to t_{end}

$$\|\mathbf{q}^n - \mathbf{q}(t_{\text{end}})\| \le C \cdot \delta t^p \cdot \frac{e^{M(t_{\text{end}} - t_0)} - 1}{M}, \tag{6.8}$$

see [297], Theorem 3.4. A corresponding result holds also for the error in the momentum $\mathbf{p}$. Again, we can control the accuracy of the solution by the selection of the time step δt and convergence order p of the method.[4] The fundamental problem is now that the error grows exponentially with the time t_{end}. In many molecular dynamics applications, we are interested in simulations over long times, where t_{end} is large compared to the Lipschitz constant M which is dominated by the highest frequency of the oscillations. A small perturbation in the initial data may therefore be amplified *exponentially* with the time t_{end} and, at the end of the simulation, only the effects of the perturbation are visible. In general this is summarized under the concept of chaotic behavior.[5] As a consequence, the results of a simulation no longer directly depend on the initial data and derived quantities such as energy and momentum may diverge with t. A simulation over long periods of time seems therefore meaningless, regardless which integration method and which accuracy is used. Thus, there is no hope to use molecular dynamics simulations as a molecular microscope to look at actual long-time trajectories of particles.

Let us illustrate this point in a numerical experiment. To study the exponential growth of errors we consider two simple simulations with 1000 particles each. We measure the distance of the particle trajectories $\mathbf{q}$ and $\hat{\mathbf{q}}$

---

[3] A function f defined on D is called Lipschitz continuous or is said to satisfy a Lipschitz condition, if there exists a constant $M \ge 0$ such that for all $x, y \in D$

$$\|f(x) - f(y)\| \le M \cdot \|x - y\|.$$

The smallest such number is called the Lipschitz constant of the function.

[4] Here, a decrease in the time step δt may also lead to a loss of accuracy since the rounding error grows with the number of time steps.

[5] In this context the largest Ljapunov exponent [297] plays the role of the Lipschitz constant M.

from the two simulations. The simulations differ only in the initial conditions for *two* particles: The velocities of these two particles differ by the factor of 10^{-10} in the $\mathbf{x}_1$ component. According to (6.8), we expect that this small perturbation in the initial conditions will be amplified exponentially in time. Thus, the particle configurations in both simulations should be essentially independent after some time.

The parameters for the experiment are given in Table 6.1. The 1000 particles in each system interact with each other by the smoothed Lennard-Jones potential (3.63) from Section 3.7.3. We measure the distance of the trajectories by $\|\mathbf{q}(t) - \hat{\mathbf{q}}(t)\|$. The results are given in Figure 6.1 (left). Indeed, we see that the distance between the two simulations grows exponentially with time, i.e., the estimate (6.8) is sharp. A small perturbation in the initial conditions is sufficient to cause a difference in the positions of the particles of the order of the size of the simulation domain. At time $t = 5$ we reach such a difference and the particles can not drift further apart during the onward simulation.

$$
\begin{array}{lll}
L_1 = 7.5, & L_2 = 7.5, & L_3 = 7.5, \\
\varepsilon = 1, & \sigma = 1, & m = 39.95, \\
N = 10 \times 10 \times 10, & m = 39.95, & T = 2000 \text{ K}, \\
r_{\text{cut}} = 2.5\,\sigma, & \delta t = 0.001, & t \in [0, 5]
\end{array}
$$

Table 6.1. Parameter values for the simulation of two particle systems with Lennard-Jones forces.

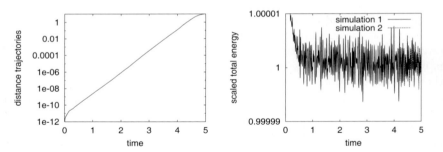

Fig. 6.1. Simulation of a particle system with 1000 particles over 5000 time steps and comparison with a slightly perturbed system. The distance of the trajectories diverges exponentially in time (left). The time average of the total energy is conserved, at least approximately (right).

The total energy of the system is shown in Figure 6.1 (right). We see that, unlike the distance of the trajectories, the total energy only oscillates. Its average value even seems to stay constant over time. In the following, we will study this effect in more detail.

Conserved Quantities. Despite the fact that the numerically computed trajectory of a particle system differs in a substantial way from the exact trajectory $(\mathbf{q}(t), \mathbf{p}(t))$ already after a short simulation time and is completely off for long simulation times, molecular dynamics techniques can still help to obtain meaningful results. It can be observed that at least certain statistical averages of the simulation like the total energy can be obtained quite accurately. Thus, reliable values for macroscopic observables in phase space can be gained even for long simulation times as we will see in the following.

In every closed mechanical system $\mathcal{H}(\mathbf{q}, \mathbf{p})$, there are a number of quantities which are conserved over time by the exact solution of the Hamilton equation.[6] Prominent examples are the total energy, the momentum and the angular momentum.[7] There exist time integration schemes that conserve one or more of these quantities exactly. One possibility to construct such an integration scheme is to compute a projection of the variables $(\mathbf{q}^{n+1}, \mathbf{p}^{n+1})$ after each time step such that, given a conserved quantity A, the projected value $(\hat{\mathbf{q}}^{n+1}, \hat{\mathbf{p}}^{n+1})$ satisfies

$$A(\mathbf{q}^n, \mathbf{p}^n) = A(\hat{\mathbf{q}}^{n+1}, \hat{\mathbf{p}}^{n+1}).$$

Hence, numerical simulations of molecular systems can be meaningful even though the computed trajectories of the particles are not accurate for long times and quickly leave the neighborhood of the exact trajectory $(\mathbf{q}(t), \mathbf{p}(t))$.

Symplectic Integrators. We now consider a particular conserved quantity which is characteristic for Hamiltonian systems, namely the volume in phase space, see also Section 3.7.2. In the space of points $(\mathbf{q}, \mathbf{p})$, the measure of a set is conserved under transport along the trajectories of the system. This property can be expressed with the (outer) differential form

$$\omega := \sum_i \mathrm{d}\mathbf{q}_i \wedge \mathrm{d}\mathbf{p}_i,$$

as $\mathrm{d}\omega = 0$, see also [296, 377]. A system with such a property is called *symplectic*. A symplectic mapping Ψ from $(\mathbf{q}, \mathbf{p})$ to $(\tilde{\mathbf{q}}, \tilde{\mathbf{p}})$,

$$\begin{pmatrix} \tilde{\mathbf{q}} \\ \tilde{\mathbf{p}} \end{pmatrix} = \Psi(\mathbf{q}, \mathbf{p}) := \begin{pmatrix} \Psi_1(\mathbf{q}, \mathbf{p}) \\ \Psi_2(\mathbf{q}, \mathbf{p}) \end{pmatrix},$$

---

[6] Conserved quantities of a Hamiltonian $\mathcal{H}$ can also be characterized by $\{A, \mathcal{H}\} = 0$ with $\{\cdot, \cdot\}$ being a Poisson bracket. $\{A, \mathcal{H}\}$ is defined as $\sum_i \frac{\partial A}{\partial \mathbf{q}_i} \frac{\partial \mathcal{H}}{\partial \mathbf{p}_i} - \frac{\partial A}{\partial \mathbf{p}_i} \frac{\partial \mathcal{H}}{\partial \mathbf{q}_i}$ which corresponds to the time derivative $\dot{A}(\mathbf{q}, \mathbf{p})$.

[7] In simple cases one can directly eliminate the conserved quantity from the system. However, eliminating or conserving one quantity exactly does not imply anything on the quality of other quantities from the simulation. Unfortunately, one cannot conserve all analytically conserved quantities numerically as well. Additionally, the relevant conserved quantities might not be known or very hard to determine for the system under study, especially for more complex systems.

can equivalently be defined by the condition

$$\begin{pmatrix} \nabla_{\mathbf{q}}\Psi_1 \ \nabla_{\mathbf{p}}\Psi_1 \\ \nabla_{\mathbf{q}}\Psi_2 \ \nabla_{\mathbf{p}}\Psi_2 \end{pmatrix}^T \begin{pmatrix} 0 & I \\ -I & 0 \end{pmatrix} \begin{pmatrix} \nabla_{\mathbf{q}}\Psi_1 \ \nabla_{\mathbf{p}}\Psi_1 \\ \nabla_{\mathbf{q}}\Psi_2 \ \nabla_{\mathbf{p}}\Psi_2 \end{pmatrix} = \begin{pmatrix} 0 & I \\ -I & 0 \end{pmatrix}. \qquad (6.9)$$

A one-step method (6.5) is called symplectic if the mapping Ψ in (6.5) from the approximation for one time t_n to that for the next point in time t_{n+1} is symplectic. A point in phase space is then transported by symplectic mappings along a trajectory.

Let us now study the mappings implied by time integration schemes in a numerical experiment. We consider a simple model of the mathematical pendulum, with the Hamiltonian

$$\mathcal{H}(\mathbf{q}, \mathbf{p}) = \mathbf{p}^2/2 - \cos \mathbf{q}$$

and the corresponding differential equation

$$\dot{\mathbf{q}} = \mathbf{p}, \qquad \dot{\mathbf{p}} = -\sin \mathbf{q}.$$

This corresponds to an ideal undamped pendulum with unit mass which is suspended with a thread at a fixed point and swings without friction in a plane. For simplicity we use units such that the gravitational constant has the numerical value one. We choose the coordinate system such that the position $\mathbf{q}$ corresponds to the angular deflection of the pendulum. The momentum $\mathbf{p}$ is then the angular velocity. If the pendulum oscillates periodically, the position $\mathbf{q}$ lies in $[\mathbf{q}_{min}, \mathbf{q}_{max}] = [-\gamma, \gamma]$ with $\gamma < \pi$. Because of the periodicity in the description of the problem, systems with positions differing by a multiple of 2π correspond to the same physical situation, and the given range for $\mathbf{q}$ could be shifted by any multiple of 2π. The momentum of the system lies in $[-2, 2]$. If the system possesses more energy, the pendulum no longer oscillates, but turns over. The position then passes through all values in $\mathbb{R}$, either monotonically increasing or monotonically decreasing. Between periodic oscillation and overturning there is a limit case with the maximal angle of π which is not reached by the system in finite time. The behavior of the system is sketched as a phase diagram in Figure 6.2, upper left. Here, several orbits of the system are drawn and marked with arrows. The periodic solutions are located close to the origin. They correspond to a periodically swinging pendulum. For small amplitudes, the trajectories are nearly circular and the potential is approximately harmonic. The thick line corresponds to the limit case and separates the phase space. Any trajectory outside corresponds to an overturning pendulum.

In our experiment, we now follow a disk in phase space for several time steps using different time integration schemes: Besides the Störmer-Verlet method (3.22), we consider the following variants of the Euler method for (6.4):

$$\Psi_{\text{explicit}}(\mathbf{q}^n, \mathbf{p}^n) = \begin{pmatrix} \mathbf{q}^n + \delta t\, m^{-1}\mathbf{p}^n \\ \mathbf{p}^n - \delta t \nabla_{\mathbf{q}} V(\mathbf{q}^n) \end{pmatrix}, \tag{6.10}$$

$$\Psi_{\text{implicit}}(\mathbf{q}^n, \mathbf{p}^n) = \begin{pmatrix} \mathbf{q}^n + \delta t\, m^{-1}\mathbf{p}^{n+1} \\ \mathbf{p}^n - \delta t \nabla_{\mathbf{q}} V(\mathbf{q}^{n+1}) \end{pmatrix}, \tag{6.11}$$

$$\Psi_{\text{symplectic}}(\mathbf{q}^n, \mathbf{p}^n) = \begin{pmatrix} \mathbf{q}^n + \delta t\, m^{-1}\mathbf{p}^{n+1} \\ \mathbf{p}^n - \delta t \nabla_{\mathbf{q}} V(\mathbf{q}^n) \end{pmatrix}. \tag{6.12}$$

These methods are of first order in time. The results of the experiment are shown in Figure 6.2. The exact solution is shown in the top right picture. To allow easier orientation, all pictures also contain several exact orbits.

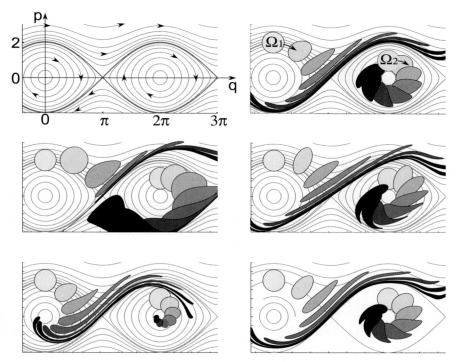

Fig. 6.2. Results of different time integration schemes for Hamiltonian flows in phase space $(\mathbf{q}, \mathbf{p})$, for the example of a mathematical pendulum. $\delta t = \pi/4$. Flow (top left), exact solution (top right) with the initial positions of the disks Ω_1 and Ω_2, explicit (middle left), implicit (bottom left), and symplectic (middle right) Euler method, Störmer-Verlet method (bottom right). The disks of initial values (light color) are deformed time step by time step (successively darker colors).

One can clearly see that the volume of the disk in phase space is growing for the explicit Euler method, and is shrinking for the implicit Euler method. Both methods are therefore not symplectic. In contrast, the symplectic Euler method (6.12) and the Störmer-Verlet method (3.22) conserve the volume in phase space and are therefore symplectic.

For one-dimensional systems, symplecticity is even equivalent to energy conservation. This is however not the case for higher-dimensional systems. In addition to the conservation of the physical structure of the Hamiltonian, the symplecticity of a time integration scheme has another consequence which is important for the interpretation of the numerical error. Backward error analysis interprets a computed trajectory as the exact solution of a slightly perturbed differential equation.[8] If the time integration scheme is symplectic, this perturbed differential equation can be represented again by a Hamiltonian $\tilde{\mathcal{H}}$. We can therefore write

$$\tilde{\mathcal{H}}(\mathbf{q},\mathbf{p}) = \mathcal{H}(\mathbf{q},\mathbf{p}) + \delta t \mathcal{H}_1(\mathbf{q},\mathbf{p}) + \delta t^2 \mathcal{H}_2(\mathbf{q},\mathbf{p}) + \dots ,$$

where the perturbed Hamiltonian $\tilde{\mathcal{H}}$ is symplectic if the original Hamiltonian is smooth[9] [78, 292, 296, 510]. However, with $\tilde{\mathcal{H}} \neq \mathcal{H}$ this implies that symplectic methods in general do not conserve energy exactly [201] if they do not reproduce the exact solution. But since the numerically computed trajectory is an exact solution of the system described by $\tilde{\mathcal{H}}$, the conserved quantities $\tilde{A}$ of $\tilde{\mathcal{H}}$ are conserved on the computed trajectory. If the perturbed Hamiltonian $\tilde{\mathcal{H}}$ and the original Hamiltonian $\mathcal{H}$ differ only slightly and if $\tilde{\mathcal{H}} \approx \mathcal{H}$ implies $\tilde{A} \approx A$, it follows that the numerically obtained A, which correspond to the conserved quantity of the original Hamiltonian $\mathcal{H}$, is at least approximately conserved along the computed trajectory. Hence, one has to show that the perturbed Hamiltonian $\tilde{\mathcal{H}}$ and the original Hamiltonian $\mathcal{H}$ differ only slightly. In special cases, the perturbed Hamiltonian $\tilde{\mathcal{H}}$ as well as its conserved quantities $\tilde{A}$ can be constructed explicitly. However, in general, the perturbed Hamiltonian $\tilde{\mathcal{H}}$ as well as other conserved quantities from the perturbed system are not available, which restricts the usefulness of backward error analysis.

Errors in Long Time Integration. Often some observable quantity A is not conserved exactly but stays close to an unknown value $\tilde{A}$. In addition, the observed A usually oscillates. Therefore, it is more sensible to consider appropriate time averages of A instead of the value of A along the trajectory of the system. The ergodic hypothesis states that, roughly speaking, an average over a long period of time corresponds to an average over statistical ensembles, see also Section 3.7.2. If the numerical method is symplectic, one

---

[8] The forward error analysis (6.8) directly considers the distance of the computed trajectory from the exact solution.

[9] This is the case for the potentials from Chapter 2, but not for potentials which are truncated or modified as in Section 3.7.3.

can hope, because of the conservation of the structure of the system by the numerical method, that one can observe some kind of "numerical" ergodicity. Thus, in a simulation that is run for a sufficiently long time, the numerically computed trajectory comes arbitrarily close to all points in phase space. Under this assumption, one can prove error estimates for exponentially large integration times. For the time interval from t_0 to

$$t_{\text{end}} = t_0 + C_1 \cdot e^{C_2/\delta t} \cdot \delta t$$

and correspondingly exponentially many time steps $n = C_1 \cdot e^{C_2/\delta t}$, the difference between the time averages of a conserved quantity for the computed trajectory and for the exact trajectory is bounded by

$$\|\langle A_{\mathbf{q}^1,\ldots,\mathbf{q}^n}\rangle_\tau - \langle A_{\mathbf{q}(t)}\rangle_\tau\| \leq C \cdot \delta t^p \tag{6.13}$$

with p being the order of the method from (6.6), see [78, 293, 295, 296, 377, 510]. Thus the error of the computed averages can be made arbitrarily small by decreasing the time step size δt. One can also try to bound the difference of the perturbed $\tilde{\mathcal{H}}$ to the original $\mathcal{H}$. These estimates imply the numerical stability of symplectic time integrators: Depending on the time step size δt, the accuracy of the computed averages of conserved quantities is high also for long integration times. We have already observed such a behavior in the simple experiment in Figure 6.1.

If δt is chosen to be larger, but below the stability threshold and thereby far enough from the resonance frequencies of the system[10], we still obtain reasonable values as results of our numerical experiments. Now, we can no longer successfully perform a backward error analysis. But at least an analysis of a model problem is often possible. As an example let us consider a simple chain of springs (the Fermi-Pasta-Ulam problem [222, 295, 296]) in which springs are alternatingly stiff and soft. The springs are modeled by harmonic potentials. In addition, the potentials for the soft springs are perturbed by a fourth-order term. Then, one can show that a certain class of integration methods does conserve energy on average over long times according to (6.13), but the exchange of energy is not represented correctly because of the slow linear terms in the energy. In this context, the time-reversibility of the almost symplectic methods is of special importance.[11]

So-called stable and quasi-periodic orbits are further aspects to be considered in the study of errors of computed trajectories. According to KAM

---

[10] Resonance is the tendency of a system to oscillate with high amplitude when excited by energy at a certain frequency. This frequency is also called eigenfrequency of the system.

[11] A method is called time-reversible or symmetric if a change of sign in time from t to $-t$ and therefore also in the momentum from $\mathbf{p}$ to $-\mathbf{p}$ allows to follow the numerical trajectory backwards. A time integration scheme such as the Störmer-Verlet method is time-reversible up to rounding errors. The properties of symplecticity and time-reversibility are independent of each other.

theory[12] one can find under certain conditions orbits in phase space which are stable against perturbations [49]. For these orbits, the exponential estimate for the error (6.8) is too pessimistic. Instead, the growth of the error is insignificant on the orbit and small in a neighborhood of the orbit. One merely obtains an estimate of $\mathcal{O}(t)$ or, in case of the two orbits drifting apart, an estimate of $\mathcal{O}(t^2)$, if the initial conditions lie on such a stable orbit. On the other hand, with initial conditions in other parts of the phase space, also chaotic behavior and therefore exponential growth of the error may occur.

Errors in the Force Computation. Up to now we have assumed that the time integration is performed in exact arithmetic. In practice however, the effect of rounding errors has to be taken into account. Furthermore, the consequences of approximations in the computation of the forces must be considered, for instance those caused by the truncation of the potentials in the linked cell method. The estimate (6.8) is valid also for such types of errors: They can lead to an exponential growth of the total error in time.

In general one can hope that the integrator remains symplectic if the *rounding errors* in its implementation are uncorrelated and therefore at least the estimate for the energy averages (6.13) still holds. But if the rounding errors depend on time or are correlated between different time steps, the computation of the potential $V(\mathbf{x}^n)$ may depend on further parameters such as $\mathbf{x}^{n-1}$. The resulting method is then no longer symplectic and one observes deviations from the averages.[13]

A further source of errors is any *approximation* in the force computation made necessary by the complexity of the evaluation of the original force term. Examples for such approximations are given by the truncation of the potentials at a radius r_{cut} in the linked cell method or the techniques still to be described in Chapter 7 (grid based methods) and Chapter 8 (tree methods) for the approximate evaluation of long-range forces. How do such approximations change the energy and the momentum of the system beyond the error estimates according to (6.13) and (6.8)? Only little is known about this. If we modify the potential in the linked cell method in such a way that it is exactly zero outside of the cutoff radius, then the truncation in (3.30) removes only terms $\mathbf{F}_{ij}$ with $\mathbf{F}_{ij} = 0$ from the summation, as presented in Section 3.7.3.[14] All other changes affect the computation of the energy so that we can no longer make any general statements about the long-time behavior

---

[12] Kolmogorov and later Arnold and Moser [49].

[13] To confront this problem, one can define ergodicity also on lattices with finitely many points. Then, all orbits consist of finitely many points in a discrete phase space. In the algorithm this is implemented by an appropriate rounding to the lattice points at each time step [201, 565]. If the rounding errors are small enough compared to the lattice constant (the distance between the points on the lattice), they disappear in the accuracy of the resulting integer arithmetic.

[14] One has to ensure the differentiability of the potential to preserve an estimate of type (6.13).

of the computed energy. This is somewhat different for the momentum. As long as we always include both $\mathbf{F}_{ij}$ and $\mathbf{F}_{ji}$ in the summation, Newton's third law implies that the total momentum is not changed. In this way, even huge errors caused by the truncation in (3.30) do not affect the conservation of the total momentum.

All in all we have different options to interpret the long-time behavior of molecular systems. First, we observed that even the smallest perturbations and errors can be amplified exponentially in time, compare (6.8) and Figure 6.1. Symplectic time integration schemes and the additional assumption of "numerical ergodicity" imply at least the conservation of certain time averages of conserved quantities. An alternative approach using KAM theory shows that, depending on the initial values, the physical process can still be stable over long times. Altogether this poses the question how far any of these additional assumptions needed for long-time stability are satisfied for realistic molecular dynamical simulations. Many real chemical and biological systems are known to be stable against various kinds of perturbations in the initial conditions and in the trajectories. This offers at least some hope that the corresponding numerical simulations could be stable as well. But as far as we know, nothing more definite can be said at the moment. Also, the effect of rounding errors and errors in the force computations on the long-term stability of a simulation is not yet sufficiently understood.

6.2 Symplectic Time Integration Methods

Methods of higher order p come to mind immediately when one thinks of the efficient integration of ordinary differential equations. There are various classes of methods such as Runge-Kutta methods, extrapolation methods, and general multi-step methods that provide higher orders in δt. We have seen in the last section that additional properties such as symplecticity and time-reversibility are important for the long-time behavior of the results of the simulation. Furthermore, to obtain a fast algorithm, it is important to keep the number of expensive force computations as small as possible. This excludes a number of implicit methods which require many evaluations of the force in the iterative solution of the resulting nonlinear systems of equations.

Multi-Step Methods. In contrast to one-step methods (6.5), multi-step methods use not only the previous point $(\mathbf{q}^n, \mathbf{p}^n, t_n)$ from the trajectory, but a longer sequence of points $(\mathbf{q}^{n-j}, \mathbf{p}^{n-j}, t_{n-j})$ with $j = 0, \ldots, s$ from the trajectory to compute a new point $(\mathbf{q}^{n+1}, \mathbf{p}^{n+1}, t_{n+1})$. In this respect, a preferred class of methods in the area of molecular dynamics is that of backward differentiation formulas (BDF) [247]. The definition of symplecticity (6.9) cannot be applied directly to multi-step methods, since we have to study a mapping of several pairs $(\mathbf{q}^i, \mathbf{p}^i)$ to the new value $(\mathbf{q}^{n+1}, \mathbf{p}^{n+1})$. However, multi-step methods can be interpreted as one-step methods in a higher-dimensional product space and symplecticity (6.9) can be generalized to such a setting. Then, it can be shown that the usual multi-step methods of order higher than the mid-point formula (which is of order two) cannot be symplectic [294, 296]. It is also easy to see that it is difficult to construct such methods which are both explicit and time-reversible since this would require the same number of force evaluations in the future δt and the past $-\delta t$. Therefore, one is left essentially only with one-step methods that could have the desired property of symplecticity. One-step methods can be written as Runge-Kutta methods. However, symplectic Runge-Kutta methods are in general implicit [297] and require the expensive solution of nonlinear systems of equations. This is the reason why we consider in the following more specific and less general time integration schemes for problems in molecular dynamics. They possess at least some of the properties of the symplectic, time-reversible and explicit Störmer-Verlet method.

Time Splitting Methods. One possibility to construct general time integration schemes is based on the idea of operator splitting. To this end, the right hand side of the differential equation is split into several parts and the parts are integrated separately. In our case we split the Hamiltonian from (6.2) into two parts $\mathcal{H}^1(\mathbf{q}, \mathbf{p})$ and $\mathcal{H}^2(\mathbf{q}, \mathbf{p})$, i.e. $\mathcal{H}(\mathbf{q}, \mathbf{p}) = \mathcal{H}^1(\mathbf{q}, \mathbf{p}) + \mathcal{H}^2(\mathbf{q}, \mathbf{p})$. If we denote the integration schemes for the two parts by $\Psi^1_{\delta t}(\mathbf{q}, \mathbf{p})$ and $\Psi^2_{\delta t}(\mathbf{q}, \mathbf{p})$, the time splitting method corresponds to their subsequent application, i.e.

$$\Psi_{\delta t}(\mathbf{q}, \mathbf{p}) := \Psi^2_{\delta t} \circ \Psi^1_{\delta t} (\mathbf{q}, \mathbf{p}) := \Psi^2_{\delta t}(\Psi^1_{\delta t}(\mathbf{q}, \mathbf{p})). \qquad (6.14)$$

By construction, this approach leads only to a first order method regardless of the order of the integration schemes for the two parts. A combination of the integration schemes for the parts in the way

$$\Psi_{\delta t}(\mathbf{q}, \mathbf{p}) := \Psi^1_{\delta t/2} \circ \Psi^2_{\delta t} \circ \Psi^1_{\delta t/2} (\mathbf{q}, \mathbf{p}) \qquad (6.15)$$

can lead to a second order method instead. This construction is called Strang splitting[15] and can also be derived from so-called Lie-Trotter factorizations [617].

If we use the natural splitting of the Hamiltonian $\mathcal{H}(\mathbf{q}, \mathbf{p}) = T(\mathbf{p}) + V(\mathbf{q})$ from (6.3) and solve the resulting subproblems with the explicit Euler method (6.10), we obtain with

$$\Psi^T_{\delta t}(\mathbf{q}^n, \mathbf{p}^n) = \begin{pmatrix} \mathbf{q}^n + \delta t \, m^{-1} \mathbf{p}^n \\ \mathbf{p}^n \end{pmatrix},$$

$$\Psi^V_{\delta t}(\mathbf{q}^n, \mathbf{p}^n) = \begin{pmatrix} \mathbf{q}^n \\ \mathbf{p}^n - \delta t \nabla_{\mathbf{q}} V(\mathbf{q}^n) \end{pmatrix},$$

---

[15] This splitting was proposed by Strang originally for partial differential equations in [590]. There, the different parts of the splitting correspond to differential operators in the different coordinate directions.

from $\Psi_{\delta t}^T \circ \Psi_{\delta t}^V$ exactly the symplectic Euler method (6.12) and from $\Psi_{\delta t/2}^V \circ \Psi_{\delta t}^T \circ \Psi_{\delta t/2}^V$ the Störmer-Verlet method (3.22). It can easily be verified that the mappings $\Psi_{\delta t}^T$ and $\Psi_{\delta t}^V$ are symplectic. This implies that the Störmer-Verlet method and the symplectic Euler method are also symplectic as compositions of symplectic mappings.

Furthermore, there are splitting formulas of higher order of the form

$$\Psi_{\delta t} := \Psi_{b_k \cdot \delta t}^2 \circ \Psi_{a_k \cdot \delta t}^1 \circ \ldots \circ \Psi_{a_2 \cdot \delta t}^1 \circ \Psi_{b_1 \cdot \delta t}^2 \circ \Psi_{a_1 \cdot \delta t}^1 \tag{6.16}$$

with coefficients a_j and b_j, $j = 1, .., k$. Such formulas use $a_j \cdot \delta t$ and $b_j \cdot \delta t$ as local time steps [235, 296, 422].

Composition Methods. It is also possible to construct time-reversible methods from symplectic methods. One then first derives the adjoint method Ψ^* from the original time integration method Ψ by reversing the time t. In this way, for example, the explicit Euler method (6.10) turns into the implicit Euler method (6.11) and vice versa. Then, the two methods Ψ^* and Ψ are used successively, i.e., they are composed to $\Psi^* \circ \Psi$ or $\Psi \circ \Psi^*$. The resulting methods are again symplectic and additionally also time-reversible [296]. However, they need twice as many function evaluations as the original method Ψ alone.

One can also subsequently use a time integration scheme with different time step sizes to construct new higher order methods. Following Yoshida [676], we can define a so-called composition method by

$$\tilde{\Psi}_{\delta t} := \Psi_{a_k \cdot \delta t} \circ \ldots \circ \Psi_{a_2 \cdot \delta t} \circ \Psi_{a_1 \cdot \delta t} \tag{6.17}$$

with coefficients a_j, $j = 1, \ldots, k$. For example, with the three values

$$a_1 = a_3 = \frac{1}{2 - 2^{1/(p+1)}}, \qquad a_2 = -\frac{2^{1/(p+1)}}{2 - 2^{1/(p+1)}}, \tag{6.18}$$

we obtain from a method Ψ of even order p a new method $\tilde{\Psi}$ of order $p + 2$ [296, 594, 676]. Such a construction can be applied repeatedly and will then lead to methods of orders $p + 4$, $p + 6$, ...

One disadvantage of (6.18) and similar sets of coefficients is that negative time steps can occur (here with a_2), and that force evaluations outside of the time interval $[t_n, t_{n+1}]$ may be required. Note that there are special composition methods which at least ensure that all force evaluations occur within the time interval of the actual time step. Note furthermore that splitting and composition methods are closely related, as are their coefficients [421, 422].

Partitioned Runge-Kutta Methods. Symplectic Runge-Kutta methods are often expensive and in general not explicit. But the splitting of the Hamiltonian from (6.3) can be used to systematically construct explicit symplectic partitioned Runge-Kutta methods. A partitioned Runge-Kutta method for (6.2) can be written as

$$Q_i = \mathbf{q}^n + \delta t \sum_{j=1}^{s} a_{ij} \nabla_{\mathbf{p}} \mathcal{H}(Q_j, P_j),$$

$$P_i = \mathbf{p}^n - \delta t \sum_{j=1}^{s} \hat{a}_{ij} \nabla_{\mathbf{q}} \mathcal{H}(Q_j, P_j),$$

$$\mathbf{q}^{n+1} = \mathbf{q}^n + \delta t \sum_{i=1}^{s} b_i \nabla_{\mathbf{p}} \mathcal{H}(Q_i, P_i), \tag{6.19}$$

$$\mathbf{p}^{n+1} = \mathbf{p}^n - \delta t \sum_{i=1}^{s} \hat{b}_i \nabla_{\mathbf{q}} \mathcal{H}(Q_i, P_i),$$

with the intermediate values Q_i and P_i and the coefficients a_{ij}, $\hat{a}_{ij}$, b_i, and $\hat{b}_i$ that specify the method. These coefficients and the number s of stages must be chosen appropriately such that the resulting method possesses desired properties. For instance, if the resulting system is of triangular form, the time integration scheme is explicit and the system of equations (6.19) is easy to solve for the unknowns Q_i and P_i. With a separable Hamiltonian (6.3), system (6.19) reduces to

$$Q_i = \mathbf{q}^n + \delta t \sum_{j=1}^{s} a_{ij} \nabla_{\mathbf{p}} T(P_j),$$

$$P_i = \mathbf{p}^n - \delta t \sum_{j=1}^{s} \hat{a}_{ij} \nabla_{\mathbf{q}} V(Q_j),$$

$$\mathbf{q}^{n+1} = \mathbf{q}^n + \delta t \sum_{i=1}^{s} b_i \nabla_{\mathbf{p}} T(P_i), \tag{6.20}$$

$$\mathbf{p}^{n+1} = \mathbf{p}^n - \delta t \sum_{i=1}^{s} \hat{b}_i \nabla_{\mathbf{q}} V(Q_i).$$

The coefficients can now be computed by collocation based on quadrature formulas.[16] We start with a Gauss-Lobatto formula in which one quadrature point is located at each end of the interval and the other points are chosen so that the quadrature is of the highest possible order. We choose the so-called Lobatto IIIA-IIIB pairs a_{ij}, $\hat{a}_{ij}$, b_i, and $\hat{b}_i$, see Tables 6.2, 6.3, and 6.4. Usually, the coefficients are arranged in Butcher tableaus [297] from which the matrices a_{ij} and $\hat{a}_{ij}$ and the vectors b_i and $\hat{b}_i$ can be read off.

In the two-stage Lobatto IIIA-IIIB pair from Table 6.2, the quadrature points lie only on the ends of $[t_n, t_{n+1}]$. Inserting the values from Table 6.2 in (6.20), we obtain $Q_1 = \mathbf{q}^n$. With one evaluation of the force, one can then obtain $P_1 \equiv P_2$. Subsequently, one determines Q_2 and $\mathbf{q}^{n+1}$ and $\mathbf{p}^{n+1}$. Thus, this method is exactly the Störmer-Verlet method (3.22).

Methods of higher order can be constructed similarly using Lobatto quadrature formulas of higher order. With the coefficients from Table 6.3,

---

[16] For ordinary differential equations, the concept of collocation consists in searching for a polynomial of degree s whose derivative coincides at s given points with the vector field of the differential equation.

$$
\begin{array}{c|cc}
a & 0 & 0 \\
& 1/2 & 1/2 \\
\hline
b & 1/2 & 1/2
\end{array}
\qquad\qquad
\begin{array}{c|cc}
\hat{a} & 1/2 & 0 \\
& 1/2 & 0 \\
\hline
\hat{b} & 1/2 & 1/2
\end{array}
$$

Table 6.2. Partitioned Runge-Kutta method based on the Lobatto IIIA-IIIB pair with two stages $s = 2$. This corresponds to the Störmer-Verlet method for the separable Hamiltonian.

$$
\begin{array}{c|ccc}
a & 0 & 0 & 0 \\
& 5/24 & 1/3 & -1/24 \\
& 1/6 & 2/3 & 1/6 \\
\hline
b & 1/6 & 2/3 & 1/6
\end{array}
\qquad
\begin{array}{c|ccc}
\hat{a} & 1/6 & -1/6 & 0 \\
& 1/6 & 1/3 & 0 \\
& 1/6 & 5/6 & 0 \\
\hline
\hat{b} & 1/6 & 2/3 & 1/6
\end{array}
$$

Table 6.3. Partitioned Runge-Kutta method based on the Lobatto IIIA-IIIB pair with three stages.

$$
\begin{array}{c|cccc}
a & 0 & 0 & 0 & 0 \\
& \frac{11+\sqrt{5}}{120} & \frac{25-\sqrt{5}}{120} & \frac{25-13\sqrt{5}}{120} & \frac{-1+\sqrt{5}}{120} \\
& \frac{11-\sqrt{5}}{120} & \frac{25+13\sqrt{5}}{120} & \frac{25+\sqrt{5}}{120} & \frac{-1-\sqrt{5}}{120} \\
& 1/12 & 5/12 & 5/12 & 1/12 \\
\hline
b & 1/12 & 5/12 & 5/12 & 1/12
\end{array}
$$

$$
\begin{array}{c|cccc}
\hat{a} & 1/12 & \frac{-1-\sqrt{5}}{24} & \frac{-1+\sqrt{5}}{24} & 0 \\
& 1/12 & \frac{25+\sqrt{5}}{120} & \frac{25-13\sqrt{5}}{120} & 0 \\
& 1/12 & \frac{25+13\sqrt{5}}{120} & \frac{25-\sqrt{5}}{120} & 0 \\
& 1/12 & \frac{11-\sqrt{5}}{24} & \frac{11+\sqrt{5}}{24} & 0 \\
\hline
\hat{b} & 1/12 & 5/12 & 5/12 & 1/12
\end{array}
$$

Table 6.4. Partitioned Runge-Kutta method based on the Lobatto IIIA-IIIB pair with four stages.

one obtains a three-stage Runge-Kutta scheme and with the values from Table 6.4, one obtains a four-stage scheme. In general such a method with s stages has order $p = 2s - 2$. All such methods are symplectic. One still obtains $Q_1 = \mathbf{q}^n$ and analogously Q_s as last term, but one has to solve a system of equations for $Q_2, \ldots, Q_{s-1}$ and $P_1, \ldots, P_s$. This also implies that the potential V has to be evaluated more often which makes these time integration schemes significantly more expensive.

There are other symplectic partitioned Runge-Kutta methods that are explicit. An example with three stages is given in Table 6.5. However, it is only accurate up to third order [297, 532]. These classes of symplectic methods are called Runge-Kutta-Nyström methods since they can be written directly for differential equations of second order $\ddot{\mathbf{x}} = f(\mathbf{x}, \dot{\mathbf{x}}, t)$ instead of treating $\mathbf{x}$ and $\dot{\mathbf{x}}$ separately in a system of first order.

Altogether, we now can construct symplectic, time-reversible and explicit methods of higher order. However, the question remains if these methods really lead to more efficient simulations than the Störmer-Verlet method.

	a					$\hat{a}$			
	0	0	0			7/24	0	0	
	2/3	0	0			7/24	3/4	0	
	2/3	$-2/3$	0			7/24	3/4	$-1/24$	
b	2/3	$-2/3$	1		$\hat{b}$	7/24	3/4	$-1/24$	

Table 6.5. An explicit, partitioned, symplectic Runge-Kutta method of third order according to Ruth [532].

For one, the force terms have to be evaluated much more often for these other methods. In addition, molecular dynamical simulations usually use time steps δt which are absolutely small but nevertheless large compared to the frequencies of the system[17] and thus the asymptotic order p of the method does not necessarily determine the local accuracy. For integrations over long times, it may even be advantageous to use the Störmer-Verlet method with a smaller time step instead of a more expensive method of higher order.

There are also attempts to construct methods with variable time steps δt for the long-time integration of Hamiltonian flows. For a number of problems it can be advantageous not to use the same time step δt for the entire calculation but to use a larger time step whenever possible. To this end, often good heuristics can be given which locally adapt the time step to the simulated problem. Note however that additional efforts are necessary to keep the resulting method time-reversible or stable in the sense of (6.13), see [136, 134, 298, 376].

6.3 Multiple Time Step Methods – the Impulse Method

Bonded and non-bonded interactions occur at different time scales. The bonded interactions in molecules which correspond to bond, angle, and torsion forces are modeled as vibrations around an equilibrium position.[18] The time integration scheme has to resolve the fastest oscillations. The very high frequencies of the bonds enforce the use of very small time steps to capture also the fastest motions. The size of the time step δt is therefore limited by the highest frequency f_{max} occurring in the considered system, i.e. $\delta t \ll 1/f_{max}$.

---

[17] The situation is different in celestial mechanics. There, often systems with a relatively small number of particles are studied, but higher accuracy is needed.

[18] The dynamics of such a system can be represented approximately as the superposition of harmonic oscillations, so-called eigenmodes. The frequencies of these modes can be computed, at least in principle. Especially for biological systems it turns out that the eigenmodes evolve on separate time scales. Bonded forces correspond to high-frequency motions, non-bonded forces correspond to motions of lower frequency. The fastest interactions limit the time step, in the case of the Störmer-Verlet method to typically 1 fs.

In general, the vibrations caused by the bond potentials are significantly faster than the vibrations caused by the angle potentials. The torsion potentials are responsible for larger deformations of the molecules and therefore act on an even slower time scale. Therefore, we are looking for time integration schemes that can exploit the different time scales and hence allow for different time steps for different classes of vibration modes depending on their frequencies. In such a way, larger time steps can be used for most of the interactions in Newton's equations. A typical representative of such methods is the impulse method, also called r-Respa method [274, 275, 620, 621, 623], which we consider now in more detail.

The idea is the following: We construct a second order splitting of the time integration scheme (6.15). To this end, we assume that the forces in Newton's equations of motion act on two time scales and write (6.1) as

$$m\ddot{\mathbf{x}} = \mathbf{F}(\mathbf{x}) = \mathbf{F}^{\text{short}}(\mathbf{x}) + \mathbf{F}^{\text{long}}(\mathbf{x})$$

with $\mathbf{F}^{\text{short}}(\mathbf{x}) = -\nabla_{\mathbf{x}} V^{\text{short}}(\mathbf{x})$ and $\mathbf{F}^{\text{long}}(\mathbf{x}) = -\nabla_{\mathbf{x}} V^{\text{long}}(\mathbf{x})$. The decomposition should be chosen in such a way that, with an appropriate stable time discretization, the time step δt^{long} for the slow term is larger than the time step δt^{short} for the fast term.[19]

The reduced problem

$$m\ddot{\mathbf{x}} = \mathbf{F}^{\text{long}}(\mathbf{x})$$

now allows for a numerically stable integration with a substantially larger time step than feasible for the whole system, typically a factor of about 4 to 5 [334]. The idea is then to evaluate $\mathbf{F}^{\text{long}}$ less often and to incorporate those evaluations in an appropriate way in the integration of the whole system.

The splitting method is based on the decomposition of the Hamiltonian into

$$\mathcal{H}(\mathbf{q}, \mathbf{p}) = V^{\text{long}}(\mathbf{q}) + \left(V^{\text{short}}(\mathbf{q}) + T(\mathbf{p})\right)$$

with the two parts V^{long} and $V^{\text{short}} + T$. Thus, the kinetic energy is treated together with the forces $\mathbf{F}^{\text{short}}$. For the two parts of the splitting, we need two appropriate time integration schemes $\Psi_{\delta t}^{\text{long}}$ and $\Psi_{\delta t}^{\text{short}}$. Here, the Störmer-Verlet method (3.22) can be directly applied for $\Psi_{\delta t}^{\text{long}}$, but $\Psi_{\delta t}^{\text{short}}$ must still be properly chosen.

This approach has been proposed for a problem from celestial mechanics in 1982 in [663], in which the reduced problem for V^{short} consists of the motion of the planets around the sun according to Kepler's theory, and V^{long} describes the interactions between the planets. The application of the method to problems in molecular dynamics has been studied in [340]. If the reduced system

---

[19] Here, we us the terms *long* and *short* to describe the different parts of the system that oscillate on different time scales. This is in anticipation of the methods in Chapters 7 and 8 where we split the forces in short and long-range parts. These parts then do indeed act on different time scales.

$$m\ddot{\mathbf{x}} = \mathbf{F}^{\mathrm{short}}(\mathbf{x})$$

can be solved analytically, the time integration scheme $\Psi_{\delta t}^{\mathrm{short}}$ corresponds to an exact solution. Then, the long-range integration scheme is just corrected appropriately in every time step by $\Psi_{\delta t}^{\mathrm{short}}$.

In general, however, the reduced problem for $\Psi_{\delta t}^{\mathrm{short}}$ cannot be integrated analytically. Therefore, we substitute for $\Psi_{\delta t}^{\mathrm{short}}$ a sequence of Störmer-Verlet steps with a smaller time step $\delta t/M$.[20] We obtain

$$\Psi_{\delta t} = \Psi_{\delta t/2}^{\mathrm{long}} \circ \Psi_{\delta t}^{\mathrm{short}} \circ \Psi_{\delta t/2}^{\mathrm{long}}$$

$$= \Psi_{\delta t/2}^{V^{\mathrm{long}}} \circ \left(\Psi_{\delta t/M}^{V^{\mathrm{short}}+T} \circ \Psi_{\delta t/M}^{V^{\mathrm{short}}+T} \circ \ldots \circ \Psi_{\delta t/M}^{V^{\mathrm{short}}+T} \right) \circ \Psi_{\delta t/2}^{V^{\mathrm{long}}} . \quad (6.21)$$

Algorithm 6.1 Impulse Method based on Velocity-Störmer-Verlet Scheme

---

```
// set initial values x⁰, v⁰
for (int n=0; n<t_end/δt; n++) {
```
$\tilde{\mathbf{v}}^n = \mathbf{v}^n + \frac{1}{2}\delta t \, m^{-1} \, \mathbf{F}^{\mathrm{long}}(\mathbf{x}^n);$
```
  for (int j=1; j<=M; j++) {
```
$\mathbf{x}^{n+j/M} = \mathbf{x}^{n+(j-1)/M} + \frac{1}{M}\delta t \, \tilde{\mathbf{v}}^{n+(j-1)/M} +$
$\qquad\qquad\qquad \frac{1}{2}(\frac{\delta t}{M})^2 \, m^{-1} \, \mathbf{F}^{\mathrm{short}}(\mathbf{x}^{n+(j-1)/M});$

$\tilde{\mathbf{v}}^{n+j/M} = \tilde{\mathbf{v}}^{n+(j-1)/M} +$
$\qquad\qquad \frac{1}{2M}\delta t \, m^{-1} \left(\mathbf{F}^{\mathrm{short}}(\mathbf{x}^{n+(j-1)/M}) + \mathbf{F}^{\mathrm{short}}(\mathbf{x}^{n+j/M}) \right);$
```
  }
```
$\mathbf{v}^{n+1} = \tilde{\mathbf{v}}^{n+1} + \frac{1}{2}\delta t \, m^{-1} \, \mathbf{F}^{\mathrm{long}}(\mathbf{x}^{n+1});$
```
}
```

In Algorithm 6.1 we use the Velocity-Störmer-Verlet method for both $\Psi^{V^{\mathrm{short}}+T}$ and $\Psi^{V^{\mathrm{long}}}$ in (6.21). Then, we obtain again a method that is both time-reversible and symplectic. This method is the so-called impulse/r-Respa time integration method. It was first proposed in [274, 275, 620, 621, 623]. Earlier papers in this direction were [231, 591]. The method can be directly generalized to more than just two different time steps.

Note that it is enough to evaluate $\mathbf{F}^{\mathrm{long}}$ just once in each long time step since the positions are modified only in the inner loop in Algorithm 6.1. For each such evaluation, we need M evaluations of $\mathbf{F}^{\mathrm{short}}$. In the implementation, both forces must be computed separately and then be stored for the use in the corresponding next step or substep. The parallelization of the method does not pose any new difficulties if the evaluation of the forces has already been parallelized.

In general, the force F can be split into the two terms $\mathbf{F}^{\mathrm{short}}$ and $\mathbf{F}^{\mathrm{long}}$ in different ways. One possibility is to split the forces according to their types

---

[20] For the trivial choice $V^{\mathrm{long}} = 0$ we obtain again the Störmer-Verlet method (3.22) with smaller time step $\delta t/M$.

so that for instance the bond and angle forces are included in the short-range part and the torsion and Lennard-Jones forces are included in the long-range part. Another possibility is, at least if there are large differences in the masses and therefore in the velocities of the particles, to split the forces into long-range or short-range according to the particle types involved. Thus, in a situation with particles of mass one and m with $m \gg 1$ one obtains a factor M of order $\sqrt{m}$ between δt^{long} and δt^{short}, see [622]. The forces can also be split into short-range and long-range forces according to distance. For this approach, forces such as Lennard-Jones or Coulomb forces have to be split into short-range and long-range parts. The different parts can be defined for instance using the cutoff function from Section 3.7.3, see [620]. Furthermore, the long-range forces can be approximated with various methods. Here, the P^3M method or the Ewald summation method from the next chapter could be appropriate [492, 495, 690]. The tree-based methods of Chapter 8 permit the introduction of additional long-range scales and coarser time steps [226, 480, 689].

The impulse method allows for a time step δt^{long} of about 4 fs for biomolecules. Studies show that time steps of 5 fs or larger cannot be attained in this way. Furthermore, resonances may occur in the method that may lead to instabilities if the frequency of the slow force evaluation is equal or close to an eigenmode of the system. An analogous problem occurs for time steps that are about half of the period of the fastest eigenmode. The "mollified" impulse method MOLLY [245] tries to avoid these problems. To this end, the potential is modified by defining it on time-averaged positions, with the time averaging taking the high-frequency vibrations into account. Different averages and extensions have been tested in [334]. There, a stable time step of about 7 fs could be attained. This is a respectable, yet somewhat disappointing result. The multiple time step method is more expensive than the simple Störmer-Verlet method and thus, depending on the particular implementation, an overall improvement of only about a factor of five in runtime is possible in practice.

Larger time steps are attainable using Langevin dynamics [318, 682, 683, 684]. In this approach, Newton's equation are augmented by friction and stochastic terms. In this way, the high-frequency modes are damped and special multiple time step methods with an even larger time step δt^{long} can be used. Here, time steps of 12 fs for water, 48 fs for a biomolecule in water, and 96 fs for a biomolecule in vacuum have been reported in [61, 62]. Further variants can be found in [548]. While Langevin dynamics methods allow for larger time steps sizes, the introduction of additional terms also changes the model substantially. This can however lead to completely different numerical results.

6.4 Constraints – the RATTLE Algorithm

In many cases, the energy transport between high-frequency and low-frequency degrees of freedom is very slow. Therefore, it is difficult to get close to an equilibrium within reasonable time. In such a case it is often advantageous to freeze the high-frequency degrees of freedom, i.e. the degrees of freedom for the bond and angle potentials. This means that the lengths of the chemical bonds and the values of the angles within the molecules are assumed to be fixed. In practice this is implemented using geometric constraints that tie bond lengths and angles to given values.[21] If the system contains degrees of freedom which are associated to those high frequencies but which only have small amplitudes or small effects on other degrees of freedom, the use of constraints allows to freeze those degrees of freedom without significant loss of accuracy. This approach works very well for bonds.

There are different methods to implement geometric constraints. One of these methods is the so-called symplectic RATTLE algorithm [43]. In this approach, the constraints are enforced by Lagrangian multipliers. Variants of this algorithm include for instance the older SHAKE algorithm [534], various methods that incorporate the constraints into the coordinate system, and a number of methods that work with soft constraints [509]. In the RATTLE algorithm, the Lagrangian multipliers are determined from a nonlinear system of equations which is solved iteratively. We first formulate the symplectic RATTLE algorithm for general Hamiltonians. Here, we follow [341, 508].

Constraints. The forces acting on each particle consist of physical forces and forces resulting from constraints. The latter forces preserve the structure of the molecule over time. We assume that the constraints are given in the form

$$\sigma^{ij}(\mathbf{x}) = \|\mathbf{x}_i - \mathbf{x}_j\|^2 - d_{ij}^2 = 0, \qquad (6.22)$$

where $\mathbf{x} = \{\mathbf{x}_1, \ldots, \mathbf{x}_N\}$ denotes the positions of the particles. Here, d_{ij} denotes the length at which the distance $r_{ij} = \|\mathbf{x}_i - \mathbf{x}_j\|$ between particles i and j is to be frozen. Such conditions are called holonomic constraints since they are independent of the time t. In total we use M different constraints. Furthermore, we denote by I the set of the index pairs of the particles involved, i.e. $(i, j) \in I$ with $i < j$ for each constraint. If the bond between atom i and j is to be frozen to the length r^0, we set $d_{ij} = r^0$. In the case of angles

---

[21] Note that the freezing of degrees of freedom changes the energy of the system. In certain circumstances, this can significantly change the results of the simulation. One method to compensate for these effects is the use of effective potentials for the remaining degrees of freedom, compare also Chapter 2. Such potentials compensate for the energy that is lost by the freezing of some degrees of freedom. Mathematically, effective potentials are derived by asymptotic analysis and homogenization, see [103, 105].

and distances between the atoms i, j, and k, for example, we have to freeze the distances between i and j, between i and k, and between j and k.

We choose an approach with Lagrangian multipliers λ^{ij} for all constraints $(i,j) \in I$, which leads to the following equations of motion of classical mechanics:

$$\begin{aligned}
m\ddot{\mathbf{x}} &= -\nabla_{\mathbf{x}}\left(V(\mathbf{x}) + \sum_{(i,j)\in I} \lambda^{ij}\sigma^{ij}(\mathbf{x}) \right) \\
&= -\nabla_{\mathbf{x}}V(\mathbf{x}) - \sum_{(i,j)\in I} \lambda^{ij}\nabla_{\mathbf{x}}\sigma^{ij}(\mathbf{x}) \\
&= \mathbf{F} + \mathbf{Z}.
\end{aligned} \tag{6.23}$$

In addition to the force $\mathbf{F} = -\nabla_{\mathbf{x}}V(\mathbf{x})$ on the particles that stems from the potential acting between the particles, now there is also the force

$$\mathbf{Z} = -\sum_{(i,j)\in I} \lambda^{ij}\nabla_{\mathbf{x}}\sigma^{ij}(\mathbf{x})$$

which stems from the constraints (6.22) that need to be satisfied. The additional M degrees of freedom λ^{ij} for $(i,j) \in I$ are to be chosen in such a way that all M constraints (6.22) are satisfied at each point in time.

In addition to (6.22), another condition is fulfilled, the so-called hidden constraint for the velocity $\dot{\mathbf{x}}$, that results from the time derivative of (6.22). This constraint is written as

$$0 = \dot{\sigma}^{ij}(\mathbf{x}) = 2\langle \mathbf{x}_i - \mathbf{x}_j, \dot{\mathbf{x}}_i - \dot{\mathbf{x}}_j \rangle \quad \forall (i,j) \in I, \tag{6.24}$$

with the inner product $\langle .,. \rangle$. The solution $\mathbf{x}(t)$ only remains on the manifold of admissible values with $\sigma^{ij}(\mathbf{x}) = 0$, $(i,j) \in I$, if the velocities $\dot{\mathbf{x}}(t)$ are tangential to the manifold and do not move particles away from this manifold.

Time Integration. We can rewrite the system (6.23) together with the constraints (6.22) in Hamiltonian form. Then, we can apply the time integration schemes from Section 6.2. To this end, we write the constraints as $\sigma(\mathbf{q}) = 0$ and the Lagrangian multipliers for time t_n as $\Lambda_n = (\lambda^{ij})_{ij\in I}^T$. Using for example the partitioned Runge-Kutta method (6.19), we obtain with $\mathbf{q} = \mathbf{x}$ and $\mathbf{p} = m\dot{\mathbf{x}}$ the system

$$\begin{aligned}
Q_i &= \mathbf{q}^n + \delta t \sum_{j=1}^{s} a_{ij} \nabla_{\mathbf{p}}\mathcal{H}(Q_j, P_j), \qquad \sigma(Q_i) = 0, \\
P_i &= \mathbf{p}^n - \delta t \sum_{j=1}^{s} \hat{a}_{ij} \left(\nabla_{\mathbf{q}}\mathcal{H}(Q_j, P_j) + \nabla_{\mathbf{q}}\sigma(Q_j)\Lambda_j \right), \\
\mathbf{q}^{n+1} &= \mathbf{q}^n + \delta t \sum_{i=1}^{s} b_i \nabla_{\mathbf{p}}\mathcal{H}(Q_i, P_i), \\
\mathbf{p}^{n+1} &= \mathbf{p}^n - \delta t \sum_{i=1}^{s} \hat{b}_i \left(\nabla_{\mathbf{q}}\mathcal{H}(Q_i, P_i) + \nabla_{\mathbf{q}}\sigma(Q_i)\Lambda_i \right).
\end{aligned} \tag{6.25}$$

The Lagrangian multipliers Λ_j have to be determined in such a way that all constraints are satisfied simultaneously. To this end, we first rewrite (6.25). For the separable Hamiltonian (6.3), this yields

$$Q_i = \mathbf{q}^n + \delta t \sum_{j=1}^{s} a_{ij} \nabla_{\mathbf{p}} T(P_j), \qquad \sigma(Q_i) = 0,$$

$$P_i = \mathbf{p}^n - \delta t \sum_{j=1}^{s} \hat{a}_{ij} \left(\nabla_{\mathbf{q}} V(Q_j) + \nabla_{\mathbf{q}} \sigma(Q_j) \Lambda_j \right),$$

$$\mathbf{q}^{n+1} = \mathbf{q}^n + \delta t \sum_{i=1}^{s} b_i \nabla_{\mathbf{p}} T(P_i),$$

$$\mathbf{p}^{n+1} = \mathbf{p}^n - \delta t \sum_{i=1}^{s} \hat{b}_i \left(\nabla_{\mathbf{q}} V(Q_i) + \nabla_{\mathbf{q}} \sigma(Q_i) \Lambda_i \right).$$

(6.26)

With the Lobatto IIIA-IIIB pair from Table 6.2, we obtain a two-stage Runge-Kutta method. In the absence of constraints, this method corresponds to the Störmer-Verlet method. With $Q_1 = \mathbf{q}^n$, $P_1 = P_2$, and the assumption that the initial values $\mathbf{q}^n$ already satisfy $\sigma(\mathbf{q}^n) = 0$, we arrive at the system of equations

$$P_2 = \mathbf{p}^n - \tfrac{\delta t}{2} \left(\nabla_{\mathbf{q}} V(\mathbf{q}^n) + \nabla_{\mathbf{q}} \sigma(\mathbf{q}^n) \Lambda_n \right),$$

$$Q_2 = \mathbf{q}^n + \delta t \nabla_{\mathbf{p}} T(P_2),$$

$$0 = \sigma(Q_2),$$

with the unknowns Q_2, P_2 and Λ_n. Substitution then yields a nonlinear system of equations in the unknown Λ_n

$$\sigma \left(\mathbf{q}^n + \delta t \nabla_{\mathbf{p}} T \left(\mathbf{p}^n - \frac{\delta t}{2} \left(\nabla_{\mathbf{q}} V(\mathbf{q}^n) + \nabla_{\mathbf{q}} \sigma(\mathbf{q}^n) \Lambda_n \right) \right) \right) = 0. \qquad (6.27)$$

From its solution Λ_n one can then obtain $\mathbf{q}^{n+1} = Q_2$.

To compute $\mathbf{p}^{n+1}$ from

$$\mathbf{p}^{n+1} = \mathbf{p}^n - \frac{\delta t}{2} \sum_{i=0}^{1} \nabla_{\mathbf{q}} V(\mathbf{q}^{n+i}) + \nabla_{\mathbf{q}} \sigma(\mathbf{q}^{n+i}) \Lambda_{n+i},$$

one additionally needs the Lagrangian multiplier Λ_{n+1} for the next time step. However, its value is not yet available at this point. Here, the hidden constraint (6.24) for the velocities and impulses comes into play. In the RATTLE algorithm, it leads to a further system of equations in $\mathbf{p}^{n+1}$ and μ_{n+1}

$$\mathbf{p}^{n+1} = \mathbf{p}^n - \tfrac{\delta t}{2} \nabla_{\mathbf{q}} \left(V(\mathbf{q}^n) + \sigma(\mathbf{q}^n) \Lambda_n + V(\mathbf{q}^{n+1}) + \sigma(\mathbf{q}^{n+1}) \mu_{n+1} \right),$$

$$0 = \dot{\sigma}(\mathbf{q}^{n+1}, m^{-1} \mathbf{p}^{n+1}),$$

(6.28)

in which μ_{n+1} plays the role of Λ_{n+1}. By way of substitution, we again obtain a system of equations in the unknown μ_{n+1}

$$\dot{\sigma}\left(\mathbf{q}^{n+1},\, m^{-1}\left(\mathbf{p}^n - \frac{\delta t}{2}\nabla_{\mathbf{q}}\left(V(\mathbf{q}^n) + V(\mathbf{q}^{n+1})\right)\right.\right. \tag{6.29}$$

$$\left.\left. + \sigma(\mathbf{q}^n)\Lambda_n + \sigma(\mathbf{q}^{n+1})\mu_{n+1}\right)\right)\right) = 0$$

with M equations and M unknowns. From its solution, one obtains $\mathbf{p}^{n+1}$ by substitution into (6.28).

Implementation. We now turn to the discussion of the practical implementation of the RATTLE algorithm. In each time step, the nonlinear systems of equations (6.27) and (6.29) for the Lagrange multipliers have to be solved. A direct solution of these systems is only possible in some very simple cases. Therefore, iterative methods for nonlinear systems of equations must normally be used. As in the Störmer-Verlet method, only one expensive force evaluation $\nabla_{\mathbf{q}}V(\mathbf{q}^{n+1})$ is needed in each time step, while the constraint function σ has to be evaluated more often in the iteration.

We again consider the special constraints (6.22). Their derivatives can be computed as

$$\nabla_{\mathbf{x}_l}\sigma^{ij} = \nabla_{\mathbf{x}_l}\left(\|\mathbf{x}_i - \mathbf{x}_j\|^2 - d_{ij}^2\right) = \begin{cases} 2(\mathbf{x}_i - \mathbf{x}_j), & \text{for } l = i, \\ 2(\mathbf{x}_j - \mathbf{x}_i), & \text{for } l = j, \\ 0, & \text{for } l \neq i, j. \end{cases} \tag{6.30}$$

This special structure can be exploited to simplify the systems of equations (6.27) and (6.29). Instead of using general iterative solvers for nonlinear equations such as Newton's method, in which the system is linearized in each step and the resulting linear system is solved, the RATTLE algorithm uses a simpler approach. Each iteration step consists of a loop over all constraints σ^{ij} that are not yet satisfied. For each unsatisfied constraint, a new approximation of the Lagrangian multiplier λ^{ij} is computed and the positions and momenta are updated according to the new approximation. Since the system is mostly close to equilibrium because the time step δt is small, this iteration usually converges.

We first consider the system of equations (6.27). With the precomputed auxiliary vectors

$$\tilde{\mathbf{p}}^{n+1/2} := \mathbf{p}^n - \frac{\delta t}{2}\left(\nabla_{\mathbf{q}}V(\mathbf{q}^n)\right),$$

$$\tilde{\mathbf{q}}^{n+1} := \mathbf{q}^n + \delta t\, m^{-1}\tilde{\mathbf{p}}^{n+1/2}, \tag{6.31}$$

the system (6.27) can be simplified to

$$\sigma\left(\tilde{\mathbf{q}}^{n+1} - \frac{\delta t^2}{2}\, m^{-1}\left(\nabla_{\mathbf{q}}\sigma(\mathbf{q}^n)\Lambda_n\right)\right) = 0.$$

If we consider the equation corresponding to the constraint $(i, j) \in I$, we obtain by (6.30)

$$\left\| \tilde{\mathbf{q}}_i^{n+1} - \tilde{\mathbf{q}}_j^{n+1} - \frac{\delta t^2}{2} \left(\frac{1}{m_i} \sum_{l:(i,l)\in I\cup I^*} (\mathbf{q}_i^n - \mathbf{q}_l^n)\lambda^{il} - \right. \right.$$
$$\left. \left. \frac{1}{m_j} \sum_{l:(l,j)\in I\cup I^*} (\mathbf{q}_j^n - \mathbf{q}_l^n)\lambda^{lj} \right) \right\|^2 = d_{ij}^2 ,$$

with the transposed index sets I^*. In the RATTLE algorithm, the left hand side is now approximated by truncating the sums to the two cases $l = j$ and $l = i$ and dropping terms corresponding to other constraints. Setting $\lambda^{ji} = \lambda^{ij}$ we then have

$$\left\| \tilde{\mathbf{q}}_i^{n+1} - \tilde{\mathbf{q}}_j^{n+1} - \frac{\delta t^2}{2} \left(\frac{1}{m_i} + \frac{1}{m_j} \right) (\mathbf{q}_i^n - \mathbf{q}_j^n)\lambda^{ij} \right\|^2 \approx d_{ij}^2 .$$

After linearizing in λ^{ij} and disregarding the δt^4-terms one obtains

$$\lambda^{ij} \approx \frac{d_{ij}^2 - \|\tilde{\mathbf{q}}_i^{n+1} - \tilde{\mathbf{q}}_j^{n+1}\|^2}{\delta t^2(\frac{1}{m_i} + \frac{1}{m_j})\langle \mathbf{q}_i^n - \mathbf{q}_j^n, \tilde{\mathbf{q}}_i^{n+1} - \tilde{\mathbf{q}}_j^{n+1} \rangle} . \tag{6.32}$$

The updates for the positions and momenta from this new approximation are then given by

$$\begin{aligned}
\tilde{\mathbf{q}}_i^{n+1} &= \tilde{\mathbf{q}}_i^{n+1} + \frac{\delta t^2}{2m_i}(\mathbf{q}_i^n - \mathbf{q}_j^n)\lambda^{ij}, \\
\tilde{\mathbf{q}}_j^{n+1} &= \tilde{\mathbf{q}}_j^{n+1} - \frac{\delta t^2}{2m_j}(\mathbf{q}_i^n - \mathbf{q}_j^n)\lambda^{ij}, \\
\tilde{\mathbf{p}}_i^{n+1/2} &= \tilde{\mathbf{p}}_i^{n+1/2} + \frac{\delta t}{2}(\mathbf{q}_i^n - \mathbf{q}_j^n)\lambda^{ij}, \\
\tilde{\mathbf{p}}_j^{n+1/2} &= \tilde{\mathbf{p}}_j^{n+1/2} - \frac{\delta t}{2}(\mathbf{q}_i^n - \mathbf{q}_j^n)\lambda^{ij}.
\end{aligned} \tag{6.33}$$

This nonlinear iteration step is repeated until the constraints are satisfied up to an acceptable accuracy. The iteration is also terminated if the number of iterations becomes too large. This can happen in the case of defective initial values which result in very large forces (when particles are too close to each other). In this case, the new positions computed without constraints can be far away from the positions in the previous time step, so that no corrections can be found that satisfy the constraints. But in most cases, dependent on the size of the time step, only few iterations are needed to satisfy the constraints up to an acceptable accuracy. However, to maintain the symplecticity of the resulting method, the system of equations (6.27) must be solved exactly.

After the successful completion of this nonlinear iteration, the new $\mathbf{q}^{n+1}$ and $\mathbf{p}^{n+1/2}$ have been computed. Now, to compute $\mathbf{p}^{n+1}$, we have to solve the system of equations (6.29). First, we precompute an auxiliary vector

$$\tilde{\mathbf{p}}^{n+1} := \tilde{\mathbf{p}}^{n+1/2} - \frac{\delta t}{2}\nabla_{\mathbf{q}}V(\mathbf{q}^{n+1}) \tag{6.34}$$

and use it to simplify (6.29) to

$$\dot{\sigma}\left(\mathbf{q}^{n+1}, m^{-1}\left(\tilde{\mathbf{p}}^{n+1} - \frac{\delta t}{2}\nabla_{\mathbf{q}}\sigma(\mathbf{q}^{n+1})\mu_{n+1} \right) \right) = 0 .$$

The constraint for the index pair $(i,j) \in I$ now yields

$$\Big\langle \mathbf{q}_i^{n+1} - \mathbf{q}_j^{n+1}, \frac{1}{m_i}\tilde{\mathbf{p}}_i^{n+1} - \frac{1}{m_j}\tilde{\mathbf{p}}_j^{n+1} -$$
$$\frac{\delta t}{2}\Big(\frac{1}{m_i}\sum_{l:(i,l)\in I\cup I^*}(\mathbf{q}_i^{n+1} - \mathbf{q}_l^{n+1})\mu^{il} - \frac{1}{m_j}\sum_{l:(l,j)\in I\cup I^*}(\mathbf{q}_j^{n+1} - \mathbf{q}_l^{n+1})\mu^{lj}\Big)\Big\rangle = 0 .$$

Again, we restrict the sum to the cases $l = j$ and $l = i$ and neglect all other constraints, set $\mu^{ji} = \mu^{ij}$, and obtain

$$\Big\langle \mathbf{q}_i^{n+1} - \mathbf{q}_j^{n+1}, \frac{1}{m_i}\tilde{\mathbf{p}}_i^{n+1} - \frac{1}{m_j}\tilde{\mathbf{p}}_j^{n+1} - \frac{\delta t}{2}\Big(\frac{1}{m_i} + \frac{1}{m_j}\Big)(\mathbf{q}_i^{n+1} - \mathbf{q}_j^{n+1})\mu^{ij}\Big\rangle \approx 0 .$$

Now, we can construct an iterative solver for this system of equations as follows: Solving this equation for μ_{ij} and using $d_{ij}^2 = \langle \mathbf{q}_i^{n+1} - \mathbf{q}_j^{n+1}, \mathbf{q}_i^{n+1} - \mathbf{q}_j^{n+1}\rangle$ – compare also (6.22) – we obtain

$$\mu^{ij} \approx \frac{2}{\delta t}\frac{\langle \mathbf{q}_i^{n+1} - \mathbf{q}_j^{n+1}, \frac{1}{m_i}\tilde{\mathbf{p}}_i^{n+1} - \frac{1}{m_j}\tilde{\mathbf{p}}_j^{n+1}\rangle}{(\frac{1}{m_i} + \frac{1}{m_j})d_{ij}^2} \tag{6.35}$$

and thus

$$\begin{aligned}\tilde{\mathbf{p}}_i^{n+1} &= \tilde{\mathbf{p}}_i^{n+1} + \frac{\delta t}{2}(\mathbf{q}_i^{n+1} - \mathbf{q}_j^{n+1})\mu^{ij},\\ \tilde{\mathbf{p}}_j^{n+1} &= \tilde{\mathbf{p}}_j^{n+1} - \frac{\delta t}{2}(\mathbf{q}_i^{n+1} - \mathbf{q}_j^{n+1})\mu^{ij}\end{aligned} \tag{6.36}$$

as corresponding updates to the other variables. These steps are used iteratively to determine $\mathbf{p}^{n+1}$. The overall method is summarized in Algorithm 6.2.

The expressions (6.32) and (6.35) can also be derived directly. To this end, one restricts the problem to a single constraint σ^{ij} and changes the positions and impulses in the direction of the vectors between $\mathbf{p}_i$ to $\mathbf{p}_j$ and between $\mathbf{q}_i$ to $\mathbf{q}_j$ until both the distance d_{ij} has the right value and the position and the velocity are orthogonal.

For the parallelization of the RATTLE Algorithm 6.2, we first assume that a parallel version of the force evaluation is available, implemented for instance as a parallel linked cell method. In addition, forces resulting from the constraints have to be taken into account in the updates for the positions and velocities. If the two particles connected by such a bond σ_{ij} both belong to the same process, the forces can be computed directly, otherwise the computations of the Lagrangian multipliers and the updates for the positions and velocities require communication between the processes. The computation of the forces caused by the constraints is relatively cheap, but data needs to be exchanged between processes in each iteration. Thus, depending on the structure of the constraints and the number of molecules in the simulation, this might suggest to compute all constraint forces for a molecule by one process. This leads to an efficient algorithm if the simulation involves many small molecules, such as many water molecules. However, if for example the simulation involves a single large (bio-)molecule or only a few such molecules, the constraint iterations and the evaluation of the constraint forces have to be parallelized as well.

Algorithm 6.2 RATTLE Time Integration for the Constrained Problem (6.22)

```
//  given initial values x, v, t  that satisfy the constraints
//  auxiliary vector x^old;
Compute forces F;
while (t < t_end) {
  t = t + delta_t;
  loop over all particles i {
      v_i = v_i + delta_t * .5 / m_i * F_i;
      x_i^old = x_i;
      x_i = x_i + delta_t * v_i;
  }
  real ε;
  do {
    ε = 0;
    loop over all constraints (i,j) {
        real r = d_{ij}^2 - ||x_i-x_j||^2;
        ε = ε + |r|;
        real lambda = r/((1/m_i+1/m_j)*⟨x_i^old-x_j^old, x_i-x_j⟩);
        x_i = x_i + (x_i^old-x_j^old) * lambda *.5 / m_i;
        x_j = x_j - (x_i^old-x_j^old) * lambda *.5 / m_j;
        v_i = v_i + (x_i^old-x_j^old) * lambda *.5 / (m_i * delta_t);
        v_j = v_j - (x_i^old-x_j^old) * lambda *.5 / (m_j * delta_t);
    }
  } while (ε > ε_{tol});
  compute forces F;
  loop over all particles i {
      v_i = v_i + delta_t * .5 / m_i * F_i;
  }
  do {
    ε = 0;
    loop over all constraints (i,j) {
        real r = ⟨x_i-x_j, v_i-v_j⟩;
        ε = ε + |r|;
        real mu = r / ((1/m_i+1/m_j)*d_{ij}^2);
        v_i = v_i + (x_i-x_j) * mu / m_i;
        v_j = v_j - (x_i-x_j) * mu / m_j;
    }
  } while (ε > ε_{tol});
  compute derived quantities such as energies, write x, v to output;
}
```

Variants. The iteration (6.32) and (6.33) without momentum updates had already been implemented in the SHAKE algorithm [534]. This algorithm omits the second iteration from the RATTLE Algorithm (6.35) and (6.36). The SHAKE algorithm is not symplectic, even though it satisfies the constraints (6.22) and the condition (6.9). It is only the inclusion of the additional constraints (6.24) for the velocities and momenta that leads to a symplectic algorithm [378].

In the case of more complex constraints, one can also use other nonlinear iterative solvers such as Newton methods [180, 587] to compute the Lagrangian multipliers. However, the iteration introduced above has the advantage that each step of the nonlinear iteration is cheap. Of course, there are also other ways to satisfy the constraints. For small problems it is often possible to find a new coordinate system in which the constraints are satisfied automatically. However, the equations of motion in such coordinate systems are more complex. For rigid bodies one possible choice are quaternion representations [34].

The freezing of degrees of freedom by the RATTLE approach eliminates forces from the system and therefore allows the use of larger time steps without compromising the stability of the method. Note finally that multiple time step methods can also be used together with the RATTLE approach [334, 492, 493, 541].[22]

---

[22] The impulse/r-Respa method together with the SHAKE algorithm allows time steps of up to 8 fs for molecular systems.

7 Mesh-Based Methods for Long-Range Potentials

In Chapters 3 and 5 we have studied so-called short-range potentials such as the Lennard-Jones potential (3.27), the Finnis-Sinclair potential (5.2), the EAM potential (5.14), and the Brenner potential (5.17). In the three-dimensional case, potentials that decay faster in r than $1/r^3$ are called short-range.[1] The relevant interactions resulting from these potentials occur only between particles that are close together. Besides those short-range potentials, there are also types of potentials in which the evolution of the simulated particle system is influenced by the interactions of particles that are far away from each other. For example, the gravitational potential (2.42) and the Coulomb potential (2.43) belong to the class of slowly decaying, long-range potentials.

The potentials V in the applications of this chapter consist of a short-range part V^{short} and an additional long-range part V^{long}, i.e.

$$V = V^{\text{short}} + V^{\text{long}}. \tag{7.1}$$

Here, V^{short} may contain all types of potentials considered up to now, such as bond, angle, torsion angle, and Lennard-Jones potentials, compare (5.38), as well as other many-body potentials. The long-range part may be[2]

$$V^{\text{long}} = \frac{1}{4\pi\varepsilon_0} \sum_{i=1}^{N} \sum_{j=i+1}^{N} q_i q_j \frac{1}{||\mathbf{x}_j - \mathbf{x}_i||} = \frac{1}{2}\frac{1}{4\pi\varepsilon_0} \sum_{i=1}^{N} \sum_{\substack{j=1 \\ j\neq i}}^{N} q_i q_j \frac{1}{||\mathbf{x}_j - \mathbf{x}_i||}. \tag{7.2}$$

This term models for instance the electrostatic potential of N point charges at the positions $\mathbf{x}_1, \dots, \mathbf{x}_N$ with the charges $q_1, \dots, q_N$.

In contrast to the rapidly decaying potentials, slowly decaying potentials cannot be truncated without a significant loss of accuracy [219, 672]. Therefore, the linked cell method cannot be applied directly. But evaluating the sum in (7.2) exactly would lead to a method with $\mathcal{O}(N^2)$ computational costs,

---

[1] In general a function $f(r)$ in $d > 2$ dimensions is called rapidly decaying if it decays faster in r than $1/r^d$. This classification into rapidly and slowly decaying functions reflects that functions which decay as $1/r^d$ or slower are not integrable over all of $\mathbb{R}^d$.

[2] $\varepsilon_0 = 8.854187817 \times 10^{-12}$ C^2/(Jm) denotes the dielectric constant.

which is generally prohibitively expensive. Instead, certain approximations of potentials such as (7.2) can be used. Here, the fundamental idea is to split the long-range part V^{long} of the potential into a smooth long-range part V^{lr} and a singular short-range part V^{sr}, i.e.

$$V^{\text{long}} = V^{\text{sr}} + V^{\text{lr}}. \tag{7.3}$$

These two parts are then treated separately with different approximation methods. For V^{sr} one can again use the linked cell method. The long-range part V^{lr} is smooth and can therefore be approximated well by so-called grid-based methods such as the P^{3}M method of Hockney and Eastwood [202, 324], the PME method of Darden et al., or the SPME method of Essmann et al. [168, 215, 374].

Long-range potentials such as (7.2) can be represented in different ways that in turn lead to different methods of approximation. Let us therefore briefly introduce these representations in the following.

Representations of the Potential. Consider a continuous charge distribution in the entire space $\mathbb{R}^3$ with a charge density ρ (charge per volume). The potential Φ induced by the charge density ρ is then

$$\Phi(\mathbf{x}) = \frac{1}{4\pi\varepsilon_0} \int_{\mathbb{R}^3} \rho(\mathbf{y}) \frac{1}{\|\mathbf{y} - \mathbf{x}\|} d\mathbf{y}. \tag{7.4}$$

Here, the potential Φ is the solution of the partial differential equation[3]

$$-\Delta\Phi(\mathbf{x}) = \frac{1}{\varepsilon_0} \rho(\mathbf{x}) \quad \text{on} \quad \mathbb{R}^3. \tag{7.5}$$

It decays like $\Phi(\mathbf{x}) \to 0$ for $\|\mathbf{x}\| \to \infty$. The equation (7.5) is called potential equation or Poisson equation. It is a classical representative of the class of elliptic partial differential equations of second order. The electrostatic energy associated with this potential is now defined as[4]

$$V = \int_{\mathbb{R}^3} \rho(\mathbf{x})\Phi(\mathbf{x})d\mathbf{x}. \tag{7.6}$$

Also, the forces on charged particles can be computed from $\Phi(\mathbf{x})$ by applying gradients.

Altogether, we have the following two formulations, which the methods for the evaluation of long-range interactions described in this and in the next chapter are based on:

---

[3] This follows from the fact that $1/r$ with $r = \|\mathbf{y} - \mathbf{x}\|$ is the fundamental solution of the Laplacian, i.e., in the sense of distributions, it holds that $-\Delta\frac{1}{r} = 4\pi\delta_0$, where δ_0 is the delta distribution with peak at $\mathbf{x} = 0$, compare also footnote 6.

[4] The existence of solutions to (7.5) or of the integrals in (7.4) and (7.6), respectively, depends on certain integrability conditions for the charge density ρ on $\mathbb{R}^3$.

Partial Differential Equation		Integral Formulation
$$-\Delta\Phi(\mathbf{x}) = \frac{1}{\varepsilon_0}\rho(\mathbf{x})$$	$\Leftrightarrow$	$$\Phi(\mathbf{x}) = \frac{1}{4\pi\varepsilon_0}\int \rho(\mathbf{y})\frac{1}{||\mathbf{y}-\mathbf{x}||}d\mathbf{y}$$

The standard approach to determine the potential Φ from the partial differential equation (7.5) consists of two steps, namely discretization and solution. In the discretization step, the continuous equation is transformed into a discrete system of equations. For example, the continuous variables in the equation can be approximated on the grid points of a given grid, as already described for some simple cases of difference quotients in Section 3.1. A finer grid then leads to a more accurate approximation of the potential. The discretization of the partial differential equation results in a linear system of equations. This system can be solved with fast direct methods such as the fast Fourier transform (FFT) [155], or with an iterative method, as for example a multigrid or multilevel method [117, 289]. The discretization of the integral expression (7.4) also leads to a linear system. However, this system does not have to be solved, but merely a matrix-vector multiplication must be computed in a fast and efficient way. To this end, the matrix-vector product is computed approximatively. Here, various methods exist, for instance the panel clustering method [289], the multilevel method [118, 119], the tree code of Barnes and Hut [58], several different multipole methods [38, 263], wavelet compression techniques [88, 165, 549], approaches that use $\mathcal{H}$-matrices [286], and pseudoskeleton approximations [257]. The tree code of Barnes and Hut and a variant of the multipole method will be introduced in more detail in Chapter 8.

The efficiency of all of these methods depends strongly on the smoothness of the functions involved. In our case, the smoothness of ρ is most important.[5] For smooth functions ρ, the associated potential Φ is also smooth. Standard discretization methods such as finite elements [111] or finite differences [284] will then lead to good approximations with known order of approximation

---

[5] The smoothness of a square integrable function can be measured by how often it is differentiable. Function classes can be defined using the order of differentiability, which leads to so-called Sobolev spaces [30]. These spaces are defined for $s \in \mathbb{N}$ as follows:

$$H^s(\Omega) := \{u \in L^2(\Omega) : \sum_{\alpha \in \mathbb{N}_0^3 : 0 \le ||\alpha||_\infty \le s} ||D^\alpha u||_{L^2}^2 < \infty\}.$$

Here, $D^\alpha u$ denotes the generalized (or weak) derivative of u and L^2 is the space of square integrable functions on Ω with the associated norm $||.||_{L^2}$. The parameter s describes the smoothness of functions in H^s. If the right hand side of the potential equation is in H^s and the boundary of Ω is sufficiently smooth, then the solution of the potential equation is in H^{s+2}. Thus, the solution is two times more (weakly) differentiable than the right hand side.

and convergence. However, in our case of a potential induced by N point charges, the charge distribution is a sum of delta distributions

$$\rho = \sum_{i=1}^{N} q_i \delta_{\mathbf{x}_i},$$

and is therefore not smooth.[6] The resulting potential Φ has singularities at the positions of the particles. Standard discretization methods for the potential equation converge only very slowly in such cases. They lead to large errors close to the positions of the particles. Nevertheless, the idea of a fast approximative solution of the potential equation can still be used: The computation of the potential is split into a smooth and a singular part. The singular part contains the interactions of a particle with its nearest neighbors. Here, the corresponding interactions can be computed with the algorithms developed in the earlier chapters. The smooth part contains the interactions of a particle with particles further away. It can be treated with the already mentioned methods for the fast solution of the Poisson equation. Altogether, this leads to algorithms that evaluate interactions between the particles with a much smaller complexity than a direct computation. Given a fixed maximal error, the complexity of such algorithms is typically $\mathcal{O}(N(\log(N))^\alpha)$, with $\alpha \geq 0$ depending on the specific method.

In this chapter, we study methods that are based on the formulation of the problem as a differential equation. Fast methods that are based on the integral formulation are discussed in Chapter 8. In the following, after a short explanation of the boundary conditions necessary for the solution of the potential equation, we explain how the potential Φ can be split appropriately into a smooth and a non-smooth part. Then, we discuss methods that rely on the discretization and fast solution of the potential equation with fast Fourier transforms or multigrid methods. As a concrete example, we present in Section 7.3 the so-called smooth particle-mesh Ewald method (SPME) in detail [168, 215, 374]. It uses the fast Fourier transform together with B-spline interpolation. Next, we explain the implementation of this method and apply it to simulate a Rayleigh-Taylor instability with a Coulomb potential, phase transitions in ionic KCl microcrystals, and water as a molecular system. Finally, in Section 7.5, we discuss the parallelization of the SPME method and study a problem from astrophysics, the formation of the large-scale structure of the universe.

---

[6] $\delta_{\mathbf{x}_i}(\mathbf{x}) := \delta(\mathbf{x}_i - \mathbf{x})$ is a distribution with $\int_\Omega f(\mathbf{y})\, \delta_{\mathbf{x}_i}(\mathbf{y})d\mathbf{y} = f(\mathbf{x}_i)$ for $\mathbf{x}_i \in \Omega, f \in L^2(\Omega)$ almost everywhere. One can think of $\delta_{\mathbf{x}_i}$ as a (generalized) function that is everywhere zero except at the point $\mathbf{x}_i$.

7.1 Solution of the Potential Equation

In this section we discuss in detail the approach based on the formulation as differential equation.

7.1.1 Boundary Conditions

Up to now, we have considered the case in which the domain is the entire space. However, discretization methods for the potential equation in general assume bounded domains. Here, one mostly finds two approaches for the boundary conditions for the potential equation: In the first one, the domain is chosen to be finite but large, for instance $\Omega = [0, a]^3$ with a large enough, and homogeneous Dirichlet boundary conditions are assumed.[7] The alternative is to extend the domain to infinity by periodicity. This approach is suitable especially for regular structures such as crystals. It also allows a simple splitting of the potential into a smooth and a singular part. For a more extensive discussion of appropriate boundary conditions and their influence on the results of the simulations, see [34, 100, 177].

In the following, we restrict ourselves to periodic systems. To this end, we extend the simulation domain in all spatial directions with periodic images, see Figure 7.1.

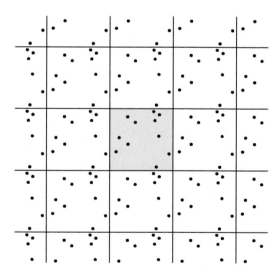

Fig. 7.1. Simulation box with particles and periodic extension to $\mathbb{R}^2$.

---

[7] Instead of Dirichlet boundary conditions, certain other non-reflecting boundary conditions can be used as well.

Every particle in the simulation box interacts with all other particles in the box as well as with all particles in the periodically translated simulation boxes, including its own periodic images. Note however that the physical quantities are only computed for the particles *in* the simulation box. Thus, the electrostatic energy of the particles in the simulation box is

$$V^{\text{long}} = \frac{1}{2} \frac{1}{4\pi\varepsilon_0} \sum_{\mathbf{n}\in\mathbb{Z}^3} \sum_{i=1}^{N} \sum_{\substack{j=1 \\ i\neq j \text{ for } \mathbf{n}=\mathbf{0}}}^{N} q_i q_j \frac{1}{||\mathbf{x}_j^{\mathbf{n}} - \mathbf{x}_i||}, \tag{7.7}$$

where the sum $\sum_{i=1}^{N}$ only runs through all particles *within* the simulation box. The sum $\sum_{\mathbf{n}} = \sum_{\mathbf{n}_1} \sum_{\mathbf{n}_2} \sum_{\mathbf{n}_3}$ runs over all periodic images of the simulation domain, and $\mathbf{x}_j^{\mathbf{n}} = \mathbf{x}_j + (n_1 \cdot L_1, n_2 \cdot L_2, n_3 \cdot L_3)$ denotes the positions of the periodic images of particle j. Here, the interaction of a particle of the simulation box with itself is excluded. However, its interaction with its periodic images is taken into account.[8] Note that this sum is not absolutely convergent, its value is therefore dependent on the order of summation.

7.1.2 The Potential Equation and the Decomposition of the Potential

Now, we consider the simulation domain $\Omega := [0, L_1[\times[0, L_2[\times[0, L_3[$, and we identify opposite sides to take the periodic extension of the simulation domain into account. In addition, a charge distribution is assumed to be given, which is *periodic* with respect to Ω on $\mathbb{R}^3$, i.e.

$$\rho(\mathbf{x} + (n_1 L_1, n_2 L_2, n_2 L_2)) = \rho(\mathbf{x}), \quad \mathbf{n} \in \mathbb{Z}^3,$$

and which also satisfies

$$\int_{\Omega} \rho(\mathbf{x}) d\mathbf{x} = 0. \tag{7.8}$$

Analogous to (7.4) and (7.5), the potential can be represented in integral form or as the solution of a potential equation. Because of the periodicity of the charge distribution ρ, the potential in Ω can be found as the solution of the potential equation

$$-\Delta\Phi = \frac{1}{\varepsilon_0} \rho\big|_{\Omega} \tag{7.9}$$

---

[8] Here, we compute the energy in the simulation box. Since, in the case $\mathbf{n} \neq \mathbf{0}$, the interactions extend beyond the simulation box, only half of the pair interactions have to be taken into account. For the interactions of the particles within the simulation box, i.e. for the case $\mathbf{n} = \mathbf{0}$, a factor of $1/2$ is necessary because of the double summation over all particle pairs. Altogether, this explains the factor $1/2$ in (7.7).

with periodic boundary conditions[9] at the boundary $\partial\Omega$ of the simulation domain Ω. Here, $|_\Omega$ denotes the restriction to the domain Ω.

For periodic charge distributions ρ, the integral (7.4) does not exist in general and we cannot directly use the representation (7.4) of the potential. But analogous to (7.4), the potentials $\Phi_{\mathbf{n}}$ which are defined on $\mathbb{R}^3$ and which are induced by charge distributions $\chi_{\mathbf{n}}\rho$ restricted to the translated simulation domain $\Omega_{\mathbf{n}} := [n_1L_1, (n_1+1)L_1[\times[n_2L_2, (n_2+1)L_2[\times[n_3L_3, (n_3+1)L_3[$, allow the representation

$$\Phi_{\mathbf{n}}(\mathbf{x}) = \frac{1}{4\pi\varepsilon_0} \int_{\mathbb{R}^3} \frac{\chi_{\mathbf{n}}(\mathbf{y})\rho(\mathbf{y})}{\|\mathbf{y}-\mathbf{x}\|} d\mathbf{y} \quad \text{for } \mathbf{n} \in \mathbb{Z}^3.$$

Here, $\chi_{\mathbf{n}}$ denotes the characteristic function of $\Omega_{\mathbf{n}}$, i.e. $\chi_{\mathbf{n}}(\mathbf{y}) = 1$ for $\mathbf{y} \in \Omega_{\mathbf{n}}$ and $\chi_{\mathbf{n}}(\mathbf{y}) = 0$ otherwise. The total potential in the simulation box Ω is then

$$\Phi(\mathbf{x}) = \sum_{\mathbf{n}\in\mathbb{Z}^3} \Phi_{\mathbf{n}}(\mathbf{x}).$$

How can these two representations for the computation of the potential be used to determine the energy and the forces in a system of N point charges? As already mentioned, the idea is to appropriately decompose the charge distribution ρ and thereby also the potential Φ and the corresponding energy (7.6) into two parts.

One proceeds as follows: A "charge cloud" $\varrho_i^{\mathbf{n}}$, spherically symmetric with respect to $\mathbf{x}_i^{\mathbf{n}}$ and with the same absolute charge but with the opposite sign, is attached to the point charge q_i at position $\mathbf{x}_i^{\mathbf{n}}$, compare Figure 7.2 (center). This charge distribution now shields the interactions that are caused by the point charges. The effect of the attached charge distributions is now removed by opposite charge distributions, compare Figure 7.2 (right). In this way, we can write the charge distribution induced by the N point charges and their periodic images

$$\rho(\mathbf{x}) = \sum_{\mathbf{n}\in\mathbb{Z}^3}\sum_{j=1}^{N} q_j\delta_{\mathbf{x}_j}^{\mathbf{n}}(\mathbf{x}) \tag{7.10}$$

as

$$\rho(\mathbf{x}) = (\rho(\mathbf{x}) - \rho^{\mathrm{lr}}(\mathbf{x})) + \rho^{\mathrm{lr}}(\mathbf{x}) = \rho^{\mathrm{sr}}(\mathbf{x}) + \rho^{\mathrm{lr}}(\mathbf{x})$$

---

[9] The solution of the potential equation with periodic boundary conditions is determined uniquely only up to a constant. Thus, if Φ is a solution, then so is $\Phi + C$ with an arbitrary constant C. Therefore, an additional constraint is necessary for a unique solution, for instance $\int_\Omega \Phi d\mathbf{x} = 0$. The missing uniqueness has however no effect on the force evaluation, since $\nabla C = \mathbf{0}$. In addition, $\frac{1}{\varepsilon_0}\int_\Omega \rho d\mathbf{x} = -\int_\Omega \Delta\Phi d\mathbf{x} = -\int_{\partial\Omega}\langle\nabla\Phi, \mathbf{n}\rangle d\Gamma = 0$ (to prove this, apply Green's formula and use the periodic boundary conditions) implies that the condition (7.8) is necessary for the solvability of (7.9). The condition $\sum_{j=1}^{N} q_j = 0$ on the electric charges then follows using (7.10).

with

$$\rho^{\mathrm{sr}}(\mathbf{x}) := \sum_{\mathbf{n} \in \mathbb{Z}^3} \sum_{j=1}^{N} q_j (\delta_j^{\mathbf{n}}(\mathbf{x}) - \varrho_j^{\mathbf{n}}(\mathbf{x})) \text{ and } \rho^{\mathrm{lr}}(\mathbf{x}) := \sum_{\mathbf{n} \in \mathbb{Z}^3} \sum_{j=1}^{N} q_j \varrho_j^{\mathbf{n}}(\mathbf{x}). \quad (7.11)$$

Here, $\varrho_j^{\mathbf{n}}(\mathbf{x})$ is the translated version of the function $\varrho(\mathbf{x})$ defined as

$$\varrho_j^{\mathbf{n}}(\mathbf{x}) = \varrho(\mathbf{x} - \mathbf{x}_j - (n_1 L_1, n_2 L_2, n_3 L_3)) \quad (7.12)$$

and $\delta_{\mathbf{x}_j^{\mathbf{n}}}(\mathbf{x}) = \delta(\mathbf{x} - \mathbf{x}_j^{\mathbf{n}})$ are delta distributions centered at $\mathbf{x}_j^{\mathbf{n}}$.

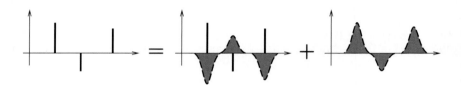

Fig. 7.2. The charge distribution of point charges (left) is split into a smoothed version (right) and the remainder (center).

The function ϱ has to satisfy the following conditions:

1. ϱ is normalized, i.e., it satisfies

$$\int_{\mathbb{R}^3} \varrho(\mathbf{x}) d\mathbf{x} = 1,$$

2. ϱ is symmetric with respect to the origin,
3. ϱ has compact support[10] or decays rapidly (compare footnote 1),
4. ϱ is a smooth function.

Condition 1 guarantees that the charge induced by $q_j \varrho_j^{\mathbf{n}}$ is equal to the charge q_j. Condition 2 implies that the functions $\varrho_j^{\mathbf{n}}$ are symmetric with respect to $\mathbf{x}_j^{\mathbf{n}}$ and therefore only depend on the distance to $\mathbf{x}_j^{\mathbf{n}}$. Conditions 1-3 together imply that the potential induced by the charge distribution $q_j(\delta_{\mathbf{x}_j^{\mathbf{n}}} - \varrho_j^{\mathbf{n}})$ is zero or at least very small outside of the (numerical) support of $\varrho_j^{\mathbf{n}}$. Condition 3 is also necessary to cause the complexity of the evaluation of ρ^{lr} to be independent of the number of particles (at least for equidistributed particles). Condition 4 guarantees that the solution of the potential equation (7.9) with ρ^{lr} as right hand side is a smooth function, compare footnote 5. Such functions can be approximated well with standard approximation methods.[11]

---

[10] The support of a function f is the closure of the set of $\mathbf{x}$ with $f(\mathbf{x}) \neq 0$. This condition therefore means that ϱ is nonzero only in a bounded domain.

[11] Instead of a single function ϱ, different charge clouds can be chosen for each particle. This is necessary especially for adaptive methods. Then, each of these charge distributions has to satisfy conditions 1-4.

Examples of appropriate choices for the shielding charge distribution ϱ are Gaussians

$$\varrho(\mathbf{x}) := \left(\frac{G}{\sqrt{\pi}}\right)^3 e^{-G^2\|\mathbf{x}\|^2} \tag{7.13}$$

or spheres with uniformly decreasing density

$$\varrho(\mathbf{x}) = \begin{cases} \frac{48}{\pi G^4}\left(\frac{G}{2} - \|\mathbf{x}\|\right), & \text{for } \|\mathbf{x}\| < \frac{G}{2}, \\ 0, & \text{otherwise,} \end{cases} \tag{7.14}$$

each with a parameter G that specifies the width of the distribution. Figure 7.3 shows the graph of the Gaussian (7.13) for different values of G. For increasing G the function becomes more and more localized and decays faster and faster. A plot with a logarithmic scale for the y axis, see Figure 7.4, shows this clearly.

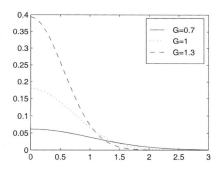

Fig. 7.3. Graphs of the charge distribution $(G/\sqrt{\pi})^3 e^{-G^2x^2}$ versus x for different values of G, linear scale.

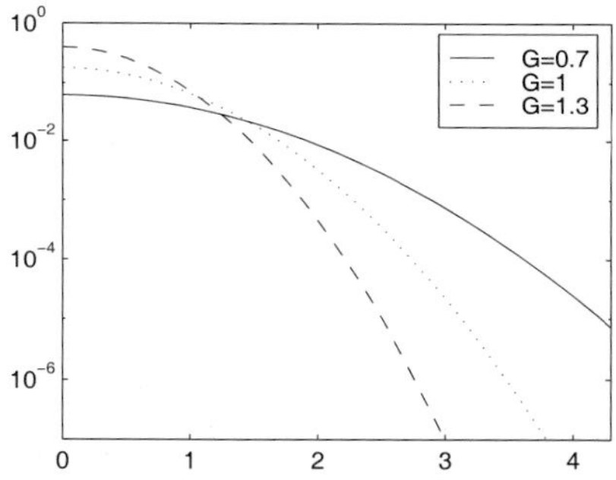

Fig. 7.4. Graph of the charge distribution $(G/\sqrt{\pi})^3 e^{-G^2x^2}$ versus x, semilogarithmic plot.

The two charge distributions ρ^{sr} and ρ^{lr}, into which the charge distribution ρ of the N point charges is split, are associated with potentials Φ^{sr} and Φ^{lr}, energies V^{sr} and V^{lr}, and forces $\mathbf{F}_i^{\text{sr}}$ and $\mathbf{F}_i^{\text{lr}}$. The total potential Φ, the total energy, and the total forces can be obtained as sums of the corresponding short-range and long-range terms. This follows from the superposition principle for the solutions of the potential equation.[12] For example, the total

---

[12] The linearity of the Laplace operator implies the following superposition principle for the solutions of the potential equation with periodic boundary conditions: If u_1 and u_2 are solutions of problems $-\Delta u_1 = f$ and $-\Delta u_2 = g$ in Ω with periodic boundary conditions on $\partial\Omega$, then the solution u of the Poisson equation with right hand side $f + g$ is given as the sum of u_1 and u_2. Thus, $-\Delta u = f + g$ holds for $u = u_1 + u_2$ in Ω.

potential is given as $\Phi = \Phi^{\mathrm{sr}} + \Phi^{\mathrm{lr}}$. The two potentials Φ^{sr} and Φ^{lr} can again be represented as solutions of potential equations

$$-\Delta\Phi^{\mathrm{sr}} = \frac{1}{\varepsilon_0}\rho^{\mathrm{sr}}\Big|_{\Omega} \quad\text{and}\quad -\Delta\Phi^{\mathrm{lr}} = \frac{1}{\varepsilon_0}\rho^{\mathrm{lr}}\Big|_{\Omega} \qquad (7.15)$$

with periodic boundary conditions on Ω.

7.1.3 Decomposition of the Potential Energy and of the Forces

The electrostatic energy (7.7) of the simulation box induced by the N point charges can now be written as

$$V = \frac{1}{2}\frac{1}{4\pi\varepsilon_0}\sum_{\mathbf{n}\in\mathbb{Z}^3}\sum_{i=1}^{N}\sum_{\substack{j=1\\i\neq j\text{ for }\mathbf{n}=0}}^{N} q_i q_j \int_{\mathbb{R}^3}\frac{\delta_{\mathbf{x}_j^{\mathbf{n}}}(\mathbf{y}) - \varrho_j^{\mathbf{n}}(\mathbf{y}) + \varrho_j^{\mathbf{n}}(\mathbf{y})}{||\mathbf{y} - \mathbf{x}_i||}d\mathbf{y}$$

$$= V^{\mathrm{sr}} + V^{\mathrm{lr}}$$

with

$$V^{\mathrm{sr}} := \frac{1}{2}\frac{1}{4\pi\varepsilon_0}\sum_{\mathbf{n}\in\mathbb{Z}^3}\sum_{i=1}^{N}\sum_{\substack{j=1\\i\neq j\text{ for }\mathbf{n}=0}}^{N} q_i q_j \int_{\mathbb{R}^3}\frac{\delta_{\mathbf{x}_j^{\mathbf{n}}}(\mathbf{y}) - \varrho_j^{\mathbf{n}}(\mathbf{y})}{||\mathbf{y} - \mathbf{x}_i||}d\mathbf{y} \qquad (7.16)$$

and

$$V^{\mathrm{lr}} := \frac{1}{2}\frac{1}{4\pi\varepsilon_0}\sum_{\mathbf{n}\in\mathbb{Z}^3}\sum_{i=1}^{N}\sum_{\substack{j=1\\i\neq j\text{ for }\mathbf{n}=0}}^{N} q_i q_j \int_{\mathbb{R}^3}\frac{\varrho_j^{\mathbf{n}}(\mathbf{y})}{||\mathbf{y} - \mathbf{x}_i||}d\mathbf{y}. \qquad (7.17)$$

If one substitutes the definition of ρ^{lr} as sum of charge clouds according to (7.11) into the integral representation (7.4), one obtains[13]

$$\Phi^{\mathrm{lr}}(\mathbf{x}) = \frac{1}{4\pi\varepsilon_0}\sum_{\mathbf{n}\in\mathbb{Z}^3}\sum_{j=1}^{N} q_j \int_{\mathbb{R}^3}\frac{\varrho_j^{\mathbf{n}}(\mathbf{y})}{||\mathbf{y} - \mathbf{x}||}d\mathbf{y}. \qquad (7.18)$$

We now evaluate this expression for all particle positions $\mathbf{x}_i \in \Omega$, multiply with the corresponding charge q_i, and compute the sum. This yields

$$\sum_{i=1}^{N} q_i \Phi^{\mathrm{lr}}(\mathbf{x}_i) = \frac{1}{4\pi\varepsilon_0}\sum_{\mathbf{n}\in\mathbb{Z}^3}\sum_{i=1}^{N}\sum_{j=1}^{N} q_i q_j \int_{\mathbb{R}^3}\frac{\varrho_j^{\mathbf{n}}(\mathbf{y})}{||\mathbf{y} - \mathbf{x}_i||}d\mathbf{y}. \qquad (7.19)$$

Then, a comparison of (7.19) with (7.17) shows that the long-range part V^{lr} of the electrostatic energy V^{long} can also be represented as

---

[13] The substitution here is entirely formal and has to be understood in the sense of the explanation on page 245.

$$V^{\text{lr}} = V^{\text{lr}}_{\text{other}} - V^{\text{lr}}_{\text{self}} = \frac{1}{2} \sum_{i=1}^{N} q_i \Phi^{\text{lr}}(\mathbf{x}_i) - \frac{1}{2} \frac{1}{4\pi\varepsilon_0} \sum_{i=1}^{N} q_i^2 \int_{\mathbb{R}^3} \frac{\varrho_i^0(\mathbf{y})}{||\mathbf{y} - \mathbf{x}_i||} d\mathbf{y}. \quad (7.20)$$

The second term on the right hand side is a correction term that removes the interaction of the particle i at position $\mathbf{x}_i$ with the charge distribution ϱ_i^0 which is included in the first term $V^{\text{lr}}_{\text{other}}$. In the following, we denote this correction as

$$V^{\text{lr}}_{\text{self}} = \frac{1}{2} \frac{1}{4\pi\varepsilon_0} \sum_{i=1}^{N} q_i^2 \int_{\mathbb{R}^3} \frac{\varrho_i^0(\mathbf{y})}{||\mathbf{y} - \mathbf{x}_i||} d\mathbf{y} \quad (7.21)$$

and call it self-energy. It holds that

$$\nabla_{\mathbf{x}_i} V^{\text{lr}}_{\text{self}} = \frac{1}{2} \frac{1}{4\pi\varepsilon_0} q_i^2 \int_{\mathbb{R}^3} \nabla_{\mathbf{x}_i} \frac{\varrho_i^0(\mathbf{y})}{||\mathbf{y} - \mathbf{x}_i||} d\mathbf{y} = 0, \quad (7.22)$$

because, according to our conditions, ϱ_i^0 is chosen to be symmetric around $\mathbf{x}_i$, and

$$\frac{\partial}{\partial t} V^{\text{lr}}_{\text{self}} = 0,$$

because $V^{\text{lr}}_{\text{self}}$ does not explicitly depend on time. Therefore, the chain rule of differential calculus implies

$$\frac{d}{dt} V^{\text{lr}}_{\text{self}} = \sum_{i=1}^{N} \nabla_{\mathbf{x}_i} V^{\text{lr}}_{\text{self}} \cdot \frac{\partial}{\partial t} \mathbf{x}_i(t) + \frac{\partial}{\partial t} V^{\text{lr}}_{\text{self}} = 0.$$

Thus, $V^{\text{lr}}_{\text{self}}$ is constant over time for given charges $q_1, \ldots, q_N$ and has to be computed just once at the beginning of each simulation. The electrostatic energy V^{long} can therefore be written as

$$V^{\text{long}} = V^{\text{sr}} + V^{\text{lr}} = V^{\text{sr}} + \frac{1}{2} \sum_{i=1}^{N} q_i \Phi^{\text{lr}}(\mathbf{x}_i) - V^{\text{lr}}_{\text{self}}. \quad (7.23)$$

Corresponding to this splitting of the energy into the two components V^{sr} and V^{lr}, the force on a particle i can be computed from the two parts

$$\mathbf{F}_i^{\text{sr}} = -\nabla_{\mathbf{x}_i} V^{\text{sr}} \quad \text{and} \quad \mathbf{F}_i^{\text{lr}} = -\nabla_{\mathbf{x}_i} V^{\text{lr}} = -\frac{1}{2} \sum_{j=1}^{N} q_j \nabla_{\mathbf{x}_i} \Phi^{\text{lr}}(\mathbf{x}_j) \quad (7.24)$$

according to $\mathbf{F}_i = \mathbf{F}_i^{\text{sr}} + \mathbf{F}_i^{\text{lr}}$. Note that the force $\mathbf{F}_i^{\text{lr}}$ does not depend on the self-energy $V^{\text{lr}}_{\text{self}}$ since the self-energy does not depend on the particle positions, compare (7.22).

Here, the expression $\nabla_{\mathbf{x}_i} \Phi^{\text{lr}}(\mathbf{x}_j)$ is an abbreviation that is to be understood in the following sense: The function $\Phi^{\text{lr}} : \Omega \rightarrow \mathbb{R}$ depends parametrically on the particle positions $\{\mathbf{x}_i\}_{i=1}^{N}$. This can also be viewed as a function of the particle positions, that means $\Phi^{\mathrm{lr}} : \Omega^N \times \Omega \to \mathbb{R}$ with $\Phi^{\mathrm{lr}} \to \Phi^{\mathrm{lr}}(\mathbf{x}_1, \ldots, \mathbf{x}_N; \mathbf{x})$. Then, the expression $\nabla_{\mathbf{x}_i} \Phi^{\mathrm{lr}}(\mathbf{x}_1, \ldots, \mathbf{x}_N; \mathbf{x})$ denotes as before the gradient $\nabla_{\mathbf{y}} \Phi^{\mathrm{lr}}(\mathbf{x}_1, \ldots, \mathbf{x}_{i-1}, \mathbf{y}, \mathbf{x}_{i+1}, \ldots, \mathbf{x}_N; \mathbf{x})$ evaluated at the position $\mathbf{x}_i$ for $\mathbf{y}$. In this sense, the expression $\nabla_{\mathbf{x}_i} \Phi^{\mathrm{lr}}(\mathbf{x})$ stands for $\nabla_{\mathbf{y}} \Phi^{\mathrm{lr}}(\mathbf{x}_1, \ldots, \mathbf{x}_{i-1}, \mathbf{y}, \mathbf{x}_{i+1}, \ldots, \mathbf{x}_N; \mathbf{x})$ evaluated at $\mathbf{y} = \mathbf{x}_i$ and $\nabla_{\mathbf{x}_i} \Phi^{\mathrm{lr}}(\mathbf{x}_j)$ stands for $\nabla_{\mathbf{y}} \Phi^{\mathrm{lr}}(\mathbf{x}_1, \ldots, \mathbf{x}_{i-1}, \mathbf{y}, \mathbf{x}_{i+1}, \ldots, \mathbf{x}_N; \mathbf{x}_j)$ evaluated at $\mathbf{y} = \mathbf{x}_i$. We will use this abbreviated notation for the remainder of the book.

7.2 The Computation of Short-Range and Long-Range Energy and Force Terms

If ϱ satisfies the conditions 1–4 then the V^{sr} from (7.16) can be treated together with any V^{short} by the linked cell method. V^{lr} can be computed according to (7.20) by first computing Φ^{lr} as the solution of the potential equation in (7.15).

In the next sections we discuss the computation of these two terms in detail.

7.2.1 Short-Range Terms – Linked Cell Method

The integral

$$\int_{\mathbb{R}^3} \frac{\varrho_j^{\mathbf{n}}(\mathbf{y})}{||\mathbf{y} - \mathbf{x}||} d\mathbf{y},$$

which occurs with $\mathbf{x} = \mathbf{x}_i$ in (7.16) and (7.17), can be split into two integrals according to

$$\int_{\mathbb{R}^3} \frac{\varrho_j^{\mathbf{n}}(\mathbf{y})}{||\mathbf{y} - \mathbf{x}||} d\mathbf{y} = \int_{B_{\mathbf{x}_j^{\mathbf{n}}}(\mathbf{x})} \frac{\varrho_j^{\mathbf{n}}(\mathbf{y})}{||\mathbf{y} - \mathbf{x}||} d\mathbf{y} + \int_{\mathbb{R}^3 \setminus B_{\mathbf{x}_j^{\mathbf{n}}}(\mathbf{x})} \frac{\varrho_j^{\mathbf{n}}(\mathbf{y})}{||\mathbf{y} - \mathbf{x}||} d\mathbf{y},$$

where $B_{\mathbf{x}_j^{\mathbf{n}}}(\mathbf{x})$ denotes the ball around $\mathbf{x}_j^{\mathbf{n}}$ with radius $||\mathbf{x}_j^{\mathbf{n}} - \mathbf{x}||$. The function $\varrho_j^{\mathbf{n}}$ is symmetric around $\mathbf{x}_j^{\mathbf{n}}$ according to condition 2. Therefore, the first summand can be written as

$$\int_{B_{\mathbf{x}_j^{\mathbf{n}}}(\mathbf{x})} \frac{\varrho_j^{\mathbf{n}}(\mathbf{y})}{||\mathbf{y} - \mathbf{x}||} d\mathbf{y} = \frac{1}{||\mathbf{x}_j^{\mathbf{n}} - \mathbf{x}||} \int_{B_{\mathbf{x}_j^{\mathbf{n}}}(\mathbf{x})} \varrho_j^{\mathbf{n}}(\mathbf{y}) d\mathbf{y}.$$

Thus, the action of such a radial charge distribution at a point $\mathbf{x}$ is the same as the action of a point charge with the value

$$\int_{B_{\mathbf{x}_j^{\mathbf{n}}}(\mathbf{x})} \varrho_j^{\mathbf{n}}(\mathbf{y}) d\mathbf{y}$$

in $\mathbf{x}_j^{\mathbf{n}}$. For the second summand, it holds

$$\int_{\mathbb{R}^3 \setminus B_{\mathbf{x}_j^n}(\mathbf{x})} \frac{\varrho_j^n(\mathbf{y})}{||\mathbf{y} - \mathbf{x}||} d\mathbf{y} = \int_{\mathbb{R}^3 \setminus B_{\mathbf{x}_j^n}(\mathbf{x})} \frac{\varrho_j^n(\mathbf{y})}{||\mathbf{y} - \mathbf{z}||} d\mathbf{y} \quad \text{for all } \mathbf{z} \in B_{\mathbf{x}_j^n}(\mathbf{x}).$$

Thus, the potential induced by the second summand is constant inside the ball $B_{\mathbf{x}_j^n}(\mathbf{x})$. Evaluating this formula at $\mathbf{z} := \mathbf{x}_j^n$ yields

$$\int_{\mathbb{R}^3 \setminus B_{\mathbf{x}_j^n}(\mathbf{x})} \frac{\varrho_j^n(\mathbf{y})}{||\mathbf{y} - \mathbf{z}||} d\mathbf{y} = \int_{\mathbb{R}^3 \setminus B_{\mathbf{x}_j^n}(\mathbf{x})} \frac{\varrho_j^n(\mathbf{y})}{||\mathbf{y} - \mathbf{x}_j^n||} d\mathbf{y}.$$

Together this results in the decomposition

$$\int_{\mathbb{R}^3} \frac{\varrho_j^n(\mathbf{y})}{||\mathbf{y} - \mathbf{x}||} d\mathbf{y} = \frac{1}{||\mathbf{x}_j^n - \mathbf{x}||} \int_{B_{\mathbf{x}_j^n}(\mathbf{x})} \varrho_j^n(\mathbf{y}) d\mathbf{y} + \int_{\mathbb{R}^3 \setminus B_{\mathbf{x}_j^n}(\mathbf{x})} \frac{\varrho_j^n(\mathbf{y})}{||\mathbf{y} - \mathbf{x}_j^n||} d\mathbf{y}. \tag{7.25}$$

The coordinate transformation $\mathbf{w} := \mathbf{y} - \mathbf{x}_j^n$ and a subsequent transformation to spherical coordinates in the integrals lead to

$$\int_{\mathbb{R}^3} \frac{\varrho_j^n(\mathbf{y})}{||\mathbf{y} - \mathbf{x}||} d\mathbf{y} = \frac{1}{||\mathbf{x}_j^n - \mathbf{x}||} \int_{B_0(\mathbf{x} - \mathbf{x}_j^n)} \varrho(\mathbf{w}) d\mathbf{w} + \int_{\mathbb{R}^3 \setminus B_0(\mathbf{x} - \mathbf{x}_j^n)} \frac{\varrho(\mathbf{w})}{||\mathbf{w}||} d\mathbf{w}$$

$$= \frac{4\pi}{||\mathbf{x}_j^n - \mathbf{x}||} \int_0^{||\mathbf{x}_j^n - \mathbf{x}||} r^2 \varrho(r) dr + 4\pi \int_{||\mathbf{x}_j^n - \mathbf{x}||}^{\infty} r \varrho(r) dr. \tag{7.26}$$

If F is now an antiderivative of $r \cdot \varrho(r)$ with $F(r) \to 0$ for $r \to \infty$,[14] i.e. $F'(r) = r \cdot \varrho(r)$, partial integration of the first integral on the right hand side of (7.26) and substitution in the second integral results in

$$\int_{\mathbb{R}^3} \frac{\varrho_j^n(\mathbf{y})}{||\mathbf{y} - \mathbf{x}||} d\mathbf{y} = \frac{4\pi}{||\mathbf{x}_j^n - \mathbf{x}||} \left(||\mathbf{x}_j^n - \mathbf{x}|| \cdot F(||\mathbf{x}_j^n - \mathbf{x}||) - \int_0^{||\mathbf{x}_j^n - \mathbf{x}||} F(r) dr \right)$$

$$- 4\pi \cdot F(||\mathbf{x}_j^n - \mathbf{x}||)$$

$$= -\frac{4\pi}{||\mathbf{x}_j^n - \mathbf{x}||} \int_0^{||\mathbf{x}_j^n - \mathbf{x}||} F(r) dr. \tag{7.27}$$

Here, we exploited the fact[15] that $r \cdot F(r) \to 0$ for $r \to 0$ in the first integral and that $F(r) \to 0$ for $r \to \infty$ in the second integral.

If one substitutes these results together with $\mathbf{x} := \mathbf{x}_i$ into (7.16), one obtains

$$V^{\mathrm{sr}} = \frac{1}{2} \frac{1}{4\pi\varepsilon_0} \sum_{\mathbf{n} \in \mathbb{Z}^3} \sum_{i=1}^N \sum_{\substack{j=1 \\ i \neq j \text{ for } n=0}}^N q_i q_j \left(\frac{1}{||\mathbf{x}_j^n - \mathbf{x}_i||} - \int_{\mathbb{R}^3} \frac{\varrho_j^n(\mathbf{y})}{||\mathbf{y} - \mathbf{x}_i||} d\mathbf{y} \right) \tag{7.28}$$

---

[14] The antiderivative is only determined up to a constant. The condition for $r \to \infty$ selects a unique antiderivative.

[15] This follows from condition 1 on ϱ in Section 7.1.2. To show it, apply L'Hôpital's rule.

$$= \frac{1}{2} \frac{1}{4\pi\varepsilon_0} \sum_{\substack{\mathbf{n} \in \mathbb{Z}^3}} \sum_{i=1}^{N} \sum_{\substack{j=1 \\ i \neq j \text{ for } \mathbf{n}=\mathbf{0}}}^{N} q_i q_j \left(\frac{1}{||\mathbf{x}_j^{\mathbf{n}} - \mathbf{x}_i||} + \frac{4\pi}{||\mathbf{x}_j^{\mathbf{n}} - \mathbf{x}_i||} \int_0^{||\mathbf{x}_j^{\mathbf{n}} - \mathbf{x}_i||} F(r) dr \right).$$

The expression in parentheses is very small for a distance between $\mathbf{x}_i$ and $\mathbf{x}_j^{\mathbf{n}}$ that is sufficiently large, because of conditions 1 and 3 from Section 7.1.2.[16] With a sufficiently large threshold parameter r_{cut}, one can therefore restrict the sum over j again to the indices j with $||\mathbf{x}_j^{\mathbf{n}} - \mathbf{x}_i|| < r_{\text{cut}}$. The resulting forces on the particles are given as negative gradients of the approximation of V^{sr} again. Therefore, the linked cell method from Chapter 3 for the case of periodic boundary conditions can be used for the efficient evaluation of the short-range part of the forces and the energies. Here, the choice of the threshold parameter r_{cut} is determined by the decay of ϱ.

If, in addition to the term V^{sr} arising from the long-range Coulomb potential, there are further short-range terms present, for example extra Lennard-Jones terms, these are to be taken into account in the computation of the short-range part of the potentials or the force as well. The cutoff radius r_{cut} is then determined by the maximum of the range of the short-range potential and the range of the densities ϱ.

7.2.2 Long-Range Terms – Fast Poisson Solvers

Since the direct evaluation of (7.17) requires $\mathcal{O}(N^2)$ operations in general, the electrostatic energy term V^{lr} and the long-range forces $\mathbf{F}_i^{\text{lr}}$ are evaluated using the representation of Φ^{lr} as the solution of the potential equation

$$-\Delta \Phi^{\text{lr}} = \frac{1}{\varepsilon_0} \rho^{\text{lr}}|_{\Omega} \qquad (7.29)$$

on Ω, compare (7.15). Here, one discretizes equation (7.29) for instance by the Galerkin method (see below) using K appropriate test basis functions ϕ_k. This results in a linear system of equations $A\mathbf{c} = \mathbf{b}$ that has to be solved efficiently. Fast Poisson solvers can be used for this task. Using the solution $\mathbf{c} = (c_0, \ldots, c_{K-1})^T$ of the linear system of equations, the solution Φ^{lr} of the potential equation can be approximated by a function Φ_K^{lr} in the form of a finite sum

$$\Phi_K^{\text{lr}} = \sum_{k=0}^{K-1} c_k \phi_k, \qquad (7.30)$$

---

[16] This can be seen for example from the representation (7.28) together with (7.25). The second summand on the right hand side of (7.25) decays very rapidly to 0 for $||\mathbf{x} - \mathbf{x}_j^{\mathbf{n}}|| \to \infty$ (this follows from the fast decay of ϱ according to condition 3) and the first summand converges rapidly to $1/||\mathbf{x} - \mathbf{x}_j^{\mathbf{n}}||$ (this follows from conditions 1 and 3 on ϱ). This implies that the expression in parentheses in (7.28) also decays rapidly to 0 for $||\mathbf{x}_i - \mathbf{x}_j^{\mathbf{n}}|| \to \infty$.

compare (7.35) in the following section. The energy term $V_{\text{other}}^{\text{lr}}$ can be computed according to (7.20) as a sum over the potential evaluated at the positions of the particles, weighted by their charges. Substituting the approximation (7.30) in (7.20), we obtain

$$V_{\text{other}}^{\text{lr}} = \frac{1}{2} \sum_{i=1}^{N} q_i \varPhi^{\text{lr}}(\mathbf{x}_i) \approx \frac{1}{2} \sum_{i=1}^{N} q_i \varPhi_K^{\text{lr}}(\mathbf{x}_i) = \frac{1}{2} \sum_{i=1}^{N} q_i \sum_{k=0}^{K-1} c_k \phi_k(\mathbf{x}_i). \quad (7.31)$$

With functions ϕ_k of bounded support, the point-wise evaluation of the approximation $\varPhi_K^{\text{lr}}$ requires only $\mathcal{O}(1)$ operations since the sum over k in (7.30) extends only over those k for which the value of ϕ_k is nonzero at the given point. The sum in (7.31) can therefore be evaluated with a complexity of $\mathcal{O}(N)$.

The forces $\mathbf{F}_i^{\text{lr}}$ on the particles can be computed directly according to (7.24)[17] from the approximation of the solution of the potential equation. We obtain[18]

$$\mathbf{F}_i^{\text{lr}} \approx -\frac{1}{2} \sum_{j=1}^{N} q_j \nabla_{\mathbf{x}_i} \varPhi_K^{\text{lr}}(\mathbf{x}_j) = -\frac{1}{2} \sum_{j=1}^{N} q_j \sum_{k=0}^{K-1} \nabla_{\mathbf{x}_i} \left(c_k \phi_k(\mathbf{x}_j) \right). \quad (7.32)$$

As for the computation of the energy, the complexity of the computation of the forces $\mathbf{F}_i^{\text{lr}}$ on N particles by (7.32) is also of order $\mathcal{O}(N)$ if the ϕ_k have local support.[19]

The computation of the long-range force and of the potential terms consists therefore of the following three steps:

1. Discretization: The potential equation (7.29) is discretized, which in turn leads to an approximation of the solution $\varPhi^{\text{lr}}$. The accuracy of this approximation depends on the type of discretization used and on the choices for its discretization parameters (such as e.g. the mesh size for a finite element method).

---

[17] For discretizations with Galerkin methods, the basis functions are (at least piecewise) differentiable. In other discretization approaches such as for example finite difference methods or collocation methods, a representation as (7.30) in terms of basis functions needs not to exist. Also, the basis functions do not have to be differentiable as is the case for finite volume methods. Then, differentiable approximations of $\varPhi_K^{\text{lr}}$ have to be generated by reconstruction or, alternatively, interpolation or possibly numerical differentiation may be employed. Note that the error introduced this way should be smaller or comparable to the other errors of the overall method.

[18] Note that the coefficients $\{c_k\}_{k=0}^{K-1}$ also depend on the positions of the particles $\{\mathbf{x}_i\}_{i=1}^{N}$. The gradient operator $\nabla_{\mathbf{x}_i}$ from (7.32) therefore does not only act on the ϕ_k, but also on the coefficients c_k, compare the remarks on page 250.

[19] In general, the straightforward use of global functions for the ϕ_k leads only to a complexity of order $\mathcal{O}(N \cdot K)$, since the sums in (7.31) and (7.32) have to be evaluated for all k.

2. Solution of the discretized problem: The discretized potential equation is solved with some (direct or iterative) method. Multilevel methods and fast Fourier transforms (FFT) are most efficiently used in this step.
3. Computation of energies and forces: The computed approximation for the potential is used to compute the energy V^{lr} and the forces $\mathbf{F}_i^{\text{lr}}$ at the positions of the particles according to (7.31) and (7.32). Here, the computation of the forces may require the computation of the derivative of the potential and an interpolation of the gradient at the particle positions.

Figure 7.5 shows in a schematic way the different steps in the computation of the long-range forces.

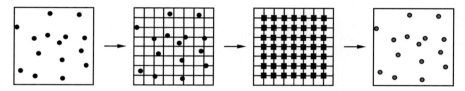

Fig. 7.5. Schematic representation of the method for the computation of the forces $\mathbf{F}_i^{\text{lr}}$, two-dimensional case. The charge density induced by the particles is approximated on a mesh. The associated potential is computed on that mesh as the solution of the potential equation. Finally, the resulting forces on the particles are computed as the gradient of the potential (after a possible interpolation from the mesh points at the positions of the particles).

In general, each of these steps contributes an error term to the total error of the approximation of the forces and the energy. The different steps for the computation of the long-range terms are interdependent and must be coordinated to obtain an efficient method for the overall problem. In addition, it is beneficial if the error for the short-range and the long-range terms are of the same order. The degrees of freedom that can be varied to achieve this goal are the form of the shielding charge distribution ϱ, the threshold parameter r_{cut}, the discretization parameters for the discretization of the potential equation, such as the mesh width, and (if necessary) the number of interpolation points for the interpolation in step 3.

We discuss in the following the three steps for the computation of the long-range terms in detail. We choose a Galerkin discretization. First, we consider the finite element method. Here, we employ a uniform mesh and use piecewise polynomials as basis functions. Their local support allows the computation of the potential and the forces in $\mathcal{O}(N)$ operations. In addition, this leads to a system of linear equations with a sparse system matrix. However, the condition number of this matrix is large, so that standard iterative methods converge only slowly. Here, multilevel methods or direct solvers based on FFT can be used instead. We also review the variants of this general approach for the computation of the long-range terms, which are commonly described in the literature. Finally, we consider the case of trigonometric basis functions in detail in the next section. Their orthogonality leads to a trivial system of equations with a diagonal system matrix. However, their support is global which generally results in a quite expensive computation of the energy and the forces. But similar to pseudo-spectral methods [258], these costs can be substantially reduced by the approximation on a uniform mesh using local basis functions and the fast Fourier transform.

Galerkin Discretization and Finite Element Method. In numerical analysis, the transition from a continuous system to a discrete system is called discretization. In Section 3.1, we already encountered the finite difference method for the time discretization of Newtons equations. We will now use the Galerkin method [111, 146] to discretize the potential equation. Let

$$(u, v) := \int_\Omega u\bar{v}d\mathbf{x}$$

be the L^2 inner product over the domain Ω. The solution Φ^{lr} of (7.29) has to satisfy the equation

$$-(\Delta\Phi^{lr}, v) = \frac{1}{\varepsilon_0}(\rho^{lr}, v)$$

for all test functions v chosen from an appropriate function space V.[20] The left hand side of the equation can be rewritten using partial integration, and one obtains the so-called weak formulation of the differential equation

$$(\nabla\Phi^{lr}, \nabla v) = \frac{1}{\varepsilon_0}(\rho^{lr}, v) \quad \text{for all } v \in V. \tag{7.33}$$

The Galerkin method now consists in the choice of a finite-dimensional subspace $V_K \subset V$ with $K = dim(V_K)$, in which the solution of the problem is approximated:

$$\text{Find } \Phi_K^{lr} \in V_K \text{ with } (\nabla\Phi_K^{lr}, \nabla v) = \frac{1}{\varepsilon_0}(\rho^{lr}, v) \quad \text{for all } v \in V_K. \tag{7.34}$$

For the numerical implementation, one selects a basis $\{\phi_0, \ldots, \phi_{K-1}\}$ of V_K. The solution of (7.34) is assumed to have the form

$$\Phi_K^{lr} = \sum_{k=0}^{K-1} c_k\phi_k, \tag{7.35}$$

compare (7.30). In this way, one obtains the system of equations

---

[20] It is necessary to have $V \subset H^1 \subset L^2$ to be able to compute the gradients on the left hand side of (7.33).

$$\sum_{k=0}^{K-1} (\nabla\phi_k, \nabla\phi_j)\, c_k = \frac{1}{\varepsilon_0}(\rho^{\mathrm{lr}}, \phi_j), \quad j = 0, \ldots, K-1, \tag{7.36}$$

or in matrix notation

$$A\mathbf{c} = \mathbf{b} \tag{7.37}$$

with $A_{jk} = (\nabla\phi_k, \nabla\phi_j)$ the entries of the matrix A, $b_j = \frac{1}{\varepsilon_0}(\rho^{\mathrm{lr}}, \phi_j)$ the components of the vector $\mathbf{b}$, and the K unknowns $c_0, \ldots, c_{K-1}$ collected in the vector $\mathbf{c}$.

The finite element method [111, 146] is a Galerkin discretization with special trial and test functions $\{\phi_j\}$ that have local support. To this end, we decompose the domain Ω into small disjoint subdomains called elements. In the two-dimensional case these elements are mostly triangles or quadrilaterals. In the three-dimensional case they are tetrahedra, cubes, bricks, pyramids, or prisms. Altogether, these elements define a mesh on Ω. In the simplest case, all elements have the same shape and size and constitute a uniform mesh. Figure 7.6 (left) shows such a decomposition of a two-dimensional domain. Here, the decomposition consists of square elements of the same size.

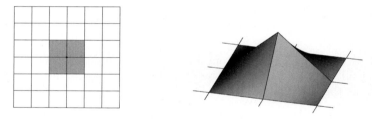

Fig. 7.6. Regular decomposition of a domain in the finite element method (left) and a bilinear basis function with its support (right).

The basis functions ϕ_k are now defined over these subdomains as piecewise polynomials that each assume the value one at their associated vertex $\mathbf{x}$ and zero at all other vertices of the mesh. The support of ϕ_k is then the union of the elements that have $\mathbf{x}$ as a vertex. In the simplest case, piecewise linear functions can be used as basis functions. An example is given in Figure (7.6) (right). The finite element method uses the approximation with piecewise polynomials of fixed order. The size of the elements and thus the global mesh size can now be decreased to obtain better approximations. Typically, this leads to errors of the form $C \cdot h^p$, where h denotes the size of the elements and p the largest obtainable order, which depends on the degree of the basis functions as well as on the smoothness of the solution.[21]

---

[21] This is the h-version of the finite element method. If one instead fixes h and changes the order of the polynomials, one obtains the so-called p-version. Simultaneously changing both h and p leads to the so-called hp-version [54, 597].

The finite element method provides an efficient interpolation (or approximation) between the positions of the particles and the mesh points in a natural way. Moreover, the evaluation of ϕ_K^{lr} at the positions of the particles directly implements an interpolation via the basis functions. In a dual way, the discretization of the right hand side in (7.36) yields an approximation of the charge distribution on the finite element mesh, smoothed by the ϱ_j. If one substitutes the definition (7.11) of the right hand side ρ^{lr} of the potential equation (7.29) into the definition of b_j, one obtains

$$b_j = \frac{1}{\varepsilon_0} \sum_{\mathbf{n} \in \mathbb{Z}^3} \sum_{i=1}^{N} q_i (\varrho_i^{\mathbf{n}}, \phi_j) \quad \text{for all } j = 0, \dots K - 1. \tag{7.38}$$

Thus, to compute any component b_j of the right hand side $\mathbf{b}$, one has to evaluate a sum over all particles, for which the integral $(\varrho_i^{\mathbf{n}}, \phi_j)$ of the product of basis function and charge cloud function is not zero (which means that the support of the basis function ϕ_j and the support of $\varrho_i^{\mathbf{n}}$ overlap). Provided that $N = \mathcal{O}(K)$ and that the particles are approximately equidistributed, one can compute the integrals in $(\varrho_i^{\mathbf{n}}, \phi_j)$ and hence b_j with $\mathcal{O}(1)$ operations since the functions ϕ_j have local support. The entire right hand side $\mathbf{b}$ can then be obtained with $\mathcal{O}(K)$ operations.

In addition, the stiffness matrix A is sparsely populated, because the local support of the basis functions implies that $(\nabla \phi_i, \nabla \phi_j)$ is zero except for the case of overlapping supports of ϕ_i and ϕ_j. This is advantageous with respect to the memory requirements of the approach. Still the issue remains how to efficiently solve the linear system of equations (7.37). It would be desirable to use a method in which the complexity increases only linearly with the number of unknowns K, so that only $\mathcal{O}(K)$ or $\mathcal{O}(K \log(K)^{\alpha})$, $\alpha > 0$, operations are necessary to solve the system of equations up to a prescribed accuracy.

The standard direct methods such as Gaussian elimination or the Cholesky method [255] have a higher memory complexity (because of fill in) and a higher time complexity than desired. Also classic iterative methods such as the Richardson method, the Jacobi method, the Gauss-Seidel method, or the SOR method [285] are in general not suitable because of their higher complexity: The convergence rate of those simple iterative methods depends on the mesh size (the dimension of the approximation space V_K); the finer the discretization, the worse the convergence rate becomes. A more careful analysis shows that the convergence rate of these methods is initially high, but decreases drastically after a few iterations. The reason for this behavior becomes clear after a Fourier analysis of the error in the iterations: The low wavelength/high frequency part of the error is reduced strongly in each step of the iteration, but the high wavelength/low frequency part is damped only weakly. Therefore, the low frequency components of the error dominate the convergence rate after a few initial steps.

The fundamental idea behind the multigrid method [117, 283], and, at least implicitly, behind other multilevel methods such as the BPX preconditioner [112, 456], is to avoid the drawbacks of the simple iteration methods by using an additional so-called coarse grid correction. This correction ensures that also the low frequency components of the error are reduced substantially in each step. For the coarse grid correction, a coarser mesh (usually with elements twice as large in all directions) is used to approximate the linear system of equations under consideration by a smaller system of equations whose solution still represents the low frequency components of the fine grid solution well. This smaller system may then again be treated by a simple iterative method. A recursive application of this idea leads to the multigrid method. Using a sequence of nested grids (and assuming that the problem under consideration satisfies certain regularity conditions), the multigrid method converges with a rate which is independent of the mesh size and therefore independent of the number K of unknowns. The runtime complexity of the solution of the Poisson equation up to a given accuracy with such multigrid or multilevel methods is then only of the order $\mathcal{O}(K)$.

In our special case of rectangular simulation domains and periodic boundary conditions, it is possible to alternatively use special direct methods based on the fast Fourier transform. These methods have a complexity of the order $\mathcal{O}(K \log(K))$ and, assuming $K = \mathcal{O}(N)$, also a complexity of the order $\mathcal{O}(N \log(N))$. We will consider such techniques in Section 7.3 in more detail.

7.2.3 Some Variants

There is a large number of possibilities to combine different discretization methods, different solvers for the discrete system, and different techniques for the computation of the forces into a method for the overall problem. In the following, we review briefly the most common approaches. They differ in the choice of the shielding charge cloud, in the discretization method, and in the computation of the forces from the potential. Often, a fast Fourier transform is used for the solution of the potential equation. A survey on these methods can be found in [211, 488, 574] and [612]. Articles that study the different sources of errors in these methods are for instance [100] (artifacts caused by boundary conditions, especially dielectric boundary conditions), [97] (artifacts in the pressure and free energy) and [570].

Particle-Particle Particle-Mesh Method (P³M). The particle-particle particle-mesh (P³M) method developed by Hockney and Eastwood [202, 324] employs spherical charge clouds of the form (7.14) to shield the point charges. The short-range terms can then be evaluated with the linked cell method. The charge distribution is interpolated by trilinear interpolation to the mesh points. The long-range terms are evaluated in Fourier space. Here, the Green's functions used can be optimized depending on the size of the system, the form of the shielding charge distributions, and the interpolation method. The forces at the mesh points are computed as differences of the values of the potential. The forces at the positions of the particles are then determined by interpolation.

The variants derived from the original P^{3}M method differ in the choice of interpolation methods, the form of the shielding charge cloud, the optimized Green's function, and the computation of the forces from the potential. Some suggestions for improvements and further developments of the method can be found in [399, 646, 660].

Particle-Mesh Ewald Method (PME). The particle-mesh Ewald method [168, 215, 374] uses the Gaussians from (7.13) as charge clouds.[22] Furthermore, higher order methods are used for interpolation (such as Lagrange interpolation or B-spline interpolation). The accuracy and efficiency of this method have been considered in [477]. A combination with Respa time stepping is studied in [485]. In the following sections of this chapter, we describe a variant in detail, the so-called smooth particle-mesh Ewald method (SPME). There, B-splines of degree $p > 2$ are used in the interpolation. Furthermore, the forces are computed as gradients of the derived approximation to the potential. Thus, the trial and test functions used in this method have to be differentiable. A comparison of the P^{3}M method, the PME method, and the SPME method can be found in [167, 179].

Fast Fourier Poisson Method. In the so-called fast Fourier Poisson method [675], the charge clouds are evaluated at the mesh points and these values are used to compute the elements b_j of the vector on the right hand side of the discretized equation. The resulting discrete Poisson problem is then solved with help of the fast Fourier transform. Energies and forces are computed from sums over the values of the approximated potential Φ_K^{lr} at the mesh points (but not at the positions of the particles).

Multigrid Methods and Adaptive Refinement. The solver in the discussed mesh-based methods is not limited to fast Fourier transform methods. Other methods for the efficient solution of the discrete potential equation can be used as well, as for instance multilevel methods.

In the case of a nonuniform particle distribution, the efficiency of FFT based techniques is substantially reduced. They require a discretization on a uniform mesh. However, the mesh size has to be fine enough to capture the nonuniformly distributed particles. To this end, adaptive methods have been developed in which the mesh points no longer have to be distributed

---

[22] The use of Gaussians to decompose the potential together with trigonometric trial and test functions first leads (without the use of fast Poisson solvers (FFT)) to the so-called classic Ewald sum [216]. For a given accuracy one can choose the number of degrees of freedom for the discretization, the threshold parameter r_{cut} in the linked cell method, and the parameter G from the Gaussian (which determines the balance between long-range and short-range terms) in such a way that the complexity is reduced from $\mathcal{O}(N^2)$ to $\mathcal{O}(N^{3/2})$, compare [228, 239, 612].

uniformly over the entire domain but which work with locally refined meshes [157, 158, 249, 473, 577], compare Figure 7.7. Here, adaptive finite element methods [53, 644] can be applied directly. For the solution of the resulting systems, adaptive multigrid methods are a natural choice [114, 115, 116, 283]. An example can be found in [343].

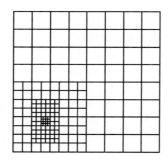

Fig. 7.7. Adaptive mesh refinement in the finite element method.

7.3 Smooth Particle-Mesh Ewald Method (SPME)

We now discuss the so-called smooth particle-mesh Ewald (SPME) method [168, 215, 374] in more detail. It is a particular variant of the approach presented in Section 7.1, which uses trigonometric functions. The components of the method are chosen as follows:

- *Gaussians* from (7.13) are used as shielding charge distributions $\varrho_i^{\mathbf{n}}$.
- *B-splines* with order > 2 are used as local basis functions ϕ_k. They are differentiable and provide approximations of higher order.
- After computing the right hand side $\mathbf{b}$ of the linear system of equations using B-splines, $\mathbf{b}$ is mapped into Fourier space via the fast discrete Fourier transform. Then, the Laplacian, which corresponds to just a diagonal scaling in Fourier space, is inverted, the result is mapped back to real space via an inverse fast Fourier transform, and the result is expressed by local basis functions.
- The force is computed according to (7.32) as the gradient of the approximation of the potential. Since the B-splines are differentiable, the gradient can be computed directly.

With such methods one can achieve a high accuracy with a complexity of $\mathcal{O}(N \log(N))$. Here, the logarithmic factor arises from the use of the fast Fourier transform.

7.3.1 Short-Range Terms

We now follow the approach described in general terms in Sections 7.2.1 and 7.2.2 for the computation of the energies and forces and explain its implementation in more detail for the SPME method. With help of the so-called error function

$$\text{erf}(x) := \frac{2}{\sqrt{\pi}} \int_0^x e^{-y^2} dy,$$

the complementary error function

$$\text{erfc}(x) := 1 - \text{erf}(x),$$

the threshold parameter r_{cut}, the distance vector $\mathbf{r}_{ij}^{\mathbf{n}} := \mathbf{x}_j^{\mathbf{n}} - \mathbf{x}_i$, the distance $r_{ij}^{\mathbf{n}} := ||\mathbf{x}_j^{\mathbf{n}} - \mathbf{x}_i||$, and with the choice of ϱ as a Gaussian according to (7.13), an approximation for (7.28) can be written explicitly as[23]

$$
\begin{aligned}
V^{\text{sr}} &\approx \frac{1}{2} \frac{1}{4\pi\varepsilon_0} \sum_{\mathbf{n}\in\mathbb{Z}^3} \sum_{i=1}^{N} \sum_{\substack{j=1 \\ i\neq j \text{ for } \mathbf{n}=0 \\ r_{ij}^{\mathbf{n}} < r_{\text{cut}}}}^{N} q_i q_j \frac{1 - \text{erf}(Gr_{ij}^{\mathbf{n}})}{r_{ij}^{\mathbf{n}}} \\
&= \frac{1}{2} \frac{1}{4\pi\varepsilon_0} \sum_{\mathbf{n}\in\mathbb{Z}^3} \sum_{i=1}^{N} \sum_{\substack{j=1 \\ i\neq j \text{ for } \mathbf{n}=0 \\ r_{ij}^{\mathbf{n}} < r_{\text{cut}}}}^{N} q_i q_j \frac{\text{erfc}(Gr_{ij}^{\mathbf{n}})}{r_{ij}^{\mathbf{n}}}.
\end{aligned}
\tag{7.39}
$$

The application of the gradient operator[24] results in

---

[23] Given $\varrho(r) = \left(\frac{G}{\sqrt{\pi}}\right)^3 \cdot e^{-G^2 r^2}$, the antiderivative F of $r \cdot \varrho(r)$ is

$$F(r) = -\frac{1}{2} \frac{G}{\pi^{3/2}} e^{-G^2 r^2}.$$

Applying (7.27), it follows that

$$
\int_{\mathbb{R}^3} \frac{\varrho_j^{\mathbf{n}}(\mathbf{y})}{||\mathbf{y} - \mathbf{x}_i||} d\mathbf{y} = -\frac{4\pi}{||\mathbf{x}_j^{\mathbf{n}} - \mathbf{x}_i||} \int_0^{||\mathbf{x}_j^{\mathbf{n}} - \mathbf{x}_i||} F(r) dr =
$$

$$
\frac{1}{||\mathbf{x}_j^{\mathbf{n}} - \mathbf{x}_i||} \frac{2}{\sqrt{\pi}} \int_0^{G\cdot||\mathbf{x}_j^{\mathbf{n}} - \mathbf{x}_i||} e^{-r^2} dr = \frac{1}{||\mathbf{x}_j^{\mathbf{n}} - \mathbf{x}_i||} \text{erf}(G \cdot ||\mathbf{x}_j^{\mathbf{n}} - \mathbf{x}_i||).
$$

[24] The gradient operator with respect to $\mathbf{x}_i^{\mathbf{0}}$ has to be applied to the total energy

$$
\frac{1}{2} \frac{1}{4\pi\varepsilon_0} \sum_{\mathbf{n},\mathbf{m}\in\mathbb{Z}^3} \sum_{\substack{i,j=1 \\ (\mathbf{n},i)\neq(\mathbf{m},j)}}^{N} q_i q_j \frac{\text{erfc}(G||\mathbf{x}_j^{\mathbf{n}} - \mathbf{x}_i^{\mathbf{m}}||)}{||\mathbf{x}_j^{\mathbf{n}} - \mathbf{x}_i^{\mathbf{m}}||}
$$

which results from the periodic extension of the simulation box to $\mathbb{R}^3$, and is infinite because of the summation over all $\mathbf{n}, \mathbf{m} \in \mathbb{Z}^3$. Here, the interactions between $\mathbf{x}_i^{\mathbf{m}}$ and $\mathbf{x}_j^{\mathbf{n}}$ appear twice. This is the reason why the factor $1/2$ no longer occurs in the computation of the force on the particles $\mathbf{x}_i^{\mathbf{0}}$ in the simulation box.

$$\mathbf{F}_i^{\mathrm{sr}} \approx -\frac{1}{4\pi\varepsilon_0}\, q_i \sum_{\substack{\mathbf{n}\in\mathbb{Z}^3}} \sum_{\substack{j=1 \\ j\neq i\ \mathrm{for}\ \mathbf{n}=0 \\ r_{ij}^{\mathbf{n}}<r_{\mathrm{cut}}}}^{N} q_j \frac{1}{(r_{ij}^{\mathbf{n}})^2}\left(\mathrm{erfc}(Gr_{ij}^{\mathbf{n}})+\frac{2G}{\sqrt{\pi}}r_{ij}^{\mathbf{n}}e^{-(Gr_{ij}^{\mathbf{n}})^2}\right)\frac{\mathbf{r}_{ij}^{\mathbf{n}}}{r_{ij}^{\mathbf{n}}}. \quad (7.40)$$

The linked cell method from Chapter 3 can then be used directly for the computation of V^{sr} and $\mathbf{F}_i^{\mathrm{sr}}$.[25]

Figure 7.8 shows the function $\mathrm{erfc}(x)/x$ (whose translated versions are combined to make up the short range term Φ^{sr}), the function $\mathrm{erf}(x)/x$ (whose translated versions are combined to make up the long range term Φ^{lr}), and the sum $\mathrm{erf}(x)/x + \mathrm{erfc}(x)/x = 1/x$. For the purpose of comparison, the fast decaying function $1/x^6$ is shown which appears in the Lennard-Jones potential.

The graphs in Figure 7.9 on the semi-logarithmic scale show clearly that $\mathrm{erfc}(x)/x$ decays very quickly, even significantly faster than $1/x^6$. For large values of x, the functions $\mathrm{erf}(x)/x$ and $1/x$ agree well with each other.

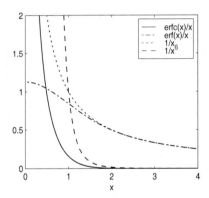

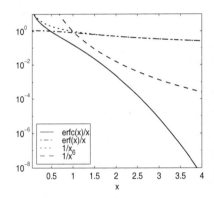

Fig. 7.8. The short-range function $\mathrm{erfc}(x)/x$, the long-range function $\mathrm{erf}(x)/x$, the Coulomb potential $1/x$ and the $1/x^6$ potential on a linear scale.

Fig. 7.9. The functions from Figure 7.8, semi-logarithmic plot.

---

[25] The computation can be accelerated using tabulation and interpolation for the exponential function, the erf function, and the erfc function. The values of these functions at several places are computed and stored beforehand in a table [75, 217]. These values can then be used directly if the function value at these positions is sought. If interim values are to be computed, they can be approximated by interpolation. This avoids the repeated and costly evaluation of the exponential function and the erf and erfc functions. However, the actual saving in runtime depends strongly on the particular implementation. Obviously, such an optimization does not reduce the order of complexity of the algorithm.

7.3.2 Long-Range Terms

Similarly to (7.39), V^{lr} can be written explicitly as

$$V^{\mathrm{lr}} = \frac{1}{2}\frac{1}{4\pi\varepsilon_0} \sum_{\mathbf{n}\in\mathbb{Z}^3} \sum_{i=1}^{N} \sum_{\substack{j=1 \\ i\neq j \text{ for } n=0}}^{N} q_i q_j \frac{\mathrm{erf}(Gr_{ij}^{\mathbf{n}})}{r_{ij}^{\mathbf{n}}}. \tag{7.41}$$

Since a direct evaluation of this sum is too expensive, the long-range terms V^{lr} of the energy are computed approximately according to (7.31) and (7.21), and the force terms $\mathbf{F}_i^{\mathrm{lr}}$ are computed approximately according to (7.32).

We have previously emphasized that the trial and test functions in the Galerkin method have to be *local* so that, on the one hand, the discretization of the right hand side according to (7.38) will not be too expensive, and that, on the other hand, the Galerkin approximation to the potential can be evaluated efficiently, compare (7.31) and (7.32). In this section, we show that the use of special *global* basis functions can be advantageous as well. The idea is to compute the inverse of the Laplace operator in Fourier space (which is easily possible since the stiffness matrix A_{trig} resulting from the use of trigonometric basis functions in the Galerkin method is diagonal, compare (7.46)), but to represent the approximation Φ_K^{lr} to the potential as a sum over local real basis functions. The coefficients of the Galerkin approximation from (7.37) (this is the approximation with local real finite element basis functions) are then determined by[26]

$$\mathbf{c} \approx T^* A_{\mathrm{trig}}^{-1} T \mathbf{b}, \tag{7.42}$$

where the matrix T expresses the change from the representation in the local real basis to the representation in the complex trigonometric basis.[27] This approach corresponds to a Galerkin discretization of the right hand side with local bases, its subsequent transformation to the Fourier space, the inversion of the Laplace operator in the Fourier space, and a subsequent back-transformation of the solution into (the dual of) the original space which is spanned by the local bases.

It is now necessary to specify the maps T and T^*. They can be implemented efficiently using fast Fourier transforms and fast inverse Fourier transforms.

---

[26] T^* is the adjoint operator to T. With the representation $T = T_1 + iT_2$, where T_1 denotes the real part and iT_2 the imaginary part of T, it holds that $T^* = (T_1 + iT_2)^* = T_1^T - iT_2^T$.

[27] Note that the spaces spanned by the local basis functions and the trigonometric basis functions are in general not the same and that therefore the maps T and T^* are in general not coordinate transforms. Rather, they correspond to the approximation of the functions from one space by the functions of another space, for instance by interpolation. In generally it only holds that $A^{-1} \approx T^* A_{\mathrm{trig}}^{-1} T$, i.e., equality does not necessarily hold. The quality of this approximation depends on how well trigonometric functions can be represented by the chosen local basis functions (and vice versa).

Discretization Using Trigonometric Functions. We associate to the index $\mathbf{k} = (k_1, k_2, k_3) \in \mathbb{Z}^3$ a new index

$$\mathbf{k}_L := \left(\frac{k_1}{L_1}, \frac{k_2}{L_2}, \frac{k_3}{L_3} \right) \in \mathbb{R}^3, \qquad (7.43)$$

which has been scaled by the dimensions of the simulation domain. Furthermore, let $|\Omega| = L_1 \cdot L_2 \cdot L_3$ again denote the volume of the simulation domain. Here, we use the complex trigonometric trial functions

$$\psi_{\mathbf{k}} = e^{2\pi i \mathbf{k}_L \cdot \mathbf{x}}, \ \mathbf{k} \in \mathcal{K} \setminus \mathbf{0}, \qquad (7.44)$$

where

$$\mathcal{K} := \left[-\left\lfloor \tfrac{K_1-1}{2} \right\rfloor, \left\lceil \tfrac{K_1-1}{2} \right\rceil \right] \times \left[-\left\lfloor \tfrac{K_2-1}{2} \right\rfloor, \left\lceil \tfrac{K_2-1}{2} \right\rceil \right] \times \left[-\left\lfloor \tfrac{K_3-1}{2} \right\rfloor, \left\lceil \tfrac{K_3-1}{2} \right\rceil \right], \quad (7.45)$$

i.e., we use the integers from the symmetric intervals around the origin of the lengths $K_i, i = 1, 2, 3$, as frequencies for our trigonometric basis functions.

The orthogonality property[28] $(\psi_{\mathbf{k}}, \psi_{\mathbf{j}}) = |\Omega| \cdot \delta_{\mathbf{k},\mathbf{j}}$ of the trigonometric functions and the relation $\nabla_{\mathbf{x}} \psi_{\mathbf{k}} = 2\pi i \mathbf{k}_L e^{2\pi i \mathbf{k}_L \cdot \mathbf{x}}$ yields the *real* diagonal matrix

$$A_{\text{trig}} = \{(\nabla_{\mathbf{x}} \psi_{\mathbf{k}}, \nabla_{\mathbf{x}} \psi_{\mathbf{j}})\}_{\mathbf{k},\mathbf{j} \in \mathcal{K} \setminus \mathbf{0}} = \text{diag}\left(\left\{ |\Omega| (2\pi)^2 ||\mathbf{k}_L||^2 \right\}_{\mathbf{k} \in \mathcal{K} \setminus \mathbf{0}} \right) \quad (7.46)$$

as the stiffness matrix from the Galerkin method, compare (7.36). If one uses trigonometric functions (7.44) as both trial and test functions, one therefore obtains the linear system of equations

$$A_{\text{trig}} \mathbf{c}^{\text{trig}} = \mathbf{b}^{\text{trig}}$$

with A_{trig} from (7.46) and $b_{\mathbf{k}}^{\text{trig}} = \frac{1}{\varepsilon_0}(\rho^{\text{lr}}, \psi_{\mathbf{k}})$, which can be solved directly. The Galerkin approximation of the potential in this discretization is then given by

$$\Phi_{\mathbf{K},\text{trig}}^{\text{lr}} = \sum_{\mathbf{k} \in \mathcal{K} \setminus \mathbf{0}} \mathbf{c}_{\mathbf{k}}^{\text{trig}} \psi_{\mathbf{k}} \qquad (7.47)$$

with

$$\mathbf{c}_{\mathbf{k}}^{\text{trig}} = \frac{1}{|\Omega|} \frac{1}{(2\pi)^2 ||\mathbf{k}_L||^2} \frac{1}{\varepsilon_0}(\rho^{\text{lr}}, \psi_{\mathbf{k}}),$$

and the right hand side satisfies the relation

---

[28] Here, $\delta_{\mathbf{k},\mathbf{j}}$ denotes the Kronecker delta. It satisfies $\delta_{\mathbf{k},\mathbf{j}} = 1$ for $\mathbf{k} = \mathbf{j}$ and $\delta_{\mathbf{k},\mathbf{j}} = 0$ otherwise.

$$
\begin{aligned}
\mathbf{b}_{\mathbf{k}}^{\text{trig}} &= \frac{1}{\varepsilon_0}(\rho^{\text{lr}}, \psi_{\mathbf{k}}) = \frac{1}{\varepsilon_0} \sum_{\mathbf{n}\in\mathbb{Z}^3} \sum_{j=1}^{N} q_j(\varrho_j^{\mathbf{n}}, \psi_{\mathbf{k}}) \\[2mm]
&= \frac{1}{\varepsilon_0}\left(\frac{G}{\sqrt{\pi}}\right)^3 \sum_{\mathbf{n}\in\mathbb{Z}^3} \sum_{j=1}^{N} q_j \int_{\Omega} e^{-G^2\|\mathbf{x}-\mathbf{x}_j^{\mathbf{n}}\|^2} e^{-2\pi i \mathbf{k}_L \cdot \mathbf{x}} d\mathbf{x} \\[2mm]
&= \frac{1}{\varepsilon_0}\left(\frac{G}{\sqrt{\pi}}\right)^3 \sum_{j=1}^{N} q_j \int_{\mathbb{R}^3} e^{-G^2\|\mathbf{x}-\mathbf{x}_j\|^2} e^{-2\pi i \mathbf{k}_L \cdot \mathbf{x}} d\mathbf{x} \\[2mm]
\underset{\mathbf{y}:=\mathbf{x}-\mathbf{x}_j}{=}\ & \frac{1}{\varepsilon_0}\left(\frac{G}{\sqrt{\pi}}\right)^3 \sum_{j=1}^{N} q_j \int_{\mathbb{R}^3} e^{-G^2\|\mathbf{y}\|^2} e^{-2\pi i \mathbf{k}_L \cdot (\mathbf{y}+\mathbf{x}_j)} d\mathbf{y} \\[2mm]
&= \frac{1}{\varepsilon_0}\left(\frac{G}{\sqrt{\pi}}\right)^3 \sum_{j=1}^{N} q_j e^{-2\pi i \mathbf{k}_L \cdot \mathbf{x}_j} \int_{\mathbb{R}^3} e^{-G^2\|\mathbf{x}\|^2} e^{-2\pi i \mathbf{k}_L \cdot \mathbf{x}} d\mathbf{x} \\[2mm]
\underset{\text{Footnote 29}}{=}\ & \frac{1}{\varepsilon_0}\left(\frac{G}{\sqrt{\pi}}\right)^3 \sum_{j=1}^{N} q_j e^{-2\pi i \mathbf{k}_L \cdot \mathbf{x}_j} \left(\frac{\sqrt{\pi}}{G}\right)^3 e^{-\pi^2 \mathbf{k}_L^2/G^2} \\[2mm]
&= \frac{1}{\varepsilon_0} e^{-\frac{\|2\pi\mathbf{k}_L\|^2}{4G^2}} \sum_{j=1}^{N} q_j e^{-2\pi i \mathbf{k}_L \cdot \mathbf{x}_j} \\[2mm]
&= \frac{1}{\varepsilon_0} e^{-\frac{\|2\pi\mathbf{k}_L\|^2}{4G^2}} \sum_{j=1}^{N} q_j \overline{\psi_{\mathbf{k}}}(\mathbf{x}_j), \quad \mathbf{k}\in\mathcal{K}\setminus\mathbf{0} \qquad (7.48)
\end{aligned}
$$

for the components of the trigonometric Galerkin discretization $\mathbf{b}^{\text{trig}}$ of the right hand side of the potential equation. The factor $e^{-\frac{\|2\pi\mathbf{k}_L\|^2}{4G^2}}$ results from the Fourier transform of the Gaussian centered at the origin and the factors $\overline{\psi_{\mathbf{k}}}(\mathbf{x}_j) = e^{-2\pi i \mathbf{k}_L \cdot \mathbf{x}_j}$ result from the translation of the Gaussian to the particle positions $\mathbf{x}_j$.

Approximation with Local Functions – B-Splines. The computation of the $b_{\mathbf{k}}^{\text{trig}}$ according to (7.48) would be very expensive, because of the global support of the functions $\psi_{\mathbf{k}}$. Therefore, we approximate the complex trigonometric functions $\psi_{\mathbf{k}}$ with a sum over local, real, and periodic functions $\phi_{\mathbf{m}}$

$$
\psi_{\mathbf{k}} \approx \sum_{\mathbf{m}=0}^{\mathbf{K}-1} \overline{t_{\mathbf{km}}}\phi_{\mathbf{m}}, \quad t_{\mathbf{km}}\in\mathbb{C}. \qquad (7.49)
$$

For example, one can choose finite element bases that are defined on a mesh.[30] The matrix T is then given by the weights $t_{\mathbf{km}}\in\mathbb{C}$, i.e., it holds that

---

[29] Use $\dfrac{1}{\sqrt{2\pi}}\displaystyle\int_{-\infty}^{\infty} e^{-x^2/2}e^{ixy}dx = e^{-y^2/2}$.

[30] Note that the running index of the sum in (7.49) is a multi-index $\mathbf{m} = (m_1, m_2, m_3)$. We use the notation $\sum_{\mathbf{m}=0}^{\mathbf{K}-1}$ for the multiple summation $\sum_{m_1=0}^{K_1-1}\sum_{m_2=0}^{K_2-1}\sum_{m_3=0}^{K_3-1}$.

$T_{\mathbf{km}} = t_{\mathbf{km}}$. The vector $\mathbf{c}$ is computed approximately by (7.42). In this way, the explicit computation of the right hand side $\mathbf{b}^{\mathrm{trig}}$ can be avoided, i.e., we use the approximation $\mathbf{b}^{\mathrm{trig}} \approx T\mathbf{b}$ instead.[31] Here, the number of functions $\phi_{\mathbf{m}}$ is chosen such that a quadratic matrix T results.

Applying the approximation from (7.49) with local bases, one obtains

$$\sum_{j=1}^{N} q_j \overline{\psi_{\mathbf{k}}}(\mathbf{x}_j) \approx \sum_{j=1}^{N} q_j \sum_{\mathbf{m}=0}^{K-1} \overline{t_{\mathbf{km}} \phi_{\mathbf{m}}}(\mathbf{x}_j)$$

$$= \sum_{\mathbf{m}=0}^{K-1} t_{\mathbf{km}} \sum_{j=1}^{N} q_j \overline{\phi_{\mathbf{m}}}(\mathbf{x}_j) = (TQ)_{\mathbf{k}}, \qquad (7.50)$$

with[32]

$$Q_{\mathbf{m}} := \sum_{j=1}^{N} q_j \overline{\phi_{\mathbf{m}}}(\mathbf{x}_j) = \underset{\substack{\phi_{\mathbf{m}}(\mathbf{x}) \in \mathbb{R}, \\ \phi_{\mathbf{m}}(\mathbf{x}) = \overline{\phi}_{\mathbf{m}}(\mathbf{x})}}{\sum_{j=1}^{N}} q_j \phi_{\mathbf{m}}(\mathbf{x}_j), \qquad \mathbf{m} \in [\mathbf{0}, \cdots, \mathbf{K} - 1]. \quad (7.51)$$

The vector Q can be understood as the interpolation of the particle point charges from the particle positions to the mesh points. For each index $\mathbf{m}$, the sum in (7.51) runs over those particles that lie within the support of $\phi_{\mathbf{m}}$. Since the support of the $\phi_{\mathbf{m}}$ is bounded, the entire vector Q for $N = \mathcal{O}(K_1 K_2 K_3)$ can be computed with a complexity of $\mathcal{O}(N)$, where the hidden constant is proportional to the size of the support of the basis functions. The matrix-vector product TQ can be implemented by way of fast Fourier transforms.

Using (7.48), one obtains all in all

$$\mathbf{b}_{\mathbf{k}}^{\mathrm{trig}} = \frac{1}{\varepsilon_0}(\rho^{\mathrm{lr}}, \psi_{\mathbf{k}}) \approx \frac{1}{\varepsilon_0} e^{-\frac{\|2\pi \mathbf{k}_L\|^2}{4G^2}} (TQ)_{\mathbf{k}}. \qquad (7.52)$$

The application of $T^* A_{\mathrm{trig}}^{-1}$ to this approximation of $\mathbf{b}^{\mathrm{trig}}$ leads to an approximation of $\mathbf{c}$ (in a representation in the local basis) with $\mathbf{c} \approx T^* A_{\mathrm{trig}}^{-1} T\mathbf{b}$.[33]

---

[31] With the approximation (7.49), it holds that

$$\mathbf{b}_{\mathbf{k}}^{\mathrm{trig}} = \frac{1}{\varepsilon_0}(\rho^{\mathrm{lr}}, \psi_{\mathbf{k}}) = \frac{1}{\varepsilon_0}\int_{\Omega} \rho^{\mathrm{lr}} \overline{\psi_{\mathbf{k}}} d\mathbf{x} \underset{(7.49)}{\approx} \frac{1}{\varepsilon_0}\int_{\Omega} \rho^{\mathrm{lr}} \sum_{\mathbf{m}=0}^{K-1} \overline{t_{\mathbf{km}} \phi_{\mathbf{m}}} d\mathbf{x}$$

$$= \sum_{\mathbf{m}=0}^{K-1} t_{\mathbf{km}} \frac{1}{\varepsilon_0}\int_{\Omega} \rho^{\mathrm{lr}} \overline{\phi_{\mathbf{m}}} d\mathbf{x} = \sum_{\mathbf{m}=0}^{K-1} t_{\mathbf{km}} \frac{1}{\varepsilon_0}(\rho^{\mathrm{lr}}, \phi_{\mathbf{m}}) = (T\mathbf{b})_{\mathbf{k}}.$$

[32] The expression $\mathbf{m} \in [\mathbf{0}, \ldots, \mathbf{K} - 1]$ is defined as $(m_1, m_2, m_3) \in \{0, \ldots, K_1 - 1\} \times \{0, \ldots, K_2 - 1\} \times \{0, \ldots, K_3 - 1\}$.

[33] Note that the result of this multiplication is real, without any complex part, even though complex intermediate values appear in the computation. This follows from the special structure of A_{trig} as a diagonal matrix with real entries.

One possible choice for these local functions in (7.49) is given by B-splines over a uniform mesh on Ω with a mesh width of

$$\mathbf{h} = (L_1/K_1, L_2/K_2, L_3/K_3). \tag{7.53}$$

The spline M_p of order[34] p is defined in one dimension by the recursion

$$M_p(x) = \frac{x}{p-1} M_{p-1}(x) + \frac{p-x}{p-1} M_{p-1}(x-1) \tag{7.54}$$

with

$$M_2(x) = \begin{cases} 1 - |x-1|, & \text{for } x \in [0,2], \\ 0, & \text{otherwise.} \end{cases} \tag{7.55}$$

The spline M_p is $p-2$ times continuously differentiable, and its derivative satisfies the recurrence relation

$$\frac{dM_p}{dx}(x) = M_{p-1}(x) - M_{p-1}(x-1). \tag{7.56}$$

Multidimensional splines are defined as the appropriate products of one-dimensional splines.

For splines of even order p, one obtains an approximation of $e^{-2\pi i \mathbf{k}_L \cdot \mathbf{x}}$ by[35]

$$e^{-2\pi i \mathbf{k}_L \cdot \mathbf{x}} \approx \sum_{\mathbf{n} \in \mathbb{Z}^3} \sum_{\mathbf{m}=0}^{\mathbf{K}-1} t_{\mathbf{km}} \cdot \prod_{d=1}^{3} M_p((\mathbf{x})_d K_d / L_d - m_d - n_d K_d), \tag{7.57}$$

where

$$t_{\mathbf{km}} := B(\mathbf{k}) \cdot \left(\prod_{d=1}^{3} e^{-2\pi i \frac{k_d m_d}{K_d}} \right) \tag{7.58}$$

and

$$B(\mathbf{k}) := \prod_{d=1}^{3} B_{K_d}(k_d) \quad \text{with} \quad B_{K_d}(k_d) = \frac{e^{-2\pi i (p-1) k_d / K_d}}{\sum_{q=0}^{p-2} e^{-2\pi i k_d q / K_d} M_p(q+1)}, \tag{7.59}$$

see also [89, 145] and [550]. Both sides of (7.57) are actually equal at the mesh points $(L_1 m_1/K_1, L_2 m_2/K_2, L_3 m_3/K_3)$ with $\mathbf{m} \in [\mathbf{0}, \ldots, \mathbf{K}-1]$. This can be verified by plugging in the coordinates of the mesh points. In this case the entries of the matrix T are given by $t_{\mathbf{km}}$ from (7.58).

---

[34] The degree of the spline function is then $p-1$.

[35] The summation $\sum_{\mathbf{n} \in \mathbb{Z}^3}$ results in a periodic spline, as necessary for this approach.

Solving the Potential Equation. With

$$Q_{\mathbf{m}} = \sum_{\mathbf{n} \in \mathbb{Z}^3} \sum_{j=1}^{N} q_j \prod_{d=1}^{3} M_p((\mathbf{x}_j)_d K_d/L_d - m_d - n_d K_d) \qquad (7.60)$$

as in (7.51) and

$$\mathrm{DF}[Q](\mathbf{k}) := \sum_{\mathbf{m}=0}^{K-1} Q_{\mathbf{m}} \cdot e^{-2\pi i \left(\frac{k_1 m_1}{K_1} + \frac{k_2 m_2}{K_2} + \frac{k_3 m_3}{K_3} \right)}, \qquad (7.61)$$

the discrete Fourier transform of Q, it holds that

$$\sum_{i=1}^{N} q_i e^{-2\pi i \mathbf{k}_L \mathbf{x}_i} \approx (TQ)_{\mathbf{k}} = B(\mathbf{k}) \mathrm{DF}[Q](\mathbf{k}), \qquad (7.62)$$

compare (7.50). Here, the action of T corresponds to a discrete Fourier transform and a multiplication by $B(\mathbf{k})$ from (7.59). In this way, one obtains the coefficients of the approximation of the potential with trigonometric functions (7.47) as

$$c_{\mathbf{k}}^{\mathrm{trig}} = \frac{1}{4\pi^2 |\Omega| \varepsilon_0} \frac{1}{||\mathbf{k}_L||^2} e^{-\pi^2 ||\mathbf{k}_L||^2 / G^2} B(\mathbf{k}) \mathrm{DF}[Q](\mathbf{k}). \qquad (7.63)$$

The factor $1/||\mathbf{k}_L||^2$ corresponds to the inverse of the Laplace operator in frequency space. The factor $e^{-\pi^2 ||\mathbf{k}_L||^2 / G^2}$ results from the Fourier transform of the Gaussians and leads to an exponential convergence of the coefficients. For large frequencies $\mathbf{k}$ the error is damped by the factor $e^{-\pi^2 ||\mathbf{k}_L||^2 / G^2}$. The reason for this very fast decay of the coefficients is given by the smoothness[36] of the right hand side ρ^{lr}.[37]

---

[36] The smoothness of a function (compare footnote 5) can be determined from the decay of its Fourier coefficients [30, 50, 616]. Then, the Sobolev spaces H^s for $s \in \mathbb{R}$ can also be defined directly with help of the Fourier transform as

$$H^s(\Omega) = \{u(\mathbf{x}) = \sum_{\mathbf{k} \in \mathbb{Z}^n} c_{\mathbf{k}} e^{-i\mathbf{k}\mathbf{x}} : \sum_{\mathbf{k} \in \mathbb{Z}^3} (1 + ||\mathbf{k}||_\infty)^{2s} \cdot |\hat{u}_{\mathbf{k}}|^2 < \infty\},$$

with the Fourier coefficients $\hat{u}_{\mathbf{k}} := \frac{1}{|\Omega|} \int_\Omega u(\mathbf{x}) e^{-2\pi i \mathbf{k}\mathbf{x}} d\mathbf{x}$ for $u \in L^1(\Omega)$. Similar relations between the smoothness of a function and the decay of its coefficients also hold for nonperiodic functions and their coefficients with respect to other multiscale bases [132, 164, 456]. In particular, the Fourier coefficients of C^∞ functions (i.e. functions which are infinitely often differentiable) decay exponentially.

[37] For $\rho = \frac{1}{\varepsilon_0} \sum_{i=1}^{N} \delta_{\mathbf{x}_i}$ instead of ρ^{lr} as right hand side in the potential equation, one would obtain the coefficients without the factors $e^{-\pi^2 ||\mathbf{k}_L||^2 / G^2}$, the summands would not be damped, and the sum could not be truncated without introducing larger errors.

The solution vector $\mathbf{c}$ is now given approximately by $\mathbf{c} \approx T^* \mathbf{c}^{\text{trig}}$ as

$$c_{\mathbf{m}} \approx \sum_{\substack{\mathbf{k} \in \mathcal{K} \\ \mathbf{k} \neq 0}} \overline{t_{\mathbf{km}} c_{\mathbf{k}}^{\text{trig}}} \tag{7.64}$$

$$= \sum_{\substack{\mathbf{k} \in \mathcal{K} \\ \mathbf{k} \neq 0}} \overline{\left(\prod_{d=1}^{3} B_{K_d}(k_d) e^{-2\pi i \frac{k_d m_d}{K_d}} \right) \frac{1}{4\pi^2 |\Omega| \varepsilon_0 \|\mathbf{k}_L\|^2} \; e^{-\pi^2 \frac{\|\mathbf{k}_L\|^2}{G^2}} B(\mathbf{k}) \mathrm{DF}[Q](\mathbf{k})}$$

$$= \sum_{\substack{\mathbf{k} \in \mathcal{K} \\ \mathbf{k} \neq 0}} \frac{1}{4\pi^2 |\Omega| \varepsilon_0 \|\mathbf{k}_L\|^2} \; e^{-\pi^2 \frac{\|\mathbf{k}_L\|^2}{G^2}} |B(\mathbf{k})|^2 \mathrm{DF}[Q](\mathbf{k}) \; e^{2\pi i (\frac{k_1 m_1}{K_1} + \frac{k_2 m_2}{K_2} + \frac{k_3 m_3}{K_3})}.$$

The last equation has the form of a Fourier series again. Therefore, we want to use a fast discrete Fourier transform for the fast evaluation of this series.

To be able to apply the FFT directly, we rewrite the sum in (7.64) so that it does not run from $-\left\lfloor \frac{K_i - 1}{2} \right\rfloor$ to $\left\lceil \frac{K_i - 1}{2} \right\rceil$, but from 0 to $K_i - 1$, $i = 1, 2, 3$. The translation invariance of the trigonometric functions[38] together with the definitions

$$d(\mathbf{0}) := 0,$$

$$d(\mathbf{k}) := \frac{1}{\varepsilon_0 |\Omega|} \frac{1}{(2\pi)^2} \frac{1}{\|\mathbf{m}\|^2} \; e^{-\frac{\pi^2 \|\mathbf{m}\|^2}{G^2}} \cdot |B(\mathbf{k})|^2, \quad \text{with} \tag{7.65}$$

$$\mathbf{m} = (m_1, m_2, m_3) \text{ with } m_d = \begin{cases} k_d / L_d & \text{for } k_d \leq K_d/2 \\ (k_d - K_d)/L_d & \text{for } k_d > K_d/2 \end{cases}, \quad d = 1, 2, 3,$$

$$a(\mathbf{k}) := d(\mathbf{k}) \cdot \mathrm{DF}[Q](\mathbf{k}), \tag{7.66}$$

implies the relations

$$\frac{1}{(2\pi)^2 |\Omega| \varepsilon_0} \sum_{\substack{\mathbf{k} = (k_1, k_2, k_3), \mathbf{k} \neq 0 \\ k_d \in \left[-\left\lfloor \frac{K_d - 1}{2} \right\rfloor, \left\lceil \frac{K_d - 1}{2} \right\rceil \right]}} \frac{1}{\|\mathbf{k}_L\|^2} \; e^{-\pi^2 \frac{\|\mathbf{k}_L\|^2}{G^2}} |B(\mathbf{k})|^2 \mathrm{DF}[Q](\mathbf{k}) \; e^{2\pi i (\sum_{d=1}^{3} \frac{k_d m_d}{K_d})}$$

$$= \sum_{\mathbf{k}=0}^{\mathbf{K}-1} d(\mathbf{k}) \mathrm{DF}[Q](\mathbf{k}) \; e^{2\pi i (\frac{k_1 m_1}{K_1} + \frac{k_2 m_2}{K_2} + \frac{k_3 m_3}{K_3})}$$

$$= \sum_{\mathbf{k}=0}^{\mathbf{K}-1} a(\mathbf{k}) e^{2\pi i (\frac{k_1 m_1}{K_1} + \frac{k_2 m_2}{K_2} + \frac{k_3 m_3}{K_3})}$$

$$= \mathrm{DF}^{-1}[a](\mathbf{m}), \tag{7.67}$$

where DF^{-1} denotes the discrete inverse Fourier transform.[39]

Altogether, we obtain for the coefficients $c_{\mathbf{m}}$ an approximation

---

[38] They satisfy $e^{2\pi i l k / K} = e^{2\pi i l (k-K)/K}$, for all $l, k, K \in \mathbb{Z}$.

[39] Note that we define the inverse transform without an additional scaling of $\frac{1}{K_1 K_2 K_3}$. In the literature, both conventions for the inverse Fourier transform can be found. Therefore, it only holds that $\mathrm{DF}^{-1}[\mathrm{DF}[Q]](\mathbf{m}) = K_1 K_2 K_3 \cdot Q(\mathbf{m})$.

$$c_{\mathbf{m}} = \mathrm{DF}^{-1}[a](\mathbf{m}) \tag{7.68}$$

and therefore

$$\Phi_{\mathbf{K}}^{\mathrm{lr}}(\mathbf{x}) = \sum_{\mathbf{n}\in\mathbb{Z}^3} \sum_{\mathbf{m}=0}^{\mathbf{K}-1} \mathrm{DF}^{-1}[a](\mathbf{m}) \cdot \prod_{d=1}^{3} M_p(K_d(\mathbf{x})_d/L_d - m_d - n_d K_d) \tag{7.69}$$

for the solution Φ^{lr} as an implementation of (7.30).

The complexity of the computation of the $c_{\mathbf{k}}$ is then $\mathcal{O}(\#\mathbf{K}\log(\#\mathbf{K})) + \mathcal{O}(p^3 N)$, with $\#\mathbf{K} = K_1 K_2 K_3$, since the sums in (7.60) only have to be computed for those points that lie in the support of M_p, and the discrete Fourier transform $\mathrm{DF}[Q]$ of Q and the discrete inverse Fourier transform $\mathrm{DF}^{-1}[a]$ can be computed using the fast Fourier transform according to Cooley and Tukey [155, 313] with a complexity of $\mathcal{O}(\#\mathbf{K}\log(\#\mathbf{K}))$. There are a number of program packages that provide very efficient implementations of the fast Fourier transforms, see for instance [240]. With $\#\mathbf{K} \sim N$, the linear system of equations can thus be solved with a total complexity of $\mathcal{O}(N\log(N))$.

The Approximation of the Energy. The self-energy (7.21) associated to the long-range potential satisfies

$$V_{\mathrm{self}}^{\mathrm{lr}} = \frac{1}{2}\frac{1}{4\pi\varepsilon_0} \sum_{i=1}^{N} q_i^2 \mathrm{erf}(0)/0 = \frac{1}{4\pi\varepsilon_0}\frac{G}{\sqrt{\pi}} \sum_{i=1}^{N} q_i^2, \tag{7.70}$$

with

$$\mathrm{erf}(0)/0 := \lim_{r\to 0} \mathrm{erf}(Gr)/r.$$

Using (7.31) together with (7.60) and (7.69), the energy term $V_{\mathrm{other}}^{\mathrm{lr}}$ can be expressed by

$$V_{\mathrm{other}}^{\mathrm{lr}} \approx \frac{1}{2} \sum_{i=1}^{N} q_i \Phi_{\mathbf{K}}^{\mathrm{lr}}(\mathbf{x}_i) = \frac{1}{2} \sum_{\mathbf{m}=0}^{\mathbf{K}-1} \mathrm{DF}^{-1}[a](\mathbf{m}) \cdot Q(\mathbf{m})$$

$$= \frac{1}{2} \sum_{\mathbf{k}=0}^{\mathbf{K}-1} a(\mathbf{k}) \cdot \overline{\mathrm{DF}[Q](\mathbf{k})} \underset{(7.66)}{=} \frac{1}{2} \sum_{\mathbf{k}=0}^{\mathbf{K}-1} d(\mathbf{k}) \cdot |\mathrm{DF}[Q](\mathbf{k})|^2 \tag{7.71}$$

and therefore in total

$$V^{\mathrm{lr}} \approx \frac{1}{2} \sum_{\mathbf{k}=0}^{\mathbf{K}-1} d(\mathbf{k}) \cdot |\mathrm{DF}[Q](\mathbf{k})|^2 - \frac{1}{4\pi\varepsilon_0}\frac{G}{\sqrt{\pi}} \sum_{i=1}^{N} q_i^2. \tag{7.72}$$

The Approximation of the Forces. Following (7.32), the application of the gradient to (7.69) yields the approximation[40]

---

[40] Note that Q appears in a, and Q contains B-splines which contain $\mathbf{x}_i$. To verify this relation, substitute a, write out the definition of Q, and differentiate.

$$\mathbf{F}_i^{\text{lr}} \approx -\frac{1}{2} \sum_{j=1}^{N} q_j \nabla_{\mathbf{x}_i} \Phi_{\mathbf{K}}^{\text{lr}}(\mathbf{x}_j) \tag{7.73}$$

$$= -q_i \sum_{\mathbf{n} \in \mathbb{Z}^3} \sum_{\mathbf{m}=0}^{K-1} \text{DF}^{-1}[a](\mathbf{m}) \cdot \nabla_{\mathbf{x}_i} \prod_{d=1}^{3} M_p(K_d(\mathbf{x}_i)_d/L_d - m_d - n_d K_d)$$

$$= -q_i \sum_{\mathbf{n} \in \mathbb{Z}^3} \sum_{\mathbf{m}=0}^{K-1} \text{DF}^{-1}[a](\mathbf{m}) \begin{pmatrix} \frac{\partial}{\partial (\mathbf{x}_i)_1} M_p(y_1^i) \cdot M_p(y_2^i) \cdot M_p(y_3^i) \\ M_p(y_1^i) \cdot \frac{\partial}{\partial (\mathbf{x}_i)_2} M_p(y_2^i) \cdot M_p(y_3^i) \\ M_p(y_1^i) \cdot M_p(y_2^i) \cdot \frac{\partial}{\partial (\mathbf{x}_i)_3} M_p(y_3^i) \end{pmatrix}, \tag{7.74}$$

with $y_d^i := (\mathbf{x}_i)_d K_d/L_d - (m_d + n_d K_d)$. In this way, only the derivatives of the interpolating functions M_p are needed to compute the gradient. This approximation can be evaluated in a complexity of $\mathcal{O}(1)$ in every point, since the sum only runs over those $\mathbf{m}$ for which the B-splines M_p are not equal to zero.

The partial derivatives of the B-splines can be computed according to the recurrence formula (7.56). This yields for instance

$$\frac{dM_p(y_1^i)}{dx_1} = \frac{K_1}{L_1} \left(M_{p-1}(y_1^i) - M_{p-1}(y_1^i - 1) \right). \tag{7.75}$$

The factor K_1/L_1 stems from the chain rule, i.e. from $dy_1^i/dx_1 = K_1/L_1$.

To determine (7.71) and (7.74), we first compute the array Q. To this end, an outer loop over all particles and an inner loop over the support of the B-splines associated to the current particle is needed. Q is computed in the body of the loops according to (7.60). Then, $\text{DF}[Q]$ is computed by a discrete fast Fourier transform, and a and $V_{\text{other}}^{\text{lr}}$ are computed subsequently in a loop over the mesh. Next, $\text{DF}^{-1}[a]$ is computed by an inverse discrete fast Fourier transform. Finally, the forces $\mathbf{F}_i^{\text{lr}}$ are computed in an outer loop over all particles and an inner loop over the supports of the associated B-splines.

In this method the force is computed directly as the gradient of the potential, and therefore the total energy is conserved up to computer accuracy, while no conservation of momentum is guaranteed.[41]

Finally, let us give a remark concerning the computation of the forces: Since the force is computed as the gradient of the potential, the described approach can only be employed for functions that are sufficiently differentiable. For B-splines, this yields the condition $p \geq 3$. If the chosen local bases $\phi_{\mathbf{m}}$ are *not* differentiable, (7.32) and (7.73) cannot be used to compute the force on a particle. Instead an approximation of $\nabla_{\mathbf{x}} \psi_{\mathbf{k}}$ by appropriate local functions $\tilde{\phi}_m$ must be found which satisfies

---

[41] Here, it is possible that more and more energy is transferred into the motion of the center of mass. This effect can be prevented if one computes in each time step the velocity $\left(\sum_{i=1}^{N} m_i \mathbf{v}_i \right) / \sum_{i=1}^{N} m_i$ of the center of mass and subtracts this velocity from the velocity of all particles.

$$\nabla_{\mathbf{x}}\psi_{\mathbf{k}}(\mathbf{x}) \approx \sum_{\mathbf{m}=\mathbf{0}}^{\mathbf{K}-1} \overline{\tilde{t}_{\mathbf{km}}\tilde{\phi}_{\mathbf{m}}}(\mathbf{x}). \tag{7.76}$$

This approximation can then be employed to approximate the forces. Using trigonometric functions and the abbreviations $f_{\mathbf{0}} := \mathbf{0}$, $f_{\mathbf{k}} := \mathbf{k}_L a(\mathbf{k})$, and $y_d := (\mathbf{x})_d K_d/L_d - m_d - n_d K_d$, the approximation of $\nabla_{\mathbf{x}} e^{2\pi i \mathbf{k}_L \cdot \mathbf{x}} = 2\pi i \mathbf{k}_L e^{2\pi i \mathbf{k}_L \cdot \mathbf{x}}$ with B-spline interpolants leads to the approximation

$$\nabla_{\mathbf{x}}\Phi_{\mathbf{K}}^{\mathrm{lr}}(\mathbf{x}) \approx 2\pi i \sum_{\mathbf{n}\in\mathbb{Z}^3}\sum_{\mathbf{m}=\mathbf{0}}^{\mathbf{K}-1} \mathrm{DF}^{-1}[\mathbf{f}](\mathbf{m}) \cdot \prod_{d=1}^{3} M_p(y_d).$$

This results in the following approximation for the force

$$\mathbf{F}_j^{\mathrm{lr}} \approx -\frac{1}{2}\sum_{r=1}^{N} q_r \nabla_{\mathbf{x}_j}\Phi_{\mathbf{K}}^{\mathrm{lr}}(\mathbf{x}_r) \approx -2\pi i q_j \sum_{\mathbf{n}\in\mathbb{Z}^3}\sum_{\mathbf{m}=\mathbf{0}}^{\mathbf{K}-1} \mathrm{DF}^{-1}[\mathbf{f}](\mathbf{m}) \cdot \prod_{d=1}^{3} M_p(y_d). \tag{7.77}$$

Both equations have to be read as vector equations. To evaluate the forces, three discrete Fourier transforms and three inverse discrete Fourier transforms are necessary (one for each spatial direction). Both, the interpolation of the charges to the mesh and the interpolation of the forces to the particle coordinates, use the same interpolation scheme. This symmetry now ensures that the momentum is conserved up to machine accuracy while no conservation of energy can be guaranteed. The sum $\sum_{\mathbf{m}=\mathbf{0}}^{\mathbf{K}-1}$ for each particle is only to be computed for the mesh points inside of the support of the spline M_p. Note that the approximations (7.74) and (7.77) do not have to lead to the same results since in general approximation and gradient operator do not commute.

Choosing the Parameters. The steps in the computation of the long-range terms are coupled with each other and have to be adapted to each other to yield an efficient global method. Thus,

– the width of the shielding charge distribution G,
– the threshold radius r_{cut}, and
– the number of degrees of freedom $\#\mathbf{K}$

have to be chosen such that the resulting global algorithm has an optimal ratio of computational complexity to accuracy. Here, the following points should be considered:

1. The parameters G and r_{cut} obviously depend on each other and have to be chosen appropriately to obtain a given accuracy.
2. The accuracy can be improved by increasing the cutoff radius r_{cut} or by using a finer mesh, i.e. an increase of $\#\mathbf{K}$.
3. A smaller value for G together with a larger value for r_{cut} leads to a decrease in the complexity of the computation of the long-range terms and an increase in the complexity of the computation of the short-range terms.

4. A larger value for G together with a smaller value for r_{cut} leads to a decrease in the complexity of the computation of the short-range terms and an increase in the complexity of the computation of the long-range terms.

A discussion on how to appropriately choose the parameters G, r_{cut}, and the number of degrees of freedom $\#\mathbf{K}$ can be found in [215].

7.3.3 Implementation of the SPME method

After the presentation of the theory of the SPME method we can now consider the actual implementation. We start with the three-dimensional variant (DIM=3) of the program for the computation of the forces and the potentials from Chapter 3, Section 3.5. This program only has to be changed in a few places. Some routines are added which implement the computation of the long-range forces and energies.

We already implemented the gravitational potential in Algorithm 3.7. If we want to use the Coulomb potential instead, the charge of the particle is needed as well. To this end, we extend the data structure 3.1 to include a variable for the charge q as shown in data structure 7.1.

Data structure 7.1 Additional Particle Data for the Coulomb Potential

```
typedef struct {
   ...            // particle data structure 3.1
   real q;        // charge
} Particle;
```

With this data structure we can already implement the computation of the short-range force terms. Besides the classical short-range forces, which are computed by force for the Lennard-Jones potential or those that result from the potentials of Chapter 5, we also have to implement the additional short-range term (7.40) from the SPME method that is caused by the shielding Gaussian charge distributions. This additional term is computed in force_sr in Algorithm 7.1. For reasons of simplicity, we omit the factor $1/(4\pi\varepsilon_0)$ here, since it can be eliminated by a proper scaling of the variables similar to Section 3.7.3. For details see Section 7.4.2. However, to allow different scalings we introduce an additional global parameter skal. The parameter G, also globally defined, describes the width of the shielding Gaussians. The functions erfc and the constant M_2_SQRTPI:$=2/\sqrt{\pi}$ are taken from the header file math.h of the math library of the programming language C.

With the programs from the linked cell method from Section 3.5 and the new function force_sr, we can already compute the short-range part of the forces and integrate Newton's equation of motion with the Störmer-Verlet method. To this end, only the new function call to force_sr has to

---

Algorithm 7.1 Short-Range Terms (7.40) of the Coulomb Potential in the SPME Method for the Linked Cell Method 3.12 and 3.15

```
real G; // parameter for the Gaussians
real skal = 1; // scaling parameter
void force_sr(Particle *i, Particle *j) {
  real r2 = 0;
  for (int d=0; d<DIM; d++)
    r2 += sqr(j->x[d] - i->x[d]);        // distance squared r2=r²ᵢⱼ
  real r = sqrt(r2);                     // distance r=rᵢⱼ
  real f = -i->q * j->q * skal *
          (erfc(G*r)/r+G*M_2_SQRTPI*exp(-sqr(G*r)))/r2;
  for (int d=0; d<DIM; d++)
    i->F[d] += f * (j->x[d] - i->x[d]);
}
```

---

be inserted into the force computation compF_LC after the call to force. Alternatively, the body of function force_sr could be appended to the body of function force to obtain *one* function force for *all* short-range forces.[42] What is still missing is the long-range portion of the forces. Besides the linked cell structure already mentioned earlier, we need an appropriate mesh for the SPME algorithm, functions for the interpolation of the particle charges to this mesh, the solution of the discrete Poisson problem on the mesh, and the evaluation of this solution at the particle positions for the computation of the forces. We will present these parts of the algorithm in the following.

In code fragment 7.1 we show the basic approach for the interpolation of the charges of the particles to a regular mesh. The details are written out in the function compQ from Algorithm 7.2.

Code fragment 7.1 Interpolation to the Mesh

```
set Q = 0;
loop over all cells ic
  loop over all particles i in cell ic {
      determine the interpolation points for the interpolant Q for the
      particle charge i->q in the SPME mesh using the macro index;
      determine the values in array Q according to (7.60);
}
```

Here, the particles are stored in the linked cell data structure which is based on a decomposition of the computational domain into a regular mesh of cells ic. This linked cell grid grid will be described as before by a variable

---

[42] Here, periodic boundary conditions have to be taken into account in the computation of the distance, compare also the remarks on page 76 in Section 3.6.4.

Algorithm 7.2 Interpolation to the Mesh

```
void compQ(Cell *grid, int *nc, fftw_complex *Q, int *K, int pmax,
           real *spme_cellh) {
  int jc[DIM];
  for (int i=0; i<K[0]*K[1]*K[2]; i++) {
    Q[i].re = 0;
    Q[i].im = 0;
  }
  for (jc[0]=0; jc[0]<nc[0]; jc[0]++)
    for (jc[1]=0; jc[1]<nc[1]; jc[1]++)
      for (jc[2]=0; jc[2]<nc[2]; jc[2]++)
        for (ParticleList *i=grid[index(jc,nc)]; NULL!=i; i=i->next) {
          int m[DIM], p[DIM];
          for (int d=0; d<DIM; d++)
            m[d] = (int)floor(i->p.x[d] / spme_cellh[d]) + K[d];
          for (p[0]=0; p[0]<pmax; p[0]++)
            for (p[1]=0; p[1]<pmax; p[1]++)
              for (p[2]=0; p[2]<pmax; p[2]++) {
                int mp[DIM];
                for (int d=0; d<DIM; d++)
                  mp[d]=(m[d]-p[d])%K[d];
                Q[index(mp,K)].re += i->p.q
                  * spline(p[0] + fmod(i->p.x[0], spme_cellh[0])
                                  / spme_cellh[0], pmax)
                  * spline(p[1] + fmod(i->p.x[1], spme_cellh[1])
                                  / spme_cellh[1], pmax)
                  * spline(p[2] + fmod(i->p.x[2], spme_cellh[2])
                                  / spme_cellh[2], pmax);
              }
        }
}
```

nc of type `int[DIM]`. Now, the SPME mesh is a new ingredient. It is in general independent of the linked cell grid structure. We index it analogously with the variable K that is also of type `int[DIM]`. The data associated to the SPME mesh are stored in the linear array Q. In this array, we keep the real values of the charges interpolated to the mesh according to (7.51), as well as the results of the Fourier transform, of the scaling in Fourier space, and of the inverse Fourier transform. We use a complex Fourier transform, and therefore Q is declared as a complex array, see below. The linked cell grid and the SPME mesh now have to be put into correspondence with each other. This is implemented by a conversion of their coordinates using our macro **index**, which we introduced in the implementation of the linked cell method on page 61.

The interpolation projects the particle charges i->q onto the SPME mesh. To implement this, we first set all values of Q to zero. In a loop over the particles, we add the contribution of each charge to the mesh. To this end, we search for the appropriate interpolation points in the SPME mesh and sum up the associated splines of order pmax weighted by the charges i->q. A particle charge has an effect on pmax different SPME mesh cells in each coordinate direction. Here, we use the mesh width **h** that is stored in the variable spme_cellh of type real[DIM].

We need a function spline for the evaluation of the splines. We implement this function for the sake of simplicity[43] in a recursive form according to (7.54) and (7.55), see Algorithm 7.3.

Algorithm 7.3 Recursive Evaluation of a B-spline (7.54)

```
real spline(real x, int p) {
  if ((x<=0.)||(x>=p)) return 0.;
  if (p==2) return 1. - fabs(x-1.);
  return (x * spline(x, p-1) + (p-x) * spline(x-1., p-1)) / (p - 1.);
}
```

In the next step, the Poisson problem with given right hand side Q has to be solved on the SPME mesh. We use the fast Fourier transform (FFT), a diagonal scaling, and the inverse fast Fourier transform. There are many different implementations for the FFT. In our examples, we employ the version 2.1.5. of the library FFTW by Frigo and Johnson [240].[44] This library and a description with usage guide and other details can be found on the website http://www.fftw.org. The complex array Q is declared as the data type fftw_complex defined in the FFTW library.

For the solution of (7.29) we need the Fourier transform $\mathrm{DF}[Q]$ of Q, a scaling of the results according to (7.66), and an inverse transform $\mathrm{DF}^{-1}[Q]$. This is implemented in the function compFFT from Algorithm 7.4 by the calls to the function fftwnd_one, the multiplication with D and a further call to the function fftwnd_one from the FFTW library.

---

[43] Since the evaluation of the spline functions is needed frequently and demands a relatively large amount of runtime in the described form, it would be advantageous to further improve this implementation later on. One can for example fix the spline order pmax, transform the recursive implementation into an iterative one, and avoid the conditional statements. The resulting Neville–Newton-like tableau can even be partially reused for the evaluation of the spline in neighboring points. In addition, the coefficients of the interpolation scheme can be precomputed once, stored in a table, and then be directly loaded from it when needed.

[44] Other FFT implementations can be found on the internet, for instance on the website http://www.netlib.org.

Algorithm 7.4 Computation of the Solution Q with FFTW

```
void compFFT(fftw_complex *Q, int *K, real *D) { // using FFTW library
  fftwnd_one(fft1, Q, NULL);           // complex in-place FFT
  for (int i=0; i<K[0]*K[1]*K[2]; i++) {
    Q[i].re *= D[i];                   // scaling by the values from D
    Q[i].im *= D[i];                   // compute energy from (7.72)
  }                                    // here as well, if needed
  fftwnd_one(fft2, Q, NULL);           // inverse complex in-place FFT
}
```

Before calling `compFFT`, we have to compute the values of the scaling factors D and to initialize the FFTW library appropriately. This is implemented in function `initFFT` in Algorithm 7.5.

Algorithm 7.5 Initialization for the FFT Solver

```
#include <fftw.h>
fftwnd_plan fft1, fft2;
void initFFT(real *D, int *K, int pmax, real *l) {
  int k[DIM];
  fft1 = fftw3d_create_plan(K[2], K[1], K[0],
                            FFTW_FORWARD, FFTW_IN_PLACE);
  fft2 = fftw3d_create_plan(K[2], K[1], K[0],
                            FFTW_BACKWARD, FFTW_IN_PLACE);
  D[0] = 0;
  for (k[0]=K[0]/2; k[0]>-K[0]/2; k[0]--)
    for (k[1]=K[1]/2; k[1]>-K[1]/2; k[1]--)
      for (k[2]=K[2]/2; k[2]>-K[2]/2; k[2]--) {
        int kp[DIM];
        for (int d=0; d<DIM; d++)
          kp[d]=(k[d]+K[d])%K[d];
        real m = sqr(k[0]/l[0])+sqr(k[1]/l[1])+sqr(k[2]/l[2]);
        if (m>0)
          D[index(kp,K)] = exp(-m*sqr(M_PI/G))*skal/
                           (m*M_PI*l[0]*l[1]*l[2])*
                           bCoeff(pmax,K[0],k[0])*
                           bCoeff(pmax,K[1],k[1])*
                           bCoeff(pmax,K[2],k[2]);
      }
}
```

In the initialization of FFTW, the two variables `fft1` and `fft2` of type `fftwnd_plan` (declared globally for reasons of simplicity) are to be set properly. Their initialization needs the direction of the Fourier transform, parameters describing the memory layout, but also the size K of the SPME mesh, since its prime factor decomposition and the order of the prime factors is needed for an efficient FFT realization. The transforms are three-dimensional and work in-place. The factors D are computed according to (7.65), using the width of the Gaussians G, the spline order pmax, the mesh width K, and the dimensions 1 of the domain.

As in the computation of the short-range forces in force_sr, we allow for a general scaling by the parameter skal. For the computation of the values of

$$|B(\mathbf{k})|^2 = \prod_{d=0}^{DIM-1} |B_{K_d}(k_d)|^2$$

with $B_{K_d}(k_d)$ from (7.59), we first note that $|e^{-2\pi i(p-1)k_d/K_d}|^2 = 1$. Hence, we only need to determine the expressions in the nominator of $B_{K_d}(k_d)$, i.e. the values of

$$\left| \sum_{q=0}^{p-2} e^{-2\pi i k_d q/K_d} M_p(q+1) \right|^{-2}. \tag{7.78}$$

They are computed in a loop from 0 to $p-2$ in the function bCoeff from Algorithm 7.6. This function calls the function spline from Algorithm 7.3 for the evaluation of M_p.

Algorithm 7.6 Computation of the Factors (7.78)

```
real bCoeff(int p, int K, int k) {
  if ((p % 2 ==1) && (2*abs(k)==K)) return 0.;
  real c=0., s=0.;
  for (int q=0; q<=p-2; q++) {
    c += spline(q + 1., p) * cos((2. * M_PI * k * q) / K);
    s += spline(q + 1., p) * sin((2. * M_PI * k * q) / K);
  }
  return 1./(sqr(c)+sqr(s));
}
```

Finally, we have to compute the long-range forces at the particle positions from the solution of the Poisson problem stored in Q. We assume that the short-range forces have already been computed and we now add the long-range forces. In code fragment 7.2 we present the basic approach.

A complete implementation can be found in the function compF_SPME in Algorithm 7.7. We traverse all particles from the linked cell structure as already in the interpolation to the mesh in code fragment 7.1 and Algorithm 7.2. The particle coordinates are converted into coordinates with respect to the SPME mesh. Now, we determine the interpolation points of the spline approximation Q of the solution of the Poisson equation and compute its gradient at the position of the particle. Again, we use splines of order pmax

Code fragment 7.2 Computing Long-Range Forces from the Solution Q

loop over all cells `ic`
  loop over all particles `i` in cell `ic` {
    determine, using the macro `index`, the interpolation points of the
    array `Q` for the evaluation of the long-range forces on the SPME mesh;
    compute the force on particle `i` according to (7.74);
}

Algorithm 7.7 Computation of the Long-Range Forces from Q

```
void compF_SPME(Cell *grid, int *nc, fftw_complex *Q, int *K, int pmax,
                real *spme_cellh) {
  int jc[DIM];
  for (jc[0]=0; jc[0]<nc[0]; jc[0]++)
    for (jc[1]=0; jc[1]<nc[1]; jc[1]++)
      for (jc[2]=0; jc[2]<nc[2]; jc[2]++)
        for (ParticleList *i=grid[index(jc,nc)]; NULL!=i; i=i->next) {
          int m[DIM], p[DIM];
          for (int d=0; d<DIM; d++)
            m[d] = (int)floor(i->p.x[d] / spme_cellh[d]) + K[d];
          for (p[0]=0; p[0]<pmax; p[0]++)
            for (p[1]=0; p[1]<pmax; p[1]++)
              for (p[2]=0; p[2]<pmax; p[2]++) {
                int mp[DIM];
                real x[DIM], s[DIM];
                for (int d=0; d<DIM; d++)
                  mp[d] = (m[d]-p[d])%K[d];
                real q = i->p.q * Q[index(mp,K)].re;
                for (int d=0; d<DIM; d++) {
                  x[d] = p[d] + fmod(i->p.x[d], spme_cellh[d])
                                / spme_cellh[d];
                  s[d] = spline(x[d], pmax);
                }
                i->p.f[0] -= q * Dspline(x[0], pmax)
                             / spme_cellh[0] * s[1] * s[2];
                i->p.f[1] -= q * s[0] * Dspline(x[1], pmax)
                             / spme_cellh[1] * s[2];
                i->p.f[2] -= q * s[0] * s[1] * Dspline(x[2], pmax)
                             / spme_cellh[2];
              }
        }
}
```

and therefore need $(\text{pmax})^3$ cells of the SPME mesh. The derivative of the spline function according to (7.56) is implemented in the function Dspline in Algorithm 7.8.[45]

Algorithm 7.8 Derivative of a B-spline (7.56)

```
real Dspline(real x, int p) {
  return spline(x, p-1) - spline(x-1., p-1);
}
```

Using these functions, we can implement the complete computation of the forces in the SPME method concisely as in code fragment 7.3. First, the short-range forces are computed with the function compF_LC of the linked cell method 3.15. There, the function force_sr of the Algorithm 7.1 has been inserted. Independently of the short-range computation, the particle charges are interpolated with compQ to the mesh, and the Poisson problem is solved with compFFT. Finally, the long-range forces on the mesh are computed with compF_SPME, then evaluated at the particle positions, and added to the short-range forces.

Code fragment 7.3 Complete Force Computation with the SPME Method

```
compF_LC(grid, nc, r_cut);
compQ(grid, nc, Q, K, pmax, spme_cellh);
compFFT(Q, K, D);
compF_SPME(grid, nc, Q, K, pmax, spme_cellh);
```

The corresponding main program in Algorithm 7.9 has several new parameters compared to the original version of the linked cell method, such as the width G of the Gaussians, the order of the splines pmax, as well as the size K[DIM] and the mesh width spme_cellh[DIM] of the SPME mesh. These parameters have to be initialized or read from a configuration file at the beginning of the program. The mesh width spme_cellh of the SPME mesh can be computed from K and the dimensions l of the domain according to (7.53).

Besides the linked cell data structure grid, the SPME mesh structures Q and D have to be initialized in the main program. There, the necessary amount of memory is allocated and the appropriate initialization routine is called. The new time integration function timeIntegration_SPME differs from timeIntegration_LC in that the extended force computation from code fragment 7.3 is used. There, the force computation function force_sr has been inserted into compF_LC.

---

[45] This function can be further improved analogously to the remarks in footnote 43.

Algorithm 7.9 Main Program of the SPME Method

```
int main() {
  int nc[DIM], K[DIM], pmax;
  real l[DIM], spme_cellh[DIM], r_cut;
  real delta_t, t_end;
  inputParameters_SPME(&delta_t, &t_end, nc, l, &r_cut, K,
                       spme_cellh, &G, &pmax);
  Cell *grid = (Cell*)malloc(nc[0]*nc[1]*nc[2]*sizeof(*grid));
  real *D = (real*) malloc(K[0]*K[1]*K[2]*sizeof(*D));
  fftw_complex *Q = (fftw_complex*) malloc(K[0]*K[1]*K[2]*sizeof(*Q));
  initFFT(D, K, pmax, l);
  initData_LC(grid, nc, l);
  timeIntegration_SPME(0, delta_t, t_end, grid, nc, l, r_cut,
                       Q, D, K, spme_cellh, G, pmax);
  freeLists_LC(grid, nc);
  free(Q); free(D); free(grid);
  return 0;
}
```

7.4 Application Examples and Extensions

In this section we show some results for the simulation of particle systems with the program described in the last section. In addition to the previously implemented short-range potentials, a Coulomb potential will be used as an interaction potential between the particles.

First, we consider two examples for an instability introduced by charges. The setting is similar to the Rayleigh-Taylor problem which we studied in Section 3.6.4, but now the dynamics of the particles is caused by the distribution of charges. Next, we run a simulation in which a salt in crystalline form (KCl) is slowly heated, melts, is cooled rapidly after melting, and then transforms into a glass state. Finally, we study the behavior of water as an example for a molecular system in which long-range forces play an important role. We discuss different models for water, describe the implementation of one specific model, and determine the self-diffusion coefficient for a small system of 216 water molecules.

7.4.1 Rayleigh-Taylor Instability with Coulomb Potential

In this section we simulate an instability induced by the charges of the particles. It arises when layers of particles with different charges are put on top of each other. Figure 7.10 (upper left) shows the initial state. The simulation domain is filled completely with particles. The particles in a layer of half the domain height (dark-shaded) carry positive charges, while all other particles carry negative charges. This unstable state resolves by a mixing of the particles from the two subdomains. The mixing process depends on the "size" of the particles (given by the parameter σ of the Lennard-Jones potential) and the charge differences. Before we describe the simulation in more detail, we discuss the introduction of dimensionless equations for the Coulomb potential.

Dimensionless Equations – Reduced Variables. As a further example of how the quantities in the equations of motion can be put in nondimensional form, we consider the mixture of two different types of particles (A or B) which interact by a Lennard-Jones and a Coulomb potential. The potential here reads (for the sake of simplicity in the nonperiodic case)

$$V = \sum_{\substack{i=1}}^{N} \sum_{\substack{j=1 \\ j>i}}^{N} \frac{1}{4\pi\varepsilon_0} \frac{q_i q_j}{r_{ij}} + \sum_{\substack{i=1}}^{N} \sum_{\substack{j=1 \\ j>i}}^{N} 4\varepsilon_{ij}\left(\left(\frac{\sigma_{ij}}{r_{ij}}\right)^{12} - \left(\frac{\sigma_{ij}}{r_{ij}}\right)^{6}\right). \quad (7.79)$$

Now, we select one type of particle (here the type A) and compute the scaling based on its parameters. Analogously to Section 3.7.3, we scale the Lennard-Jones potential by the values $\tilde{\sigma} = 2.22$ Å, $\tilde{\varepsilon} = 1.04710^{-21}$ J, and $\tilde{m} = 1$ u, compare (3.55). Furthermore, the charges are scaled by

$$q_i' = q_i/\tilde{q}$$

with $\tilde{q} = 1$ e.[46] If one now defines $\varepsilon_0' := \dfrac{4\pi\varepsilon_0\tilde{\sigma}\tilde{\varepsilon}}{\tilde{q}^2}$, one obtains by substitution into (7.79) the scaled potential energy

$$V' = \frac{V}{\tilde{\varepsilon}_1} = \sum_{\substack{i=1}}^{N} \sum_{\substack{j=1 \\ j>i}}^{N} \frac{1}{\varepsilon_0'} \frac{q_i' q_j'}{r_{ij}'} + \sum_{\substack{i=1}}^{N} \sum_{\substack{j=1 \\ j>i}}^{N} 4\varepsilon_{ij}'\left(\left(\frac{\sigma_{ij}'}{r_{ij}'}\right)^{12} - \left(\frac{\sigma_{ij}'}{r_{ij}'}\right)^{6}\right).$$

Applying the gradient operator then results in the scaled forces

$$\mathbf{F}_i' = -\sum_{\substack{j=1}}^{N} \frac{1}{\varepsilon_0'} \frac{q_i' q_j'}{(r_{ij}')^2} \frac{\mathbf{r}_{ij}'}{r_{ij}'} - \sum_{\substack{j=1 \\ j\neq i}}^{N} 24\varepsilon_{ij}'\left(2\left(\frac{\sigma_{ij}'}{r_{ij}'}\right)^{12} - \left(\frac{\sigma_{ij}'}{r_{ij}'}\right)^{6}\right)\frac{\mathbf{r}_{ij}'}{(r_{ij}')^2}.$$

In the two-dimensional example that we consider in the following[47], the two types of particles differ only in their charges so that ε_{ij}' as well as σ_{ij}' are eliminated after the re-scaling.

Figure 7.10 shows the results of a simulation for 14000 particles with the parameter values (after scaling) of Table 7.1. At the beginning of the simulation, the particles are placed on a regular grid of 200×70 grid points and are perturbed by a small thermal motion. The particles in the middle layer carry a charge of q_A', while the other particles carry a charge of q_B'.

---

[46] The atomic mass unit is u $= 1.6605655 \cdot 10^{-27}$ kg, and the elementary charge is e $= 1.6021892 \cdot 10^{-19}$ C.

[47] The algorithm simulates the example in three dimensions; to this end, the third coordinate of the position is set to a constant value, and the third coordinate of the velocity is set to zero.

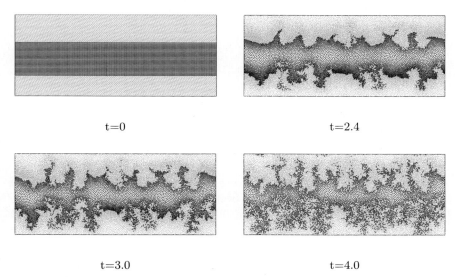

t=0   t=2.4

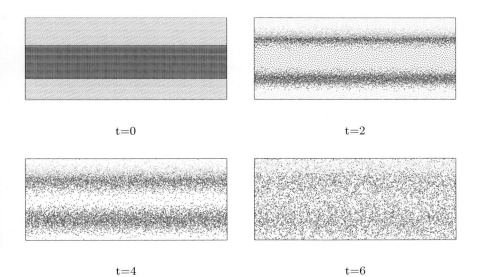

t=3.0   t=4.0

Fig. 7.10. Instability with Coulomb potential, $\sigma'_A = \sigma'_B = 1$, time evolution of the particle distribution.

t=0   t=2

t=4   t=6

Fig. 7.11. Instability with Coulomb potential, $\sigma'_A = \sigma'_B = 0.4$, time evolution of the particle distribution.

$$L'_1 = 144, \qquad L'_2 = 60,$$
$$\varepsilon'_A = \varepsilon'_B = 1, \qquad \sigma'_A = \sigma'_B = 1,$$
$$q'_A = 0.5 \ , \qquad q'_B = -0.5,$$
$$m'_A = m'_B = 23, \qquad N = 14000,$$
$$r'_{cut} = 6, \qquad \delta t' = 0.001,$$
$$T' = 0.1, \qquad G' = 0.175,$$
$$h' = 1, \qquad p_{max} = 4,$$
$$skal = 992.573$$

Table 7.1. Parameter values for the simulation of an instability with Coulomb potential.

Fibrous, mushroom-shaped structures emerge, similar to the simulation of the Rayleigh-Taylor instability from Section 3.6.4. The results of the simulation depend strongly on the values of the parameters σ'_A and σ'_B. Figure 7.11 shows the result of another simulation with $\sigma'_A = \sigma'_B = 0.4$; all other parameters are kept unchanged. In this case the particles mix diffusively since their smaller "size" allows them to pass each other more easily.

7.4.2 Phase Transition in Ionic Microcrystals

Now, we consider a three-dimensional example. We study the melting of salt crystals (KCl) and the subsequent emergence of a glass state after cooling. Here, the K^+ and Cl^- ions introduce non-negligible long-range terms into the potential.

The melting of salt crystals has already been simulated in [33] and the emergence of a glass phase has been observed in [324]. Despite the very short time spans that can be treated in molecular dynamical simulations up to now (in the range of picoseconds to microseconds), such computations help to better understand the mechanisms of melting (especially the microscopical details of the melting progress) and they can contribute to a theory of melting, compare [238, 299]. As potentials for the simulation of the phase transition in salts, the Born-Mayer-Huggins and the Tosi-Fumi potential [241] have been used. Parameters for the potentials for different salts can be found in [543]. The behavior of KCl has been simulated with the Tosi-Fumi potential in [383]. For a comparison of the results of the simulations with results of experiments see [543]. The behavior of KCl for all of its possible phase transitions was studied in a simulation in [41].

As potential we use a Coulomb interaction potential together with a short-range repulsive term

$$U(r_{ij}) = \frac{1}{4\pi\varepsilon_o} \frac{q_i q_j}{r_{ij}} \left(1 + \text{sgn}(q_i q_j) \frac{2^8}{9} \left(\frac{\sigma_{ij}}{r_{ij}} \right)^8 \right),$$

compare [468]. The force between two ions i and j with distance r_{ij} is given by

$$\mathbf{F}_{ij} = -\frac{q_i q_j}{4\pi\varepsilon_0} \frac{1}{r_{ij}^2} \left(1 + \mathrm{sgn}(q_i q_j) 2^8 \left(\frac{\sigma_{ij}}{r_{ij}} \right)^8 \right) \frac{\mathbf{r}_{ij}}{r_{ij}}. \qquad (7.80)$$

The parameters σ_{ij} are computed with the Lorentz-Berthelot mixing rule (3.37).

In our simulation we scale by $\tilde{\sigma} = 1\ r_B$, $\tilde{q} = 1$ e, $\tilde{m} = 1$ u and $\tilde{\varepsilon} = e^2/(4\pi\varepsilon_0 r_B)$. Here, $r_B = 0.52917721$Å is the Bohr radius. This scaling differs somewhat from the ones introduced up to now, since r_B is used instead of σ_K or σ_{Cl} in the scaling. With this scaling one obtains the scaled force

$$\mathbf{F}'_{ij} = -\frac{q'_i q'_j}{(r'_{ij})^2} \left(1 + \mathrm{sgn}(q'_i q'_j) 2^8 \left(\frac{\sigma'_{ij}}{r'_{ij}} \right)^8 \right) \frac{\mathbf{r}'_{ij}}{r'_{ij}}$$

and the scaled pair terms of the potential energy

$$U'(r'_{ij}) = \frac{q'_i q'_j}{r'_{ij}} \left(1 + \mathrm{sgn}(q'_i q'_j) \frac{2^8}{9} \left(\frac{\sigma'_{ij}}{r'_{ij}} \right)^8 \right).$$

We consider a cubic simulation box with periodic boundary conditions which contains in its center a small cubic microcrystal consisting of 12^3 ions (alternatingly $K+$ and $Cl-$ ions) at equilibrium with a temperature of 10 K. This crystal is heated up to a temperature of 2000 K by first scaling the velocities for 25 time steps by a factor of $\beta = 1.001$ and then simulating the system for 600 time steps without any scaling (equilibration). When the temperature of 2000 K is reached, the system is cooled down to a temperature of 10 K by scaling the velocities for 25 time steps by a factor $\beta = 0.999$ and then equilibrating the system for 600 time steps without any scaling. The parameter values used in the simulation are given in Table 7.2.

$$
\begin{array}{lll}
L'_1 = 144, & L'_2 = 144, & L'_3 = 144, \\
\sigma'_K = 2.1354, & \sigma'_{Cl} = 2.9291, & \\
m'_K = 38.9626, & m'_{Cl} = 35.4527, & \\
q'_K = 1, & q'_{Cl} = -1, & \\
r'_{cut} = 24, & \delta t' = 1.0, & \\
h' = 4, & G' = 0.1, & \\
N = 12^3, & p_{max} = 4, & skal = 1
\end{array}
$$

Table 7.2. Parameter values of the simulation of the melting of KCl.

In this simulation one can observe both the phase transition from solid to fluid and the transition into a glass phase. Figure 7.12 shows the particle distribution before the melting (left), after the melting in a fluid state (middle), and in a glass state (right).

Fig. 7.12. KCl before the melting (left), $t = 5.2$ ps, after the melting in a fluid state (middle), $t = 144.6$ ps, and in a glass state (right), $t = 309.8$ ps, NVE ensemble.

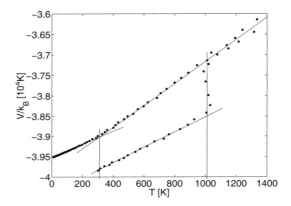

Fig. 7.13. Graph of the intrinsic energy over the temperature from a simulation of the melting and of the formation of a glass state for KCl.

In Figure 7.13 we show the intrinsic energy versus the temperature. The values of V and T (and of the kinetic energy E_{kin}) are computed as averages over the last 250 time steps of an equilibration phase (which consists of 625 time steps). One can clearly recognize that a phase transition occurs during heating. The cooling does not restore the crystal state, but it transforms the salt into a glass state. The melting point in our simulation is about 1010 K, the glass formation point is about 310 K, and the heat of fusion is 1400 K. For macroscopic samples, [324] reports the physical values 1045 K for the melting point and 1580 K for the heat of fusion. The melting point in the simulation is lower since the ratio of surface to volume is significantly larger for the macroscopic sample and the crystal starts melting at the surface first. The specific heat can be computed from the slope of the line drawn in Figure 7.13. Here, the specific heat of the liquid is larger than the specific heat of the solid.

7.4.3 Water as a Molecular System

In this section we consider the simulation of molecules with a Coulomb term in the interaction potential. To this end, we focus on water as an example. Here, we use a slightly extended version of the program developed above: If namely some specified pair interactions are not to be considered in the molecule, we have to correct the electrostatic energy of the entire system appropriately. It is then advantageous to strictly distinguish between the intermolecular and intramolecular terms in the electrostatic energy.

The potential induced by a Gaussian charge $q_j \left(G/\sqrt{\pi} \right)^3 e^{-G^2 \|\mathbf{x} - \mathbf{x}_j^\mathbf{n}\|^2}$ at the point $\mathbf{x}_j^\mathbf{n}$ is

$$\Phi_j^{\mathbf{n},\text{Gaussian}}(\mathbf{x}) = \frac{q_j}{4\pi\varepsilon_0} \frac{\text{erf}(G\|\mathbf{x}_j^\mathbf{n} - \mathbf{x}\|)}{\|\mathbf{x}_j^\mathbf{n} - \mathbf{x}\|},$$

compare (7.41). The energy of the particles within a molecule[48] is then

$$V_{\text{self}}^{\text{molecule}} = \frac{1}{2} \sum_{\mathbf{n} \in \mathbb{Z}^3} \sum_{i=1}^{N} \sum_{\substack{j=1, j \neq i \\ (i,j) \in \text{molecule}}}^{N} q_i \Phi_j^{\mathbf{n},\text{Gaussian}}(\mathbf{x}_i)$$

$$= \frac{1}{2} \frac{1}{4\pi\varepsilon_0} \sum_{\mathbf{n} \in \mathbb{Z}^3} \sum_{i=1}^{N} \sum_{\substack{j=1 \\ j \neq i \text{ for } \mathbf{n}=\mathbf{0} \\ (i,j) \in \text{molecule}}}^{N} q_i q_j \frac{\text{erf}(Gr_{ij}^\mathbf{n})}{r_{ij}^\mathbf{n}}. \qquad (7.81)$$

The sum in this equation here runs over all pairs of atoms within the molecules. In total, we obtain an electrostatic energy of

$$V_{\text{Coulomb}} = V_{\text{other}}^{\text{lr}} - V_{\text{self}} + V^{\text{sr}} - V_{\text{self}}^{\text{molecule}} + V^{\text{molecule}} \qquad (7.82)$$

with the terms

$$V^{\text{sr}} = \frac{1}{2} \frac{1}{4\pi\varepsilon_0} \sum_{\mathbf{n} \in \mathbb{Z}^3} \sum_{i=1}^{N} \sum_{\substack{j=1 \\ j \neq i \text{ for } \mathbf{n}=\mathbf{0} \\ (i,j) \notin \text{molecule}}}^{N} q_i q_j \frac{\text{erfc}(Gr_{ij}^\mathbf{n})}{r_{ij}^\mathbf{n}},$$

with $V_{\text{self}}^{\text{molecule}}$ and V_{self} as defined in (7.81) and (7.70), respectively, $V_{\text{other}}^{\text{lr}}$ as defined in (7.31), and V^{molecule} as defined in (5.38). The associated forces are obtained as gradients of the energies.

On this basis, we study in the following some properties of water. Water is the most abundant liquid on earth, it is important in biology, biochemistry, physical chemistry, and plays a central role in many of the processes

---

[48] To keep the notation simple, we here assume that the molecule is contained entirely within the simulation box. For molecules that cross the boundary and therefore extend over one or more neighboring images of the simulation box, this periodicity has to be taken into account in the inner sum over the parts of the molecule.

studied there. However, the geometric and electronic structure of its molecule makes water a very complex many-particle system. The water molecule consists of two hydrogen atoms and one oxygen atom. It assumes a tetrahedral form, with the oxygen atom in the center, the two hydrogen atoms at two of the vertices, and charge clouds with negative charges at the other two vertices, compare Figure 7.14. The charge clouds result from the way hydrogen and oxygen bond in the water molecule. Simply speaking, oxygen has eight negatively charged electrons, with two belonging to the inner (and thereby complete) electron shell and six belonging to the outer electron shell. The outer shell can have up to eight electrons, however. In the bond with the two hydrogen atoms, the electrons of the hydrogen atoms are attracted to the oxygen atom, since the outer electron shell of oxygen tends to be filled completely. This implies a higher probability of finding the hydrogen electrons close to the nucleus of the oxygen atom than to find them close to the positively charged hydrogen nuclei they are associated with. This is the reason the water molecule is polar: It has two clouds of negative charge closer to the oxygen atom, and the hydrogen nuclei possess a corresponding positive charge. The angle between the two approximately 1 Å long hydrogen-oxygen bonds is at about 105 degrees. This is somewhat smaller than the 109.5 degree angle in a perfect tetrahedron.

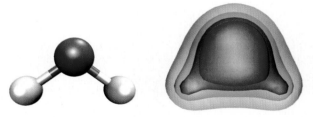

Fig. 7.14. A water molecule in ball-and-stick representation (left) and the probability density function for the electrons (right).

The flexible triangular structure and the characteristic asymmetry of the electron cloud of the water molecule makes different interactions with other water molecules or polar groups possible. In such interactions between water molecules, a positively charged hydrogen atom of one water molecule forms a so-called hydrogen bond with the negatively charged oxygen atom of another water molecule. Two such bonds are possible between the two hydrogen atoms of a water molecule and the oxygen atoms of two other water molecules. Two other such bonds are possible between its oxygen atom and the hydrogen atoms of two other water molecules. In this way, many different networks and clusters of water molecules of different size can emerge that again interact with each other and with other dissolved molecules. Note that the number of hydrogen bonds per water molecule in its fluid form varies between three and six, with an average of 4.5. The tetragonal form of its molecule gives water a structure that is much more loosely packed than the structure of most other fluids, such as oil or liquid nitrogen.

For a realistic molecular dynamical simulation of water, we need a model that represents the polarity of the water molecules and allows for hydrogen bonds just like the "real" water molecule does. Let us first consider models that are both stiff and not polarizable. Then, two kinds of interactions between the atoms (which are assumed to be arranged with constant distances and angles) of a H_2O molecule have to be taken into account: The electrostatic forces and the van der Waals forces, which are described by Coulomb potentials and Lennard-Jones potentials, respectively. Other contributions from dipole moments and quantum mechanical effects are not treated separately, but enter into the choice of the parameters of the potentials.

The first simulation of water molecules was performed by Rahman and Stillinger in the early seventies of the last century [499]. They studied the dynamics of $6 \times 6 \times 6 = 216$ water molecules in a square box for five picoseconds. The simulation reproduced, at least qualitatively, important physical properties of water such as the diffusion rate, the evaporation rate, and radial distribution functions.

In the meantime, an entire series of different water models has been developed and studied. One can classify them according to the number of modeled interaction sites. The simplest model consists of three point charges to which effective pair potential functions are attached. The starting point was the TIPS3 (transferable intermolecular potential with 3 sites) model [345]. The three point charges are given by the oxygen atom and the two hydrogen atoms. The hydrogen atoms carry positive fractional charges and the oxygen atom carries an opposite charge of twice the magnitude. These charges enter into Coulomb interactions. An additional Lennard-Jones force acts between the oxygen atoms of the water molecules. An improved set of parameters, which were adapted to the case of liquid H_2O, resulted in the SPC model (simple point charge) [81]. An extension thereof lead to the SPC/E model (extended simple point charge) [79] and a different reparametrization of the TIPS3 model for the case of liquid H_2O resulted in the TIP3P model [347].

Fig. 7.15. Positions of the centers of mass and charge for four- and five-site models of water.

Better results can be obtained by models that use four interaction sites. Here, the three-site model is only changed with respect to the position of the negative charge of the oxygen atom. It is slightly moved on the bisector of the *H-O-H* angle away from the oxygen atom towards the hydrogen atoms, see Figure 7.15 (left). Such a model had been first proposed by Bernal and Fowler [86]. TIPS2 [346] and TIP4P [347] are variants with changed geometry and improved parameters.

Five interaction sites are used in the model of Ben-Naim and Stillinger [77], as well as in the improved models ST2 [583], ST4 [306] and TIP5P [404]. Here, the negative charge of the oxygen atom is distributed to two sites so that the charges form an equilateral tetrahedron. The mass of the oxygen atom is located in its center, see Figure 7.15 (right). To evaluate the resulting potential, one has to compute seventeen distances, which is relatively expensive when compared to the ten distances of a four-site model and the nine distances of a three-site model. Therefore, four-site models are most-often used in practice because they are more economical than five-site models but deliver more accurate results than the simpler three-site models. A more detailed description of these different models together with parameter sets, references to the literature, and an accuracy comparison can be found for instance in [79, 347, 404]. Newer comparisons with experimental data are given in [575].

Even though the geometry and the parameters of the potentials of these models have been continuously adapted over time, it is still not possible to achieve a good approximation of *all* the properties of water that can be measured in experiments. To this end, we have to turn to more sophisticated water models which are either polarizable or where the charges fluctuate, as for instance in the WK [655], TIP4P-FQ [518], POL5 [582] and SWFLEX [638] models. In these models, the charges and therefore the potentials depend on the neighboring atoms, modeling electron clouds that can deform and adapt to their environment.

The parameters of the potentials are usually chosen by a comparison of data from experiments with results from Monte-Carlo simulations in which the force is truncated at a sufficiently large radius r_{cut} and which is then evaluated with the linked cell method. The results of the simulations depend on the cutoff radius r_{cut} [633] and on the chosen ensemble [405]. Other approaches use the PME method [215], reaction fields [633], or the Ewald summation method for water modeled by SPC and TIP3P models [100, 219]. A study of boundary conditions that resemble a dielectric environment of the observed system such as water, can be found for the SPC water model in [400].

In the TIP3P model the electric charge of the oxygen atom is chosen as -0.834 e, and the charges of the hydrogen atoms are correspondingly chosen to be 0.417 e, where e denotes the elementary charge. The geometry of the water molecule is given by the fixed distance of the hydrogen and oxygen atoms and the fixed angle of the H-O-H bonds. In this model, the distance is chosen to be 0.957 Å and the angle is chosen to be 104.52 degrees. Also, with this fixed geometry, Lennard-Jones interactions are only computed between oxygen atoms. Instead of the original nine degrees of freedom, the entire water molecule then has only six degrees of freedom, which can be parametrized by the center of mass and three angles.

It is both advantageous in our context and results in a simpler algorithm, to move the hydrogen and oxygen atoms separately and to use the original nine degrees of freedom. The distances and angles are fixed with the appropriate bond potentials, as described in Section 5.2.2. Furthermore, we use Lennard-Jones forces with the appropriate mixing rule between all atoms from different molecules. Note that this is different to the TIPS3 model in which only the oxygen atoms from different molecules interact in this way. The parameters for our model are given in Table 7.3.

distance potential O-H	$r_0 = 0.957$ Å,	$k_b = 450$ kcal/mol,
angle potential H-O-H	$\theta_0 = 104.52$ degrees,	$k_\theta = 55$ kcal/mol,
Coulomb potential	$q_H = 0.417$ e,	$q_O = -0.834$ e,
Lennard-Jones potential	$\varepsilon_H = 0.046$ kcal/mol,	$\sigma_H = 0.4$ Å,
	$\varepsilon_O = 0.1521$ kcal/mol,	$\sigma_O = 3.1506$ Å,
	$m_H = 1.0080$ u,	$m_O = 15.9994$ u

Table 7.3. Parameter values for the TIP3P-C water model.

An analogous model with a slightly different set of parameters is used in CHARMM [125]. We call the new water model TIP3P-C. Even though it is still a three-site model, it already has fluctuating charges and can be more easily implemented in our context. But because of the bond potential, we have to use smaller time steps ($\delta t = 0.1$ fs) than with a stiff water model or with frozen bonds ($\delta t = 1$ fs).

Now, we simulate the self-diffusion of water. To this end, we place 216 TIP3P-C water molecules into a periodic box with a length of 18.77 Å. The particle system is first equilibrated in the NVE ensemble to a temperature of 300 K and a density of 0.97 g/cm^3. The simulation then proceeds at constant temperature. We compute all bond and angle terms explicitly, i.e., we do not freeze any degrees of freedom. The SPME method from Section 7.3 is applied for the long-range Coulomb terms. We use a time step of 0.1 fs. The values of the parameters for the simulation are summarized in Table 7.4.

In our simulation we compute an approximation to the self-diffusion coefficient, which is given as the limit $t \to \infty$ of

$$D(t) = \sum_{i=1}^{N} \frac{d_i^2(t)}{6Nt},$$

$$L_1 = 18.77 \text{ Å}, \quad L_2 = 18.77 \text{ Å}, \quad L_3 = 18.77 \text{ Å},$$
$$\text{N} = 216 \text{ H}_2\text{O}, \quad T = 300 \text{ K}, \quad \rho = 0.97 \text{ g/cm}^3,$$
$$r_{\text{cut}} = 9.0 \text{ Å}, \quad \delta t = 0.1 \text{ fs}, \quad t_{\text{end}} = 100 \text{ ps},$$
$$h = 1 \text{ Å}, \quad G = 0.26 \text{ Å}^{-1}, \quad p = 6$$

Table 7.4. Parameter values for the simulation of water with the TIP3P-C model.

compare also (3.61). Here, $d_i(t)$ is the distance[49] from the center of mass of the molecule i at time t to its initial position $\mathbf{x}_i(t_0)$ at time t_0. In our example, the value of D is measured every 0.5 ps. After every 10 ps the current configuration is used as the new initial position for the next measurement, i.e., t_0 is reset. In this way, it can be checked whether the values vary only statistically or whether the system has not yet been equilibrated for a sufficiently long time span. Typically the value for D is taken at the end of such a series of measurements, since then the particle system is as much equilibrated as possible. We also determine the potential energy of the system.

Figure 7.16 shows two different views of the spatial distribution of the water molecules at a certain time during the simulation. Here, a clustering can be observed. Figure 7.17 shows the evolution of the computed approximation

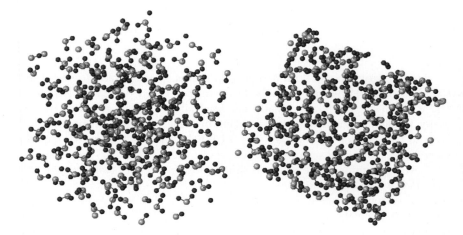

Fig. 7.16. Two different views of 216 water molecules simulated with the TIP3P-C model.

---

[49] For the case of periodic boundary conditions as in our simulations, this distance must be measured from the actual position of the particle. If a particle leaves the simulation domain at one side of the simulation domain and enters it from the opposite side, we therefore correct the value of its *initial position* $\mathbf{x}(t_0)$ accordingly.

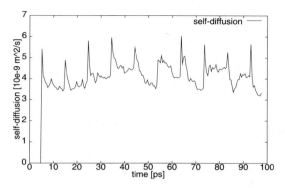

Fig. 7.17. Self-diffusion coefficient for the TIP3P-C water model.

to the self-diffusion coefficient D over time. Its measurement starts after 5 ps. The restart of the computation after every 10 ps results in a jump of the graph. At the end of the simulation, the computed approximation of D is about $4.1 \cdot 10^{-9}$ m^2/s. The potential energy varies in a 2% range around the value of -9.63 kcal/mol. These values are given in Table 7.5 together with values from other references for comparison.

model	reference	diffusion [10^{-9} m^2/s]	reference	E_{pot} [kcal/mol]
ST2	[583]	4.2	[583]	-9.184
TIP5P	[404]	2.62	[404]	-9.86
POL5/TZ	[582]	1.81	[582]	-9.81
TIP4P	[347]	3.29	[633]	-9.84
TIP4P (reaction field)			[633]	-10.0
TIP4P-FQ	[518]	1.9	[638]	-9.89
WK	[655]	1.1		
SPC	[655]	3.3	[633]	-9.93
SPC/E	[405]	2.49	[633]	-11.2
TIP3P (linked cell)	[347]	5.19	[633]	-9.60
TIP3P (PME)	[215]	5.1	[215]	-9.5
TIP3P-C (PME)		4.1		-9.63
data from experiments	[365]	2.30	[633]	-9.96

Table 7.5. Simulation results using the TIP3P-C water model in comparison with other models for normal conditions 10^5 Pa and 300 K or 307.5 K (depending on reference).

Our computed potential energy is close to the results for the TIP3P model and tends towards the values obtained with the TIP4P and TIP5P models and the experimentally observed result. For the diffusion coefficient, our value differs somewhat from that of the TIP3P model. Note here that the value measured in experiments at 300 K is $2.3 \cdot 10^{-9}$ m^2/s [365]. This value is reproduced relatively well by the more expensive four-center and five-center models.[50] Also our TIP3P-C model, which is more expensive than the original TIP3P model, reproduces these results better and yields a somewhat more accurate, i.e. lower value for the diffusion coefficient. However, the small number of particles still causes relatively large statistical fluctuations in the measured values.

In addition to the water molecules one can also take further atoms and molecules into account, like peptides or proteins [251]. This leads to the simulation of structures dissolved in an aqueous solution. We discuss some experiments in this direction in the Sections 9.1, 9.2 and 9.3.

7.5 Parallelization

We now turn to the parallelization of the program from Section 7.3. Our parallel SPME code will be based on our parallelized linked cell method, extended by a parallel version of the computation of the long-range forces. The parallelization of the program modules responsible for the evaluation of the short-range terms in the forces and the potential, and for the motion of the particles has already been described in Chapter 4.

The choice of suitable parallelization strategies for mesh-based methods for the computation of long-range forces depends on the components used in the method. A discussion for different computer architectures can be found for instance in [221, 606, 679]. In the following, we again consider parallel computers with distributed memory and we use a domain decomposition technique as parallelization strategy. We moreover rely on the communication library MPI, see also Chapter 4 and Appendix A.3.

We do not discuss the parallelization of the fast Fourier transform here [140, 595], instead we opt for a modular implementation so that readers can apply a parallel FFT of their choice. However, the data distribution of this FFT implementation has to match that of our domain decomposition approach. This is the case for instance for the parallel implementation fft_3d of Plimpton [15], which can be based on the FFTW library [240].

7.5.1 Parallelization of the SPME Method

Now, we discuss the parallelization of the SPME method in more detail. To this end, the following steps have to be implemented:

---

[50] More accurate simulation results can be obtained for systems containing a small number of water molecules with ab initio methods and the DFT approach [138, 555, 578].

- The sums (7.60) in the computation of the array Q have to be computed in parallel.
- The vector a from (7.66) has to be computed in parallel. This includes in particular the computation of a parallel fast Fourier transform of Q.
- A parallel fast inverse Fourier transform of a has to be computed.
- The sum (7.71) in the computation of the long-range energy terms has to be computed in parallel. First, partial sums are computed by each process. Then, the results of each process are added up. This can be directly implemented in a communication step with MPI_Allreduce.
- The sums (7.74) in the computation of the long-range force terms have to be computed in parallel.

The first three steps correspond to the solution of the potential equation. The last two steps implement the computation of the long-range energy terms and the long-range force terms.

Domain Decomposition as Parallelization Strategy. For the parallelization, we again use the domain decomposition strategy. Here, we build on the decomposition (4.2) of the simulation domain Ω into the subdomains $\Omega_{\texttt{ip}}$ with the multi-indices $\texttt{ip}$ to distribute the data to the processes and thereby to the processors, compare Figure 7.18. It has already been used in the parallelization of the linked cell method.

The data exchange necessary for the parallel execution of the linked cell code is already known from Sections 4.2 and 4.3. In addition, we assume that the numbers $\texttt{K[d]}$ of degrees of freedom of the SPME mesh used in the discretization of the potential equation are (componentwise) integer multiples of $\texttt{np[d]}$. This assumption guarantees that the decomposition of the simulation domain in the linked cell method is also a valid decomposition of the SPME mesh and can therefore be used to distribute the SPME mesh to the processes, compare Figure 7.18.

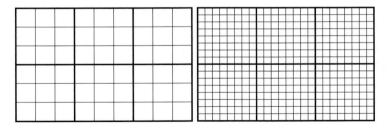

Fig. 7.18. The decomposition of the domain Ω into subdomains $\Omega_{\texttt{ip}}$ implies a distribution of the cells in the linked cell method (left) as well as a distribution of the degrees of freedom of the SPME mesh (right) to the processes.

The subdomains Ω_{ip} are again decomposed into $\prod_{d=0}^{DIM-1} K[d]/np[d]$ mesh cells each. The necessary computations for the long-range terms in the forces and the energy are performed by each process, but only on the local data assigned to it. However, at certain points during the computation, each process needs data that are assigned to neighboring processes or that have been computed by neighboring processes. Therefore, in each time step in the computation of the long-range force and energy terms, these data have to be exchanged between processes that handle neighboring subdomains.

Communication between the processes is thus necessary at the following three points of the algorithm:

− After the computation of Q according to (7.60).
− During the fast Fourier transform and the fast inverse Fourier transform.
− Before the computation of the forces according to (7.74).

To compute Q and the forces on some of the particles in its subdomain, each process needs values within a border neighborhood of width $p_{max} - 1$ across the lower, front, and left boundary from neighboring subdomains. This results from our chosen definition of B-splines in (7.55). To store this data of the border neighborhoods, each subdomain is extended by $p_{max} - 1$ cell rows in these directions. If the data in the border cells are copied to the process and thus are locally available, the process can compute the array Q and the forces on the particles in its subdomain independently of other processes. This works for all processes and subdomains, and allows for the parallel computation of the forces on the particles. In the following, we consider the three points of communication separately.

Computation of Q. According to (7.60), the array Q is computed as follows

$$Q_{\mathbf{m}} = \sum_{\mathbf{n} \in \mathbb{Z}^3} \sum_{j=1}^{N} q_j \prod_{d=1}^{3} M_p((\mathbf{x}_j)_d K_d/L_d - m_d - n_d K_d).$$

The multi-index $\mathbf{m}$ in this formula runs over all SPME mesh points. A particle at the position $\mathbf{x}$ influences the value of Q at all mesh points $\mathbf{m}$ with

$$0 \leq \frac{(\mathbf{x})_d}{h_d} - m_d < p_{max}, \text{ with the mesh width } h_d = \frac{L_d}{K_d}, d \in \{1, 2, 3\}.$$

These are its so-called support points. If the particles and meshes are distributed to the processes, each process computes the sum only over the particles associated to it.

It may happen at the left, lower, or front local boundary that the mesh points $\mathbf{m}$ needed in the summation belong to other processes. The corresponding border neighborhoods of the subdomain have been shown schematically in Figure 4.6 already. But in this way, each process also computes parts of the sum that are associated to other processes. For our definition of the M_p

in (7.54), (7.55), the border neighborhoods have a width of $p_{\max} - 1$, and are situated next to the left, lower, and front boundary of the subdomain. To compute Q completely, these data are needed on neighboring processes and thus have to be sent there.

The data to be exchanged between processes lie at the left, lower, and front border of the subdomain. The exact memory layout of those layers of mesh cells of width `pmax-1` depends on the parallel FFT used. Similar to the parallelization of the linked cell method, one could extend the local Q array by a border neighborhood. However, the FFT library – FFTW – that we chose in the sequential case uses a *linearized* array to describe the three-dimensional grid. It is easy to distinguish between border neighborhood cells and interior cells of the grid if one uses a multi-index `ic`, but it is difficult to distinguish them using the linear cell indices computed by `index`. The parallel FFT is only computed on the data inside the subdomain, and therefore the FFT would have to differentiate explicitly between border neighborhood cells and interior cells. Essentially this means that the FFT would have to work internally with an analogous index computation. An alternative approach is not to extend the array for Q explicitly, but to introduce new arrays for the border neighborhoods. Then, the loops in the computation of the interpolants and in the evaluation of the forces have to be split into loops over the mesh points over the different arrays. This approach is the more flexible one since the parallel FFT does not have to be changed.

The actual communication of data then consists in the transport of these border neighborhoods, as already shown in the Figures 4.9 and 4.10 for the particle moving in the linked cell method. The received data are *added to* the local values. We have to send and receive mesh data, similarly to how we already sent and received particle data in Algorithm 4.7. But unlike the number of particles there, the number of values to be sent is now determined a priori and can be computed from the mesh structure. Furthermore, only one communication in each direction is necessary. In this way, also values in the corners are sent to the correct processes. Again, one has to handle the special cases for subdomains next to the boundary $\partial\Omega$ of the simulation domain Ω which are caused by periodic (or other) boundary conditions. At the end of these communication steps, all values of Q are completely determined, and they are distributed to all processes according to the domain decomposition layout.

Fast Fourier Transform, Scaling, and Inverse Transform. The parallel fast Fourier transform can now be applied to the distributed data. Of course, data has to be exchanged also during the Fourier transform, as well as during the fast inverse Fourier transform. However, a parallel FFT does in general not use the additional border neighborhood arrays that we introduced for the evaluation of the splines, it uses other, internally defined data and communication structures. Therefore, it is sufficient to call the parallel FFT with the array Q that is distributed across the processes. We will not discuss the implementation of the parallel FFT in more depth. Instead we refer to the literature and the available software packages [15, 240]. However, the distribution of the data in the parallelized FFT should match with the one in our domain decomposition approach. We use in the following the parallel FFT routine fft_3d developed by Plimpton in [15].

After the execution of the parallel Fourier transform, the Fourier transformed data have to be multiplied with the factor (7.65). The different multiplications are independent of each other and do not require any communication.

Finally, a parallel fast inverse Fourier transform has to be applied to the scaled data. After this step, the result $\mathrm{DF}^{-1}[a]$ of the inverse Fourier transform is stored in distributed form across all processes.

Computation of the Forces. The summation over the mesh points in the computation of the force on one particle in a subdomain can be restricted to the support of the associated spline M_p. In the parallel execution of the program, each process computes the forces on all those particles that lie within its subdomain Ω_{ip}. In this computation the process needs the values of $\mathrm{DF}^{-1}[a]$ also in a border neighborhood of width $p_{\max} - 1$. Correspondingly, the computation of the forces needs this data from the neighboring processes which are associated to the front, lower, and left neighboring subdomain. The data are stored in the previously introduced additional arrays for the border neighborhoods. We already presented an analogous data exchange in the Figures 4.7 and 4.8 for the force computation in the linked cell method and we implemented it in Algorithm 4.6. In-between the communication steps, the received data have again to be copied into the mesh data structure to ensure that they can be transmitted further in the next communication step.

Altogether, we obtain the parallel Algorithm 7.10 for the computation of the long-range terms.

Algorithm 7.10 Parallel SPME Algorithm

---
compute the scaling factors D;
compute the interpolant of the charge by each process for its subdomain Ω_{ip} including border neighborhoods;
exchange data from the border neighborhoods and add them to the local Q;
perform a parallel FFT;
compute the electrostatic energy and a;
perform a parallel inverse FFT;
exchange data of the border neighborhood;
compute the forces locally by each process;

Note that our definition of the splines M_p in (7.54) and (7.55) offers the advantage that data only have to be sent to the left, lower, and front neighboring processes and vice versa, and not also to the right, upper, and back neighboring processes. This significantly reduces the number of communication steps, but not the amount of data to be transported.

7.5.2 Implementation

Subdomain. We start with the description of the subdomain Ω_{ip} associated to a process. The type `SubDomainSPME` is declared in data structure 7.2. It collects all the data that a process needs for its computations.

Data structure 7.2 Subdomains, Cells and Neighboring Processes of Ω_{ip}

```
typedef struct {
    struct SubDomain lc;  //  data structure 4.1 linked cell
    int K[DIM];           //  SPME mesh cells
    int K_lower_global[DIM]; //  global index of the first mesh point
                          //  of the subdomain
    int K_start[DIM];     //  width of border neighborhood, also
                          //  smallest local index inside the subdomain
    int K_stop[DIM];      //  first index following the upper border
                          //  neighborhood of the subdomain
    int K_number[DIM];    //  number of mesh points in the subdomain
                          //  including border neighborhoods
    real spme_cellh[DIM]; //  mesh width of the SPME mesh
} SubDomainSPME;
```

In code fragment 7.4, the appropriate values for the subdomain are computed from the values for the entire domain Ω together with the process number.[51]

The next step towards a parallel SPME method is the adaptation of the sequential code to this new generalized domain decomposition. The first step is the replacement of the sequential linked cell method for the short-range force terms with its parallel version (see Chapter 4). For the treatment of the long-range force terms we have to adapt the code to the new subdomains: We change all loops running over all mesh points in compQ_SPME, etc., into loops that run only over the local subdomain. Here, we do not only have to appropriately change the stopping criteria for the loops, but we also have to take into account that the data are distributed over two arrays Q and Q_boundary; i.e., every code segment accessing memory in the form `for (int d=0; d<DIM; d++) kp[d]=(k[d]-p[d])%K[d];` `Q[index(kp,K)].re = ...` has to be changed appropriately.

---

[51] Compare also data structure 4.1 and code fragment 4.1 for the analogous construction for the grid in the linked cell method.

Code fragment 7.4 Initialization of the Data Structure SubDomainSPME

```
void inputParameters_SPMEpar(real *delta_t, real *t_end, int pmax,
                             SubDomainSPME *s) {
  inputParameters_LCpar(delta_t, t_end, &(s->lc));
  ...  // set s->K
  for (int d=0; d<DIM; d++) {
    s->spme_cellh[d] = s->lc.l[d] / s->K[d];
    s->K_start[d] = pmax-1;
    s->K_stop[d] = s->K_start[d] + (s->K[d]/s->lc.np[d]);
    s->K_number[d] = (s->K_stop[d] - s->K_start[d]) + s->K_start[d];
    s->K_lower_global[d] = s->lc.ip[d] * (s->K[d]/s->lc.np[d]);
  }
}
```

Main Program. The changes in the main program are minimal, see Algorithm 7.11. The routine inputParameters_SPMEpar, as described in code fragment 7.4, determines the subdomain for each process. Memory is allocated for the arrays D and Q as well as for the border neighborhood array Q_boundary of Q. Furthermore, the parallel FFT library has to be initialized with the routine initFFTpar before time integration is started.

In the routine timeIntegration_SPMEpar for the time integration, which is given in code fragment 7.5, one only has to add the calls to the routines compQpar for the parallel computation of the interpolant, compFFTpar for the parallel solution of the Poisson equation, and compF_SPMEpar for the parallel evaluation of the long-range force terms.

Exchange of Boundary Data and Parallel Evaluation of Forces. The code fragments 7.6 and 7.7 present the implementation of the parallel interpolation and the parallel evaluation of the forces. As in the parallel linked cell method, a further communication step is introduced *after* the computation of the interpolant and *before* the evaluation of the forces. The corresponding border neighborhood of the subdomain Ω_{ip} is shown in Figure 4.6 (right). The communication after the computation of the interpolant, in which the data in the border neighborhood is sent, was already shown in Figure 4.10. It has to be implemented appropriately in the routine compQ_comm. Data are transported in exactly opposite order (see Figure 4.8) in the routine compF_SPME_comm, which is executed before the computation of the long-range force terms in compF_SPME. There, data from the left, lower, or front border have to be sent to the corresponding neighbors.

Communication. The entire communication between neighboring processes[52] is again to be implemented in a central routine like sendReceiveCell which allows to realize the communication patterns of the Figures 4.8 and

---

[52] Except for the communication in the parallel FFT.

Algorithm 7.11 Main Program of the Parallel SPME Method

```
int main(int argc, char *argv[]) {
  int N, pnc, pmax;
  real r_cut;
  real delta_t, t_end;
  SubDomainSPME s;
  int ncnull[DIM];
  MPI_Init(&argc, &argv);
  inputParameters_SPMEpar(&delta_t, &t_end, &N, &s, &r_cut, &G, &pmax);
  pnc = 1;
  for (int d = 0; d < DIM; d++)
    pnc *= s.lc.ic_number[d];
  Cell *grid = (Cell*) malloc(pnc*sizeof(*grid));
  pnc = 1;
  for (int d = 0; d < DIM; d++)
    pnc *= s.K_stop[d]-s.K_start[d];
  real *D = (real*) malloc(pnc*sizeof(*D));
  fft_type *Q = (fft_type*) malloc(pnc*sizeof(*Q));
                // data type fft_type depends on the chosen FFT library
  fft_type *Q_boundary = (fft_type*) malloc(.... *sizeof(*Q));
                // arrays for the border neighborhood of the Q array,
                // its memory layout has to be adapted to the memory layout
                // of the chosen parallel FFT library
  initFFTpar(D, s.K, pmax, s.lc.l); // possibly further parameters
                                    // depending on the chosen parallel
                                    // FFT library
  initData_LC(N, grid, &s);
  timeIntegration_SPMEpar(0, delta_t, t_end, grid, Q, Q_boundary, D,
                          &s, r_cut, pmax);
  for (int d = 0; d < DIM; d++)
    ncnull[d] = 0;
  freeLists_LC(grid, ncnull, s.lc.ic_number, s.lc.ic_number);
  free(grid); free (Q); free (D);
  MPI_Finalize();
  return 0;
}
```

4.10. With the help of this routine, the routines compF_SPME_comm and compQ_comm can easily be implemented. Since the data is associated to the mesh structure, the routine sendReceiveGrid implemented here is simpler than the routine sendReceiveCell from algorithm 4.4. However, one has to take into account the decomposition of the SPME-specific vector Q into the two arrays Q and Q_boundary when sending and receiving data.

Code fragment 7.5 Time Integration

```
timeIntegration_SPMEpar(real t, real delta_t, real t_end, Cell *grid,
                        fft_type *Q, fft_type *Q_boundary, real *D,
                        SubDomainSPME *s, real r_cut, int pmax) {
  compF_LCpar(grid, &s->lc, r_cut);
  compQpar(grid, s, Q, Q_boundary, pmax);
  compFFTpar(Q, s->K, D);
  compF_SPMEpar(grid, s, Q, Q_boundary, pmax);
  while (t < t_end) {
    t += delta_t;
    compX_LC(grid, &s->lc, delta_t);
    compF_LCpar(grid, &s->lc, r_cut);
    compQpar(grid, s, Q, Q_boundary, pmax);
    compFFTpar(Q, s->K, D);
    compF_SPMEpar(grid, s, Q, Q_boundary, pmax);
    compV_LC(grid, &s->lc, delta_t);
    compoutStatistic_LCpar(grid, s, t);
    outputResults_LCpar(grid, s, t);
  }
}
```

Code fragment 7.6 Parallel Interpolation of the Charge Distribution

```
compQpar(Cell *grid, SubDomainSPME *s, fft_type *Q,
         fft_type *Q_boundary, int pmax) {
  compQ(grid, s, Q, Q_boundary, pmax); // version adapted to s
                                       // and Q_boundary
  compQ_comm(Q, Q_boundary, s, pmax);
}
```

Code fragment 7.7 Parallel Evaluation of the Long-Range Force Terms

```
compF_SPMEpar(Cell *grid, SubDomainSPME *s, fft_type *Q,
              fft_type *Q_boundary, int pmax) {
  compF_SPME_comm(Q, Q_boundary, s, pmax);
  compF_SPME(grid, s, Q, Q_boundary, pmax); // version adapted to s
                                            // and Q_boundary
}
```

7.5.3 Performance Measurements and Benchmarks

In this section we analyze the parallel scaling of the components of our code for the computation of the long-range terms in the forces and potentials. We employ the potential

$$U(r_{ij}) = \frac{1}{4\pi\varepsilon_0} \frac{q_i q_j}{r_{ij}} + 4\varepsilon_{ij} \left(\left(\frac{\sigma_{ij}}{r_{ij}} \right)^{12} - \left(\frac{\sigma_{ij}}{r_{ij}} \right)^6 \right)$$

for a model problem. We study the melting of a salt akin to sodium chloride. We scale the variables as in the KCl example from Section 7.4.2 and obtain the parameters from Table 7.6.

$$m_1' = 22.9898, \quad m_2' = 35.4527,$$
$$\sigma_{11}' = 4.159, \quad \sigma_{22}' = 7.332,$$
$$\varepsilon_{11}' = 75.832, \quad \varepsilon_{22}' = 547.860,$$
$$q_1' = 1, \quad q_2' = -1,$$
$$r_{cut}' = 24, \quad G' = 0.1,$$
$$h' = 4.0, \quad p = 4$$

Table 7.6. Parameter values for the benchmark problem: Melting of salt.

We use this problem as a benchmark to study the properties of our parallelized SPME method. All computations have been carried out on a PC cluster, compare Section 4.4 and [557]. Here, when the number of particles is increased, the size of the simulation domain is increased by the same factor so that the particle density remains constant. Analogously, the number of mesh points for the solution of the potential equation is increased and the mesh width is kept constant. Table 7.7 lists the values used for the number of particles, the length of the domain, and the number of mesh points.

	particles									
	1728	4096	8000	17576	32768	64000	140608	262144	592704	1191016
domain length	96	144	192	240	288	360	480	576	768	960
mesh points	24^3	36^3	48^3	60^3	72^3	90^3	120^3	144^3	192^3	240^3

Table 7.7. Numbers of particles, numbers of mesh points, and the length of the domain for the benchmark problem.

Tables 7.8 and 7.9 show the runtimes for one time step for the computation of the long-range and the short-range force terms, respectively.[53] Table 7.10 shows the runtimes for a fast Fourier transform.

The corresponding parallel efficiency and speedup for computations with 262144 atoms are shown in the Tables 7.11 and 7.12. Both the computations for the short-range terms and the computations for the long-range terms show very good parallel scaling. The slight degradation in the parallel efficiency of the computation of the long-range terms stems from the less efficient

---

[53] Here we have used the implementation in which the splines and their derivatives are computed recursively using the routines `spline` and `Dspline`. With the optimizations discussed in footnote 43, one can speed up the runtime by a factor of two to three for `pmax=4`, depending on the particular problem. However, the parallel scaling of the parallel algorithm changes only minimally.

runtime mesh points	processors							
	1	2	4	8	16	32	64	128
13824	**1.10**	0.55	0.28	0.14				
46656	2.68	**1.36**	0.68	0.35				
110592	5.36	2.77	**1.39**	0.72	0.36			
216000	11.62	5.99	3.05	**1.59**	0.83			
373248	21.62	11.14	5.70	2.97	**1.47**	0.74		
729000	42.20	21.30	10.81	5.67	3.05	**1.56**		
1728000	**93.79**	48.55	24.59	13.30	6.63	3.38	**1.68**	
2985984	179.23	**90.13**	45.50	24.39	12.27	6.32	3.13	**1.63**
7077888			**103.91**	55.45	28.10	14.61	7.38	3.86
13824000				**111.60**	55.97	29.31	14.92	7.82
23887872					**107.40**	55.64	28.33	15.11
56623104						**121.59**	63.96	34.25
110592000							**126.31**	67.54
242970624								**156.30**

Table 7.8. Parallel runtimes (in seconds) for the long-range force terms for one time step.

runtime particles	processors							
	1	2	4	8	16	32	64	128
1728	**1.32**	0.67	0.35	0.18				
4096	2.28	**1.16**	0.61	0.31				
8000	3.99	1.97	**1.02**	0.52	0.27			
17576	9.32	4.70	2.44	**1.24**	0.82			
32768	18.31	9.21	4.75	2.40	**1.22**	0.62		
64000	35.78	18.81	10.30	5.49	2.38	**1.27**		
140608	**74.61**	37.46	19.26	9.74	4.94	2.50	**1.24**	
262144	144.26	**73.87**	37.88	19.14	9.62	4.84	2.42	**1.29**
592704			**82.74**	41.76	20.97	10.53	5.26	2.67
1191016				**85.35**	42.66	21.77	10.76	5.51
2299968					**89.90**	44.07	21.85	11.02
4913000						**90.25**	44.43	22.95
9800344							**89.39**	45.52
21024576								**91.38**

Table 7.9. Parallel runtimes (in seconds) for the short-range force terms for one time step.

parallelization of the fast Fourier transform. Table 7.13 shows the parallel efficiency and the speedup for the fast Fourier transform for a mesh with 144^3 mesh points. Here, the high communication complexity of the parallel fast Fourier transform causes a decrease in the parallel efficiency with increasing numbers of processors. However, the runtime of the fast Fourier transform is relatively small compared to the total time needed for the computation of

runtime mesh points	processors							
	1	2	4	8	16	32	64	128
13824	**0.018**	0.0085	0.0057	0.0045				
46656	0.071	**0.043**	0.023	0.017				
110592	0.19	0.13	**0.078**	0.046	0.022			
216000	0.36	0.27	0.17	**0.11**	0.14			
373248	0.65	0.49	0.32	0.21	**0.094**	0.053		
729000	1.26	0.92	0.64	0.47	0.27	**0.15**		
1728000	**3.62**	2.43	1.51	1.23	0.57	0.33	**0.15**	
2985984	5.99	**4.25**	2.75	1.92	1.00	0.60	0.29	**0.22**
7077888			**6.30**	4.15	2.42	1.46	0.76	0.54
13824000				**7.95**	4.80	3.10	1.64	1.15
23887872					**8.35**	5.45	2.78	1.43
56623104						**12.71**	7.15	5.78
110592000							**13.68**	7.56
242970624								**19.44**

Table 7.10. Parallel runtimes (in seconds) for the FFT.

	processors							
	1	2	4	8	16	32	64	128
speedup	1.000	1.999	3.939	7.348	14.607	28.359	57.262	109.957
efficiency	1.000	0.994	0.985	0.919	0.913	0.886	0.895	0.859

Table 7.11. Speedup and parallel efficiency for one time step of the simulation of a material akin to $NaCl$ with 144^3 mesh points and 262144 atoms, long-range terms.

	processors							
	1	2	4	8	16	32	64	128
speedup	1.000	1.953	3.809	7.537	14.996	29.806	59.611	111.829
efficiency	1.000	0.976	0.952	0.942	0.937	0.931	0.931	0.874

Table 7.12. Speedup and parallel efficiency for one time step of the simulation of a material akin to $NaCl$ with 262144 atoms, short-range terms.

	processors							
	1	2	4	8	16	32	64	128
speedup	1.0000	1.4094	2.1782	3.1198	5.9900	9.9833	20.6552	27.2273
efficiency	1.0000	0.7047	0.5445	0.3900	0.3744	0.3120	0.3227	0.2127

Table 7.13. Speedup and parallel efficiency of the FFT with 144^3 mesh points.

the long-range force terms, see Table 7.10, so that this decrease in efficiency does not have a dominant effect on the total runtime.

Figure 7.19 shows the speedup and the parallel efficiency for the long-range terms (including the parallel FFT), the short-range terms, and the parallel FFT. One clearly sees the decrease in efficiency for the parallel FFT. However, for our moderate numbers of processors, this decrease does not strongly impact the efficiency of the entire computation of the long-range terms. Further studies of the scalability of this type of algorithm can be found in [160].

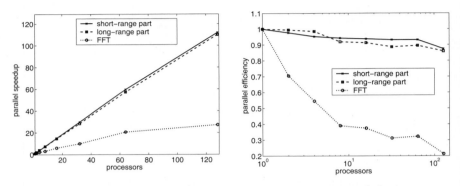

Fig. 7.19. Speedup and parallel efficiency of the computation of the long-range terms (including parallel FFT), the short-range terms, and the parallel FFT.

7.6 Example Application: Large-Scale Structure of the Universe

The parallel version of our SPME code allows the simulation of problems with larger numbers of particles. In the following, we consider an example from astrophysics – a simulation of the evolution of the large-scale structure of the universe.

Our Milky Way, which consists of approximately 200 billion stars, is only one galaxy among millions of galaxies in the universe. The galaxies that we can observe from earth are all moving away from us. The farther they are away, the faster they move away. The entire universe expands since its creation in the big bang, in which the universe arose from an unimaginably hot and dense primeval mass. This expansion of the universe is slowed down by the gravitational forces among the matter in the universe. If the average mass density is smaller than a certain critical value, the gravitational force cannot stop the expansion, and the universe will expand forever (the so-called open universe). If the average mass density is larger than the critical value, the expansion will be stopped by gravitation and the universe will start to collapse at some point (the so-called closed universe) [188, 303].

In its simplest form, the big bang theory [387] assumes that mass and radiation are distributed uniformly in the universe. This theory accounts for the existence of the cosmic background radiation and for the existence of light elements, but it cannot account for the large-scale structures that are observed in the universe. Galaxies are mostly aggregated in so-called galaxy clusters. Up to now, about 10000 of these clusters are known. These galaxy clusters again form so-called superclusters. Here, a kind of bubble structure emerges, with galaxy clusters at the surface of the bubbles and almost empty interiors. Most cosmologists assume that these observable structures have formed under the influence of gravitation from small initial fluctuations in the density of the universe. The universe expanded more slowly in regions with higher relative density, so that the relative density of those regions increased further [459, 472].

Theories about the evolution of the structure of the universe cannot be tested experimentally. They have to be simulated instead. A numerical simulation of the evolution of the structure of the universe essentially needs three components: Assumptions on the cosmological model (masses, densities, etc.), a model for the fluctuations at the beginning of the simulation (initial conditions), and a method to solve the equations of motions that control the movement of the masses.

We assume in the following that the masses in the universe move according to Newton's equation of motion

$$\dot{\mathbf{x}}_i = \mathbf{v}_i \\ \dot{\mathbf{v}}_i = \mathbf{F}_i/m_i, \quad i = 1, \ldots, N,$$

where $\mathbf{F}_i$ denotes the gravitational force from the gravitational potential (2.42). Initial conditions are needed to solve these equations of motion.

Initial Conditions. To obtain initial conditions for cosmological simulations, the so-called Zel'dovich approximation [681] is most often used. There, one specifies a spectrum which is modified by so-called transfer functions that describe the circumstances during the early evolution of the universe (as implicitly given by the model). The precise form of the transfer functions depends on the specific cosmological model that is used. How can we now obtain initial conditions from these data for our numerical simulations? The idea is to compute a density distribution from the data and then to position particles (equipped with velocities) according to the computed distribution. To this end, a number of codes have been developed that produce such initial conditions. Mostly, a mass distribution is created that differs only slightly from a regular grid. The strength of the perturbations from the regular grid strongly depends on the specific model.

Initial conditions are usually specified in comoving coordinates. Such coordinates are standard in simulations of the evolution of the structure of the universe. This simplifies the computations since the coordinate system already takes into account the expansion of the universe.

Formulation of the Problem in Comoving Coordinates. The universe appears to be homogeneous and isotropic if averaged over sufficiently large regions of space so that the inhomogeneities resulting from the structures are smoothed out. This mean density $\bar{\rho}(t)$ then only depends on time but no longer on position. Every point in the universe can thus be chosen as the origin from which all other masses move away. The assumption of homogeneity then implies that the expansion of the universe corresponds to a radial motion that can be characterized by an expansion factor

$$a(t) = \frac{r_i(t)}{r_i(0)},$$

where $r_i(t)$ denotes the distance of a particle i at time t from an arbitrarily chosen center. This describes by which factor the distance of a mass from the origin has changed relative to the distance of the mass at a fixed time $t = 0$. This factor does not depend on the position, since space is assumed to be homogeneous. The factor $a(t)$ is typically chosen to be of the form $a(t) \sim t^n$ with $n < 1$. The definition of the expansion factor $a(t)$ directly implies that the escape velocity is proportional to the distance from the origin (Hubble's law). The factor $a(t)$ obeys the so-called Friedman equation

$$\dot{a}(t)^2 - \frac{8}{3}\pi\frac{G\bar{\rho}(0)}{a(t)} = -k \tag{7.83}$$

with k the constant of integration (the curvature of space). This equation can be derived from the homogeneity of space and the law of gravity [324]. Here, $\bar{\rho}(0)$ is the mean density at time $t = 0$. If $k < 0$ holds, then the gravitational force cannot stop the expansion (and it holds that the kinetic energy is larger than the potential energy), and the universe will expand forever (open universe). However, if $k > 0$ holds (and the kinetic energy is smaller than the potential energy), the expansion is stopped by gravitation and the universe starts to collapse at some point in time (closed universe). In the special case $k = 0$, the universe expands, but is closed.

The positions scaled by the expansion factor $a(t)$ can now be used as the new coordinate system $\mathbf{x}_i = \mathbf{x}_i^{\text{old}}/a(t)$. In this coordinate system, the equations of motion read

$$\begin{aligned}\dot{\mathbf{x}}_i &= \mathbf{v}_i, \\ \dot{\mathbf{v}}_i &= \mathbf{F}_i/m_i - \gamma\mathbf{v}_i,\end{aligned} \quad i = 1,\ldots,N,$$

with $\gamma(t) = 2H(t)$ and the so-called Hubble constant

$$H(t) = \frac{\dot{a}(t)}{a(t)}.$$

The force $\mathbf{F}_i$ is now given as the gradient[54] of the solution Φ of the potential equation

---

[54] Since we work in the new, comoving, coordinates, the chain rule produces an additional factor of $1/a^3$ in the force.

$$\Delta\Phi(\mathbf{x}, t) = 4\pi G_{\text{Grav}}(\rho(\mathbf{x}, t) - \rho_0), \tag{7.84}$$

i.e. $\mathbf{F}_i = -\frac{1}{a^3}\nabla_{\mathbf{x}_i}\Phi(\mathbf{x}_i)$. Here, $\rho(\mathbf{x}, t)$ denotes the mass density at the point $\mathbf{x}$ at time t and ρ_0 denotes the mean mass density (which is constant over time in the new comoving coordinate system).

The change of coordinates has two main effects:

- An additional friction term, which depends on $H(t)$, is introduced into the equations of motion and leads to a slowdown in the expansion of the universe.
- Locally negative mass densities $\rho(\mathbf{x}, t) - \rho_0$ can occur on the right hand side of the potential equation (7.84), since the mean density is subtracted from the density $\rho(\mathbf{x}, t)$. This now guarantees that the potential equation with periodic boundary conditions has a solution: The integral over the entire space satisfies $\int \rho(\mathbf{x}, t) - \rho_0 d\mathbf{x} = 0$, which is necessary for the solvability of the potential equation with periodic boundary conditions, compare (7.8).

Time Integration. In this application of our code, we use the leapfrog version of the Störmer-Verlet method from Section 3.1, compare (3.20) and (3.21). The velocities of the particles at half-step are given according to (3.20) as

$$\mathbf{v}_i^{n+1/2} = \mathbf{v}_i^{n-1/2} + \frac{\delta t}{m_i}(\mathbf{F}_i^n - \gamma^n m_i \mathbf{v}_i^n). \tag{7.85}$$

The right hand side is evaluated here at time t_n. To evaluate all velocities at the times $t_{n+1/2}$ and $t_{n-1/2}$, we use the central difference $\mathbf{v}_i^n \approx (\mathbf{v}_i^{n-1/2} + \mathbf{v}_i^{n+1/2})/2$. We obtain[55]

$$\mathbf{v}_i^{n+1/2} = \frac{1 - \gamma^n \delta t/2}{1 + \gamma^n \delta t/2}\mathbf{v}_i^{n-1/2} + \frac{\delta t}{1 + \gamma^n \delta t/2}\mathbf{F}_i^n.$$

For the new positions of the particles, (3.21) yields

$$\mathbf{x}_i^{n+1} = \mathbf{x}_i^n + \delta t \mathbf{v}_i^{n+1/2}.$$

The forces $\mathbf{F}_i$ are computed from the solution of the potential equation (7.84). We split the computation of the forces into two parts according to the SPME method from Section 7.3; one part, which we obtain as a direct sum, and another part, which we obtain from the approximate solution of the potential equation with a smoothed right hand side.[56]

---

[55] The term $\mathbf{F}_i^n$ is multiplied with an additional factor $1/a^3$ in the new coordinates, compare footnote 54.

[56] In earlier sections we have always used the Coulomb potential as a long-range potential, with a right hand side of $\frac{1}{\varepsilon_0}\rho$, where ρ is the charge density. We now consider the gravitational potential, which is similar in form to the Coulomb potential. However, the constants on the right hand side are different, which has to be taken into account in the code.

In addition, equation (7.83) has to be solved for $a(t)$ at the same time as the other equations. From $a(t)$ one obtains $H(t)$, and finally γ at time t_n, which is needed for the computation of the velocities of the particles according to (7.85).

Smoothing of the Potential. The gravitational potential is a purely attractive potential, which leads to some numerical problems. For instance, singularities occur that can strongly limit the quality of the simulations. As a remedy, instead of the term $1/r_{ij}$ in the potential, one uses the expression $1/(r_{ij} + \varepsilon)$ with a small parameter ε. The short-range energy and force terms are smoothed in this way and now read

$$V^{\mathrm{sr}} = -\frac{1}{2} G_{\mathrm{Grav}} \sum_{i=1}^{N} \sum_{\mathbf{n} \in \mathbb{Z}^3} m_i \sum_{\substack{j=1 \\ i \neq j \text{ for } \mathbf{n}=0 \\ r_{ij}^{\mathbf{n}} < r_{\mathrm{cut}}}}^{N} m_j \frac{\mathrm{erfc}(G(r_{ij}^{\mathbf{n}} + \varepsilon))}{r_{ij}^{\mathbf{n}} + \varepsilon}$$

and

$$\mathbf{F}_i^{\mathrm{sr}} = \frac{1}{a^3} G_{\mathrm{Grav}} m_i \sum_{\mathbf{n} \in \mathbb{Z}^3} \sum_{\substack{j=1 \\ j \neq i \text{ for } \mathbf{n}=0 \\ r_{ij}^{\mathbf{n}} < r_{\mathrm{cut}}}}^{N} m_j \frac{1}{(r_{ij}^{\mathbf{n}} + \varepsilon)^2} \left(\mathrm{erfc}(G(r_{ij}^{\mathbf{n}} + \varepsilon)) \right.$$
$$\left. + \frac{2G}{\sqrt{\pi}} (r_{ij}^{\mathbf{n}} + \varepsilon) e^{-(G(r_{ij}^{\mathbf{n}} + \varepsilon))^2} \right) \frac{\mathbf{r}_{ij}^{\mathbf{n}}}{r_{ij}^{\mathbf{n}}}.$$

Example. We present some results of simulations with 32^3 and with 64^3 particles. For simplicity, we limit ourselves to the case $\dot{a}(t) = \frac{1}{a(t)^{1/2}}$, which is a special case of (7.83) resulting from the scaling of time by $H(0)$. Then, it holds that

$$a(t) = \sqrt{\frac{3}{2}}(t + t_0)^{2/3} \quad \text{and} \quad \dot{a}(t) = \sqrt{\frac{2}{3}}(t + t_0)^{-1/3},$$

and therefore

$$H(t) = \frac{\dot{a}(t)}{a(t)} = \frac{2}{3} \frac{1}{t + t_0}.$$

Figure 7.20 shows the distribution of the particles at the beginning of the simulation in the case of 32^3 particles. To determine initial conditions in our example, we use the Fortran77 code *Cosmics* [87]. Figure 7.21 shows the particle distribution at the end of the simulation. One can see that the particles agglomerate in some subdomains and that larger structures emerge. The results of a simulation with 64^3 particles and slightly changed initial conditions are shown in Figure 7.22. A color coded representation of such particle densities in space was already shown in Figure 1.3. Again, one can see that structures emerge. Further results for the simulation of the large-scale structure of the universe can be found for instance in [16, 17, 18, 353].

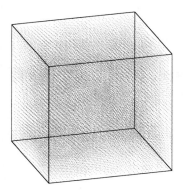
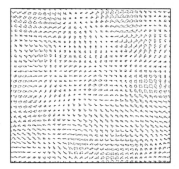

Fig. 7.20. Initial configuration with 32^3 particles and its projection to two dimensions.

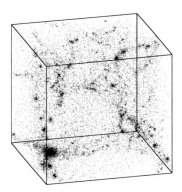
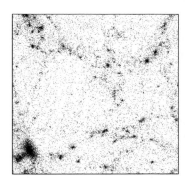

Fig. 7.21. Final configuration with 32^3 particles and its projection to two dimensions.

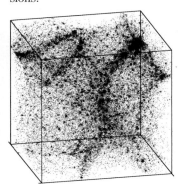
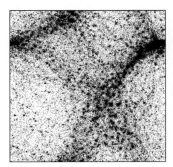

Fig. 7.22. Final configuration with 64^3 particles and its projection to two dimensions.

8 Tree Algorithms for Long-Range Potentials

In Chapter 7 we described mesh-based methods for long-range potentials using Coulomb and gravitational potentials as examples. These methods rely on a representation of the potential Φ as a solution of the Poisson equation (7.5). Such methods work well as long as the corresponding potentials are of type $1/r$ and particles are distributed approximately uniformly. In the case of a nonuniform particle distribution, i.e., when particles accumulate in only some parts of the simulation domain, the efficiency of methods that work with uniform meshes is substantially reduced. The mesh has to be fine enough to resolve the inhomogeneous particle distribution. Such situations arise particularly often in astrophysics, but they also occur in many molecular dynamical simulations in biochemistry. This has led to the development of adaptive methods, which use no longer a uniform mesh over the entire domain, but which work with locally refined meshes, compare Subsection 7.2.3.

Alternatively, so-called tree algorithms can be employed when nonuniform particle distributions are present. Consider for example a continuous charge distribution in the entire space $\mathbb{R}^3$ with a charge density ρ (charge per volume). Then, tree algorithms are based on the representation of the potential Φ as

$$\Phi(\mathbf{x}) = \frac{1}{4\pi\varepsilon_0} \int_{\mathbb{R}^3} \rho(\mathbf{y}) \frac{1}{||\mathbf{y} - \mathbf{x}||} d\mathbf{y}, \qquad (8.1)$$

compare also (7.4). They use tree-like hierarchical decompositions of the simulation domain to adaptively approximate the density distribution ρ of the particles. These hierarchical decompositions implement splittings of the simulation domain into near and far fields. Often decompositions associated with octrees are used for this purpose. This leads to algorithms which achieve a complexity of order $\mathcal{O}(N \log N)$ or even $\mathcal{O}(N)$ for a given accuracy. Due to their natural adaptivity, tree algorithms allow the computation of the potentials and the forces with a high accuracy also for inhomogeneous particle distributions.

Tree algorithms can also be used directly for long-range potentials other than $1/r$-potentials. They admit the treatment of ionized systems, which occur for instance in biomolecular problems. Tree algorithms also allow the evaluation of modified gravitational potentials, such as the regularized Plummer potential or the Biot-Savart kernel for vortex methods.[1]

In this chapter we consider tree algorithms in more detail. First, we present the adaptive Barnes-Hut algorithm. We then parallelize it by means of space-filling curves, see Section 8.4. In Section 8.5 we extend our method to higher orders of approximation and introduce a variant of the fast multipole method. We also discuss some applications from astrophysics.

8.1 Series Expansion of the Potential

We recall first the integral form of the potential given in (8.1) (compare also (7.4) in Chapter 7), written for general kernels G as

$$\Phi(\mathbf{x}) = \int_\Omega G(\mathbf{x}, \mathbf{y})\rho(\mathbf{y})d\mathbf{y} \tag{8.2}$$

with ρ being the particle density in the domain Ω.

Taylor Expansion. Under the assumption that the integral kernel G is, up to a singularity at $\mathbf{x} = \mathbf{y}$, differentiable sufficiently often, we can expand it in $\mathbf{y}$ around $\mathbf{y}_0$ up to terms of degree p as long, as $\mathbf{x}$ does not lie on the line segment $[\mathbf{y}, \mathbf{y}_0]$. We obtain

$$G(\mathbf{x}, \mathbf{y}) = \sum_{\|\mathbf{j}\|_1 \le p} \frac{1}{\mathbf{j}!} G_{\mathbf{0}, \mathbf{j}}(\mathbf{x}, \mathbf{y}_0)(\mathbf{y} - \mathbf{y}_0)^{\mathbf{j}} + R_p(\mathbf{x}, \mathbf{y}), \tag{8.3}$$

with the multi-indices $\mathbf{j} = (j_1, j_2, j_3)$ and the abbreviations and definitions $\frac{d^{\mathbf{j}}}{d\mathbf{z}^{\mathbf{j}}} = \frac{d^{j_1}}{d\mathbf{z}_1^{j_1}} \frac{d^{j_2}}{d\mathbf{z}_2^{j_2}} \frac{d^{j_3}}{d\mathbf{z}_3^{j_3}}$, $\mathbf{j}! := j_1! \cdot j_2! \cdot j_3!$, $\|\mathbf{j}\|_1 = j_1 + j_2 + j_3$, and $\mathbf{y}^{\mathbf{j}} = \mathbf{y}_1^{j_1} \cdot \mathbf{y}_2^{j_2} \cdot \mathbf{y}_3^{j_3}$. This expression generalizes the standard one-dimensional Taylor expansion to higher dimensions. Here,

$$G_{\mathbf{k}, \mathbf{j}}(\mathbf{x}, \mathbf{y}) := \left[\frac{d^{\mathbf{k}}}{d\mathbf{w}^{\mathbf{k}}} \frac{d^{\mathbf{j}}}{d\mathbf{z}^{\mathbf{j}}} G(\mathbf{w}, \mathbf{z}) \right]_{\mathbf{w}=\mathbf{x}, \mathbf{z}=\mathbf{y}}$$

denotes the mixed $(\mathbf{k}, \mathbf{j})$-th derivative of G evaluated at the point $(\mathbf{x}, \mathbf{y})$. The remainder $R_p(\mathbf{x}, \mathbf{y})$ can be written in Lagrange form as

---

[1] Since mesh-based methods rely on the representation of the potential Φ as a solution of a differential equation (like the Poisson equation), they do not allow to treat such potentials in a straightforward way. While in general one can find for a given differential operator a corresponding integral kernel (for the fundamental solution), the converse does not have to be true. In general, the corresponding operator is only a pseudo-differential operator that still needs an integral representation.

$$R_p(\mathbf{x}, \mathbf{y}) = \sum_{\|\mathbf{j}\|_1 = p+1} \frac{1}{\mathbf{j}!} G_{0,\mathbf{j}}(\mathbf{x}, \mathbf{y}_0 + \zeta \cdot (\mathbf{y} - \mathbf{y}_0)) \cdot (\mathbf{y} - \mathbf{y}_0)^{\mathbf{j}} \qquad (8.4)$$

with a function ζ, $0 \le \zeta \le 1$. Besides the term $(\mathbf{y} - \mathbf{y}_0)^{\mathbf{j}}$, the remainder contains the $(p+1)$-th derivatives of $G(\mathbf{x}, \mathbf{y})$ with respect to $\mathbf{y}$ on the line segment $[\mathbf{y}_0, \mathbf{y}]$. If we substitute the expansion (8.3) into the integral form (8.2), we obtain

$$\Phi(\mathbf{x}) = \int_\Omega \rho(\mathbf{y}) \sum_{\|\mathbf{j}\| \le p} \frac{1}{\mathbf{j}!} G_{0,\mathbf{j}}(\mathbf{x}, \mathbf{y}_0)(\mathbf{y} - \mathbf{y}_0)^{\mathbf{j}} d\mathbf{y} + \int_\Omega \rho(\mathbf{y}) R_p(\mathbf{x}, \mathbf{y}) d\mathbf{y}$$

$$= \sum_{\|\mathbf{j}\| \le p} \frac{1}{\mathbf{j}!} G_{0,\mathbf{j}}(\mathbf{x}, \mathbf{y}_0) \int_\Omega \rho(\mathbf{y})(\mathbf{y} - \mathbf{y}_0)^{\mathbf{j}} d\mathbf{y} + \int_\Omega \rho(\mathbf{y}) R_p(\mathbf{x}, \mathbf{y}) d\mathbf{y}.$$

Let us introduce so-called *moments*

$$M_{\mathbf{j}}(\Omega, \mathbf{y}_0) := \int_\Omega \rho(\mathbf{y})(\mathbf{y} - \mathbf{y}_0)^{\mathbf{j}} d\mathbf{y}. \qquad (8.5)$$

Using them, we obtain the expression

$$\Phi(\mathbf{x}) = \sum_{\|\mathbf{j}\|_1 \le p} \frac{1}{\mathbf{j}!} M_{\mathbf{j}}(\Omega, \mathbf{y}_0) G_{0,\mathbf{j}}(\mathbf{x}, \mathbf{y}_0) + \int_\Omega \rho(\mathbf{y}) R_p(\mathbf{x}, \mathbf{y}) d\mathbf{y}. \qquad (8.6)$$

Near Field and Far Field. Now we will construct an algorithm for the fast computation of the energy and the force using the approximation

$$\Phi(\mathbf{x}) \approx \sum_{\|\mathbf{j}\|_1 \le p} \frac{1}{\mathbf{j}!} M_{\mathbf{j}}(\Omega, \mathbf{y}_0) G_{0,\mathbf{j}}(\mathbf{x}, \mathbf{y}_0). \qquad (8.7)$$

We use an idea that is similar to the use of the splitting $V^{\text{short}} + V^{\text{long}}$ of the potential into near field and far field in (7.1) in the last chapter. First, for any given $\mathbf{x}$, we partition the entire integration domain Ω into a near region Ω^{near} and a far region Ω^{far} with $\Omega = \Omega^{\text{near}} \cup \Omega^{\text{far}}$ and $\Omega^{\text{near}} \cap \Omega^{\text{far}} = \emptyset$. Then, we further decompose the far region into a set of disjoint, convex subdomains Ω_ν^{far}, with a "center" $\mathbf{y}_0^\nu \in \Omega_\nu^{\text{far}}$ associated with each of them. This decomposition then satisfies[2]

$$\Omega = \Omega^{\text{near}} \cup \bigcup_\nu \Omega_\nu^{\text{far}}. \qquad (8.8)$$

Here, we choose the decomposition of the far field in such a way that

---

[2] The decomposition is dependent on $\mathbf{x}$, i.e., it must be chosen appropriately for each $\mathbf{x}$. We will see later that we can derive an appropriate decomposition for all $\mathbf{x}$ from a single tree decomposition of the entire domain using hierarchical methods.

$$\frac{diam}{\|\mathbf{x} - \mathbf{y}_0^\nu\|} \leq \theta \tag{8.9}$$

holds with a given constant $\theta < 1$ for all $\Omega_\nu^{\mathrm{far}}$ in the decomposition, where

$$diam := \sup_{\mathbf{y} \in \Omega_\nu^{\mathrm{far}}} \|\mathbf{y} - \mathbf{y}_0^\nu\|. \tag{8.10}$$

Figure 8.1 shows the situation for one subdomain in the far field for a discrete particle distribution.

Fig. 8.1. Interaction of a particle at $\mathbf{x}$ with a set of particles in $\Omega_\nu^{\mathrm{far}}$ around the center $\mathbf{y}_0^\nu$.

Now, we first apply to (8.2) the decomposition of the domain in near field and far field(s) according to (8.8). Then, we use the approximation (8.7) in each subdomain of the far region, with $\Omega_\nu^{\mathrm{far}}$ as the domain of integration for the moments (8.5). Thus, for a fixed $\mathbf{x}$, we obtain

$$
\begin{aligned}
\Phi(\mathbf{x}) &= \int_\Omega \rho(\mathbf{y}) G(\mathbf{x}, \mathbf{y}) d\mathbf{y} \\
&= \int_{\Omega^{\mathrm{near}}} \rho(\mathbf{y}) G(\mathbf{x}, \mathbf{y}) d\mathbf{y} + \int_{\Omega^{\mathrm{far}}} \rho(\mathbf{y}) G(\mathbf{x}, \mathbf{y}) d\mathbf{y} \\
&= \int_{\Omega^{\mathrm{near}}} \rho(\mathbf{y}) G(\mathbf{x}, \mathbf{y}) d\mathbf{y} + \sum_\nu \int_{\Omega_\nu^{\mathrm{far}}} \rho(\mathbf{y}) G(\mathbf{x}, \mathbf{y}) d\mathbf{y} \\
&\approx \int_{\Omega^{\mathrm{near}}} \rho(\mathbf{y}) G(\mathbf{x}, \mathbf{y}) d\mathbf{y} + \sum_\nu \sum_{\|\mathbf{j}\|_1 \leq p} \frac{1}{\mathbf{j}!} M_{\mathbf{j}}(\Omega_\nu^{\mathrm{far}}, \mathbf{y}_0^\nu) G_{\mathbf{0},\mathbf{j}}(\mathbf{x}, \mathbf{y}_0^\nu) \quad (8.11)
\end{aligned}
$$

with the corresponding *local* moments

$$M_{\mathbf{j}}(\Omega_\nu^{\mathrm{far}}, \mathbf{y}_0^\nu) = \int_{\Omega_\nu^{\mathrm{far}}} \rho(\mathbf{y})(\mathbf{y} - \mathbf{y}_0^\nu)^{\mathbf{j}} d\mathbf{y}. \tag{8.12}$$

Error Estimates. Using (8.4), the corresponding *relative* local approximation error for $\Omega_\nu^{\mathrm{far}}$ is given for a fixed $\mathbf{x}$ as

$$e_\nu^{\mathrm{rel}}(\mathbf{x}) := \frac{e_\nu^{\mathrm{abs}}(\mathbf{x})}{\Phi_\nu(\mathbf{x})}$$

with

$$e_\nu^{\mathrm{abs}}(\mathbf{x}) := \int_{\Omega_\nu^{\mathrm{far}}} \rho(\mathbf{y}) \sum_{\|\mathbf{j}\|_1=p+1} \frac{1}{\mathbf{j}!} G_{0,\mathbf{j}}(\mathbf{x}, \mathbf{y}_0^\nu + \zeta \cdot (\mathbf{y} - \mathbf{y}_0^\nu))(\mathbf{y} - \mathbf{y}_0^\nu)^{\mathbf{j}} d\mathbf{y}, \quad (8.13)$$

$$\Phi_\nu(\mathbf{x}) := \int_{\Omega_\nu^{\mathrm{far}}} \rho(\mathbf{y}) G(\mathbf{x}, \mathbf{y}) d\mathbf{y}. \tag{8.14}$$

We now assume that $\Omega_\nu^{\mathrm{far}}$ is convex[3] and that G and its $(p+1)$th derivative behave somewhat similar to the $1/r$-potential and its $(p+1)$th derivative. More precisely, we assume that for $\mathbf{j}$ with $\|\mathbf{j}\|_1 = p+1$ it holds that

$$|G_{0,\mathbf{j}}(\mathbf{x}, \mathbf{y})| \le c \cdot \frac{1}{\|\mathbf{x} - \mathbf{y}\|^{\|\mathbf{j}\|_1+1}}, \quad c \cdot \frac{1}{\|\mathbf{x} - \mathbf{y}\|} \le G(\mathbf{x}, \mathbf{y}). \tag{8.15}$$

Then, for positive densities ρ and positive[4] kernels G, the following estimate for the local relative approximation error holds:

$$e_\nu^{\mathrm{rel}}(\mathbf{x}) \le \mathcal{O}(\theta^{p+1}). \tag{8.16}$$

To show this, we use first that

$$|e_\nu^{\mathrm{rel}}(\mathbf{x})| \le \frac{\displaystyle\int_{\Omega_\nu^{\mathrm{far}}} \rho(\mathbf{y}) \sum_{\|\mathbf{j}\|_1=p+1} \frac{1}{\mathbf{j}!} |G_{0,\mathbf{j}}(\mathbf{x}, \mathbf{y}_0^\nu + \zeta(\mathbf{y} - \mathbf{y}_0^\nu))| \cdot |(\mathbf{y} - \mathbf{y}_0^\nu)^{\mathbf{j}}| d\mathbf{y}}{\displaystyle\int_{\Omega_\nu^{\mathrm{far}}} \rho(\mathbf{y}) d\mathbf{y} \cdot g_{\min}^\nu(\mathbf{x})}$$

$$\le c \frac{g_{\max}^{\nu,p+1}(\mathbf{x}) \cdot diam^{p+1}}{g_{\min}^\nu(\mathbf{x})},$$

where

$$g_{\max}^{\nu,p+1}(\mathbf{x}) := \sup_{\mathbf{y}\in\Omega_\nu^{\mathrm{far}}} \max_{\|\mathbf{j}\|_1=p+1} \frac{1}{\mathbf{j}!} |G_{0,\mathbf{j}}(\mathbf{x}, \mathbf{y})|,$$

$$g_{\min}^\nu(\mathbf{x}) := \inf_{\mathbf{y}\in\Omega_\nu^{\mathrm{far}}} G(\mathbf{x}, \mathbf{y}).$$

In the derivation of this estimate we have used the positivity of ρ and G and the property $\mathbf{y}_0^\nu + \zeta(\mathbf{y}_0^\nu - \mathbf{y}) \in \Omega_\nu^{\mathrm{far}}$, which follows from the convexity of $\Omega_\nu^{\mathrm{far}}$.

Inequality (8.9) implies the relations

$$\frac{1}{\|\mathbf{x} - \mathbf{y}\|} \ge \frac{1}{\|\mathbf{x} - \mathbf{y}_0^\nu\| + \|\mathbf{y}_0^\nu - \mathbf{y}\|} \ge \frac{1}{\|\mathbf{x} - \mathbf{y}_0^\nu\| + \theta\|\mathbf{x} - \mathbf{y}_0^\nu\|} = \frac{1}{1+\theta} \frac{1}{\|\mathbf{x} - \mathbf{y}_0^\nu\|},$$

$$\frac{1}{\|\mathbf{x} - \mathbf{y}\|} \le \frac{1}{\|\mathbf{x} - \mathbf{y}_0^\nu\| - \|\mathbf{y}_0^\nu - \mathbf{y}\|} \le \frac{1}{\|\mathbf{x} - \mathbf{y}_0^\nu\| - \theta\|\mathbf{x} - \mathbf{y}_0^\nu\|} = \frac{1}{1-\theta} \frac{1}{\|\mathbf{x} - \mathbf{y}_0^\nu\|}.$$

---

[3] Later on, we will exclusively consider decompositions into cubic subdomains $\Omega_\nu^{\mathrm{far}}$. Such subdomains are convex.

[4] An analogous result is valid for negative densities ρ or negative kernels G, but ρ or G must not change sign.

Combining these estimates with (8.15), we obtain

$$g_{\max}^{\nu,p+1}(\mathbf{x}) \le c \cdot \frac{1}{\|\mathbf{x} - \mathbf{y}_0^\nu\|^{p+2}} \quad \text{and} \quad c \cdot \frac{1}{\|\mathbf{x} - \mathbf{y}_0^\nu\|} \le g_{\min}^\nu(\mathbf{x}).$$

This results in

$$|e_\nu^{\mathrm{rel}}(\mathbf{x})| \le c \frac{g_{\max}^{\nu,p+1}(\mathbf{x}) \cdot diam^{p+1}}{g_{\min}^\nu(\mathbf{x})} \le c \cdot \frac{\|\mathbf{x} - \mathbf{y}_0^\nu\| \cdot diam^{p+1}}{\|\mathbf{x} - \mathbf{y}_0^\nu\|^{p+2}}$$

$$= c \cdot \left(\frac{diam}{\|\mathbf{x} - \mathbf{y}_0^\nu\|} \right)^{p+1} \le c\theta^{p+1}.$$

Here, c denotes a generic constant depending on p. Thus, we have shown the estimate (8.16).

For the *global* relative error

$$e^{\mathrm{rel}}(\mathbf{x}) = \sum_\nu e_\nu^{\mathrm{abs}}(\mathbf{x}) / \sum_\nu \Phi_\nu(\mathbf{x}),$$

(8.13) and (8.14) then imply the bound

$$|e^{\mathrm{rel}}(\mathbf{x})| \le \frac{\sum\limits_\nu |e_\nu^{\mathrm{abs}}(\mathbf{x})|}{\sum\limits_\nu \Phi_\nu(\mathbf{x})} = \frac{\sum\limits_\nu \frac{|e_\nu^{\mathrm{abs}}(\mathbf{x})|}{\Phi_\nu(\mathbf{x})} \Phi_\nu(\mathbf{x})}{\sum\limits_\nu \Phi_\nu(\mathbf{x})} = \frac{\sum\limits_\nu |e_\nu^{\mathrm{rel}}(\mathbf{x})| \Phi_\nu(\mathbf{x})}{\sum\limits_\nu \Phi_\nu(\mathbf{x})}$$

$$\le \frac{\sum\limits_\nu c\,\theta^{p+1} \Phi_\nu(\mathbf{x})}{\sum\limits_\nu \Phi_\nu(\mathbf{x})} = \frac{c\,\theta^{p+1} \sum\limits_\nu \Phi_\nu(\mathbf{x})}{\sum\limits_\nu \Phi_\nu(\mathbf{x})} = c\,\theta^{p+1},$$

where c denotes again a generic constant depending on p. The condition (8.9) for the decomposition of the far field for the point $\mathbf{x}$ therefore allows to control the global relative approximation error at point $\mathbf{x}$. Relation (8.9) also implies a geometric condition for the far field decomposition: The closer the subdomain $\Omega_\nu^{\mathrm{far}}$ is to $\mathbf{x}$, the smaller the subdomain has to be in order to satisfy (8.9). This is shown in Figure 8.2 for several subdomains from a far field decomposition.

Similar relative error bounds for a number of other types of potentials can be obtained with the same approach. Furthermore, there are also absolute error bounds for nonpositive kernels and nonpositive charge densities [289]. They are useful to obtain estimates for the forces instead of the potentials, since the derivative of the kernel is often nonpositive.

Note that Taylor expansion is not the only possible suitable series expansion. If one uses other coordinate systems instead of Cartesian coordinates, one obtains other series expansions. For instance, for spherical coordinates

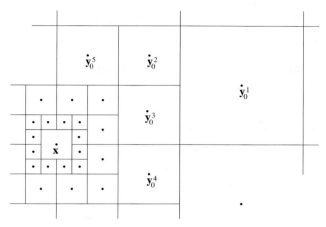

Fig. 8.2. Cells $\Omega_\nu^{\mathrm{far}}$ of different sizes that partition the far field Ω^{far} for $\mathbf{x}$ and satisfy the condition (8.9) with $\theta = 0.4472$ in two dimensions.

one obtains series of spherical harmonics [260]. Another possibility is the use of plane waves [260] or piecewise Lagrange polynomials [121].

Now, the central question is the following: How do we construct suitable decompositions of the far field as efficiently as possible? Note here that it is not sufficient to determine a far field decomposition for a single given $\mathbf{x}$. In fact, decompositions for the far field for *all* particle positions $\mathbf{x}_i$, $i = 1, \ldots N$, have to be found, the subdomains of which *all* have to satisfy the condition (8.9) for a given accuracy θ. Furthermore, since the moments $M_{\mathbf{j}}(\Omega_\nu^{\mathrm{far}}, \mathbf{y}_0^\nu)$ are rather expensive to compute, they should not only be usable for the evaluation of the potential at a single point $\mathbf{x}$, but be reusable in the computation for other particle positions.[5] Therefore, it is crucial that the moments of larger subdomains can be computed from those of the smaller subdomains subdividing them. This is achieved for example with a recursive decomposition of the entire domain Ω into a sequence of smaller and smaller cubic or cuboid subdomains. Altogether, such an approach allows an efficient approximative computation of the potential and the forces. The resulting structures can be described with geometric trees that will be introduced in the next section.

---

[5] Let m_i denote again the mass of particle i. The identification of the respective far fields and their decompositions into subdomains for a given mass distribution $\rho = \sum_{i=1}^N m_i \delta_{\mathbf{x}_i}$ is equivalent to finding partitions of the set of particles into subsets that on the one hand satisfy (8.9) and on the other hand minimize the complexity for the evaluation of the approximation of the potential. This second task can be formulated abstractly as a discrete optimization problem which, however, is very expensive to solve. The tree algorithms described in the following are good heuristics for the efficient approximate solution of this optimization problem.

8.2 Tree Structures for the Decomposition of the Far Field

First, we define trees in an abstract way. To this end, we introduce some concepts and notations: A graph is given as a set of vertices/nodes and a set of edges. Here, an edge is a connection between two vertices. A path is a sequence of different vertices in which any two successive vertices are connected by an edge. A tree is a graph in which there is exactly one path connecting any two vertices. One of the vertices of the tree is designated as the root of the tree. Then, there is exactly one path from the root to each other vertex in the tree.

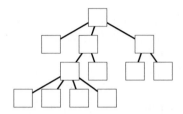

Fig. 8.3. An example of a tree of depth three.

Contrary to biological intuition, we represent a tree with its root on top and growing in the downward direction, see Figure 8.3. The edges are directed downward from the root to the other nodes. A node that lies directly below another node in the tree and is connected to it is called son node of that node. The node above is correspondingly called father node. Further terms such as brother, grandson or grandfather can be defined in similar fashion. The first generation of nodes, directly connected to the root by one edge, belongs to level one. Every further generation of nodes, connected to the root by paths of length l, belongs to level l. The depth of a tree is the maximum of the levels of all nodes of the tree, i.e. the length of the longest path starting at the root.

We can distinguish the nodes in the tree according to whether they have sons, in which case they are called inner nodes, or not, in which case they are called leaf nodes. A subtree consists of one node of the original tree that has been designated as the root of the subtree, of all of its descendants, and of the edges of the original tree between them.

Quadtree and Octree. In the following, we combine the abstract definition of trees with geometric information and use them to provide the splitting of the domain into near field and far field and the decomposition of the far field into subdomains for each particle i. The tree is then associated to a recursive partition of the entire domain into cells of different sizes. We proceed as follows: First, we assign the entire cubic or cuboid domain Ω and all the particles contained in it to the root of the tree. Then, we partition Ω into disjoint subdomains and assign them to the son nodes of the root. Every subdomain is then again partitioned into smaller subdomains and assigned to the son nodes in the next generation in the tree. This way, we proceed recursively. The recursion is terminated when there is either one particle only or no particle in the corresponding subdomain, compare Figures 8.5 and 8.7.

The question is now how to split a subdomain into smaller subdomains. A simple approach is to split a subdomain in each coordinate direction into two equal parts. For a square domain, this results in four equally sized smaller squares, for a cubic domain this results in eight equally sized smaller cubes with their edges being half as long as those of the original cube. An example for the two-dimensional case is shown Figure 8.4.

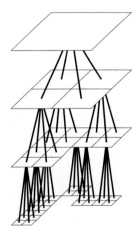

Fig. 8.4. An example for a quadtree.

The inner nodes of the tree constructed in this way have four sons which correspond to the quadrants of the local coordinate system. We obtain a so-called quadtree. In the three-dimensional case, the inner nodes of the tree have eight sons which correspond to the eight octants of the local coordinate system. Therefore, this type of tree is also called octree, or more accurately, PR octree (for point region octree) [540].

There are several other ways to decompose the domain into cells. Binary trees, i.e. trees with exactly two sons for each inner node, correspond to a bisection of the domain. This bisection can be carried out perpendicular to a coordinate axis so that in general rectangular instead of square cells result [47], see Figure 8.5. Furthermore, the decomposition does not always have to produce cells of equal volume. It might make more sense to decompose a cell depending on ρ, such that the number of particles in the resulting subcells is approximately the same. The volume of cells on the same level will then in general be different. In addition, cells can also be shrunk to exclude parts of the domain in which the particle density ρ vanishes, i.e. parts of the domain without particles [386], see Figure 8.5 (middle). For the sake of simplicity we will restrict ourselves in the following to PR octrees.

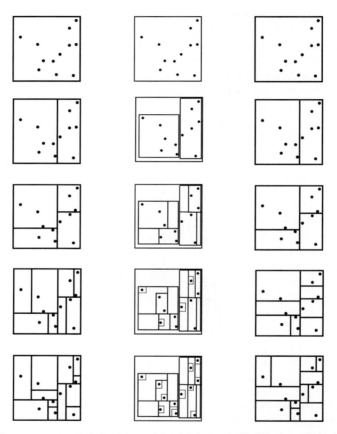

Fig. 8.5. Decomposition of the domain for different tree variants. Left: In each step the x_1 and x_2 axis are split alternatingly so that approximately the same number of particles is contained in each resulting subdomain. Middle: Shrunken version. After each subdivision, the resulting subdomains are shrunk so that they are just large enough to contain all particles in the subdomain. Right: In each step the longer coordinate axis is split so that approximately the same number of particles is contained in each resulting subdomain.

Recursive Computation of the Far Field. We now use the octree structure determined from a given set of particles to find a decomposition of the domain Ω into far field and near field for all particles and to furthermore find a decomposition of the far field into subdomains that satisfy the condition (8.9). Using this decomposition, we then compute the approximation (8.11), where we also try to use as few cells as possible. We proceed as follows: Every inner node of the tree represents a subdomain of Ω. To each subdomain we associate a designated expansion point $\mathbf{y}_0$. For instance, this could be the center of the cell or the center of mass of the particles contained in the cell. Furthermore, for each vertex, we know the size of the corresponding subdomain. From this, we can determine the size *diam* of the circumscribed sphere. We now start for a given particle i with particle position $\mathbf{x}_i$ in the root of the tree and descend recursively to the son nodes. There, we compute the ratio $diam/\|\mathbf{x}_i - \mathbf{y}_0^\nu\|$. If this value is smaller than or equal to the given θ, we terminate the recursion and have found a cell that locally satisfies (8.9). Otherwise, we descend the tree to the son nodes and again compute the associated ratios, etc. When the recursive procedure is finished, we have determined a decomposition (8.8) of the domain for $\mathbf{x}_i$, in which the subdomains of the far field all satisfy (8.9). This recursive procedure can be executed for each particle separately. Note that in general we will obtain different decompositions for different particles. However, the decompositions for *all* particles are contained in *one* tree.

Recursive Computation of the Moments. We still have to determine the moments $M_\mathbf{j}(\Omega_\nu^{\mathrm{far}}, \mathbf{y}_0^\nu)$ from (8.12) that are needed for the computation of the approximation (8.11). One could be tempted to compute these values directly by numerical integration over the density, or, in the case of a discrete density, by a sum over the particles. This, however, is not efficient. Instead we employ the hierarchical tree structure to compute *all* moments for all subdomains which are associated to the nodes of the tree. We also use the tree structure to store the moments in the respective nodes of the tree for their multiple use in a subsequent evaluation of (8.11). Note that it is possible to compute the moments for a father cell from the moments of its son cells. To this end, the following properties of the moments are useful:

For two disjoint subdomains $\Omega_1 \cap \Omega_2 = \emptyset$ and the *same* expansion point $\mathbf{y}_0$ it holds that

$$M_\mathbf{j}(\Omega_1 \cup \Omega_2, \mathbf{y}_0) = M_\mathbf{j}(\Omega_1, \mathbf{y}_0) + M_\mathbf{j}(\Omega_2, \mathbf{y}_0), \qquad (8.17)$$

which follows directly from the properties of the integral. A translation of the expansion point within a subdomain Ω_ν from $\mathbf{y}_0$ to $\hat{\mathbf{y}}_0$ changes the moments to

$$M_\mathbf{j}(\Omega_\nu, \hat{\mathbf{y}}_0) = \int_{\Omega_\nu} \rho(\mathbf{y})(\mathbf{y} - \hat{\mathbf{y}}_0)^\mathbf{j} d\mathbf{y}$$

$$= \sum_{i \leq j} \binom{j}{i} \int_{\Omega_\nu} \rho(\mathbf{y})(\mathbf{y} - \mathbf{y}_0)^i (\mathbf{y}_0 - \hat{\mathbf{y}}_0)^{j-i} d\mathbf{y}$$

$$= \sum_{i \leq j} \binom{j}{i} (\mathbf{y}_0 - \hat{\mathbf{y}}_0)^{j-i} M_i(\Omega_\nu, \mathbf{y}_0) , \tag{8.18}$$

where $\mathbf{i} \leq \mathbf{j}$ has to be understood component-wise and $\binom{j}{i}$ is defined by $\prod_{d=1}^{\text{DIM}} \binom{j_d}{i_d}$.[6] Given an expansion of the moments $M_j(\Omega_\nu^{\text{son}_\mu}, \mathbf{y}_0^{\nu,\text{son}_\mu})$ of the son cells $\Omega_\nu^{\text{son}_\mu}$ of a cell Ω_ν, we can then compute an expansion of the moments $M_j(\Omega_\nu, \mathbf{y}_0^\nu)$ of the father cell $\Omega_\nu = \bigcup_\mu \Omega_\nu^{\text{son}_\mu}$ according to (8.17) and (8.18) without having to compute all the integrals or sums again, see also Figure 8.6. The computation of all moments can then proceed recursively starting from all leaves of the tree and ending in the root.

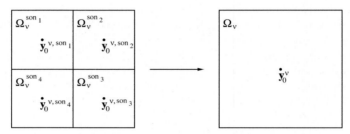

Fig. 8.6. The moments $M_j(\Omega_\nu, \mathbf{y}_0^\nu)$ of the father cell Ω_ν are computed from the moments $M_j(\Omega_\nu^{\text{son}_\mu}, \mathbf{y}_0^{\nu,\text{son}_\mu})$ of the son cells $\Omega_\nu^{\text{son}_\mu}$. Here, we have chosen the centers of the cells as expansion points.

It is still open how to compute the values for the leaves of the tree. By construction, a subdomain associated to a leaf of the tree contains (besides the trivial case[7]) exactly one particle $\mathbf{x}_i$ with mass m_i. For a discrete particle density $\rho(\mathbf{x}) = \sum_{j=1}^{N} m_j \delta_{\mathbf{x}_j}$ the moment is then given as[8]

$$M_j(\Omega_\nu^{\text{leaf}}, \mathbf{y}_0^\nu) = \int_{\Omega_\nu^{\text{leaf}}} \rho(\mathbf{y})(\mathbf{y} - \mathbf{y}_0^\nu)^j d\mathbf{y}$$

$$= \int_{\Omega_\nu^{\text{leaf}}} m_i \delta_{\mathbf{x}_i}(\mathbf{y} - \mathbf{y}_0^\nu)^j d\mathbf{y} = m_i(\mathbf{x}_i - \mathbf{y}_0^\nu)^j.$$

The integrals (or sums) for the near field approximation are evaluated in a similar fashion. Altogether, we see that the computation of all moments in

---

[6] We have used here the binomial formula $(x - a)^p = \sum_{i=0}^{P} \binom{p}{i} x^i a^{p-1}$.

[7] If a leaf contains no particle at all, the associated moment is trivially zero. We will not explicitly store such empty leaves later.

[8] In the case of a continuous density ρ, one has to use an appropriate quadrature method to compute the integral [289].

the tree constitutes a successive summation process. The moments associated with particular subdomains are partial sums and are stored in the associated nodes for later use.

To fully devise a concrete numerical method for the approximate evaluation of the potentials and the associated forces, we have to specify in detail how the tree is constructed from the set of particles, how the tree is stored, how the cells and the expansion points are chosen, and how far field and near field are determined in the force evaluation. We will discuss possible choices in more detail in the following sections.

8.3 Particle-Cluster Interactions and the Barnes-Hut Method

The simplest form of a tree-like method for the approximative computation of potentials and forces traces back to Barnes and Hut [58].[9] It had been originally developed for astrophysical problems. In general, very large numbers of particles and inhomogeneous density distributions occur in such applications. The interaction between the particles is modeled by the gravitational potential

$$U(r_{ij}) = -G_{\text{Grav}} \frac{m_i m_j}{r_{ij}} \tag{8.19}$$

or modifications thereof.[10] The Barnes-Hut method uses octrees. As already discussed, the simulation domain is recursively split into subdomains of equal size (cells) until each cell contains at most one particle. The cells are then mapped to vertices in the tree. The inner vertices represent cells with several particles, so-called clusters. For an inhomogeneous distribution of particles, such an approach results in so-called unbalanced trees, see for example Figure 8.4.

The method of Barnes and Hut can be interpreted as a special case of the approximation by the Taylor series expansion (8.11), see below. It is based on the idea that the effect of the gravitation of many particles in a cell far away is essentially the same as the effect of the gravitation of one large particle in the center of mass of the cell. Therefore, the many interactions with these many particles can be modeled by one interaction with a so-called pseudoparticle. The position of the pseudoparticle is the center of mass of the particles in the cell and the mass of the pseudoparticle is the total mass of all the particles in the cell.

---

[9] For some earlier developments, see [47].

[10] In the modeling of the dynamics of galaxies, various modifications of the gravitational potential are used. Examples are the potentials of Plummer [486], Jaffe [338], Hernquist [316], or Miyamoto and Nagai [435].

8.3.1 Method

The method consists of three main elements: The construction of the tree, the computation of the pseudoparticles and the computation of the forces. Pseudoparticles represent cells with more than one particle and are therefore associated with inner vertices of the tree, while the real particles are only stored in the leaves of the tree. The geometric coordinates of a pseudoparticle are given by the average of the coordinates of all particles in the associated cell weighted by their masses. The mass of the pseudoparticle is the total mass of all particles in its cell. Both values can be computed recursively for all vertices of the tree in one traversal starting from the leaves according to

$$m^{\Omega_\nu} := \sum_{\mu=1}^{8} m^{\Omega_\nu^{\text{son},\mu}},$$

$$\mathbf{y}_0^\nu := \sum_{\mu=1}^{8} \frac{m^{\Omega_\nu^{\text{son},\mu}}}{\sum_{\gamma=1}^{8} m^{\Omega_\nu^{\text{son},\gamma}}} \mathbf{y}_0^{\text{son},\mu} = \frac{1}{m^{\Omega_\nu}} \sum_{\mu=1}^{8} m^{\Omega_\nu^{\text{son},\mu}} \mathbf{y}_0^{\text{son},\mu}. \qquad (8.20)$$

Here, as in the previous section, Ω_ν denotes a cell associated to a vertex in the tree, $\Omega_\nu^{\text{son},\mu}$ denotes the eight son cells associated to that vertex (if they exist), $\mathbf{y}_0^\nu$ and $\mathbf{y}_0^{\text{son},\mu}$, respectively, denote the associated expansion points, and m^{Ω_ν} and $m^{\Omega_\nu \text{son},\mu}$ denote the associated masses. The expansion point of a (nontrivial) leaf, which contains by construction only one particle, and its mass, are given as the position and the mass of that particle.[11]

For the computation of the forces, the algorithm descends, for each given particle i with position $\mathbf{x}_i \in \Omega$, along the edges of the tree starting at the root until the visited cells satisfy the selection criterion (θ criterion)

$$\frac{diam}{r} \geq \theta, \qquad (8.21)$$

where $diam$ is defined in (8.10) and r denotes the distance of the associated pseudoparticle from the position $\mathbf{x}_i$ of particle i. Then, the interactions of particle i with the pseudoparticles associated to these resulting vertices are computed and added to the global result. In the case that the descent ends in a leaf, which by definition only contains one particle, the interaction between the two particles is computed directly and added to the global result. The set of leaves reached in this way constitutes the near field Ω^{near} of the decomposition. An example for the descent in a tree controlled by (8.21) is shown in Figure 8.7.

---

[11] In a simpler variant of the Barnes-Hut method, the geometric cell centers are used as expansion points, as already suggested in Figure 8.6. For strongly inhomogeneous particle distributions, however, as they occur in the simulation of collisions of galaxies, such an approach can negatively effect the accuracy of the obtained results.

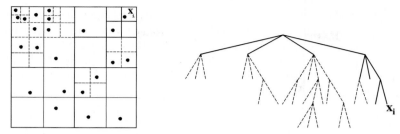

Fig. 8.7. Decomposition of the domain Ω by the quadtree (left) and the parts of the quadtree traversed in the computation of the force for $\mathbf{x}_i$ with $\theta = \frac{1}{2}$ (right).

There are several variants of the criterion (8.21), differing in the definition of the distances *diam* and r between the particle and the cell, see also Figure 8.8. These differences can be significant for the convergence of the method, especially in the case of extremely inhomogeneous particle distributions. In practice, one often uses the length of the edge of the cell for *diam* since it can be computed very inexpensively. This corresponds to an upper bound of (8.10) (up to a constant), which (for an appropriately changed choice of θ) does not change the error estimates from Section 8.1. There are a number of other selection criteria that have been developed from expansions of the relative error or from local error estimates, for examples see [58, 538].

As already mentioned, the method of Barnes and Hut can be interpreted as a special case of the approximation by the Taylor series expansion from the last section, compare (8.11), where the degree of approximation is chosen

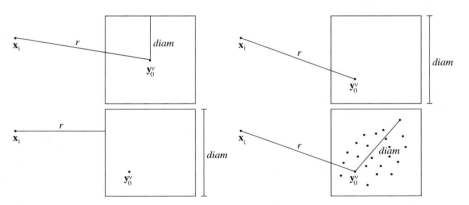

Fig. 8.8. Some variants for the definition of *diam* and r in the Barnes-Hut method.

to be $p = 0$.[12] The error of the method therefore only depends on the control parameter θ. The smaller the chosen θ, the more accurate is the evaluation by the truncated series expansion. However, this also increases the complexity of the method since there is a direct correlation between the size of θ and the number of cells in the decomposition. In [537] it has been shown that, for the case of almost uniformly distributed particles, the number of cells is bounded by $C \log N/\theta^3$, with N being the number of particles and C being a constant. In total, the method then has a complexity of $\mathcal{O}(\theta^{-3} N \log N)$. For $\theta \to 0$ the method degenerates to the original $\mathcal{O}(N^2)$ summation in the near field since only leaves of the tree are selected in the decomposition.

8.3.2 Implementation

In the following, we implement the tree method of Barnes and Hut for the approximate evaluation of the forces. To this end, we reuse the data structure 3.1 from Section 3.2 for a particle with mass, position, velocity, and force. To implement a tree we have to define the nodes of the tree and the edges that connect the nodes. Memory for one particle is allocated for each node of the tree. The connections between nodes are implemented as pointers, compare also the discussion for linked lists in Section 3.5. Here, it is enough to provide pointers from the father node to the son nodes. For each node we need `POWDIM:=` 2^{DIM} pointers, i.e., for a quadtree we need four pointers and for an octree eight pointers. In addition, we store the position and the size of the subdomain which is associated to a node. To this end, we declare the structure `Box`, see data structure 8.1, in which we store the position of the left lower front vertex `lower[DIM]` and the position of the right upper back vertex `upper[DIM]` of the boundary of the respective cell[13]. These coordinates of course satisfy `lower[d]` $<$ `upper[d]` component-wise.

Data structure 8.1 Geometric Cell in a Tree

```
typedef struct Box {
  real lower[DIM];
  real upper[DIM];
} Box;
```

---

[12] The quality of the approximation is indeed even better in the center of mass since the first moments vanish identically, see also (8.25). This, however, holds only for positive masses, i.e. particularly in astrophysical applications. In the case of molecules, the electrostatic charge can be both positive and negative. Then, the original Barnes-Hut method does not perform as well, since for the values of θ chosen in practice, its lower quality of approximation strongly affects the energy conservation.

[13] One could also compute the boundaries of the subdomains recursively in a tree traversal later in the algorithms and save the memory for `Box`.

The implementation of a vertex in a tree is given in data structure 8.2.[14]

Data structure 8.2 Node of a Tree

```
typedef struct TreeNode {
  Particle p;
  Box box;
  struct TreeNode *son[POWDIM];
} TreeNode;
```

A tree is then represented as follows: The root of the tree is a `TreeNode`. Its address is stored in `root`. Each node has pointers to all its sons. Pointers that are not used and do not point to any valid address are set to `NULL`. A node with all pointers set to `NULL` does not have any sons and is therefore a leaf.

Various operations can now be implemented on this tree structure. Among them are different traversals of all nodes in the tree, the insertion and deletion of nodes, and the search for certain nodes in the tree.

Tree Traversal. We first discuss how a given tree is traversed and a desired operation is performed for all its nodes. The simplest implementation is a recursive one: The tree traversal function calls itself but always with a different argument. The first time the function is called with the root node `root` as argument, the next time it is called with a son node of the root as argument, etc. Thus, the recursion descends in the tree until the leaf nodes are reached. During the descent (the way to the leaves) or during the ascent (the way back), after the son nodes have been processed, operations can be performed on each node.

Algorithm 8.1 Abstract Post-Order Traversal of the Nodes of a Tree via the Function `FUNCTION`

```
void FUNCTION(TreeNode *t) {
  if (t != NULL) {
    for (int i=0; i<POWDIM; i++)
      FUNCTION(t->son[i]);
    Perform the operations of the function FUNCTION on *t ;
  }
}
```

---

[14] There are several approaches to store trees. For example, if the sons of a node are connected by a linked list or are defined as an array of fixed size, it is enough to store one pointer to the sons instead of `POWDIM` pointers. One can also completely avoid pointers and implement trees with associative data structures (so-called hash techniques) [357].

This is shown abstractly in Algorithm 8.1 for a general function FUNCTION. Here, an operation is only performed on each node on the way back. This results in a so-called *post-order* traversal. Later on, we will insert specific function names and operations into this abstract algorithm.

We can traverse all the nodes of the tree with a call FUNCTION(root) and perform the appropriate operations on them. An example for such a post-order traversal of a tree is shown in Figure 8.9.

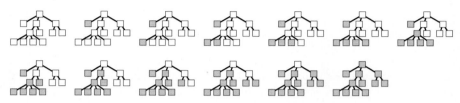

Fig. 8.9. A post-order tree traversal (from left to right).

Analogously, an operation could be performed only on the way to the leaf nodes. We then obtain a so-called *pre-order* traversal. The operations on *t are now inserted before the for loop. An example for such a pre-order traversal of a tree is shown in Figure 8.10.[15]

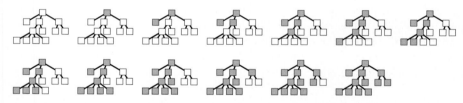

Fig. 8.10. A pre-order tree traversal (from left to right).

An important variant of the post-order traversal results if the operations are only performed on the leaves of the tree. This can be implemented by a simple conditional statement. We will use this variant later to traverse all particles. By construction, particles are only associated to leaves of the tree. An example for such a traversal of the leaf nodes of a tree is given in Figure 8.11.

---

[15] Recursive pre-order or post-order traversal of the tree is a so-called *depth-first* approach. The function descends first into the depth of the tree. An alternative is the so-called *breadth-first* approach, in which all nodes on one level are processed first before the next level is visited.

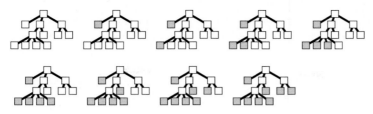

Fig. 8.11. A traversal of all leaves of a tree (from left to right).

Building the Tree. We now consider the building of the tree. In principle, there is a variety of different types of trees that differ in the number of son nodes and the order in which the nodes are sorted into the tree. In the following, we use a geometric octree. The sons of a node then describe the subcubes which are created by bisection of all edges of the cube corresponding to the node. The tree is built by successive insertion of the particles.

We start with the root of the tree. For a given particle and a given node of the tree, we determine the son tree into which the particle has to be inserted. This is easily possible from the particle coordinates and the geometric data of the node. We recursively descend the tree until we encounter the situation that the son tree into which the particle would belong does not yet exist (i.e., the corresponding pointer to the son node is NULL). Here, we have to distinguish two cases: Either the current node is an inner node or it is a leaf node. In the first case, we allocate memory for a new tree node, insert it as the appropriate son node of the current node, and fill it with the data of the particle. In the second case, the current node is a leaf node and thus already represents a particle. This node is now turned into an inner node. We again allocate memory for a new tree node, insert it as a son node into the current node, and fill it with the data for the particle to be inserted. The particle that was stored beforehand in the leaf node now has to be inserted into the created subtree. This approach is implemented in Algorithm 8.2.[16]

Figure 8.12 shows how a particle is inserted into a tree by way of this algorithm.

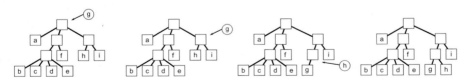

Fig. 8.12. Insertion of a particle **g** into a tree (from left to right).

---

[16] Here, instead of implementing the insertion recursively, we could implement the insertion with loops as well. Such an implementation would be somewhat faster.

Algorithm 8.2 Insertion of a Particle into an Existing Tree (`t != NULL`)

```
void insertTree(Particle *p, TreeNode *t) {
   determine the son b of t in which particle p is located;
   compute the boundary data of the subdomain of the son node
   and store it in t->son[b].box;
   if (t->son[b] == NULL) {
      if (*t is a leaf node) {
         Particle p2 = t->p;
         t->son[b] = (TreeNode*)calloc(1, sizeof(TreeNode));
         t->son[b]->p = *p;
         insertTree(&p2, t);
      } else {
         t->son[b] = (TreeNode*)calloc(1, sizeof(TreeNode));
         t->son[b]->p = *p;
      }
   } else
      insertTree(p, t->son[b]);
}
```

One can also delete a leaf node from a tree in a similar way. Then, one has to erase the leaf node and possibly also further inner nodes in a process inverse to the insertion. Figure 8.13 shows an example.

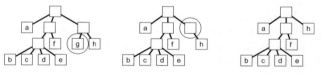

Fig. 8.13. Deletion of a leaf node g from a tree (from left to right).

In the function `insertTree`, we allocate new memory as we did in the code fragment 3.1 in Section 3.2, but we also have to set all son node pointers `son[]` to NULL. This can be implemented with the C library function `calloc`, which does not only allocate the new memory but also automatically initializes it to zero. To determine whether a node is a leaf node, we test if all pointers to son nodes are NULL. Finally, we have to determine the number `b` of the son cell which contains the particle. To this end, we compare the coordinates of the particle with the coordinates of the boundary vertices which are stored in the data structure `TreeNode`.

We start with the boundary data of the entire simulation domain which we store in `root->box` in the root `root` of the tree. The cell which corresponds after subdivision to the new `*t` is then computed from `t->box.lower` and t->box.upper. The index b of the subcell and its boundary data can be computed for instance as in Algorithm 8.3.

Algorithm 8.3 Determining the Index and the Boundary Data of a Son Cell

```
int sonNumber(Box *box, Box *sonbox, Particle *p) {
  int b = 0;
  for (int d=DIM-1; d>=0; d--) {
    if ( p->x[d] < .5 * (box->upper[d] + box->lower[d]) ) {
      b = 2*b;
      sonbox->lower[d]  = box->lower[d];
      sonbox->upper[d]  = .5 * (box->upper[d] + box->lower[d]);
    }
    else {
      b = 2*b+1;
      sonbox->lower[d]  = .5 * (box->upper[d] + box->lower[d]);
      sonbox->upper[d]  = box->upper[d];
    }
  return b;
}
```

We now use the function insertTree to build the tree for a given set of particles with the positions $\{x_i\}_{i=1}^{N}$. We initialize root so that root->p contains the first particle, and all its pointers to son nodes are set to NULL. All other particles are then inserted with insertTree, see Algorithm 8.4. The data for the particles either have to be read before from a file or have to be generated in an appropriate way.

Algorithm 8.4 Building the Particle Tree for the Barnes-Hut Method

```
void initData_BH(TreeNode **root, Box *domain, int N) {
... // read particle data from a file or generate such data
  *root = (TreeNode*)calloc(1, sizeof(TreeNode));
  (*root)->p = (first particle with number i=1);
  (*root)->box = *domain;
  for (int i=2; i<=N; i++)
    insertTree(&(particle number i), *root);
}
```

We have now constructed the tree in such a way that every leaf node contains exactly one particle.[17] The algorithm has a complexity of order

---

[17] Compare also footnote 7.

$\mathcal{O}(N \log N)$ for approximately uniformly distributed[18] particles, as can be easily verified.

Computing the Values for the Pseudoparticles. Now, the inner nodes of the tree, which resemble the so-called pseudoparticles, have to be filled with the coordinates of the center of mass and the sum of the mass of the corresponding particles according to (8.20). To this end, we start with the leaf nodes and ascend recursively to the root. For the implementation we use a post-order tree traversal. The values for the pseudoparticles are then computed recursively as shown in Algorithm 8.5 with a call to compPseudoParticle(root).

Algorithm 8.5 Computation of the Values for Pseudoparticles Using a Post-Order Traversal

```
void compPseudoParticles(TreeNode *t) {
   called recursively as in Algorithm 8.1;
   // start of the operation on *t
   if (*t is not a leaf node) {
      t->p.m = 0;
      for (int d=0; d<DIM; d++)
         t->p.x[d] = 0;
      for (int j=0; j<POWDIM; j++)
         if (t->son[j] != NULL) {
            t->p.m += t->son[j]->p.m;
            for (int d=0; d<DIM; d++)
               t->p.x[d] += t->son[j]->p.m * t->son[j]->p.x[d];
         }
      for (int d=0; d<DIM; d++)
         t->p.x[d] = t->p.x[d] / t->p.m;
   }
   // end of the operation on *t
}
```

After this tree traversal, every node of the tree contains a mass and a set of coordinates, which either correspond to an actual particle or a pseudoparticle.

Computing the Forces. In the force computation one determines for each particle in the tree an approximation of the sum of the forces which corresponds to the interaction with all other particles. We use a tree traversal over all leaf nodes. There, for each leaf node, we again traverse the tree descending

---

[18] The complexity can increase to $\mathcal{O}(N^2)$ for extremely degenerated particle distributions. However, such particle distributions only rarely occur in practice. Here, geometrically more flexible subdivision strategies can be an alternative, compare Figure 8.5.

recursively from the root until the criterion (8.21) is satisfied for the given θ, compare Algorithm 8.6.

Algorithm 8.6 Loop over all Particles for the Force Computation

```
void compF_BH(TreeNode *t, real diam) {
   called recursively as in Algorithm 8.1;
   // start of the operation on *t
   if (*t is a leaf node) {
     for (int d=0; d<DIM; d++)
       t->p.F[d] = 0;
     force_tree(t, root, diam);
   }
   // end of the operation on *t
}
```

We start the force computation with the root of the tree and with the size of the simulation domain which we store in the variable `diam`. We can then compute the current cell size `diam` in the recursion in a similar way as the cell coordinates `t->box`. Algorithm 8.7 uses the function `force` from Algorithm 3.7 to compute the gravitational force between a particle `t1->p` and another (pseudo-)particle `t->p`. For the sake of simplicity, we declare the control parameter `theta` as a global variable.

Algorithm 8.7 Force Computation in the Barnes-Hut Algorithm

```
real theta;
void force_tree(TreeNode *tl, TreeNode *t, real diam) {
   if ((t != tl) && (t != NULL)) {
     real r = 0;
     for (int d=0; d<DIM; d++)
       r += sqr(t->p.x[d] - tl->p.x[d]);
     r = sqrt(r);
     if ((*t is a leaf node) || (diam < theta * r))
       force(tl->p, t->p);
     else
       for (int i=0; i<POWDIM; i++)
         force_tree(p, t->son[i], .5 * diam);
   }
}
```

The value `r` is here computed as the distance between the particle `t1->p` and the (pseudo-)particle `t->p`. If a modification of the distance function is wanted as shown in Figure 8.8, it can be implemented here.

Time Integration. For a complete implementation of the Barnes-Hut method we also have to implement a time integration scheme and the resulting transport of the particles. We do this in the routine timeIntegration_BH. A step in the Störmer-Verlet time stepping method (or any other desired explicit scheme) can be implemented with a tree traversal[19] over all particles, in which we call the update routines of Algorithm 3.5 from Section 3.2 for the particles, see Algorithm 8.8. The positions and velocities are updated in compX_BH and compV_BH.[20]

Algorithm 8.8 Part of a Störmer-Verlet Time Step for a Tree of Particles

```
void compX_BH(TreeNode *t, real delta_t) {
    called recursively as in Algorithm 8.1;
    // start of the operation on *t
    if (*t is a leaf node) {
        updateX(t->p, delta_t);
    // end of the operation on *t
}
void compV_BH(TreeNode *t, real delta_t) {
    called recursively as in Algorithm 8.1;
    // start of the operation on *t
    if (*t is a leaf node) {
        updateV(t->p, delta_t);
    // end of the operation on *t
}
```

After the update, every particle has new coordinates and velocities. Thus, some of the particles in the existing tree may no longer be at the appropriate position in the tree. Since the particles only move for a short distance in each time step and therefore most of the particles will still be in the right cell after the time step, it makes sense to only modify the tree from the previous time step, rather than to construct it anew from scratch. This can be accomplished as follows: We introduce for each particle an additional label moved that shows whether the particle has already been moved or not. We also introduce a label todelete which shows if the particle has already been sorted into its new position. To this end, we extend the particle data structure 3.1 appropriately.[21] Then, the re-sorting of the particles can be implemented as shown in Algorithm 8.9.

---

[19] Analogously, one can implement the routines outputResults_BH and compoutStatistic_BH using tree traversals.

[20] Here, the boundary conditions have to be taken into account appropriately.

[21] Alternatively, we could store the label moved in the array entry F[0] which is not needed in this phase of the algorithm. The labelling of a particle with todelete can be implemented efficiently by setting the mass of the particle to zero.

Algorithm 8.9 Re-Sorting of the Particles in the Tree

```
void moveParticles_BH(TreeNode *root) {
  setFlags(root);
  moveLeaf(root,root);
  repairTree(root);
}
void setFlags(TreeNode *t) {
  called recursively as in Algorithm 8.1;
  // start of the operation on *t
    t->p.moved = false;
    t->p.todelete = false;
  // end of the operation on *t
}
void moveLeaf(TreeNode *t, TreeNode *root) {
  called recursively as in Algorithm 8.1;
  // start of the operation on *t
  if ((*t is a leaf node)&&(!t->p.moved)) {
    t->p.moved=true;
    if (t->p outside of cell t->box) {
      insertTree(&t->p, root);
      t->p.todelete = true;
    }
  } // end of the operation on *t
}
void repairTree(TreeNode *t) {
  called recursively as in Algorithm 8.1;
  // start of the operation on *t
  if (*t is not a leaf node) {
    int numberofsons = 0;
    int d;
    for (int i=0; i<POWDIM; i++) {
      if (t->son[i] != NULL) {
        if (t->son[i]->p.todelete)
          free(t->son[i]);
        else {
          numberofsons++;
          d = i;
        }
      }
    }
    if (0 == numberofsons) // *t is an "empty" leaf node and can be deleted
      t->p.todelete = true;
    else if (1 == numberofsons) {
      // *t adopts the role of its only son node and
      //   the son node is deleted directly
      t->p = t->son[d]->p;
      free(t->son[d]->p);
    }
  } // end of the operation on *t
}
```

The labels are initialized with the function setFlags in the first traversal of the tree.[22] The second traversal moves the particles which are at the wrong position to the correct leaves of the tree by recursively inserting them into the tree with the function insertTree, see function moveLeaf in Algorithm 8.9.[23] The particles that have been moved by moveLeaf should not be deleted immediately, in order to still allow the checking of all other nodes. Instead, it is simpler and faster to clean up the entire tree in a third step as shown in repairTree. There, leaf nodes that no longer contain particles are removed recursively in a post-order traversal. Furthermore, inner nodes are removed as well if they only have one son node. This is achieved by copying the data from the son node to the inner node and by deleting the empty son node afterwards. This procedure is continued automatically on the next higher level in the post-order tree traversal.

The main program for a particle simulation with the Barnes-Hut force computation is given in Algorithm 8.10. The routine timeIntegration_BH for the time integration with the Störmer-Verlet method can be adapted from Algorithm 3.2 from Section 3.2. Now, just compX_BH, compF_BH and compV_BH are called correspondingly. In addition, the tree has to be re-sorted by a call to Algorithm 8.9 after the particles have been moved. Finally, the routine freeTree_BH frees the memory in a recursive post-order tree traversal.

Algorithm 8.10 Main Program

```
int main() {
  TreeNode *root;
  Box box;
  real delta_t, t_end;
  int N;
  inputParameters_BH(&delta_t, &t_end, &box, &theta, &N);
  initData_BH(&root, &box, N);
  timeIntegration_BH(0, delta_t, t_end, root, box);
  freeTree_BH(root);
  return 0;
}
```

---

[22] If todelete is implemented by a mass of zero, then all nodes already have a non-zero mass after the computation of the pseudoparticles.

[23] Instead of the insertion of a particle starting at the root, one could also ascend in the tree, test if the particle belongs to a cell, and then insert it into that cell. However, this is often more expensive than the simple insertion starting at the root.

8.3.3 Applications from Astrophysics

Tree codes such as the method of Barnes and Hut have been developed especially for problems from astrophysics. In the early works [47, 342], the idea of hierarchical clusters had already been implemented, but the search for neighbors was very expensive, since arbitrarily structured trees had been used. The Barnes-Hut method avoids this search for neighbors by using octrees instead of general trees. It has been used in many forms in astrophysics [315, 538]. Typical problems studied with it are the formation and collision of galaxies and the formation of planets and protoplanetary systems. But it was also used to study theories of the formation of the moon [353] or the collision of the comet Shoemaker-Levy-9 with Jupiter [538].

In the following, we consider the formation of galaxies. There are essentially two different fundamental types of galaxies: The elliptical type, in which the stars are distributed within an ellipsoid and have three-dimensional orbits, and the structurally richer spiral type, in which the stars are distributed within a flat disk and all rotate in the same direction around a common center. Here, spiral arms are formed. Elliptical galaxies develop from star clusters with low global angular momentum, while spiral galaxies develop from star clusters with higher global angular momentum.[24] Spiral galaxies consist of approximately 90% stars and 10% atomic hydrogen gas. Elliptical galaxies contain even less gas. Therefore, we disregard the gas in our simulation.[25] Furthermore, we assume that the galaxies can be modeled as a collision-free[26] system.

We choose a spiral galaxy for our simulation. At the beginning, we randomly distribute particles/stars with a constant uniform density inside a sphere with radius one. The stars interact by way of the gravitational potential (8.19). For simplicity, all stars have the same mass. We set $m_i = 1/N$, which leads to a total system mass of one. The initial velocities of the stars are chosen in such a way that the sphere formed by the particles rotates as a rigid body around an axis through the center of the sphere. Here, the velocities are set in such a way that the centrifugal force approximately equals the gravitational force (as in Kepler orbits) and therefore the sphere does not implode or explode.

---

[24] Some few galaxies belong to neither of the two types. It is believed that they were formed by a collision of conventional galaxies.

[25] Astrophysical simulations of gas-rich galaxies are most often performed with a combination of the Barnes-Hut method with the smoothed particle hydrodynamics (SPH) method [196, 315].

[26] Collision-free systems consist of very many – typically 10^{10} to 10^{12} – particles. The time evolution of the system is then determined by the particle density instead of the exact positions of the particles. Collision systems consist of few particles, typically in the order of several hundred. The trajectories then strongly depend on the exact positions of all particles. For a discussion see for instance [324].

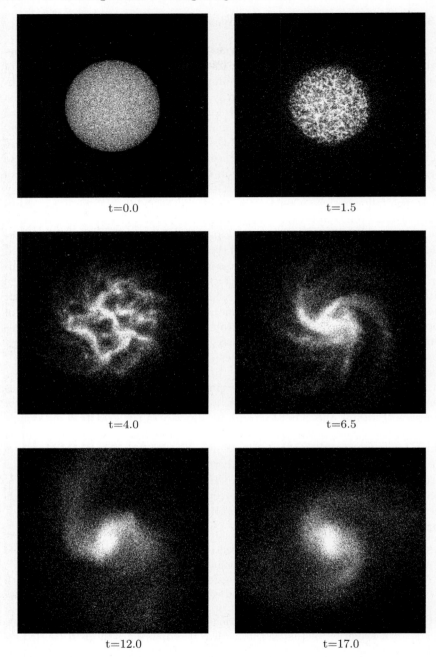

Fig. 8.14. Formation of a spiral galaxy, evolution of the particle distribution over time, view parallel to the axis of rotation.

Figure 8.14 shows the formation of a spiral structure at different points in time in such a simulation with 100000 stars. We used here $\delta t = 0.001$ as step size for the time integration and we chose a θ of 0.6. In the upper left, we see the spherically homogeneous initial configuration. Already after a small number of rotations, the configuration evolves into a structure with first smaller and then ever larger inhomogeneities. A swirly pattern forms next, which finally transforms into a stable spiral structure with two spiral arms. The conservation of angular momentum prevents strong contractions perpendicular to the axis of rotation and leads to the spiral structure in which the inner stars rotate quickly around the center and the outer stars rotate more slowly. Parallel to the axis of rotation, gravitation leads to a strong contraction which results in a thin, disk-like structure. Note that the chosen projection for the three-dimensional data parallel to the axis of rotation in Figure 8.14 does not reveal this structure in the figure. It turns out that the evolution of the system over time is strongly dependent on the angular momentum of the initial configuration, whereas the exact initial positions of the particles have only little impact.

8.4 Parallel Tree Methods

We now discuss how the Barnes-Hut method can be parallelized on a system with distributed memory. Here, an appropriate distribution of the data to the processes and thereby to the processors is of crucial importance to obtain an efficient method. In the previous chapters we always assumed an approximately uniform distribution of the particles in the simulation domain and could therefore use a uniform decomposition of the domain in the parallelization of the linked cell method and the SPME method. In this way, approximately the same number of cells and, therefore, particles were assigned to each process. Tree methods however adapt to inhomogeneous particle distributions. Thus, a simple uniform decomposition of the domain will in general lead to performance losses in the parallelization. Such losses are caused by the resulting *load imbalance*. As an example, consider the irregular distribution of particles shown in Figure 8.15.

A partition of the domain into the four subdomains would lead to an unbalanced assignment of particles to processes and therefore to an unbalanced distribution of computations to the processors. This load imbalance impairs the parallel efficiency of the method. Additionally, since the particles move according to Newton's laws over time, their (optimal) assignment to processes should change dynamically over time as well.

For the distribution of the nodes of a tree to several processes one cannot use domain decomposition for the upper nodes closest to the root. We first have to descend a certain number of levels in the tree to reach a point at which the number of nodes/subtrees is larger than or equal to the number

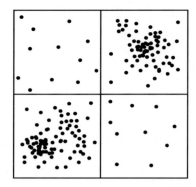

Fig. 8.15. A non-uniform particle distribution and its static partition into four equally-sized subdomains.

of processes. At that point, one or several subtrees (and thus their corresponding subdomains and particles) can be assigned to each process. The number of particles per process should be approximately the same to avoid a load imbalance. All operations that can be executed independently on the subtrees are then executed by the appropriate process. Altogether, such a distribution of subtrees can be interpreted as a decomposition of the domain which is adapted to the particle density. However, in the Barnes-Hut method, computational operations are executed not only for the leaf nodes, which correspond to particles, but also for the inner nodes, which correspond to pseudoparticles. Here, the operations on the inner nodes of the subtrees cause no problems, but the operations that work on the coarse upper part of the global tree which connects the subtrees have to be treated separately. Furthermore, the distribution of the tree to the processes has to be updated over time to retain load balance, since the particles move in each time step.

Hence, the distribution of the particles to the processes is no longer as straightforward as in the grid-based or cell-based algorithms. In the following, we use a partition strategy which exploits the structure of the current octree. On the one hand, we attempt to balance the computational load, and on the other hand, we try to assign complete subtrees to each process so that the amount of communication is minimized. To this end, we present an implementation of trees for parallel computers with distributed memory that works with keys and thus allows a simple assignment of the data to the processes. In addition, we use a heuristic based on space-filling curves to achieve good load balancing. Such methods have been first proposed by Salmon and Warren [538, 539, 651, 652, 653, 654] in computational astrophysics. Meanwhile, they are also applied successfully in other application areas [556, 693].

8.4.1 An Implementation with Keys

Key-Based Domain Decomposition. We now partition the tree and assign subtrees to different processes. On a parallel computer with distributed memory, an implementation of a tree data structure based on pointers as introduced in the last section leads to the following problem: Pointers and memory addresses can be used for the computation in the memory of *one* process but are in general meaningless for the other processes. What is missing is an absolute addressing scheme for all nodes of the tree for all processes. We can obtain such an addressing scheme by encoding the address of a node with integer keys. Every *possible* node is then assigned a unique number. In this way the set of all possible nodes of an octree – or, equivalently, of the corresponding geometric cells – is mapped to the natural numbers. A particular tree is then described by the set of keys that are associated to existing nodes.

The challenging question is how to choose the keys such that the corresponding decomposition of the domain and the associated tree nodes to different processes results in an overall efficient method. This includes the condition that entire subtrees should lie in the same process and that as little communication as possible is needed in the parallel algorithm. We will reach this goal in several steps. First, we introduce a very simple level by level mapping of tree nodes to keys and discuss a first decomposition which gives a distribution to processes. Then, we modify the resulting keys to obtain decompositions, where whole subtrees are distributed to processes. Finally, we use space-filling curves to obtain decompositions that lead to an even more localized decomposition for which the parallel communication is substantially reduced.

Let us first introduce a simple level by level mapping of tree nodes to keys. Starting with the root, *all* possible nodes of each new generation are numbered consecutively. This is implemented by a mapping that encodes in the key the path from the root of the tree to the node to be numbered. We already assigned a local number (from 0 to 7) to the direct sons of a node of the octree in function sonNumber (in Algorithm 8.3). We also used these numbers as indices for the son nodes in the data structure son. Based on these local numbers, we can now describe all nodes of the tree: We start at the root and encode the path to a node by listing the local numbers of the relevant son nodes on that path. We then concatenate the numbers and obtain an integer. This is the *path key* of the node to be numbered. For simplicity we use the binary representation for the numbers of the son nodes and also for their concatenation in the path key. The root node of the tree is encoded by the number 1.[27] An example with a quadtree is given in Figure 8.16. If we have to descend from the root first into the first son node (01), then into the third

---

[27] This serves as stop bit and is necessary to guarantee the uniqueness of the key values.

son node (11), then into the zeroth son node (00), and finally into the second son node (10) to reach the desired node, then this node is associated with the path key 101110010. We proceed analogously in the three-dimensional case for octrees. In that case, three bits are added to the key in each level of the tree.

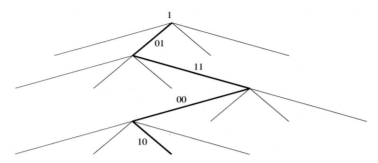

Fig. 8.16. A path in the tree and the associated path key.

This definition of a path key allows to determine the key of the father node of a given node by deleting the last three bits (in the case of an octree) of the path key of the given node. Conversely, all sons, grandsons, and further descendants of a tree node can be recognized, since the leading bits of their keys correspond exactly to the key of the tree node in question.

The use of keys for each tree node that are global, unique, and of integer type, allows a simple description of the decomposition of a given tree and, correspondingly, of a given domain. To decompose the tree to several processes, we now just have to distribute the keys that correspond to the nodes of the tree to the processes. The set of all possible keys, i.e. the set of all integers up to a certain length in bits,[28] is decomposed into P subsets that are assigned to the P processes available for computations. The keys are natural numbers and can therefore be *sorted* according to their magnitude. A decomposition of the set of sorted keys into P subsets can then be given by interval limits according to

$$0 = \text{range}_0 \leq \text{range}_1 \leq \text{range}_2 \leq \ldots \leq \text{range}_P = \text{KEY\_MAX}. \qquad (8.22)$$

Using this decomposition, we assign to each process i all keys $\mathtt{k}$ in the half-open interval

$$[\text{range}_i, \ \text{range}_{i+1}) \ = \ \Big\{ \mathtt{k} \mid \text{range}_i \leq \mathtt{k} < \text{range}_{i+1} \Big\},$$

see Figure 8.17.

---

[28] The maximal value of a key **KEY_MAX** is a constant which depends on the compiler available on the parallel computer. This constant limits the maximal refinement depth of the trees that can be used in the code.

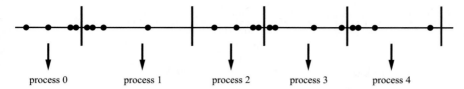

Fig. 8.17. Distribution of the particles to the processes according to the values `range[0]` to `range[P]`.

We thus have a unique procedure to determine which tree node will be stored in which process. The procedure is deterministic and can be used at any time by any process while always giving the same result.

For the implementation, we declare the data type `keytype` for the keys as `unsigned long` in the code fragment 8.1.[29]

Code fragment 8.1 Definition of the Key Data Type and the Subdomain Data Structure

```
typedef unsigned long keytype;
#define KEY_MAX ULONG_MAX
const int maxlevel = (sizeof(keytype)*CHAR_BIT - 1)/DIM;
typedef struct {
  int myrank;
  int numprocs;
  keytype *range;
} SubDomainKeyTree;
```

To this end, the interval limits `range` are defined in the data structure `SubDomainKeyTree` which also contains the number of the processes participating in the parallel computation and the local process number. Given a key k, one can then compute its process number as in Algorithm 8.11.[30] With this simple approach, every process can determine in which (possibly different) process a certain tree node is stored.

---

[29] The bit length depends on the computer and its operating system. In the case of a 32 bit architecture, up to 10 tree levels can be handled, in the case of a 64 bit architecture, up to 21 tree levels can be managed. Some 32 bit systems offer a 64 bit wide data type called `long long` which again allows 21 tree levels. Further extensions could be implemented using a multi precision library which would allow longer integer data types.

[30] The complexity of this implementation is $\mathcal{O}(P)$. It can be improved to $\mathcal{O}(\log P)$ if a binary search is used instead of a linear search on the components of `range` sorted in ascending order.

Algorithm 8.11 Mapping a Key to a Process Number

```
int key2proc(keytype k, SubDomainKeyTree *s) {
  for (int i=1; i<=s->numprocs; i++)
    if (k >= s->range[i])
      return i-1;
  return -1; // error
}
```

The question remains if the use of the path key for the decomposition of the domain and the associated tree nodes results in an efficient method. The assignment of nodes to processes should be done in such a way that entire subtrees lie in the same process.[31] This would lead to efficient tree traversals and thus to efficient tree methods. However, the path key numbering introduced above rather leads to a horizontal than a vertical ordering of the tree nodes, as can be seen in Figure 8.18. Consequently, entire subtrees will not lie in the same process after the partition. We therefore have to suitably transform the path key to obtain a better ordering.

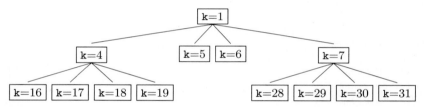

Fig. 8.18. Horizontal ordering in the tree using the path keys k for an example quadtree. The keys are given here in decimal notation and not in binary notation.

A transformation that will lead to a vertical ordering of the tree nodes is the following: We first remove the leading bit (the original key of the root node). The remaining bits are shifted to the left until the maximal bit length of the key type is reached.[32] We denote the resulting bit word as *domain key*. In the domain key ordering the tree nodes are sorted vertically and we can identify subtrees and assign them to processes using the simple interval decomposition (8.22). Applied to the two-dimensional example from Figure 8.16, the path key is transformed into the domain key as follows

---

[31] In addition, the decomposition should be carried out such that as little communication as possible is needed in the parallel algorithm. To achieve this, we will later modify the keys again and use space-filling curves to obtain decompositions that result in a more efficient method.

[32] This transformation needs $\mathcal{O}(1)$ operations if the refinement level of the tree is known. Otherwise, $\mathcal{O}(\texttt{maxlevel})$ operations are needed.

$$\underbrace{00000000000000000000001}\, \underbrace{01110010}_{\text{path}} \longmapsto$$

$$\underbrace{01110010}_{\text{path}}\,000000000000000000000000, \tag{8.23}$$

if a 32 bit key type is used.

If the keys for the nodes of *one* level of the tree are mapped in this way, they are all different, the keys for nodes of *different* levels can however happen to be the same: The zeroth son of a node has the same key as the node itself. Only then the (new) keys of the other sons follow. This holds recursively in the entire tree.[33] We show the effect of the transformation from path key to domain key in a two-dimensional example in Figure 8.19. There, we place 65536 randomly distributed particles in a circle and determine the associated quadtree as well as the path key and the domain key for each particle (leaf node) of the tree. We then connect the points representing the particles with lines in the order given by the two different keys. On the left, we see the polygonal line resulting from the path key ordering, on the right, we see the polygonal line resulting from the domain key ordering. The right polygonal line clearly shows a stronger local character and is shorter. If the domain key

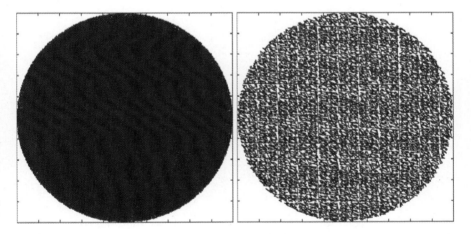

Fig. 8.19. The linear particle ordering induced by the path key (left panel) and by the domain key (right panel) for 65536 particles randomly placed in a two-dimensional circle. One can clearly see the improved locality of the ordering induced by the domain key.

---

[33] The transformed numbers of the nodes are no longer unique for the nodes of the tree, but they are still unique for the leaf nodes. This is sufficient to compute the domain decomposition and the partitioning of the particle set with the intervals $[\text{range}_i,\ \text{range}_{i+1})$.

ordering is used for the parallelization in the decomposition of the particle set into subsets using **range** values, one thus obtains a better parallelization. The resulting geometric domain decomposition is more connected and more compact, and less data has to be exchanged between processes.

Note that the description of the data distribution with the intervals $[\text{range}_i, \text{range}_{i+1})$ defines a minimal upper part of the tree that has to be present in all processes as a copy to ensure the consistency of the distributed global tree. The leaf nodes of this common coarse tree are the coarsest tree cells for which all possible descendants are stored in the same process, compare Figure 8.20. The values of the domain keys of all possible descendants of a leaf node of this coarse tree lie in the same interval $[\text{range}_i, \text{range}_{i+1})$ as the domain key of this node (if the value range_{i+1} was chosen matching with the coarse tree). These leaf nodes of the common coarse tree are also the roots of the local subtrees which are associated to single processes. Such a given coarse tree can then be used to determine at any time which particle belongs to which process.

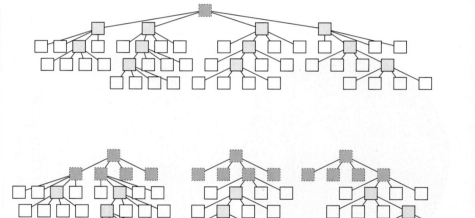

Fig. 8.20. A tree (top) distributed to three processes (bottom). The nodes of the common coarse tree (dark grey with dashed border) are kept as copies in all processes, while the nodes representing particles (white) and further pseudoparticles (light grey) are only kept in one process.

Altogether, we now have three different types of tree nodes: Leaf nodes, which are nodes without sons in which particle data are stored, inner tree nodes that do not belong to the common coarse tree and in which pseudoparticles are stored, and tree nodes that belong to the common coarse tree and which describe the domain decomposition. To simplify the implementation of the parallel tree operations later, only pseudoparticle data is stored in the last kind of nodes, and never particle data. We tag the different types of nodes explicitly. This is implemented in code fragment 8.2.

Code fragment 8.2 Definition of the Types of Nodes in the Tree

```
typedef enum { particle, pseudoParticle, domainList } nodetype;

typedef struct {
   ...
   nodetype node;
} TreeNode;
```

Here, we extended the data structure `TreeNode` by a flag `node`. We mark nodes of the common coarse tree which are kept as copies in all processes with the type `domainList`, since they describe the geometric distribution of the computational domain to the processes. The leaf nodes are marked with type `particle` and the remaining nodes are flagged with type `pseudoParticle`.

An important procedure for the parallelization is the computation and flagging of the `domainList`-nodes from the `range` values contained in data structure `SubDomainKeyTree`. This is implemented in detail in Algorithm 8.12.

Algorithm 8.12 Computation of the `domainList` Flags Using `range` Values

```
void createDomainList(TreeNode *t, int level, keytype k,
                      SubDomainKeyTree *s) {
  t->node = domainList;
  int p1 = key2proc(k,s);
  int p2 = key2proc(k | ~(~0L << DIM*(maxlevel-level)),s);
  if (p1 != p2)
    for (int i=0; i<POWDIM; i++) {
      t->son[i] = (TreeNode*)calloc(1, sizeof(TreeNode));
      createDomainList(t->son[i], level+1,
                       k + i<<DIM*(maxlevel-level-1), s);
    }
}
```

We here use bit operations such as the complement and the bit-wise OR to manipulate the domain keys appropriately. First, we look at the key of the node itself, and thus at the minimal key of all son nodes in the subtree. By filling all the remaining digits with 1, we also look at the maximal key of all son nodes. Now, if both keys are mapped to the same process, the monotonicity of `key2proc` ensures that the entire subtree is mapped to the same process. However, if they are mapped to different processes, we have to descend deeper into the tree. `createDomainList` is called as shown in code fragment 8.3.

Code fragment 8.3 Initialization of the Tree

```
root = (TreeNode*)calloc(1, sizeof(TreeNode));
createDomainList(root, 0, 0, s);
```

We now adapt the sequential routines for the tree operations from the last section to the parallel tree data structures according to the type of the node (i.e. according to the flag `node`). Here in particular, the nodes flagged with `domainList` may not be deleted in `repairTree`. Furthermore, `insertTree` should not insert particle data into `domainList` nodes. Thus, when inserting a node in `insertTree`, particles cannot be inserted directly into a `domainList` node, but have to be inserted (recursively) into its son nodes. In this way, some particle might lie in deeper levels in the parallel tree with `domainList` nodes as in the sequential case. However, the numerical result of the Barnes-Hut method does not change.[34]

The Barnes-Hut method computes not only particle-particle interactions but also particle-pseudoparticle interactions. This can lead to the situation that a process has to access pseudoparticle data on a different process in the parallel computation. Thus, the process has to insert certain pseudoparticles from other processes into its local tree to be able to compute the force on its particles. Therefore, we have to extend the routine `insertTree` in such a way that we can insert a `domainList` node into a given tree. Then, the associated decomposition of the domain changes. To this end, pseudoparticles can be simply re-flagged as `domainList` there. However, if we encounter a real particle, we have to create a `domainList` node and insert the particle as a son node.

There are several possibilities to provide the parallel program with initial data: All particles could be created in process zero or are read from a file into process zero. For large numbers of particles, this approach is limited by the available memory of process zero. Therefore, we want to be able to create or read the particles for several processes in parallel. In general, the initial parallel distribution of the particles will then be different from the distribution that would result from our domain decomposition. Hence, the particles have to be redistributed to the appropriate processes.

Altogether, we proceed as follows: We assume that the values of `range` are given. First, every process creates a local tree using `createDomainList`. Particles created or read by the process are inserted into this local tree. With

---

[34] Alternatively, one could also flag particles with `domainList`. However, this would lead to more case distinctions in the code.

help of the `domainList` flag, we can decide to which process the particle truly belongs. Every process now traverses its local tree, removes particles that are not assigned to it, and sends these particles to the appropriate process. After this operation, the tree should be cleaned up as in the sequential case, i.e., its structure should be adapted to the actual distribution of the particles. Subsequently, the process receives the particles from the other processes and sorts them into its local tree. This approach is implemented schematically in Algorithm 8.13. For simplicity, we store the particles to be sent to process `to` in a particle list `plist[to]`, an instance of the abstract list data type `ParticleList`.[35]

Algorithm 8.13 Sending Particles to Their Owners and Inserting Them in the Local Tree

```
void sendParticles(TreeNode *root, SubDomainKeyTree *s) {
  allocate memory for s->numprocs particle lists in plist;
  initialize ParticleList plist[to] for all processes to;
  buildSendlist(root, s, plist);
  repairTree(root); // here, domainList nodes may not be deleted
  for (int i=1; i<s->numprocs; i++) {
    int to = (s->myrank+i)%s->numprocs;
    int from = (s->myrank+s->numprocs-i)%s->numprocs;
    send particle data from plist[to] to process to;
    receive particle data from process from;
    insert all received particles p into
      the tree using insertTree(&p, root);
  }
  delete plist;
}


void buildSendlist(TreeNode *t, SubDomainKeyTree *s,
                   ParticleList *plist) {
  called recursively as in Algorithm 8.1;
// start of the operation on  *t
  int proc;
  if ((*t is a leaf node) &&
      ((proc = key2proc(key(*t), s)) != s->myrank)) {
    // the key of *t can be computed step by step in the recursion
      insert t->p into list plist[proc];
      mark t->p as to be deleted;
    }
  // end of the operation on *t
  }
```

---

[35] This list data type might be implemented as the list data type from Chapter 3.

Here, we use the sequential implementation of `insertTree` which is extended to handle the field `node`. In this way, we have distributed the particles to the processes according to the information given in `range`. All nodes that are marked as `particle` or `pseudoParticle` are contained in *one* subtree that belongs completely to *one* process. The root of the tree and the nodes on the coarse levels of the tree, which are all marked with `domainList`, have to be treated separately for some operations, since son nodes may belong to other processes. In the actual implementation, the content of the particle lists first has to be copied into an array, which then can be handed to the message passing library MPI for parallel communication. Also, the right amount of memory has to be allocated for `buffer`. This can be implemented for instance by a prior message that only communicates the length of the following message. Alternatively, one can use other MPI commands to determine the length of an incoming message, to allocate the right amount of memory, and then to receive and process the message.

Computing the Values for the Pseudoparticles. As in the sequential case one first computes the values for the pseudoparticles and subsequently determines the forces. We split both routines into separate parts for communication and computation. The computation of the values of the pseudoparticles for a subtree completely owned by a process can proceed as in the sequential case from the leaf nodes to the root of the subtree. We then exchange the values of these `domainList` nodes, which are the leaf nodes of the global coarse tree, among all processes. Finally, we compute the values of the pseudoparticles in the coarse `domainList` tree by *all* processes at the same time, in a redundant fashion. This approach is presented in Algorithm 8.14.

We can use a global sum over all processes for the communication[36] if we transfer and sum the first moments for each node, i.e. `m*x[i]`, instead of the coordinates. Each process then contributes either zero or the value already computed for the node and receives afterwards the global sum needed for the operations on the coarse `domainList` tree. In this way, the pseudoparticles assume exactly the same values as in the sequential case. In summary, we compute the values on the subtrees completely independently, communicate data globally, and then finish the computation on the coarse `domainList` tree independently and redundantly for each process.

Force Computation. The force on a particle will be computed by the process to which the particle has been assigned. To this end, the sequential implementation of `compF_BH` is changed as shown in Algorithm 8.15. The algorithm uses the routine `force_tree` (see Algorithm 8.7) locally in each process. For this, the necessary data from other processes have to be stored

---

[36] When implementing this with `MPI_Allreduce`, one has to copy the data into an array first.

Algorithm 8.14 Parallel Computation of the Values of the Pseudoparticles

```
void compPseudoParticlespar(TreeNode *root, SubDomainKeyTree *s) {
  compLocalPseudoParticlespar(root);
  MPI_Allreduce(..., {mass, moments} of the lowest domainList nodes,
                MPI_SUM, ...);
  compDomainListPseudoParticlespar(root);
}


void compLocalPseudoParticlespar(TreeNode *t) {
  called recursively as in Algorithm 8.1;
  // start of the operation on *t
    if ((*t is not a leaf node)&&(t->node != domainList)) {
      // operations analogous to Algorithm 8.5
    }
  // end of the operation on *t
}


void compDomainListPseudoParticlespar(TreeNode *t) {
  called recursively as in Algorithm 8.1 for the coarse domainList-tree;
  // start of the operation on *t
    if (t->node == domainList) {
      // operations analogous to Algorithm 8.5
    }
  // end of the operation on *t
}
```

Algorithm 8.15 Adapting the Sequential Routine to the Parallel Implementation

```
void compF_BH(TreeNode *t, real diam, SubDomainKeyTree *s) {
  called recursively as in Algorithm 8.1;
  // start of the operation on *t
  if ((*t is a leaf node) && (key2proc(key(*t), s) == s->myrank)) {
    // the key of *t can be computed step by step in the recursion
    for (int d=0; d<DIM; d++)
      t->p.F[d] = 0;
    force_tree(t, root, diam);
  // end of the operation on *t
  }
}
```

as copies in the local process. Because of the θ criterion used in the Barnes-Hut method, the recursion does not descend through entire subtrees that lie in other processes, but terminates earlier, depending on the distance to the particle. If we ensure that all pseudoparticles and particles necessary in the computation of the forces are copied to the local process, the sequential version of `force_tree` can be used and will give the same result in the parallel case as in the sequential case. To this end, we have to determine which data are needed and we have to exchange this data between the processes appropriately and efficiently.

The difficulty is that the local process cannot determine which nodes of the other trees it has to request since it does not know the exact structure of the entire tree nor the data for all processes. We therefore proceed as follows: Every process determines which nodes *another* process will request from it during the force computation and sends the appropriate data to the appropriate process. This can be implemented in the following way: Each process p0 (i.e. `myrank == p0`) stores all global `domainList` nodes in addition to its own nodes. To a `domainList` node `td` of a different process p1, there exists an associated subtree below it in process p1. Then, all of the particles and pseudoparticles in this subtree lie in the geometric cell `td->box`. Therefore, when process p0 tests for its particles and pseudoparticles `t->p`, whether process p1 might possibly need them in its force computation, it can assume that all particles from p1 lie in the cell of `*td`. If now an interaction is excluded even for the minimal distance from `t->p.x` to the cell `td->box` due to the θ criterion, process p1 will not need the particle `t->p` in its computation. The same holds for all descendants of `*t` since the size `diam` of the associated cells is smaller than that of `*t`. This is shown in 8.21 for an example.

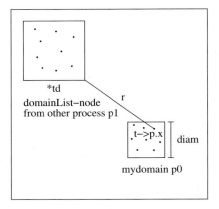

Fig. 8.21. Extension of the θ criterion to geometric cells and thus to entire subtrees.

In summary, we employ a communication routine before the parallel force computation in which every process sends a number of particles and pseudoparticles to other processes. These processes sort these particles into their own tree. For this, we need the routine `symbolicForce` which is given in Algorithm 8.16.

Algorithm 8.16 Determining Subtrees that are Needed in the Parallel Force Computation

```
void symbolicForce(TreeNode *td, TreeNode *t, real diam,
                   ParticleList *plist, SubDomainKeyTree *s) {
  if ((t != NULL) && (key2proc(key(*t), s) == s->myrank)) {
    // the key of *t can be computed step by step in the recursion
    insert t->p into list plist;
    real r = smallest distance from t->p.x to cell td->box;
    if (diam >= theta * r)
      for (int i=0; i<POWDIM; i++)
        symbolicForce(td, t->son[i], .5 * diam, plist, s);
  }
}
```

Algorithm 8.17 Parallel Force Computation

```
void compF_BHpar(TreeNode *root, real diam, SubDomainKeyTree *s) {
  allocate memory for s->numprocs particle lists in plist;
  initialize ParticleList plist[to] for all processes to;
  compTheta(root, s, plist, diam);
  for (int i=1; i<s->numprocs; i++) {
    int to = (s->myrank+i)%s->numprocs;
    int from = (s->myrank+s->numprocs-i)%s->numprocs;
    send (pseudo-)particle data from plist[to] to process to;
    receive (pseudo-)particle data from process from;
    insert all received (pseudo-)particles p into
    the tree using insertTree(&p, root);
  }
  delete plist;
  compF_BH(root, diam);
}

void compTheta(TreeNode *t, SubDomainKeyTree *s, ParticleList *plist,
               real diam) {
  called recursively as in Algorithm 8.1;
  // start of the operation on *t
  int proc;
  if ((*t is a domainList node) &&
      ((proc = key2proc(key(*t), s)) != s->myrank))
    // the key of *t can be computed step by step in the recursion
    symbolicForce(t, root, diam, &plist[proc], s);
  // end of the operation on *t
}
```

The computation of the distance of a cell to a particle can be implemented with appropriate case distinctions. One has to test whether the particle lies left, right, or inside the cell along each coordinate direction. The particles to be sent are collected in lists. It could happen that a (pseudo-)particle is inserted into the list several times, if several cells td are traversed for one process. Such duplicate (pseudo-)particles should be removed before the communication step. This can be implemented easily via sorted lists.

The actual parallel force computation is shown in Algorithm 8.17. After the particles have been received by a process, they have to be inserted into its local subtree. Here, the pseudoparticles should be inserted first, each one on its appropriate level, and then the particles should be inserted. This ensures the proper insertion of the subtree. An ordering into pseudoparticles and particles can be guaranteed for example by sorting the (pseudo-)particles according to decreasing mass. This sorting can be implemented in the routine `symbolicForce` in the insertion into the send lists. Son nodes are inserted automatically after father nodes into the local subtree.

Main Program. The remaining parts needed to complete the parallel program can be implemented in a straightforward way. After the force computation, copies of particles from other processes have to be removed. The routine for the time integration can be reused from the sequential case. It only processes all particles that belong to the process. Particles are moved in two phases. First, the sequential routine is used to re-sort particles that have left their cell in the local tree. Afterwards, particles that have left the process have to be sent to other processes. We have already implemented this in the

---

Algorithm 8.18 Parallel Main Program

```
int main(int argc, char *argv[]) {
  MPI_Init(&argc, &argv);
  TreeNode *root;
  Box box;
  SubDomainKeyTree s;
  real delta_t, t_end;
  inputParameters_BHpar(&delta_t, &t_end, &box, &s);
  root = (TreeNode*)calloc(1, sizeof(TreeNode));
  createDomainList(root, 0, 0, &s);
  initData_BHpar(&root, &box, &s);
  timeIntegration_BHpar(0, delta_t, t_end, root, &box, &s);
  outputResults_BHpar(root, &s);
  freeTree_BHpar(root);
  free(s.range);
  MPI_Finalize();
  return 0;
}
```

routine `sendParticles`. Whenever the tree is cleaned up, one has to ensure that nodes of the type `domainList` are not be deleted.

Altogether, this results in the main program from Algorithm 8.18. The `domainList` nodes are created and initialized with `createDomainList(root, 0, 0)` in `initData_BHpar`. Then, as in the sequential case, particles of the local process are inserted into its subtree. We leave further details to the reader.

8.4.2 Dynamical Load Balancing

A still open question is how to choose the values of **range**. If the particles are geometrically equidistributed, we can simply use a *uniform* partition of the set of key values.[37] However, if the particles are not uniformly distributed, such a static and fixed choice of the values in **range** is unsatisfactory. Instead, the values should be adapted to the particle data which are changing over time in the simulation. Here, values of **range** are sought that lead to an almost uniform distribution of the particles to the processes.

One possible approach is described in code fragment 8.4. Note that the tree, particles, `domainList` nodes, and **range** values have to be consistent when this code is called. This is for instance the case after a call to `sendParticles`. In this code fragment, we first determine the current load distribution and then compute a new, balanced load distribution.

Code fragment 8.4 Determining Current and New Load Distribution

```
long c = countParticles(root);
long oldcount[numprocs], olddist[numprocs+1];
MPI_Allgather(&c, &oldcount, MPI_LONG);
olddist[0] = 0;
for (int i=0; i<numprocs; i++)
  olddist[i+1] = olddist[i] + oldcount[i];
long newdist[numprocs+1];
for (int i=0; i<=numprocs; i++)
  newdist[i] = (i * olddist[numprocs]) / numprocs;
```

Here, we first count the number of particles of each process in the routine `countParticles` using a post-order tree traversal. After a global communication with `MPI_Allgather`, we then know how many particles are owned by each single process. If we now number the particles in increasing order

---

[37] Since we want to keep the depth of the `domainList` tree and thereby the number of `domainList` nodes small, we can choose values of **range** that correspond to nodes of as high a level as possible. These are nodes with domain keys that have been filled with as many zeros as possible from the right. Altogether, this is achieved by an appropriate rounding instead of a straightforward integer division.

of their keys, we know furthermore that process `i` owns the particles with numbers `olddist[i]` to `olddist[i+1]-1`. This is the case since the particles are distributed according to the old `range` values to the processes and these numbers are monotonically increasing. The total number of particles is `olddist[numprocs]`. If we now distribute the particles uniformly to the processes, i.e., assign each process `olddist[numprocs]/numprocs` particles up to rounding, each process `i` will receive the particles with numbers `newdist[i]` to `newdist[i+1]-1`. In this way, each process can determine which particles it has to send to which process and which particles it has to receive from which process.[38]

We could in principle transfer all particles right away in the routine of code fragment 8.4. However, we would save some work if we reused the routines `sendParticles` and `createDomainList`. We then only have to determine the new values of `range`. The values `range[0]=0` and `range[numprocs]=MAX_KEY` can be set correctly by each process independently, but the other `range` values cannot be determined independently by any process. Here, we proceed as follows: Each process runs through its particles and counts them. If a `newdist`-limit is reached, the corresponding key of the current particle is chosen as new value for `range`. In this way, every process knows exactly one new `range` value. If the `range` values have been initialized with zero, computing the maximum over all processes over the `range` values is then enough to obtain the correct `range` values for all processes. This global communication can then be implemented with a call to `Allreduce(range, MPI_MAX)`.

To find the keys for the particles sought, the process has to walk through its particles in increasing key order. Because of the construction of the key, this can be achieved simply with a post-order traversal of the leaf nodes of the tree. The first particle encountered by the process has the number `olddist[myrank]`. Starting with this number, the process can count along. The details are given in code fragment 8.5.

Afterwards, one has to delete the old `domainList` flags in the tree, create new ones with `createDomainList`, and transport the particles with `sendParticles`. The number of particles to be sent depends strongly on the load distribution. If the load is redistributed in every time step, usually only few particles have to be transported.

The presented approach can be interpreted as a parallel sorting method. In a first step, each process has a bucket into which all particles within its subdomain are sorted. The transporting of the particles to the right processes and the sorting into their bucket can then also be interpreted as local sorting.

---

[38] It could possibly happen that a process has to send all of its particles to other processes and will receive completely different particles. This would be the case if $[\text{olddist}_i, \text{olddist}_{i+1}) \cap [\text{newdist}_i, \text{newdist}_{i+1}) = \emptyset$. However, this would only occur when there is a very large load imbalance which generally does not happen in practice.

Code fragment 8.5 Determining New **range** Values from the Load Distribution

```
for (int i=0; i<=numprocs; i++)
  range[i] = 0;
int p = 0;
long n = olddist[myrank];
while (n > newdist[p])
  p++;
updateRange(root, &n, &p, range, newdist);
range[0]       = 0;
range[numprocs] = MAX_KEY;
MPI_Allreduce(..., range, MPI_MAX,...);

void updateRange(TreeNode *t, long *n, int *p,
                 keytype *range, long *newdist) {
  called recursively as in Algorithm 8.1;
  // the key of *t can be computed step by step in the recursion
  // start of the operation on *t
  if (*t is a leaf node) {
    while (*n >= newdist[*p]) {
      range[*p] =  key(*t);
      (*p)++;
    }
    (*n)++;
  }
  // end of the operation on *t
}
```

Analogous approaches can be found in algorithms like **bucket sort** or one-stage **radix sort** [357].

The described method can be generalized to an equidistribution of the computational load. Then, one distributes estimated computational costs per particle. The vectors **olddist** and **newdist** there no longer contain the number of particles, but for instance the accumulated load or the number of interactions in the previous time step.

8.4.3 Data Distribution with Space-Filling Curves

In the last subsection, we distributed the particles uniformly to the processes. However, the question still remains how much communication is needed in the resulting parallel tree method. This depends on the number of processes and the number of particles. If we fix both numbers, we directly see that the amount of **domainList** nodes plays a certain role in the parallel computation of the pseudoparticles, especially if the **domainList**-subtrees are not very deep. Furthermore, the communication in the parallel force computation is even more involved. The reason is that often relatively large subtrees have to be sent and received here. The question is therefore, how effective our domain decomposition technique is with respect to the amount of data to be communicated.

This question cannot be answered exactly, both for our as well as for many other domain decomposition heuristics. Let us first assume equidistributed particles and balanced trees. Then, it is rather easy to see how to improve our heuristics: If we run through the particles according to their domain keys in increasing order and connect them by line segments, we obtain a so-called discrete Lebesgue curve, compare Figure 8.22 for the two-dimensional case. If we decompose the computational domain into several subdomains, the curve will be subdivided as well. The same statement holds the other way around, i.e., a partitioning of the curve induces a certain domain decomposition. The subdomain described by a piece of the curve is then assigned to one process. Since the Lebesgue curve has "jumps", these subdomains might consist of several disconnected pieces. This is not a problem in itself, but it is an indication that the geometric boundary of a subdomain produced in this way might be relatively large. However, the size of the subdomain boundaries has a strong influence on the communication load, i.e., on how many particles have to be sent in the force computation. Hence, to improve the domain decomposition with respect to the resulting communication effort, we need to find keys such that the resulting curve leads to subdomains with small geometric boundary.

One way to improve our domain decomposition is to use a so-called Hilbert curve instead of the Lebesgue curve, which we implicitly employed up to now. A partitioning of the Hilbert curve will always lead to connected subdomains, see Figures 8.28 and 8.29. Both the Lebesgue curve and the Hilbert curve are examples of space-filling curves. They were discovered in 1890 and 1891 by Peano [470] and Hilbert [321]. The aim was to construct surjective mappings of a line segment, e.g. the unit interval $[0, 1]$, to a two-dimensional surface, for instance $[0, 1]^2$. An introduction to the theory of space-filling curves can be found in [535].

A space-filling curve $K : [0, 1] \to [0, 1]^2$ is given as the limit of a sequence of curves $K_n : [0, 1] \to [0, 1]^2, n = 1, 2, 3, \ldots$ Every curve K_n connects the centers of the 4^n squares that are created by successive subdivision of the unit square by line segments in a certain order. The curve K_{n+1} results from the curve K_n as follows: Each square is subdivided, the centers of the newly created four smaller squares are connected in a specific given order, and all the 4^n groups of 4^{n+1} centers of smaller squares are connected in the order given by the curve K_n. In this sense, the curve K_{n+1} refines the curve K_n.

Hilbert and Lebesgue curves differ in the ordering of the centers in each refinement step. In the case of the Lebesgue curve, the same order is used everywhere, as shown in the upper left of Figure 8.22. For the Hilbert curve the order is chosen in such a way that, whenever two successive centers are

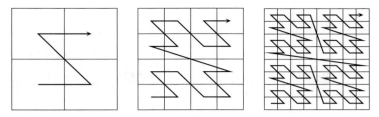

Fig. 8.22. Three steps in the construction of the Lebesgue curve.

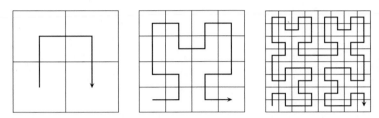

Fig. 8.23. Three steps in the construction of the Hilbert curve.

connected by a straight line, only the common edge of the two squares is crossed. The construction is made clearer in Figure 8.23. One can show that the sequence K_n for Hilbert's curve converges uniformly to a curve K, which implies that the limit curve K is continuous. For the Lebesgue curve, the sequence only converges pointwise and the limit is discontinuous.

The construction can be generalized to arbitrary space dimensions DIM, i.e. to curves $K : [0,1] \rightarrow [0,1]^{\text{DIM}}$. Such a Hilbert curve is shown for the three-dimensional case in Figures 8.24 and 8.25.

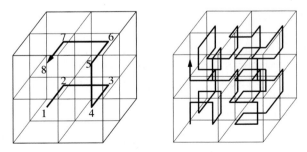

Fig. 8.24. Construction of a three-dimensional Hilbert curve.

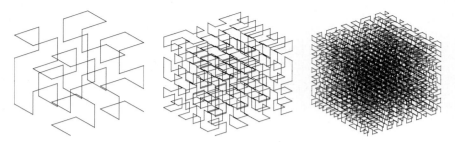

Fig. 8.25. A number of refinement steps of a three-dimensional Hilbert curve.

We implement the replacement of the Lebesgue curve by the Hilbert curve with a second transformation of the path key.[39] Thus, we still compute the path key during the recursion which is mapped by shifting to the corresponding domain key according to (8.23). This key corresponds to the ordering of the Lebesgue curve. The domain key is subsequently transformed into the corresponding *Hilbert key* using the function `Lebesgue2Hilbert` from Algorithm 8.19.

Algorithm 8.19 Transformation of a Lebesgue Key into a Hilbert Key. (Tables for DIM=2.)

```
const unsigned char DirTable[4][4] =
  { {1,2,0,0}, {0,1,3,1}, {2,0,2,3}, {3,3,1,2} };
const unsigned char HilbertTable[4][4] =
  { {0,3,1,2}, {0,1,3,2}, {2,3,1,0}, {2,1,3,0} };

keytype Lebesgue2Hilbert(keytype lebesgue) {
  keytype hilbert = 1;
  int level = 0, dir = 0;
  for (keytype tmp=lebesgue; tmp>1; tmp>>=DIM, level++);
  for (; level>0; level--) {
    int cell = (lebesgue >> ((level-1)*DIM)) & ((1<<DIM)-1);
    hilbert = (hilbert<<DIM) + HilbertTable[dir][cell];
    dir     = DirTable[dir][cell];
  }
  return hilbert;
}
```

---

[39] One can also compute the Hilbert keys directly during the descent in the tree. Then, one has to take into account how the local ordering of the Hilbert curve depends on the current position. The fast computation of Hilbert keys is also used in computer graphics and coding theory, see for instance [131].

For clarity, this routine is presented for the two-dimensional case. To extend it to the three-dimensional case, one just has to exchange the tables `DirTable` and `HilbertTable` with those from code fragment 8.6.

Code fragment 8.6 Tables for Algorithm 8.19 for DIM=3.

```
const unsigned char DirTable[12][8] =
 { { 8,10, 3, 3, 4, 5, 4, 5}, { 2, 2,11, 9, 4, 5, 4, 5},
   { 7, 6, 7, 6, 8,10, 1, 1}, { 7, 6, 7, 6, 0, 0,11, 9},
   { 0, 8, 1,11, 6, 8, 6,11}, {10, 0, 9, 1,10, 7, 9, 7},
   {10, 4, 9, 4,10, 2, 9, 3}, { 5, 8, 5,11, 2, 8, 3,11},
   { 4, 9, 0, 0, 7, 9, 2, 2}, { 1, 1, 8, 5, 3, 3, 8, 6},
   {11, 5, 0, 0,11, 6, 2, 2}, { 1, 1, 4,10, 3, 3, 7,10} };
const unsigned char HilbertTable[12][8] =
 { {0,7,3,4,1,6,2,5}, {4,3,7,0,5,2,6,1}, {6,1,5,2,7,0,4,3},
   {2,5,1,6,3,4,0,7}, {0,1,7,6,3,2,4,5}, {6,7,1,0,5,4,2,3},
   {2,3,5,4,1,0,6,7}, {4,5,3,2,7,6,0,1}, {0,3,1,2,7,4,6,5},
   {2,1,3,0,5,6,4,7}, {4,7,5,6,3,0,2,1}, {6,5,7,4,1,2,0,3} };
```

The function `Lebesgue2Hilbert` works level by level, starting at the coarsest level which corresponds to the root of the tree. On each level, the variable `cell` is set to the number of the cell into which the algorithm will descend, numbered in the Lebesgue order. The variable `dir` keeps track which of the four possible refinement orders is used for the refinement, as shown in Figure 8.26.[40]

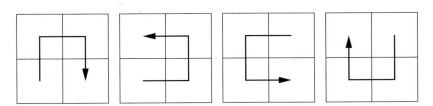

Fig. 8.26. Refinement orders 0 to 3 for the two-dimensional Hilbert curve.

From the variable `dir` of the refinement order and the cell `cell` one can now, by means of table `HilbertTable`, determine the cell number in the Hilbert ordering into which one descends. This table thus describes which cell in the Lebesgue order corresponds to which cell in the Hilbert order with respect to the refinement order `dir`. One obtains for instance for `dir=0` the mapping $0 \mapsto 0$, $1 \mapsto 3$, $2 \mapsto 1$, $3 \mapsto 2$, which leads to the

---

[40] There are of course more than four different orderings but the others are not needed here.

entry {0,3,1,2} in `HilbertTable`, compare also Figure 8.27. The number `HilbertTable[dir][cell]` constructed in this way is appended at the end of the current Hilbert key. The refinement order for the next refinement step is then determined using the table `DirTable`. This approach is repeated until all levels have been processed. The 4×4 tables of the two-dimensional case have to be replaced by corresponding 12×8 tables in the three-dimensional case, see code fragment 8.6.

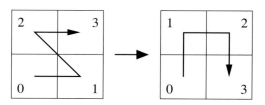

Fig. 8.27. Mapping of the Lebesgue order to the Hilbert order for refinement order 0. The cells $\{0, 1, 2, 3\}$ are mapped to the cells $\{0, 3, 1, 2\}$.

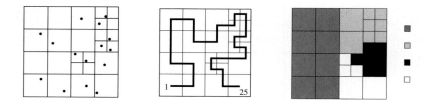

Fig. 8.28. Mapping of the cells to the processes. The Hilbert curve is split into segments according to the **range** values. The domain decomposition created by this splitting is shown at the right.

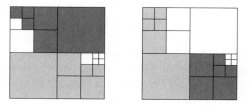

Fig. 8.29. Domain decomposition using the Lebesgue curve (left, disconnected subdomains) and the Hilbert curve (right, connected subdomains).

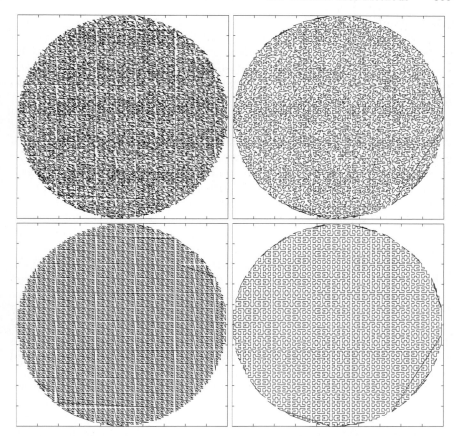

Fig. 8.30. Particle ordering induced by the domain key (left) and the Hilbert key (right) for 65536 randomly (top) and uniformly (bottom) distributed particles in a circle in two dimensions. One clearly sees the improved locality of the Hilbert curve as compared to the Lebesgue curve.

To apply the space-filling Hilbert curve in our program, we have to insert the transformation to the Hilbert key at any place where the number of the process is computed from the domain key of a cell by `key2proc`. In addition, in the load balancing, we have to traverse all particles in order of increasing Hilbert keys. Here, it is no longer enough to just use a post-order tree traversal of all leaf nodes of the tree. Instead, we have to test in the recursion which son node possesses the lowest Hilbert key. For this, we compute the Hilbert keys of all (four or eight) son nodes, sort them, and then descend in the tree in the correct order.

8.4.4 Applications

Our parallelized program has two advantages. First, it results in substantially faster running times than the sequential version. Second, we can now use the larger distributed main memory of the parallel computer, which allows us to run substantially larger application problems.

Collision of Two Spiral Galaxies. We consider the collision of two spiral galaxies. Numerical simulations can here recreate the specific structures of interacting systems that can be observed in space and give hints on how they evolved.[41] To obtain the initial configuration we put two spheres with a generic radius of one into our simulation box. The stars are distributed randomly with a constant density inside the spheres. Each galaxy contains 500000 particles that interact by the gravitational potential (8.19). All stars have, for simplicity, the same mass and we set $m_i = 1/N$, which normalizes the total mass of the system to one. The initial velocities are chosen so that each sphere rotates like a rigid body around an axis through the center of the sphere which is also parallel to the x_3 axis. The velocities are again chosen so that the centrifugal force is approximately equal to the gravitational force (Kepler orbits). In addition we move the right upper sphere with a velocity of -0.1 in the x_1 direction and move the left lower sphere with a velocity of 0.1 in the x_1 direction. Without gravitation the two systems would just pass each other. With gravitation the systems attract each other, rotate around each other, and finally merge.

Figure 8.31 shows the formation of two spiral galaxies and their collision at different times. We have again chosen $\delta t = 0.001$ as time step for the time integration. We use the value 0.6 for θ. In the upper left, we see the initial configuration consisting of two homogeneous spheres. As in the experiment in Section 8.3.3, the homogeneous spheres turn into spiral structures with two spiral arms each. The two spiral galaxies move towards each other because of gravitation and start to rotate around each other.[42] The high intrinsic rotation leads to the formation of two large spiral arms. Finally, the two galactic cores merge and a larger spiral galaxy with two arms is formed. Further results of simulations of the collision of galaxies can be found on Barnes' website [19].

In Figure 8.32 we see the distribution[43] of the domain cells that is created by the partitioning of the particles using a Hilbert curve. The cells assigned to

---

[41] Barnes [19] explains the approach as follows: "The rules of this game are to build models of isolated galaxies, place them on approaching orbits, and evolve the system until it matches the observations; if the model fails to match the observations, one adjusts the initial conditions and tries again."

[42] This rotation would not occur if the two galaxies would collide head-on.

[43] We show essentially the cells of the leaf nodes of the tree. In the graphical representation we have limited the depth of the tree to 6 levels and thereby fixed the size of the smallest cells.

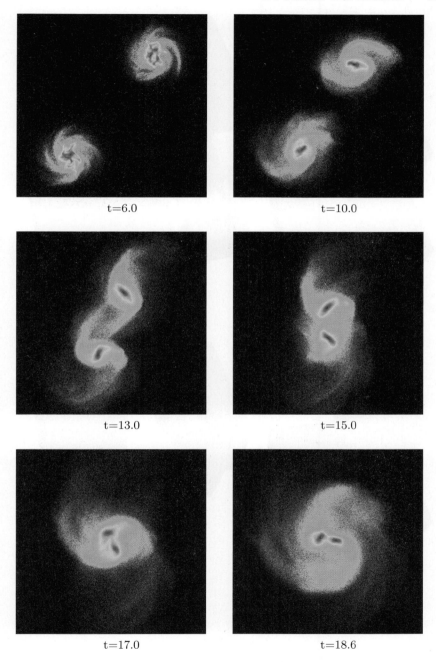

t=6.0

t=10.0

t=13.0

t=15.0

t=17.0

t=18.6

Fig. 8.31. Collision of two spiral galaxies, time evolution of the particle distribution, view parallel to the common axis of rotation. The particle densities are shown as color-map.

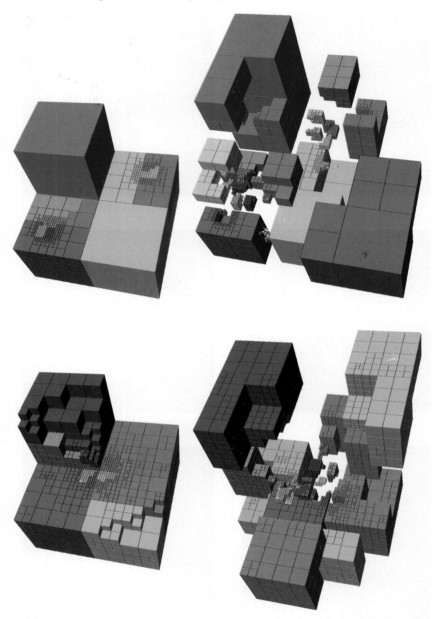

Fig. 8.32. Tree adaptivity and color-mapped data distribution. Section of the distribution (left) where some cells were removed for better visibility. Exploded view of the entire distribution (right). Collision of two spiral galaxies: Two galaxies (top), merged single galaxy (bottom) (at two different points of time in the simulation.)

a single processor are all shown in the same color. The distribution of the data is dynamical, it changes over time and follows the particles of the colliding galaxies. The load balance is maintained by rebalancing via the dynamical partitioning with a Hilbert curve.

Speedup and Parallel Efficiency. Table 8.1 and Figure 8.33 show the speedup and the parallel efficiency for the computation for one time step of the entire Barnes-Hut method for 32768, 262144, and 2097152 particles with up to 256 processors of a Cray T3E-1200. Here, the particles were uniformly distributed within the simulation domain. θ was set again to a value of 0.6. For the case of 2097152 particles, the main memory of a single processor is not sufficiently large to hold the data. We therefore use the definition $S(P) = 4 \cdot T(4)/T(P)$ for the speedup and thus normalize to 4 processors.

	32768 particles		262144 particles		2097152 particles	
proc.	speedup	efficiency	speedup	efficiency	speedup	efficiency
1	1.00	1.000	1.00	1.000		
2	1.96	0.981	1.99	0.998		
4	3.84	0.960	3.92	0.980	4.00	1.000
8	7.43	0.928	7.83	0.978	7.98	0.998
16	12.25	0.765	14.68	0.917	16.02	1.000
32	18.15	0.567	26.08	0.815	31.89	0.997
64	26.87	0.420	45.71	0.714	61.49	0.961
128	34.36	0.268	70.93	0.554	104.41	0.816
256	35.21	0.138	112.65	0.440	194.96	0.762

Table 8.1. Speedup and parallel efficiency of the Barnes-Hut method.

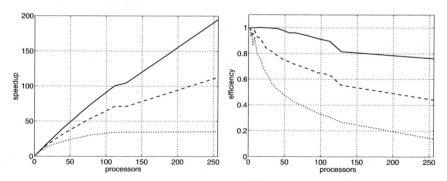

Fig. 8.33. Speedup and parallel efficiency of the parallel Barnes-Hut method on a Cray T3E-1200. Dotted line: 32768 particles, dashed line: 262144 particles, solid line: 2097152 particles.

For the smaller number of 32768 particles we observe a saturation starting at around 16 processors. With increasing number of processors, the processors are assigned less and less computations but the communication load remains approximately constant. For the case of 262144 particles we see good results up to approximately 64 processors, before saturation occurs. For the largest case of 2097152 particles good speedups and efficiencies were achieved for up to 256 processors. We observe an almost linear behavior of the speedups.[44] Analogous results are obtained in the case of colliding galaxies due to our dynamical load balancing.

8.5 Methods of Higher Order

In the previous sections we considered the Barnes-Hut method in detail. It is an example for an approximation (8.11) of the potential using a Taylor expansion of the kernel function up to degree $p = 0$. In the following, we now discuss methods of higher order. There, the series expansion (8.6) is not truncated after the first term but higher order terms are included into the computation.[45] Still, the algorithmic approach of the Barnes-Hut method can be used. Also the parallelization is similar. Essentially only the force computation routine in Algorithm 8.7 and the computation of the pseudoparticles in Algorithm 8.5 have to be changed. There, the appropriate equation with higher degree p

$$\Phi(\mathbf{x}) \approx \int_{\Omega^{\text{near}}} \rho(\mathbf{y})G(\mathbf{x},\mathbf{y})d\mathbf{y} + \sum_{\nu} \sum_{\|\mathbf{j}\|_1 \leq p} \frac{1}{\mathbf{j}!} M_{\mathbf{j}}(\Omega_\nu^{\text{far}}, \mathbf{y}_0^\nu) G_{\mathbf{0},\mathbf{j}}(\mathbf{x}, \mathbf{y}_0^\nu) \quad (8.24)$$

has to be used for the force computation instead of (8.11) with $p = 0$. Here, the local moments are again given by $M_{\mathbf{j}}(\Omega_\nu^{\text{far}}, \mathbf{y}_0^\nu) = \int_{\Omega_\nu^{\text{far}}} \rho(\mathbf{y})(\mathbf{y} - \mathbf{y}_0^\nu)^{\mathbf{j}} d\mathbf{y}$ from (8.11).

In the Barnes-Hut method we just precomputed the zeroth moments and stored them in the inner nodes of the tree, i.e. in the pseudoparticles. Now, we also have to determine the higher moments $M_{\mathbf{j}}(\Omega_\nu^{\text{far}}, \mathbf{y}_0^\nu)$ with $\|\mathbf{j}\|_1 \leq p$. To use the tree approach also for the higher moments, we again need appropriate transfer operations according to (8.17) and (8.18) that allow to compute the moments of a node from the moments of its son nodes. For the Barnes-Hut method this was simple, we just summed the masses to compute the values for the pseudoparticles. For higher values of p, additional loops over p are necessary.

---

[44] The kink at 128 processors in Figure 8.33 is probably caused by the memory architecture of the parallel machine. We suspect cache effects.

[45] Also other expansions can be used instead of the Taylor series. The modular version of the method presented here is independent of the specific series expansion, it can be employed for the Taylor series as well as for expansions into spherical harmonics which were used in the original fast multipole method.

As in the Barnes-Hut method we choose the position of the pseudoparticles as the center of mass

$$\mathbf{y}_0^\nu := \frac{\int_{\Omega_\nu^{\mathrm{far}}} \rho(\mathbf{z})\mathbf{z}d\mathbf{z}}{\int_{\Omega_\nu^{\mathrm{far}}} \rho(\mathbf{z})d\mathbf{z}},$$

compare also (8.20). Then, it holds for the first moments $M_{\mathbf{j}}(\Omega_\nu^{\mathrm{far}}, \mathbf{y}_0^\nu)$ with $\|\mathbf{j}\|_1 = 1$ that

$$M_{\mathbf{j}}(\Omega_\nu^{\mathrm{far}}, \mathbf{y}_0^\nu) = \int_{\Omega_\nu^{\mathrm{far}}} \rho(\mathbf{y}) \left(\mathbf{y} - \frac{\int_{\Omega_\nu^{\mathrm{far}}} \rho(\mathbf{z})\mathbf{z}d\mathbf{z}}{\int_{\Omega_\nu^{\mathrm{far}}} \rho(\mathbf{z})d\mathbf{z}} \right)^{\mathbf{j}} d\mathbf{y}$$

$$= \int_{\Omega_\nu^{\mathrm{far}}} \rho(\mathbf{y})\mathbf{y}^{\mathbf{j}} d\mathbf{y} - \frac{\int_{\Omega_\nu^{\mathrm{far}}} \rho(\mathbf{y})d\mathbf{y} \int_{\Omega_\nu^{\mathrm{far}}} \rho(\mathbf{z})\mathbf{z}^{\mathbf{j}} d\mathbf{z}}{\int_{\Omega_\nu^{\mathrm{far}}} \rho(\mathbf{z})d\mathbf{z}}$$

$$= \int_{\Omega_\nu^{\mathrm{far}}} \rho(\mathbf{y})\mathbf{y}^{\mathbf{j}} d\mathbf{y} - \int_{\Omega_\nu^{\mathrm{far}}} \rho(\mathbf{z})\mathbf{z}^{\mathbf{j}} d\mathbf{z} = 0. \qquad (8.25)$$

Thus, the first moments, the so-called dipole moments, vanish. This shows that the original Barnes-Hut method is already of order two.[46]

The overall algorithm again splits into three parts: Building the tree, computing the moments, and computing the force. Algorithms to build the tree have already been discussed and can be reused without changes. The two other parts of the algorithm now have to be extended to the higher order terms.

8.5.1 Implementation

To modify our implementation of the Barnes-Hut method to the case $p > 0$, we have to extend the routine for the computation of the pseudoparticles by the computation of the higher moments. In the routine for the force computation we have to take the additional terms of the Taylor series into account. First, we have to be able to store the higher moments. To this end, we extend the data structure `Particle` by a field `moments`. There, we retain the values of the moments with respect to the monomials, ordered for instance as follows:

$$\{ 1,$$
$$x_1, x_2, x_3,$$
$$x_1^2, x_1 x_2, x_1 x_3, x_2^2, x_2 x_3, x_3^2,$$
$$x_1^3, x_1^2 x_2, x_1^2 x_3, x_1 x_2^2, x_1 x_2 x_3, x_1 x_3^2, x_2^3, x_2^2 x_3, x_2 x_3^2, x_3^3,$$
$$\dots \}.$$

---

[46] We also could have chosen the center of the cell as the position of the pseudoparticle. The position then would not have to be explicitly saved because it can be determined easily. However, the dipole moments then no longer vanish in general and have to be computed and stored.

To this end, we define the constant DEGREE for the polynomial degree p of the expansion of the potentials and the constant MOMENTS for the number of the associated coefficients in the code fragment 8.7.[47]

Code fragment 8.7 Definition of Higher Moments

```
#define DEGREE 2
#if DIM==2
#define MOMENTS (((DEGREE+1)*(DEGREE+2))/2)
#else
#define MOMENTS (((DEGREE+1)*(DEGREE+2)*(DEGREE+3))/6)
#endif

typedef struct {
  ...
  real moments[MOMENTS];
} Particle;
```

The moments can now be obtained in different ways: First, one could compute the moments directly for each cell in the tree which represents a cluster of particles. To this end, one would sum over all particles belonging to the cell, in a coordinate system which has its origin at the pseudoparticle. For each cell, the total complexity of this operation is proportional to the number of particles contained in the cell. With N particles approximately equidistributed in the entire domain, one would obtain a balanced tree and one would need $\mathcal{O}(N \log N)$ computational operations to compute all moments. With respect to N, the computation of the moments has therefore the same complexity as the force computation. Note that the constant in front of the $N \log N$-term depends on p and its size plays an important role in practice.

However, the moments can also be obtained more efficiently in $\mathcal{O}(N)$ operations by computing the moments of a cell recursively from the moments of its son cells. To this end, we translate the moments from the coordinate system of the son cells to the coordinate system of the father cell and then just add these vectors.

To translate a monomial, we can use the binomial formula

$$(x - a)^p = \sum_{i=0}^{p} \binom{p}{i} x^i a^{p-i}.$$

---

[47] If we still store the total mass in mass, instead of in moments[0], and if we put the pseudoparticles at the center of mass, then the dipole moments vanish and we can directly start with the storage of the moments with quadratic terms, i.e. the quadrupole moments. The vector moments is then DIM+1 entries shorter.

The necessary binomial coefficients $\binom{p}{i}$ can be computed beforehand via Pascal's triangle and can be stored in a table. This is implemented in Algorithm 8.20.

Algorithm 8.20 Tabulating the Binomial Coefficients

```
int binomial[DEGREE+1][DEGREE+1];

void compBinomial() {
  for (int i=0; i<=DEGREE; i++) {
    binomial[0][i] = 1;
    binomial[i][i] = 1;
    for (int j=1; j<i; j++)
      binomial[j][i] = binomial[j-1][i-1] + binomial[j][i-1];
  }
}
```

We first consider how to re-expand polynomials of several variables in a coordinate system with a different origin. This can be implemented by re-expanding the polynomial along each coordinate direction, one direction after the other. Here, we use the binomial formula for each single monomial. Since it takes a complexity proportional to p to re-expand each monomial, and since there are $\mathcal{O}(p^3)$ polynomials up to degree p, the overall re-expansion has a total complexity of $\mathcal{O}(p^4)$ computational operations.[48]

Admittedly, we do not have to transform the polynomials but the moments. To this end, we can again use the binomial formula. Here, we do not consider each monomial in the source coordinate system and transform it to the target coordinate system, but instead we expand the monomials in the target coordinate system and match them with the source polynomial. Thus, if a term x^i occurs on the right side of the binomial formula, the coefficient $\binom{p}{i}$ "mediates" between the old moment associated to x^i and the new moment associated to x^p. The transformation of the polynomials can be interpreted as a linear mapping. The transformation of the moments then corresponds to the adjoint of that linear mapping. With this interpretation we can also translate the moments in a complexity of $\mathcal{O}(p^4)$.[49]

Algorithm 8.21 gives a routine that shifts the vector moments of moments along the x_1 axis by a and stores the result in m. To implement arbitrary translations in three dimensions, we also need the corresponding shifts along the x_2 and x_3 axis. For simplicity we have used the moment vectors as three-dimensional arrays in code fragment 8.8 to be able to express derivatives

---

[48] Other transformations such as rotations are more expensive. A general linear mapping would require $\mathcal{O}(p^6)$ operations.

[49] In the case $p = 0$, the adjoint mapping is again (the multiplication with) the mass. The translation does not change the mass.

with respect to the three coordinate directions.[50] These indices still have to be mapped to the linearly addressed vector of moments.

Code fragment 8.8 Indexing Moments by Numbers of Directional Derivatives

```
int dm[DEGREE+1][DEGREE+1][DEGREE+1] =        //  for DIM=3
   {{{ 0, 3, 9},{ 2, 8,-1},{ 7,-1,-1}},        //  and DEGREE=2
    {{ 1, 6,-1},{ 5,-1,-1},{-1,-1,-1}},
    {{ 4,-1,-1},{-1,-1,-1},{-1,-1,-1}}};
```

Algorithm 8.21 Shifting Moments of a Three-Dimensional Taylor Series by a in x_1 Direction

```
void shiftMoments_x0(real* moments, real a, real* m) {
  for (int j=0; j<=DEGREE; j++)
    for (int k=0; k<=DEGREE-j; k++) {
      for (int i=0; i<=DEGREE-j-k; i++)
        m[dm[i][j][k]] = 0;
      for (int i=0; i<=DEGREE-j-k; i++) {
        real s = moments[dm[i][j][k]];
        for (int l=i; l<=DEGREE-j-k; l++) {
          m[dm[l][j][k]] += s * binomial[i][l];
          s *= a;
        }
      }
    }
}
```

To correctly start the recursion for the computation of the moments of the pseudoparticles, we need the moments of the particles in their original coordinate systems. The zeroth moment of a particle is the mass of the particle. All other moments are set to zero. The moments of the son nodes are shifted and added to compute the moments for each pseudoparticle. The corresponding algorithm is given in 8.22.

Now, we are only missing a routine for the computation of the forces. For each particle we compute an approximation of the force acting on that particle. Here, the recursive descent in the tree is still controlled by the geometric θ criterion. But now, the values of the higher moments have to be taken into account in the actual computation of the forces. The force acting on a particle at position $\mathbf{x}$ is as always given as the negative gradient of the potential $\mathbf{F}(\mathbf{x}) = -\nabla\Phi(\mathbf{x})$. Using the Taylor expansion of the potential as in

---

[50] Here, unused indices are set to -1 for clarity.

Algorithm 8.22 Computing Pseudoparticles and Higher Moments (replaces `compPseudoParticles` from Algorithm 8.5)

```
void compMoments(TreeNode *t) {
  called recursively as in Algorithm 8.1;
  // start of the operation on *t
  for (int i=0; i<MOMENTS; i++)
    t->p.moments[i] = 0;
  if (*t is a leaf node)
    t->p.moments[0] = p.m;
  else {
    determine the coordinates of the pseudoparticle t->p.x;
    for (int j=0; j<POWDIM; j++)
      if (t->son[j] != NULL)
        t->p.moments +=
        shift moments t->son[j]->p.moments
          by (t->p.x - t->son[j]->p.x) using Algorithm 8.21;
  }
  // end of the operation on *t
}
```

Algorithm 8.23 Computing the Force Between Particles and Pseudoparticles up to Degree p=DEGREE

```
void force(Particle *p, Particle *q) { // particle p, pseudoparticle q
  for (int i=0; i<=DEGREE; i++)
    for (int j=0; j<=DEGREE-i; j++)
      for (int k=0; k<=DEGREE-i-j; k++) {
        real tmp = fact[i] * fact[j] * fact[k] *
                   p->m * q->moments[dm[i][j][k]];
        p->F[0] -= tmp * PotentialDeriv(p->x, q->x, i+1, j  , k  );
        p->F[1] -= tmp * PotentialDeriv(p->x, q->x, i  , j+1, k  );
        p->F[2] -= tmp * PotentialDeriv(p->x, q->x, i  , j  , k+1);
      }
}

real PotentialDeriv(real xp[3], real xq[3], int d1, int d2, int d3) {

  return (G_{(0,0,0),(d1,d2,d3)}(xp, xq));

}
```

(8.11), one has to evaluate a sum of moments and derivatives of the kernel G in the far field. We have already computed the moments for all pseudoparticles in the routine `compPseudoParticles`. The force can now be computed as before using the routines `compF_BH` and `force_tree`. The only change required is in the evaluation of the force between particles and pseudoparticles in routine `force` to also take higher moments into account now. This is implemented in Algorithm 8.23. Here, an array `fact[DEGREE+1]` is used to store the precomputed factorials. It has to be initialized appropriately beforehand.

This approach works for arbitrary kernels G that satisfy the properties from Section 8.1. The evaluation of the different derivatives using `PotentialDeriv` is generally quite expensive. But for a fixed, *given* G, as for instance for the gravitational potential, and for a fixed degree `DEGREE`, the force can be written out as an explicit formula and can thus be directly coded. This can speed up the force evaluation substantially.

Finally, let us consider the quality of the results of the Barnes-Hut method for different values of p. In Figure 8.34 we plot the resulting relative error of the potential over θ for a case with a very inhomogeneous distribution of particles as it occurs in the simulation of the collision of two spiral galaxies from Section 8.4.4 at time $t = 17$.

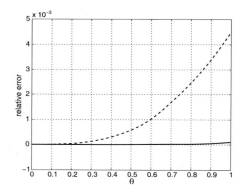

Fig. 8.34. Dependence of the relative error on θ for the Barnes-Hut method with $p = 0$ (dashed line) and $p = 2$ (solid line) for the simulation of the formation of a spiral galaxy.

In the case $p = 0$ we obtain a method of second order because of (8.25). In the case $p = 2$ we obtain a method of third order. One clearly sees the substantially improved error behavior for $p = 2$ even for relatively large values of θ. Using a method of higher order thus allows the choice of substantially larger values for θ, which significantly decreases the computing time.

8.5.2 Parallelization

Only a few changes are necessary in the parallelization of the Barnes-Hut method of higher order as compared to the parallelization of the original method. Note here that our simple approach with the summation over the masses with `Allreduce` from the original method only works for the computation of the moments of the pseudoparticles if the moments are already correctly translated. This results however in additional costs in the computations. Instead, it is more efficient to use `Allgather` to exchange the moments and coordinates of the `domainList` nodes and to translate the moments locally for each process separately before the summation. The computation of the forces does not change. When the (pseudo-)particles are communicated, their moments now have to be communicated as well. They are then used in the local computation of the force. The tree selection criterion involving θ has not been changed by the higher order of approximation, so that the parallel symbolic computation of the forces can be reused without changes as well.

8.6 Cluster-Cluster Interactions and the Fast Multipole Method

The method of Barnes and Hut, which has been discussed up to now, relies on the idea to substitute as many particle-particle interactions in the far field as possible by interactions between a particle and a pseudoparticle. This allows to reduce the complexity of the naive approach from $\mathcal{O}(N^2)$ to $\mathcal{O}(N \log N)$ if the particles are approximately uniformly distributed. In the following, we discuss an extension of the method that allows us to reach an optimal linear complexity, i.e. the order $\mathcal{O}(N)$.

The essential idea is the following: Instead of directly computing an approximation of the interaction for each single particle with far field pseudoparticles, we now take interactions between different pseudoparticles, the so-called cluster-cluster interactions,[51] into account. Their multiple use then further reduces the number of necessary computations in an approximate evaluation of the potential and forces.

8.6.1 Method

We start again from the expansion of the kernel function $G(\mathbf{x}, \mathbf{y})$ into a Taylor series as in Section 8.1. Now, G is not only expanded up to degree p in the variable $\mathbf{y}$ around a point $\mathbf{y}_0^\nu$, but also in the variable $\mathbf{x}$ around the point $\mathbf{x}_0^\mu$. We obtain

$$G(\mathbf{x}, \mathbf{y}) = \sum_{\|\mathbf{k}\|_1 \leq p} \sum_{\|\mathbf{j}\|_1 \leq p} \frac{1}{\mathbf{k}!\mathbf{j}!} (\mathbf{x} - \mathbf{x}_0^\mu)^{\mathbf{k}} (\mathbf{y} - \mathbf{y}_0^\nu)^{\mathbf{j}} G_{\mathbf{k},\mathbf{j}}(\mathbf{x}_0^\mu, \mathbf{y}_0^\nu) + \hat{R}_p(\mathbf{x}, \mathbf{y}). \quad (8.26)$$

The remainder term $\hat{R}_p(\mathbf{x}, \mathbf{y})$ will be studied more closely in the context of the error estimate in Section 8.6.3.

---

[51] The cells belonging to each pseudoparticle, together with the particles contained in them, are called clusters, see [289].

By substituting (8.26) into the integral representation of the potential (8.2), we obtain the interaction $\Phi_{\Omega_\nu}(\mathbf{x})$ of a subdomain $\Omega_\nu \subset \Omega$ with a point $\mathbf{x} \in \Omega$ as

$$
\begin{aligned}
\Phi_{\Omega_\nu}(\mathbf{x}) &= \int_{\Omega_\nu} G(\mathbf{x}, \mathbf{y}) \rho(\mathbf{y}) d\mathbf{y} \\
&\approx \int_{\Omega_\nu} \sum_{\|\mathbf{k}\|_1 \le p} \sum_{\|\mathbf{j}\|_1 \le p} \frac{1}{\mathbf{k}! \mathbf{j}!} G_{\mathbf{k},\mathbf{j}}(\mathbf{x}_0^\mu, \mathbf{y}_0^\nu)(\mathbf{x} - \mathbf{x}_0^\mu)^{\mathbf{k}} (\mathbf{y} - \mathbf{y}_0^\nu)^{\mathbf{j}} \rho(\mathbf{y}) d\mathbf{y} \\
&= \sum_{\|\mathbf{k}\|_1 \le p} \frac{1}{\mathbf{k}!} (\mathbf{x} - \mathbf{x}_0^\mu)^{\mathbf{k}} \sum_{\|\mathbf{j}\|_1 \le p} \frac{1}{\mathbf{j}!} G_{\mathbf{k},\mathbf{j}}(\mathbf{x}_0^\mu, \mathbf{y}_0^\nu) M_{\mathbf{j}}(\mathbf{y}_0^\nu, \Omega_\nu), \qquad (8.27)
\end{aligned}
$$

where the moments $M_{\mathbf{j}}(\mathbf{y}_0^\nu, \Omega_\nu)$ are given again by (8.12). The expansion of the kernel function G in $\mathbf{x}$ around $\mathbf{x}_0^\mu$ thus induces an expansion of $\Phi_{\Omega_\nu}(\mathbf{x})$ around $\mathbf{x}_0^\mu$, in which the *coefficients*

$$
\sum_{\|\mathbf{j}\|_1 \le p} \frac{1}{\mathbf{j}!} G_{\mathbf{k},\mathbf{j}}(\mathbf{x}_0^\mu, \mathbf{y}_0^\nu) M_{\mathbf{j}}(\mathbf{y}_0^\nu, \Omega_\nu)
$$

of this expansion are given by the interaction between Ω_μ and Ω_ν. This allows to convert the cluster-cluster interaction between Ω_μ and Ω_ν to $\mathbf{x}$, see also Figure 8.35. Thus, for all particles contained in the cell associated to Ω_μ, the interaction between Ω_μ and Ω_ν has to be computed only once and can then be used for all $\mathbf{x} \in \Omega_\mu$ after an appropriate conversion. In this way a certain amount of computational operations can be saved and the resulting method has a complexity of order $\mathcal{O}(N)$.

Fig. 8.35. Interaction of a cluster of particles around the center $\mathbf{x}_0^\mu$ with the particles from a distant cluster of particles around the center $\mathbf{y}_0^\nu$.

The multipole method applies this principle in a hierarchical fashion: Initially, interactions between large clusters are computed. These interactions are converted to the next smaller clusters, and interactions are computed on the next lower level of clusters. This approach is repeated while one descends down the tree. Here, on each level of the tree, each cluster inherits the interactions of its father cluster and interacts itself with other clusters. Finally, every particle (as a leaf node in the tree) receives the complete interaction with all other particles.

For this hierarchical approach we need an appropriate decomposition of the domain $\Omega \times \Omega$. We again start from the quadtree or octree decomposition of the domain Ω as described in Section 8.2. It therefore holds that

$$\Omega = \bigcup_{\nu \in I} \Omega_\nu, \tag{8.28}$$

where I is an index set for the occurring indices ν. Each cell Ω_ν represents a pseudoparticle, if it corresponds to an inner node of the tree, or a single particle, if it corresponds to a leaf node of the tree. Every Ω_ν is again assigned a "center" $\mathbf{x}_0^\nu = \mathbf{y}_0^\nu \in \Omega_\nu$.[52] Usually the center of mass is chosen for $\mathbf{x}_0^\nu$ and $\mathbf{y}_0^\nu$, respectively.[53]

From the decomposition of the domain Ω one immediately obtains a decomposition of $\Omega \times \Omega$:

$$\Omega \times \Omega = \bigcup_{(\mu, \nu) \in I \times I} \Omega_\mu \times \Omega_\nu. \tag{8.29}$$

The tree decompositions of Ω and $\Omega \times \Omega$ are not disjoint. To be able to cover the total interaction

$$\int_\Omega \int_\Omega G(\mathbf{x}, \mathbf{y}) \rho(\mathbf{x}) \rho(\mathbf{y}) d\mathbf{x} d\mathbf{y}$$

exactly once for later formulae, we select a subset $J \subset I \times I$ so that

$$\Omega \times \Omega = \bigcup_{(\mu, \nu) \in J} \Omega_\mu \times \Omega_\nu \quad \cup \quad \Omega_{\text{near}} \tag{8.30}$$

form a disjoint partitioning of the domain (up to the boundaries of the cells). For the error estimate in Section 8.6.3 one needs that each pair $(\mu, \nu) \in J$ satisfies the selection criterion (θ criterion)

$$\frac{\|\mathbf{x} - \mathbf{x}_0^\mu\|}{\|\mathbf{x}_0^\mu - \mathbf{y}_0^\nu\|} \leq \theta \quad \text{and} \quad \frac{\|\mathbf{y} - \mathbf{y}_0^\nu\|}{\|\mathbf{x}_0^\mu - \mathbf{y}_0^\nu\|} \leq \theta \quad \text{for all } \mathbf{x} \in \Omega_\mu, \mathbf{y} \in \Omega_\nu, \tag{8.31}$$

compare also (8.21). Since this criterion cannot be satisfied along the diagonal, a near field $\Omega_{\text{near}} \subset \Omega \times \Omega$ remains, which yet can be chosen so small that it does not substantially affect the computing time needed. The θ criterion of the multipole method corresponds to that of the Barnes-Hut method. The difference is that, for the treatment of cluster-cluster interactions, $\Omega \times \Omega$

---

[52] It is possible to choose different centers $\mathbf{x}_0^\nu \neq \mathbf{y}_0^\nu$ for the two sides of $\Omega \times \Omega$. In general, this does not lead to a better method. We will thus not pursue this variant here. Nevertheless, we will use the notation $\mathbf{x}_0^\nu$ for the left side and $\mathbf{y}_0^\nu$ for the right side for clarity, since they fit with the used variable names $\mathbf{x}$ and $\mathbf{y}$.

[53] In the original multipole method of Greengard and Rokhlin, the geometric center of the cell was chosen.

is now decomposed instead of Ω and the old θ criterion is to be fulfilled on both, the $\mathbf{x}$ side *and* the $\mathbf{y}$ side. An example of a decomposition satisfying the criterion for the one-dimensional case is shown in Figure 8.36. The algorithmic construction of such a decomposition is explained in more detail in Section 8.6.2. Again, as already discussed in Section 8.3.1 for the Barnes-Hut method, the selection criterion (8.31) can be replaced by a modified version.

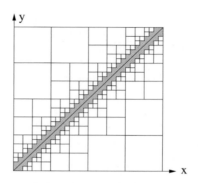

Fig. 8.36. Decomposition of $\Omega \times \Omega$ in one space dimension. The near field close to the diagonal is shaded in grey.

This hierarchical domain decomposition can now be used for the computation of the interactions. To this end, we consider the interaction of a point $\mathbf{x} \in \Omega$ with the whole domain Ω. It consists of the re-expanded interactions between all Ω_μ and Ω_ν with $\mathbf{x} \in \Omega_\mu$ and $(\mu, \nu) \in J$. We obtain

$$\Phi(\mathbf{x}) = \sum_{\substack{(\mu,\nu)\in J \\ \mathbf{x}\in\Omega_\mu}} \int_{\Omega_\nu} G(\mathbf{x}, \mathbf{y})\rho(\mathbf{y})d\mathbf{y}$$

$$\approx \sum_{\substack{(\mu,\nu)\in J \\ \mathbf{x}\in\Omega_\mu}} \sum_{\|\mathbf{k}\|_1 \leq p} \frac{1}{\mathbf{k}!}(\mathbf{x} - \mathbf{x}_0^\mu)^{\mathbf{k}} \sum_{\|\mathbf{j}\|_1 \leq p} \frac{1}{\mathbf{j}!} G_{\mathbf{k},\mathbf{j}}(\mathbf{x}_0^\mu, \mathbf{y}_0^\nu) M_{\mathbf{j}}(\mathbf{y}_0^\nu, \Omega_\nu)$$

$$=: \tilde{\Phi}(\mathbf{x}), \tag{8.32}$$

where $\tilde{\Phi}(\mathbf{x})$ denotes the approximation after the remainder term has been dropped. This sum can now be decomposed corresponding to the tree hierarchy of Ω. To this end, let μ_0 be the root of the tree, i.e. $\Omega_{\mu_0} = \Omega$. Furthermore, denote recursively μ_l as the son of μ_{l-1} which contains $\mathbf{x}$, until μ_L is finally a leaf node and Ω_{μ_L} just contains the particle $\mathbf{x}$. Therefore, L denotes the level of the tree on which the leaf node occurs that contains $\mathbf{x}$. It depends on the particle $\mathbf{x}$ and can be different for different particles. With this notation, one can rewrite (8.32) as

$$\tilde{\Phi}(\mathbf{x}) = \sum_{l=0}^{L} \sum_{\|\mathbf{k}\|_1 \le p} \frac{1}{\mathbf{k}!} (\mathbf{x} - \mathbf{x}_0^{\mu_l})^{\mathbf{k}} W_{l,\mathbf{k}}, \tag{8.33}$$

where the terms

$$W_{l,\mathbf{k}} := \sum_{\nu:(\mu_l,\nu)\in J} \sum_{\|\mathbf{j}\|_1 \le p} \frac{1}{\mathbf{j}!} G_{\mathbf{k},\mathbf{j}}(\mathbf{x}_0^{\mu_l}, \mathbf{y}_0^{\nu}) M_{\mathbf{j}}(\mathbf{y}_0^{\nu}, \Omega_\nu) \tag{8.34}$$

describe the direct interactions of Ω_{μ_l}. Using the recursively defined coefficients

$$K_{0,\mathbf{k}} := W_{0,\mathbf{k}}, \qquad K_{l,\mathbf{k}} := W_{l,\mathbf{k}} + \sum_{\substack{\|\mathbf{m}\|_1 \le p \\ \mathbf{m} \ge \mathbf{k}}} \frac{1}{(\mathbf{m}-\mathbf{k})!} (\mathbf{x}_0^{\mu_l} - \mathbf{x}_0^{\mu_{l-1}})^{\mathbf{m}-\mathbf{k}} K_{l-1,\mathbf{m}},$$

$$\tag{8.35}$$

one can rewrite (8.33) as follows:

$$\tilde{\Phi}(\mathbf{x}) = \sum_{l=0}^{L} \sum_{\|\mathbf{k}\|_1 \le p} \frac{1}{\mathbf{k}!} (\mathbf{x} - \mathbf{x}_0^{\mu_l})^{\mathbf{k}} W_{l,\mathbf{k}}$$

$$= \sum_{l=1}^{L} \sum_{\|\mathbf{k}\|_1 \le p} \frac{1}{\mathbf{k}!} (\mathbf{x} - \mathbf{x}_0^{\mu_l})^{\mathbf{k}} W_{l,\mathbf{k}} + \sum_{\|\mathbf{k}\|_1 \le p} \frac{1}{\mathbf{k}!} (\mathbf{x} - \mathbf{x}_0^{\mu_1} + \mathbf{x}_0^{\mu_1} - \mathbf{x}_0^{\mu_0})^{\mathbf{k}} K_{0,\mathbf{k}}$$

$$= \sum_{l=1}^{L} \ldots + \sum_{\|\mathbf{k}\|_1 \le p} \sum_{\substack{\|\mathbf{m}\|_1 \le p \\ \mathbf{m} \le \mathbf{k}}} \frac{1}{\mathbf{k}!} \binom{\mathbf{k}}{\mathbf{m}} (\mathbf{x} - \mathbf{x}_0^{\mu_1})^{\mathbf{m}} (\mathbf{x}_0^{\mu_1} - \mathbf{x}_0^{\mu_0})^{\mathbf{k}-\mathbf{m}} K_{0,\mathbf{k}}$$

$$= \sum_{l=1}^{L} \ldots + \sum_{\|\mathbf{k}\|_1 \le p} \frac{1}{\mathbf{k}!} (\mathbf{x} - \mathbf{x}_0^{\mu_1})^{\mathbf{k}} \sum_{\substack{\|\mathbf{m}\|_1 \le p \\ \mathbf{m} \ge \mathbf{k}}} \frac{1}{(\mathbf{m}-\mathbf{k})!} (\mathbf{x}_0^{\mu_1} - \mathbf{x}_0^{\mu_0})^{\mathbf{m}-\mathbf{k}} K_{0,\mathbf{m}}$$

$$= \sum_{l=2}^{L} \ldots + \sum_{\|\mathbf{k}\|_1 \le p} \frac{1}{\mathbf{k}!} (\mathbf{x} - \mathbf{x}_0^{\mu_1})^{\mathbf{k}}$$

$$\cdot \left(W_{1,\mathbf{k}} + \sum_{\substack{\|\mathbf{m}\|_1 \le p \\ \mathbf{m} \ge \mathbf{k}}} \frac{1}{(\mathbf{m}-\mathbf{k})!} (\mathbf{x}_0^{\mu_1} - x_0^{\mu_0})^{\mathbf{m}-\mathbf{k}} K_{0,\mathbf{m}} \right)$$

$$= \sum_{l=2}^{L} \ldots + \sum_{\|\mathbf{k}\|_1 \le p} \frac{1}{\mathbf{k}!} (\mathbf{x} - \mathbf{x}_0^{\mu_1})^{\mathbf{k}} K_{1,\mathbf{k}}$$

$$= \sum_{l=3}^{L} \ldots + \sum_{\|\mathbf{k}\|_1 \le p} \frac{1}{\mathbf{k}!} (\mathbf{x} - \mathbf{x}_0^{\mu_2})^{\mathbf{k}} K_{2,\mathbf{k}}$$

$$= \ldots$$

$$= \sum_{\|\mathbf{k}\|_1 \le p} \frac{1}{\mathbf{k}!} (\mathbf{x} - \mathbf{x}_0^{\mu_L})^{\mathbf{k}} K_{L,\mathbf{k}}. \tag{8.36}$$

This computation suggests an algorithm on each level l as follows: The interactions inherited from the father cluster μ_{l-1} are re-expanded from its center $\mathbf{x}_0^{\mu_{l-1}}$ to the respective center $\mathbf{x}_0^{\mu_l}$ of the son cluster. Then, the direct interactions $W_{l,\mathbf{k}}$ of the current level are added. This is also reflected in the definition of the coefficient $K_{l,\mathbf{k}}$ as the sum of the direct interactions $W_{l,\mathbf{k}}$ of the cluster and the re-expanded contributions of the father cluster. Starting from the root, all terms $K_{l,\mathbf{k}}$ are computed by descending in the tree until the leaf nodes are reached. These leaf nodes consist only of one particle $\mathbf{x}_0^{\mu_L}$, for which one obtains the interaction as

$$\tilde{\Phi}(\mathbf{x}_0^{\mu_L}) = K_{L,\mathbf{0}}. \tag{8.37}$$

We now need the negative gradient of the potential to compute the force vector. For this we could substitute $G_{\mathbf{e}_d,\mathbf{0}}$ for G in the computation above and repeat the computation for each component.[54] However, it can be shown that the dth component of the force is given by

$$-\frac{\partial}{\partial(\mathbf{x})_d}\Phi(\mathbf{x}_0^{\mu_L}) = K_{L,\mathbf{e}_d}, \tag{8.38}$$

if one substitutes the potential for G. But this is just equivalent to a series expansion of $G_{\mathbf{e}_d,\mathbf{0}}$ up to degree $p-1$ instead of p. Therefore, the computation has to be executed only once for the potential G and we obtain at the same time an approximation of the potential up to degree p and an approximation of the force up to degree $p-1$.

8.6.2 Implementation

We use the higher-order particle-cluster method from Section 8.5 as a starting point for the implementation of the cluster-cluster algorithm. Since the two methods just differ in the computation of the interactions, i.e. in the computation of the potential and the forces, only the implementation of the computation of the interactions is discussed in detail.

The arguments in the previous Section 8.6.1 lead to the following algorithmic approach for the computation of the potential and the force: First, the moments $M_{\mathbf{j}}$ are computed in a post-order traversal (that means ascending from the leaf nodes to the root). Then, the coefficients $K_{l,\mathbf{k}}$ are determined in a pre-order traversal (that means descending from the root to the leaf nodes) and added to the interactions. We already described the recursive computation of the moments with a post-order traversal for the Barnes-Hut method, see Algorithm 8.22. Here, a new algorithmic element was given by the decomposition of $\Omega \times \Omega$ and the computation of the coefficients $K_{l,\mathbf{k}}$. Both tasks must now be realized in a common routine. We use for the implementation a function force_fmm with a node t as first argument and a list L of nodes as

---

[54] $\mathbf{e}_d$ here denotes the dth unit vector.

the second argument. The node t plays the role of the cluster Ω_μ, and the list L describes a set of clusters Ω_ν. The function force_fmm is now called with the root node as the first argument and a list just containing the root node as the second argument. The function tests for each cluster t2 from the list L, whether it satisfies the θ criterion (8.31) together with t. If that is the case, the interaction of the two clusters is computed, which corresponds to the addition of the appropriate part of the $W_{l,\mathbf{k}}$ term in (8.35). If that is not the case, one has to refine further. The refinement is symmetric on both sides, i.e., one looks at both the son nodes of t and the son nodes of t2. For this task, the appropriate son nodes of t2 are written into a list L2 and force_fmm is called recursively for each son node of t. Here, L contains just the set of clusters for which interactions have to be computed with t, either on the same or on a finer level. Figure 8.37 shows the so-called interaction sets $\{\Omega_\mu \times \Omega_\nu, (\mu, \nu) \in J\}$ for a two-dimensional example.

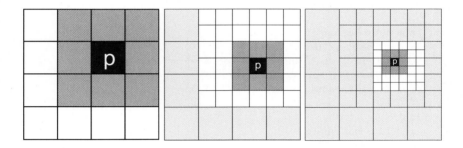

Fig. 8.37. Interaction sets (black×white) in the fast multipole method for a particle p. Interactions with the white cells are computed on this level of the tree, interactions with the light gray cells already have been computed, and interactions with the dark cells will be computed later.

If a leaf of the tree is reached on either side, one cannot refine anymore on this side, and one must descend in the tree asymmetrically. If leaf nodes are reached on both sides, the interaction is computed in any case, independently of the result of the θ criterion. This is shown in Figure 8.38. If t is a leaf node and if all interactions are processed, i.e. L2 is empty, the computation of the terms $K_{l,\mathbf{k}}$ is finished and the force is added to t->p.F according to (8.38).[55]

---

[55] There are variants of the method in which more than one particle (for instance a fixed maximal amount of particles) is stored in a leaf node of the tree, i.e., in the tree construction one no longer refines until all particles belong to different cells of the tree. Then, leaf nodes can also carry corresponding pseudoparticles. With an approximation of higher-order, the pseudoparticle of a leaf node can then be used for the interaction with the field of several particles at a larger distance.

Algorithm 8.24 Domain Decomposition and Computation of the Direct Interaction in the Multipole Method for DIM=3

```
void force_fmm(TreeNode *t, list L, real diam, real *K) {
  if (t==NULL) return;
  list L2;   // create empty list L2
  for all elements t2 of L {
      if (diam < theta * ( distance of t->p.x and t2->p.x))
         compInteract_fmm(&(t->p), &(t2->p), K);
      else
        if (t2 is not a leaf node)
          for (int i=0; i<POWDIM; i++)
            append t2->son[i] to list L2
        else
          if (t is not a leaf node)
            append t2 to list L2
          else
            compInteract_fmm(&(t->p), &(t2->p), K);
  }
  if (t is not a leaf node)
    for (int i=0; i<POWDIM; i++) {
      compute K2 as re-expansion of K from t->p.x
        to t->son[i]->p.x according to (8.35)
      force_fmm(t->son[i], L2, diam/2, K2);
    }
    else
      if (L2 is not empty)
        force_fmm(t, L2, diam/2, K);
      else {
        t->p.F[0]  -= K[dm[1][0][0]];
        t->p.F[1]  -= K[dm[0][1][0]];
        t->p.F[2]  -= K[dm[0][0][1]];
      }
}


void compInteract_fmm(Particle *p, Particle *p2, real *K) {
  for (int k=0; k<DEGREE; k++)
    for (int j=0; j<DEGREE-k; j++)
      for (int i=0; i<DEGREE-k-j; i++)
        for (int k2=0; k2<DEGREE; k2++)
          for (int j2=0; j2<DEGREE-k2; j2++)
            for (int i2=0; i2<DEGREE-k2-j2; i2++)
              K[dm[i][j][k]] += PotentialDeriv2(p->x, p2->x, i,
                  j, k, i2, j2, k2) * p2->moments[dm[i2][j2][k2]] *
                  fact[i2] * fact[j2] * fact[k2];
}
```

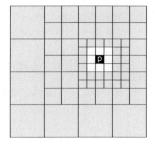

Fig. 8.38. Direct summation in the fast multipole method of particles in the white cells on the finest level of the tree.

The near field Ω_{near} no longer appears explicitly. The reason is that the algorithm refines the domain $\Omega \times \Omega$ until all particles away from the diagonal belong to the far field. The near field then only consists of the diagonal which is automatically treated by particle-particle interactions. We describe the function `force_fmm` with its case distinctions in Algorithm 8.24. The implementation of the lists is left to the reader. They could be implemented as linked lists, for instance. We do not store the terms K explicitly in the `Particle` structure but we pass them as arguments in the recursion. The decomposition of $\Omega \times \Omega$ is not stored explicitly either, but created by the recursion.

The function `force_fmm` uses the function `compInteract_fmm` for the computation of the direct interactions. This function in turn calls the function `PotentialDeriv2`, which computes the derivative of the potential G analogously to function `PotentialDeriv` in Algorithm 8.23, but now with respect to both of its arguments.[56] For specific potentials, such as for $G(\mathbf{x}, \mathbf{y}) = \frac{1}{\|\mathbf{y}-\mathbf{x}\|}$, and for fixed degree `DEGREE` of the expansion, one should exploit specific properties of the potential, as for instance radial symmetry, and optimize the routine `compInteract_fmm` accordingly. This allows to further reduce the complexity of the method with respect to p.

8.6.3 Error Estimate

The error estimate proceeds analogously to the error estimate for the Barnes-Hut method. The remainder term $\hat{R}_p(\mathbf{x}, \mathbf{y})$ of the expansion (8.26) is given by

$$\hat{R}_p(\mathbf{x}, \mathbf{y}) = \sum_{(\mathbf{k},\mathbf{j}) \in I_p} \frac{1}{\mathbf{k}!\mathbf{j}!}(\mathbf{x} - \mathbf{x}_0^\mu)^{\mathbf{k}}(\mathbf{y} - \mathbf{y}_0^\nu)^{\mathbf{j}} G_{\mathbf{k},\mathbf{j}}(\tilde{\mathbf{x}}_{\mathbf{k},\mathbf{j}}, \tilde{\mathbf{y}}_{\mathbf{k},\mathbf{j}}) \qquad (8.39)$$

with $I_p = \{(\mathbf{k}, \mathbf{j}) : \mathbf{k} = \mathbf{0}, \|\mathbf{j}\|_1 = p + 1 \text{ or } \|\mathbf{k}\|_1 = p + 1, \|\mathbf{j}\|_1 \leq p\}$. For all multi-indices $(\mathbf{k}, \mathbf{j}) \in I_p$, it holds that $\|\mathbf{k} + \mathbf{j}\|_1 \geq p + 1$. If $\mathbf{x}, \mathbf{x}_0^\mu \in \Omega_\mu$ and

---

[56] For the sake of efficiency, the different derivatives should be written out explicitly for a given G and implemented directly.

$\mathbf{y}, \mathbf{y}_0^\nu \in \Omega_\nu$, and if the cells Ω_μ and Ω_ν are convex,[57] then the points $\tilde{\mathbf{x}}_{\mathbf{k,j}}$ and $\tilde{\mathbf{y}}_{\mathbf{k,j}}$ at which the remainder term is evaluated, are also contained in Ω_μ and Ω_ν, respectively.

We assume again that G and ρ are positive functions and that G and its derivatives are equivalent to the $1/r$ potential and its derivatives up to multiplicative constants, compare (8.15). Then, it holds for all $\tilde{\mathbf{x}}, \tilde{\mathbf{y}}$ by virtue of the θ criterion (8.31) that

$$G_{\mathbf{k,j}}(\tilde{\mathbf{x}}, \tilde{\mathbf{y}}) \le c\|\tilde{\mathbf{x}} - \tilde{\mathbf{y}}\|^{-\|\mathbf{k+j}\|_1 - 1} \le c\big((1 + 2\theta)\|\mathbf{x}_0^\mu - \mathbf{y}_0^\nu\|\big)^{-\|\mathbf{k+j}\|_1 - 1},$$
$$G(\mathbf{x}, \mathbf{y}) \ge c\|\mathbf{x} - \mathbf{y}\|^{-1} \ge c(1 - 2\theta)\|\mathbf{x}_0^\mu - \mathbf{y}_0^\nu\|^{-1}. \tag{8.40}$$

Using the notation
$$\Omega_{\text{near}}^{\mathbf{x}} := \{\mathbf{y} : (\mathbf{x}, \mathbf{y}) \in \Omega_{\text{near}}\}, \tag{8.41}$$

the relative error of $\tilde{\Phi}(\mathbf{x})$ can then be estimated as follows:

$$\left|\frac{\Phi(\mathbf{x}) - \tilde{\Phi}(\mathbf{x})}{\Phi(\mathbf{x})}\right| \le \frac{\displaystyle\sum_{\substack{(\mu,\nu)\in J \\ \mathbf{x}\in\Omega_\mu}} \int_{\Omega_\nu} |\hat{R}_p(\mathbf{x}, \mathbf{y})|\rho(\mathbf{y})d\mathbf{y}}{\displaystyle\sum_{\substack{(\mu,\nu)\in J \\ \mathbf{x}\in\Omega_\mu}} \int_{\Omega_\nu} G(\mathbf{x}, \mathbf{y})\rho(y)d\mathbf{y} + \int_{\Omega_{\text{near}}^{\mathbf{x}}} G(\mathbf{x}, \mathbf{y})\rho(y)d\mathbf{y}}$$

$$\le c\frac{\displaystyle\sum_{(\mathbf{k,j})\in I_p} \sum_{\substack{(\mu,\nu)\in J \\ \mathbf{x}\in\Omega_\mu}} \int_{\Omega_\nu} \|\mathbf{x} - \mathbf{x}_0^\mu\|^{\|\mathbf{k}\|_1} \|\mathbf{y} - \mathbf{y}_0^\nu\|^{\|\mathbf{j}\|_1} \|\mathbf{x}_0^\mu - \mathbf{y}_0^\nu\|^{-\|\mathbf{k+j}\|_1 - 1}\rho(\mathbf{y})d\mathbf{y}}{\displaystyle\sum_{\substack{(\mu,\nu)\in J \\ \mathbf{x}\in\Omega_\mu}} \int_{\Omega_\nu} \|\mathbf{x}_0^\mu - \mathbf{y}_0^\nu\|^{-1}\rho(y)d\mathbf{y}}$$

$$\le c\frac{\displaystyle\sum_{(\mathbf{k,j})\in I_p} \sum_{\substack{(\mu,\nu)\in J \\ \mathbf{x}\in\Omega_\mu}} \int_{\Omega_\nu} \theta^{\|\mathbf{k+j}\|_1} \|\mathbf{x}_0^\mu - \mathbf{y}_0^\nu\|^{-1}\rho(\mathbf{y})d\mathbf{y}}{\displaystyle\sum_{\substack{(\mu,\nu)\in J \\ \mathbf{x}\in\Omega_\mu}} \int_{\Omega_\nu} \|\mathbf{x}_0^\mu - \mathbf{y}_0^\nu\|^{-1}\rho(y)d\mathbf{y}} \le c\theta^{p+1}. \tag{8.42}$$

The relative error for the multipole method is therefore of order $\mathcal{O}(\theta^{p+1})$ as well.

8.6.4 Parallelization

The parallelization of the Barnes-Hut method has already been described in Section 8.4. The multipole method differs only in the force computation. As in the parallel Barnes-Hut algorithm, it is done locally by each process after the domain decomposition. Again, one has to make sure that all needed data

---

[57] The cells in our decompositions into cubical subdomains are convex.

from other processes are available locally as a copy. To this end, we proceeded as follows for the Barnes-Hut method in Section 8.4.1: Each process checks which particles or pseudoparticles each other process will need in the force computation later. To test whether the θ criterion is satisfied, the distance between the pseudoparticle and the particle of the other process (whose exact position is only known to the other process) is estimated by the minimal distance of all cells which belong to this other process. These cells are given by the `domainList` nodes which are stored and thus known to all processes. It is possible that more data are sent in this way than necessary. However, in any case all the data possibly needed by the other processes will then be available there.

This approach might fail for the multipole method with an arbitrary domain decomposition that satisfies the θ criterion (8.31), since the distance $\|\mathbf{x} - \mathbf{x}_0^\mu\|$ for the criterion is not known locally. But the implementation suggested in Section 8.6.2 uses a *symmetric* criterion to decide whether to descend or not. Therefore, the distances $\|\mathbf{x} - \mathbf{x}_0^\mu\|$ and $\|\mathbf{y} - \mathbf{y}_0^\nu\|$ are symmetric and the θ criterion for the Barnes-Hut method is viable for the multipole method as well. The method only descends asymmetrically if one of the two sides is a leaf node. If the side which belongs to the other process (corresponding to Ω_μ) is a leaf node, the criterion is tested with the smaller cell diameter of the local pseudoparticle (as can be seen in Algorithm 8.24), and the method will succeed. If the side which belongs to the local process is a particle, the particle is communicated in any case and thus all necessary data is exchanged in this case as well. Therefore, the domain decomposition suggested in Section 8.6.2 combined with the symmetric criterion allows to reuse the parallelized version of the Barnes-Hut method without modification for the multipole method.

Further reading for parallel cluster-cluster algorithms and parallel fast multipole methods can be found in [196, 261, 454, 500, 501, 564, 680].

8.7 Comparisons and Outlook

In the previous section, we introduced the cluster-cluster method using an expansion into Taylor series. This way, one can achieve a complexity of order $\mathcal{O}(\theta^{-3}p^6 N)$ for uniformly distributed particles in the *three-dimensional* case.[58] The factor p^6 stems from the computation of the direct interactions

---

[58] For this complexity bound one assumes a uniform tree in which the particles are already stored in the leaf nodes. The construction of the tree and the sorting of the particles into the tree are neglected. In the case of uniformly distributed particles the sorting of the particles into the tree can be implemented by a bucket or radix sort in $\mathcal{O}(N)$ operations [357]. In the case of extremely non-uniformly distributed particles, both the cluster-cluster method and the fast multipole method can degenerate to a method of order $\mathcal{O}(N^2)$, see also [36].

between any two clusters. The computation of the moments and the conversion of the coefficients takes $\mathcal{O}(p^4 N)$ operations. Other expansions besides Taylor series can be used in the method as well. The original fast multipole method of Greengard and Rokhlin [38, 139, 260, 263, 264, 265, 525] employed spherical harmonics, which appear naturally in the treatment of the Coulomb potential and the gravitational potential. Then, since the kernel is radially symmetric, only $\mathcal{O}(p^2)$ moments are necessary, the computation of the direct interactions between two clusters costs therefore only $\mathcal{O}(p^4)$, and the total complexity of the method is $\mathcal{O}(\theta^{-3} p^4 N)$. Several implementations with slightly modified spherical harmonics are described in [206, 474, 650, 659]. Further explanations and related methods can be found in [66, 185, 262, 481]. Adaptive variants are discussed in [143, 454]. A generalization to more general kernel functions can be found in [599]. Greengard, Rokhlin and others have presented a new version of the fast multipole method in [207, 266, 329] which decreases the complexity to $\mathcal{O}(\theta^{-3} p^2 N)$ by using plane waves.[59] A related approach uses the technique of Chebyshev economization [398]. One can find an error estimate for the fast multipole method with a slightly improved constant in [478, 479]. Note that the multipole method as described in the previous section only works for the nonperiodic case. However, it can be generalized to the periodic case in a similar fashion as the Ewald method, see [84, 96, 370]. A comparison with the PME method can be found in [488, 574].

An independent direction of development started with the panel clustering method suggested by Hackbusch and Novak in [289] for boundary integral equations, see also [253, 290]. It uses the expansion into Taylor series. Closely related to this approach are the pseudoskeleton approximation and its generalizations [72, 256, 257, 368]. There, one uses special interpolation methods as well as low rank approximations which are constructed by singular value decompositions. Further developments along this line are the panel clustering method of variable order by Hackbusch and Sauter [424, 545, 600], in which lower values of p are used deep into the tree and higher values of p are used close to the root. This allows to further reduce the complexity while maintaining the same order in the error. In the best case one can achieve a complexity of order $\mathcal{O}(\theta^{-3} N)$, independent of p. The panel clustering approach has lead in the meantime to the theory of $\mathcal{H}$ and $\mathcal{H}^2$ matrices [101, 102, 287, 286, 288], which offers, among other things, a fast and efficient approximate matrix-vector product for a wide class of matrices and their inverses.

Finally, Brandt [118, 119, 121, 542] suggested approaches that are based directly on multigrid techniques. They can be interpreted as panel clustering

---

[59] The complexity of the method is actually $\mathcal{O}(\theta^{-3} p^3 N)$, see [266]. But if the tree is constructed in such a way that $s = 2p$ particles lie in each leaf node and these particles interact directly with each other and with particles from neighboring leaf nodes, one can eliminate the leading term in the complexity and obtains a method of complexity $\mathcal{O}(\theta^{-3} p^2 N)$.

methods or multipole methods that use Lagrange interpolation polynomials as expansion systems.

A fair comparison of the different methods is not easy. Which approach is faster depends on the number of particles, their distribution in the simulation domain, and finally on the particular implementation of the particular method. Already the difference in complexity between the two techniques presented in the two last sections helps to clarify this point: The Barnes-Hut method of higher order has a complexity of order $\mathcal{O}(\theta^{-3} p^4 N \log N)$, the cluster-cluster method has a complexity of $\mathcal{O}(\theta^{-3} p^6 N)$. Assuming the same constant, they have the same complexity for $\log N = p^2$, i.e. for $N = 8^{p^2}$. For larger values of p, the number N of the particles has to become extremely large, before the cluster-cluster method gives an advantage. Take as an example the case $p = 4$. Then, the number of particles has to be $N > 8^{16} = 281474976710656$ for the cluster-cluster method to be faster than the particle-cluster method. This shows the importance of the variants of the cluster-cluster method mentioned above. They allow to reduce the order of complexity in p. Similar approaches can also be used for the particle-cluster method to reduce its complexity in p. A comparison of the complexities of the parallelized versions of the different method is strongly machine-dependent and is even harder [94]. We will therefore refrain from any further discussion of the subject here.

9 Applications from Biochemistry and Biophysics

Genetic engineering and biotechnology have become a more and more important subject within the last decade. Therefore, we now want to give a perspective on the various problems from biochemistry and biophysics which can be treated and studied with the molecular dynamics methods that we described in this book. Applications include the general dynamics of proteins and their conformations, the formation of membrane structures, the determination of inhibitor-ligand binding energies, and the study of the folding and unfolding of peptides, proteins, and nucleic acids.

Such problems are at the front of current research. They are challenging and mostly require long running times on parallel computers. Additionally, the molecular dynamics techniques that we discussed up to now, often need to be adapted and modified to fit the specific problem at hand. Moreover, the data for the potential and the setup for the specific experiment mostly require specialist knowledge which is beyond the scope of this book. Therefore, we will not treat the following applications in depth discussing all the relevant details. However, we want to provide at least some insight into the various possibilities molecular dynamics methods have to offer in this area of research, to give readers further suggestions for their own activities and projects, and to direct readers to the appropriate literature for further studies.

Properties of biomolecules should be considered in aqueous solutions. Often the aqueous solution also contains additional salts. The presence and concentration of those salts can cause drastic changes in a biomolecule. Electrostatic effects of the molecule and the surrounding water result in long-range forces that need to be considered in the numerical simulations. However, even today many simulations still use only the linked cell method (compare Chapters 3 to 5) together with a slightly larger cutoff radius. In general, this approach is not appropriate (and will presumably lead to wrong results). Instead, techniques for long-range potentials like the SPME method from Chapter 7 or one of the tree methods from Chapter 8 need to be used.

However, due to the time complexity of the methods, only relatively fast physical processes can be studied by molecular dynamics simulations with current algorithms and computers. Nowadays, on large parallel computers, processes up to the microsecond range are within reach, see [192].

9.1 Bovine Pancreatic Trypsin Inhibitor

One of the biomolecules that have been studied most extensively with molecular dynamics methods is the trypsin inhibitor of the bovine pancreas, in short BPTI for bovine pancreatic trypsin inhibitor. It is a small, monomeric, spherical molecule consisting of 910 atoms and 58 amino acids, which can be found in the pancreas and the body tissues of cattle. It inhibits the action of trypsin on foodstuff. Trypsin is a molecule that breaks peptide bonds between amino acids in proteins. BPTI has been very well studied experimentally and there is a wealth of measured data. The first simulation of BPTI was published in 1977 [419]. Other early work can be found in [351, 352, 382, 635, 636, 664]. Since then, BPTI serves as a popular model for the test of new numerical methods. By now, it is a standard task in student lab courses to study the molecular dynamics of BPTI on a computer. This way, one can learn much about its structure and its stability under different conditions (such as temperature and solution). A simulation in vacuum already allows the analysis of the trajectory, the study of the movement of side chains and subdomains of the structure, as well as first comparisons with the results of x-ray diffraction and NMR (nuclear magnetic resonance) experiments.

The coordinates for BPTI are available from the Brookhaven protein data bank [85], for instance under the name 1bpti. The protein structures stored there often stem from a dehydrated state, since most proteins have to be dehydrated to allow analysis by, e.g. x-ray crystallography. Their structure is therefore given in the crystalline phase, see also Section 5.2.3. However, the geometry and form of the molecule in an aqueous solution can differ substantially from the dehydrated one. It is known that already the incorporation of a single water molecule can stiffen a protein [406, 665]. The method of molecular dynamics can now be used to relax such a protein molecule, after water molecules have been added, to simulate the resulting structure in aqueous solution at different temperatures, to study the protein's movements, and to compare the results with results from simulations in vacuum [162, 305].

For the numerical simulation of BPTI, we use the molecular coordinates from the Brookhaven protein data bank and add the missing hydrogen atoms according to chemical rules (using HyperChem [14]), see Section 5.2.3. The parameters for the bond, angle, and torsion potentials for the molecule can be obtained from CHARMM [125]. Then, we additionally place 5463 TIP3P-C water molecules, see Section 7.4.3, into our cubic simulation box with a size of 56.1041 Å and relax the entire system. Afterwards, we heat the system step-by-step up to room temperature (300 K). We finally let the BPTI molecule relax again in the aqueous solution using the NVE ensemble at a constant temperature for 40 ps. To this end, we employ our parallel SPME method from Chapter 7 coupled with the routines for the evaluation of the fixed bond potentials from Chapter 5.2.2 for the overall simulation. We use a time step of 0.1 fs in the Störmer-Verlet method, i.e., we propagate the system for a total of 400000 time steps. Figure 9.1 shows snapshots of the molecule in ball-stick

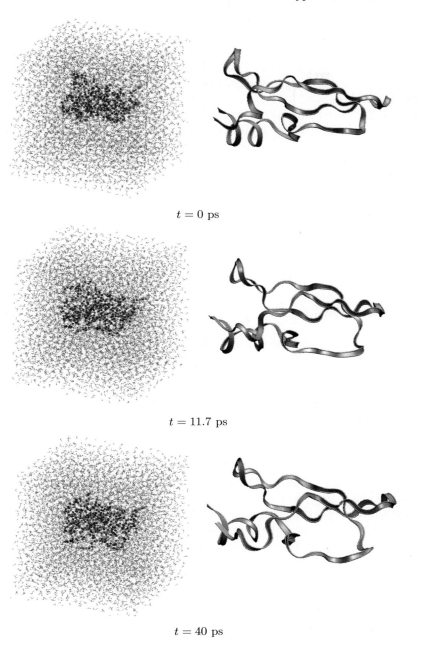

$t = 0$ ps

$t = 11.7$ ps

$t = 40$ ps

Fig. 9.1. Simulation of a BPTI molecule in aqueous solution, evolution over time. Ball-stick representation (left), ribbon representation without water molecules (right).

representation (left) and in ribbon representation (right), produced by VMD [332], at the times $t = 11.7$ ps and $t = 40$ ps. For clarity, the water molecules are shown in a smaller scale. It can be seen how the molecule relaxes under the influence of the water molecules and changes its form.

Further studies of BPTI with molecular dynamics methods can be found for instance in [523, 674]. Often a simplified united atom model and also the SHAKE method are employed. As already mentioned, mostly variants of the linked cell method or of Verlet's neighbor-list algorithm with somewhat larger cutoff radius are used to take electrostatics at least approximately into account [246]. However, this is in general not accurate enough. A comparison of the conventional cutoff method, the P^3M method and a variant of the multipole method is given in [561, 562]. There, a BPTI-water system with 23531 atoms was considered. Additionally, the differences between the electrostatics of the system for periodic and for nonperiodic experiments are discussed. Studies of the PME method and the Ewald summation technique are found in [215]. Furthermore, detailed studies of the structural stability and of the unfolding process of BPTI can be found in [128, 163, 547].

9.2 Membranes

Membranes are thin layers of molecules. They appear in a number of applications in material sciences, such as in graphite oxides [576], polymers [142], silicates [487], or zeolites [563]. They also occur in oil-water emulsions with amphiphilic molecules such as surfactants [443], lipids or detergents, where they form as primary structures together with spherical micelles and vesicles [250].

Biomembranes are of particular interest. They typically consist of a double layer of lipids into which different proteins are embedded. Such biomembranes are generally just a few nanometers thick. A cubic centimeter of biological tissues contains membranes with a surface area of approximately 10^5 cm^2. They include the plasma membrane, which delineates the cell, and a large number of intracellular membranes that enclose the nucleus, the mitochondria, the Golgi apparatus, and the organelles of the cell such as the endoplasmic reticulum, the lysosomes, the endosomes, and the peroxisomes. Membranes are therefore fundamental building blocks that give structure to biological material. In addition, many important receptor molecules are embedded in the lipid double layer of a membrane. Here, the lipid environment has an effect on the structure and the properties of such molecules. The permeation of smaller molecules (intermediate catabolic products, endogenous components such as peptides, active substances from pharmaceuticals) across the lipid double layer is also of great importance. Therefore, an understanding of the functionalities and the mechanisms of membranes is central to biochemistry and biomedicine. Here, simulations with molecular dynamics methods can make a relevant contribution, at least to some extent.

The first numerical simulations can be found in [362] for a monolayer membrane and in [630, 631] for a bilayer membrane made from decane chains. There, a simplified model for the alkanes was used, in which each methyl or methylene group is represented as one particle (united atom model) but realistic torsion potentials are employed between the groups. Another simple model goes back to [161]. It features two types of particles, i.e. oil-like and water-like particles. A tenside molecule is then modeled as a short chain of such particles connected by harmonic potentials. The two different types of particles interact by a Lennard-Jones potential. It turns out that, for a certain range of temperatures, this simple approach already leads to a demixing of the oil and water particles and to the formation of a stable boundary layer between the two fluid phases. Moreover, single or double membranes develop depending on the concentration of the tenside molecules. The formation of micelles and vesicles can also be observed. Further details can be found in [161, 213, 567].

With increasing computer power, more and more complex models could be used in the numerical simulations and, thus, the results of simulations became more and more accurate. By now, models are employed that take every single atom into account. The challenge was and still is the study of a realistic biological bilayer membrane in full atomic detail. One can find a current survey of molecular dynamics simulations of lipid bilayer membranes in [82], [609], and [611]. The material explored best is DPPC (dipalmitoylphosphatidylcholine). In addition there are also a few studies of DLPE, DMPC, and DPPS[1]. Lately, systems with unsaturated lipids have been considered more intensively. Such systems are of special interest since most biological membranes contain mixtures of proteins and unsaturated lipids. [310] examined a POPC (palmitoyloleoylphosphatidylcholine) system with 200 lipids. Furthermore, DOPC [331], DOPE [331], and DLPE [214, 687] are subjects of intensive molecular dynamics studies. Here, it is interesting that phospholipids such as DPPC carry a large dipole moment. The dipoles arrange themselves preferentially parallel to the boundary layer and interact with the water molecules. This causes a substantial electric field which is equalized by the orientation of the surrounding water molecules. Furthermore, order parameters, atomic and electronic density profiles as well as radial distribution functions can be determined, from which one can compute the hydration numbers around the lipid headgroups.

Altogether, the structure of a double membrane is well-described by a four region model which was first proposed by Marrink and Berendsen [411], see Figure 9.2. In the first region one finds water molecules with orientations influenced by the headgroups and tails of the lipids. In a second region, the density of the water decreases to almost zero and the density of the lipid

---

[1] DLPE is an abbreviation for dilaureoylphosphatidylethanolamine, DMPC abbreviates dimyristoylphosphatidylcholine, and DPPS abbreviates dipalmitoylphosphatidylserine.

becomes maximal. Furthermore, all atoms of the headgroups and some parts of the tail methylene group are located here. The water molecules in this phase are all part of hydration shells of the phospholipid headgroups. In the third region, the lipid chains are increasingly aligned, similar to a soft polymer. This region is the main barrier for the permeation of the membrane by small proteins. The fourth region – the center of the bilayer – is completely hydrophobic and has a relatively low density which is comparable to decane. Larger hydrophobic molecules could be dissolved in this region. From here onward, the third, second, and first region follow in reverse order, see Figure 9.2.

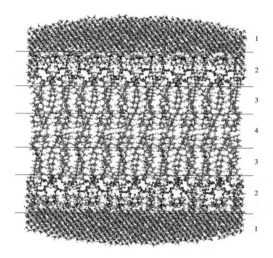

Fig. 9.2. POPC double membrane with the different regions of the four region model.

The simulation of biological bilayer membranes poses the following problem: How do we find a good structure that can be used as initial data for a simulation? Starting from a crystalline structure of the membrane, the simulation would need far too long to obtain the equilibrated liquid-crystalline phase, in which the actual studies are to be performed. The time scale on which biomembranes fully form is very long. Thus, to obtain a reasonable initial structure one could instead select lipids from a library of structures and arrange them on a grid in such a way that certain properties agree with data from physical experiments, such as order parameters observed in NMR measurements [643, 666]. Alternatively, one could select only one or a few lipid structures, arrange them on a grid, and randomly rotate and translate them perpendicularly to the membrane to obtain a certain given surface roughness. Good structures from earlier simulations also could be used as initial structures for new simulations. Quite a number of biomembranes have been studied until now and their data can be downloaded from web sites [20, 21, 22, 23].

Another question is the choice of ensemble to be used in a simulation. One can find experiments in the NPT as well as the NVT ensemble. It seems that the NPT ensemble has become widely accepted as the standard approach for simulations of biomembranes.

As a numerical example for the simulation of a membrane, let us study the evolution of a POPC double membrane. We use data from Heller [23, 308, 310]. To produce the initial configuration for the experiment, 200 POPC molecules are arranged in a box of size 84 Å×96 Å×96 Å in a water bath of 4526 H_2O molecules. This leads to a total number of 40379 particles. We fix a temperature of 300 K for the simulation, use periodic boundary conditions, and employ the parameters for the potentials from CHARMM v27 [401]. We apply our SPME method from Chapter 7 coupled with the routines for the short-range force computation (here with a cutoff radius of 12 Å) from Chapters 3 and 5. Figure 9.3 shows the initial configuration (left) and the result after a simulation time of 19.5 ps (right). One can observe how some water molecules penetrate the membrane layer and how the molecules align with each other. More detailed studies can be found in [23, 308, 310].

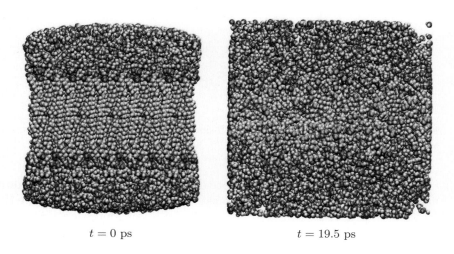

$t = 0$ ps $t = 19.5$ ps

Fig. 9.3. Simulation of a POPC double membrane from Figure 9.2.

Further numerical studies consider the temperature dependency of the gel and fluid phases of membranes [214, 220, 619, 642], the mixing of different lipids [521], the diffusion and permeation of lipid membranes [589], the transport of ions through membranes, and the formation of pores in membranes [608]. Numerical experiments to determine adhesive forces in double membranes are also of special interest. To this end, a spring with a harmonic potential is attached to the atom of the headgroup of a lipid and the spring is pulled. The force necessary to pull the lipid from the membrane can be measured over time and the conformation of the lipid being pulled out can be traced dynamically. Both depend strongly on the velocity with which the spring is pulled. This idea goes back to [276] and was applied to phospholipid membranes in [412, 580]. Studies of surface tension can be found in [413]. Finally, the interaction of a lipid membrane with smaller molecules and proteins such as phosolipase A_2 [688], bacteriorhodopsin [203], alamethicin [610] or cholesterol [572, 573] has been studied intensively.

9.3 Peptides and Proteins: Structure, Conformation, and (Un-)Folding

Amino acids (monopeptides) are the basic modules of life. They are aminocarbon acids which are formed by an amino group (-NH_2) and a carboxyl group (-$COOH$). In addition, there is also a central CH group and a specific side chain, see Figure 9.4. This side chain characterizes the amino acid and can consist of further carbon, amino or hydrosulfide groups.

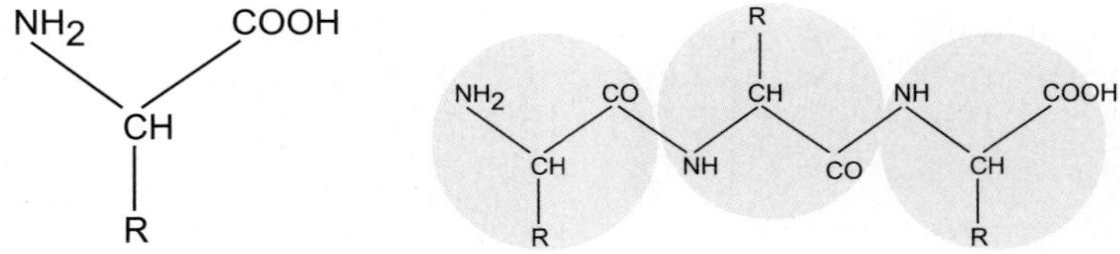

Fig. 9.4. Structure of an amino acid with side chain R (left) and primary structure of a tripeptide (right).

There are 20 different naturally occuring amino acids. Each has its own name and is usually represented by an associated three letter code. Figure 9.5 shows the chemical structural formula and a ball-stick representation of cysteine.

Amino acids can be joined together in polymer chains similar to alkanes or polyamides. Depending on the number and types of the involved amino acids one speaks of peptides, proteins (long-chain polypeptides), or nucleic acids (DNA or RNA). The amino acids are connected in these chains by peptide bonds, where the amino and carboxyl groups are reduced to ($-CO-NH-$), releasing a molecule of water H_2O, see also Figure 9.4. Hence, a peptide or a protein can be described by its sequence of amino acids in the chain. This is the so-called *primary structure*. An example is given in Figure 9.6.

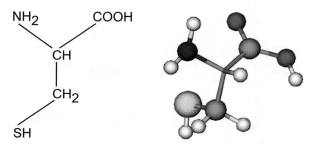

Fig. 9.5. The amino acid cysteine $SH–CH_2–CH(NH_2)–COOH$. Structural formula and three-dimensional ball-stick representation.

Lys-Val-Phe-Gly-Arg-Cys-Glu-Leu-Ala-Ala-Ala-Met-Lys-Arg-His-Gly-Leu-Asp-
Asn-Tyr-Arg-Gly-Tyr-Ser-Leu-Gly-Asn-Try-Val-Cys-Ala-Ala-Lys-Phe-Glu-Ser-
Asn-Phe-Asn-Thr-Gln-Ala-Thr-Asn-Arg-Asn-Thr-Asp-Gly-Ser-Thr-Asp-Tyr-Gly-
Ile-Leu-Gln-Ile-Asn-Ser-Arg-Try-Try-Cys-Asp-Asn-Gly-Arg-Thr-Pro-Gly-Ser-Arg-
Asn-Leu-Cys-Asn-Ile-Pro-Cys-Arg-Ala-Leu-Leu-Ser-Ser-Asp-Ile-Thr-Ala-Ser-Val-
Asn-Cys-Ala-Lys-Lys-Ile-Val-Ser-Asp-Gly-Asp-Gly-Met-Asn-Ala-Try-Val-Ala-Try-
Arg-Asn-Arg-Cys-Lys-Gly-Thr-Asp-Val-Gln-Ala-Try-Ile-Arg-Gly-Cys-Arg-Leu

Fig. 9.6. Primary structure of lysozyme.

Many peptides and – because of their length – all proteins assume well-defined three-dimensional structures in space.[2] This so-called *secondary structure* results from the arrangement of the side chains of consecutive amino acids due to regular hydrogen bonds between the peptide bonds. Two frequently occurring structures are the α-helix and the β-sheet. An α-helix is formed when the chain of amino acids twists around itself in a regular fashion. Then, a cylinder is formed in which every peptide bond is connected to other peptide bonds by way of hydrogen bonds. In this arrangement, the side chains of the amino acids point to the outside. An example is shown in Figure 9.7.

Fig. 9.7. α-helix in ball-stick representation and in ribbon representation.

---

[2] Some peptides assume stable forms based on secondary structure, others do not assume any definite stable form but assume random coil configurations instead.

A β-sheet is formed when two peptide chains lie side by side and each peptide chain forms a hydrogen bond with the corresponding peptide bond in the other chain. The sheet is folded like an accordion, the side chains of the amino acids are aligned almost vertically up and down, see Figure 9.8.

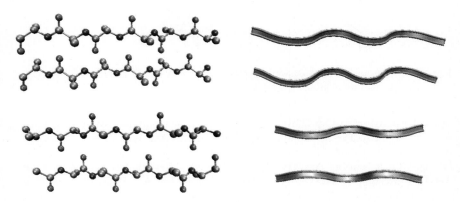

Fig. 9.8. β-sheet in ball-stick-representation and in ribbon representation. Parallel structure (top) and anti-parallel hairpin structure (bottom). The two strands still have to be connected at one end by a bend.

Proteins also have more advanced spatial structures. The *tertiary structure* refers to the three-dimensional conformation of the protein. It describes the position of all atoms and thereby the relative spatial position of the basic configurations from the secondary structure. It is determined by many types of interactions between the different amino acids. These interactions include hydrogen bonds, ionic bonds between positively and negatively charged groups in the side chains, hydrophobic bonds in the interior of the proteins, and disulfide bridge bonds.[3] Disulfide bridges can also connect two amino acid chains of different lengths. Proteins that consist of two or more chains are said to have a *quaternary structure*. It describes the shape and the relative spatial position of the polypeptide chains. The chains can be identical copies of the same protein or can be different proteins with different amino acid sequences. Examples for the quaternary structure of proteins can be seen in Figure 9.9.

Ribonucleic acids (RNA) and desoxyribosenucleic acid (DNA) are formed in a similar chain-like way from four different nucleotides. They also assume higher-order structures in space, as for instance the famous double helix according to Watson and Crick [656].

Peptides, proteins, ribonucleic and desoxyribosenucleic acids fulfil specific *functions* in the cell. Peptides, for instance, regulate the activity of other

---

[3] Disulfide (bridge) bonds are formed by the oxidation of the SH-groups of two cysteine side chains to cystine side chains.

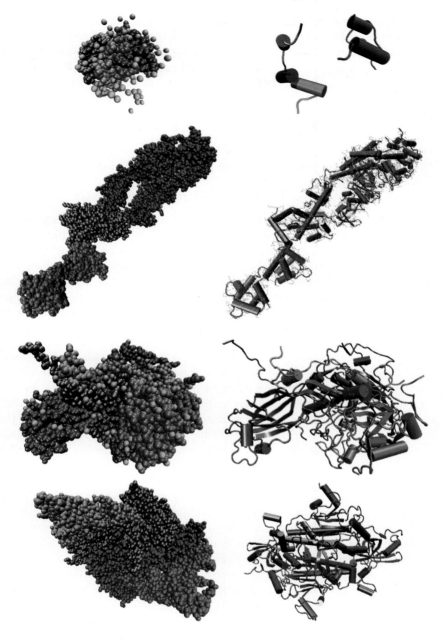

Fig. 9.9. Quarternary structure of some proteins in atomic (left) and cartoon (right) representation. From top to bottom: Insulin (two protein chains), myosin (three protein chains), rhinovirus 14 (four protein chains), aminoacyl-tRNA synthetase (two protein chains with two active pieces of tRNA).

molecules and proteins by interactions with a target molecule. In addition, there are peptides with hormonal or antibiotic properties. Enzymes, a subclass of proteins, catalyze certain biochemical reactions. DNA serves to store and RNA serves to translate genetic information. These specific functions are made possible only by the native spatial structure of the different biopolymers. There is a strong direct connection between the function of a biopolymer and its dynamical behavior.

Some of the greatest challenges of molecular biology and biophysics are

- the prediction of the structure of biological macromolecules given their primary structure (that is the sequence of basic modules such as amino acids or nucleotides) and
- the study of the folding pathway of biological macromolecules [420].

If the folding pathway could be followed in a realistic way by simulation from the primary structure to the tertiary or even quaternary structure, the problem of structure prediction would be solved. If furthermore it would be possible to predict conformational or structural changes depending on, e.g. solvent, salt concentration, temperature, and other nearby macromolecules, one could gain a crucial understanding of many processes in the cell. Structure prediction would also have practical benefits: Protein molecules with improved or even new properties could be constructed, which is of great interest to pharmaceutics and biotechnology.

Dynamical behavior of proteins occurs on different time scales: Conformational changes of parts of the molecule happen in the nanosecond to microsecond range, reactions between different proteins and changes in the quaternary structure take milliseconds to tenths of seconds, and the entire folding process can take seconds or even minutes. Hence, the main problem in the simulation of the folding of proteins is to reach sufficiently long simulation times. Molecular dynamics methods on current computers are not yet able to attain such time scales. The longest simulations of proteins in aqueous solution up to date reach the microsecond range. However, this is only possible with simplified models. Even then, such computations require running times of half a year or more on large parallel computers.

One way to simplify the model is to freeze high frequency modes for example by using the SHAKE method, compare Section 6.4. The elimination of degrees of freedom that are responsible for the extremely high frequencies in the simulation then allows the selection of larger time steps. There are more advanced approaches in which the model is reduced to its essential dynamics by "normal mode analysis" (a singular value decomposition is used to project the time evolution of the system to the low-frequency modes of the motion) [37, 44, 127, 601, 669, 682, 683]. Improved time stepping techniques, such as multiple time step methods, can also help to some extent, compare Section 6.3. Then, for specific problems, time steps can be increased up to about 8 fs. However, these techniques are also more expensive than for example the simple Störmer-Verlet method and cannot be parallelized as efficiently. According to our experience, one thus can only achieve a total speedup by a factor of three to four. Additionally, simple cutoff techniques for the electrostatic interaction could be used at the price of dramatically decreased accuracy. Overall, these techniques are not yet able to solve the time scale problem of the molecular dynamics method even on the fastest computers available today.

Under some circumstances, stochastic approaches might be more promising. In such techniques, Newton's equations of motion are supplemented by stochastic friction terms. These terms have a damping effect on the high frequency modes of the dynamical system and therefore allow substantially larger time steps. Two classes of such techniques are Langevin dynamics[4] [318, 682, 683, 684] and Brownian dynamics [242, 243, 403]. However, the simulation results are no longer deterministic but stochastic and the dynamics of the system may be changed. Additionally, there is the question how accurate the results are in the sense of statistical physics. Langevin dynamics is often used in practice to approximate the effect of a surrounding solvent on a protein without explicitly taking the many degrees of freedom of the solvent into account.

Another approach to reduce the complexity is to treat the solvent implicitly by a macroscopic model. The biomolecule and a thin layer of water molecules and ions are handled with the molecular dynamics method, while the effect of the surrounding fluid is modeled with a continuous model. For example, the electrostatics of the surrounding water domain can be described with Poisson-Boltzmann models [45, 312, 664].

Still, scientists are intensively using molecular dynamics techniques to simulate the behavior of peptides and proteins. At least a preliminary understanding of protein folding can be gained from the simulation of peptide folding. The formation of small scale structures can be observed in peptides in the microsecond range.[5] For example, the folding of a small part of protein, the head piece of the villin molecule, was simulated in aqueous solution in [192]. After a fast "burst" phase with hydrophobic collapse, the precursor of a helix was formed and a tertiary structure was observed. This computation with a simulation time of more than one microsecond is the longest simulation of a peptide in solution to date. In fact, a structure was obtained that was very close to the known native conformation.

Another focus of recent work is the study of the stability of the conformations of peptides in aqueous solution at different temperatures [169, 172, 175, 476, 567, 569, 592, 632, 658]. Changes in the conformation of small parts

---

[4] In Langevin dynamics, two terms are added to the equation of motion $m\dot{\mathbf{v}}_i = \mathbf{F}_i - \gamma_i\mathbf{v}_i + \mathbf{R}_i(t)$: A pure friction term $-\gamma_i\mathbf{v}_i$, which depends on the velocity, and a stochastic noise term $\mathbf{R}_i(t)$, which just depends on time, with a vanishing mean over time.

[5] However, it will be a very, very long way to the complete folding of a DNA molecule given the data from the recently completed human genome project.

of peptides, such as the β hairpin structure, have been studied intensively in [99, 460, 522]. The results lead to a model in which the bend is formed first, hydrogen bonds then close the hairpin, and finally hydrophobic interactions of the side chains stabilize the hairpin. The reversible formation of secondary structures for a small peptide in methanol has been studied in a number of simulations for a time span of 50 nanoseconds in [170, 171]. It was observed that the accessible space of conformation quickly narrows to a few clusters. Additionally, different folding pathways could be determined depending on the temperature.

There is currently an intensive discussion on whether the simulation of the process reverse to folding, the unfolding of a protein, can yield information about folding pathways. Unfortunately, even for long simulation times, complete unfolding cannot be observed yet. Therefore, the idea is to accelerate the unfolding process by the application of an external force or other constraints. In [319, 320] the dynamics of a protein complex, the streptavidin-biotin system, was first studied under external forcing.[6] In such a simulation one can observe several phases of separation and unfolding. In the simulations of the streptavidin-biotin system, the dominant hydrogen bonds were broken first, new hydrogen bonds were established next, which then finally were broken as well. The separation of dinitrophenyl hapten from a fragment of a monoclonal antibody was studied with this method in [320]. There, several different phases of the separation and quite complex separation patterns were observed. This technique has also been applied under the name "steered molecular dynamics" to the avidin-biotin complex [335], the titin immunoglobulin domain [391, 392, 393, 415], and further proteins [389]. Whether, and if so, how far the spontaneous unfolding corresponds to such a forced unfolding is still an open question at this time. Nevertheless, this simulation approach can complement experiments with atomic force microscopes and optical tweezers [513, 335].

In "targeted molecular dynamics" (TMD) one similarly intervenes into the dynamics of a protein to shorten the running time needed for the simulation. Here, one introduces a reaction coordinate which connects the initial state and a given target conformation of the molecule. Typically, this is the mean distance between the positions of the atoms in the initial state and the target state. This distance is then slowly decreased to zero using appropriate additional constraints for the system, while allowing the other degrees of freedom of the molecule and of the aqueous solution to relax freely. For larger proteins the radius of gyration of the molecule can also be used as

---

[6] Essentially, this is a computer simulation of the functionality of an atomic force microscope. For a few years now, such a microscope offers the possibility to study the behavior of single molecules or molecule complexes under local application of force [519]. Analogously, in the simulations a spring with a harmonic potential is attached to a part of the molecule complex, a tractive force is exerted over the spring, and the molecule is pulled apart step-by-step until it breaks.

the reaction coordinate. In such a way it is possible to study the transition between conformations and to analyze chemical reactions. This method was applied to the G protein Ha-ras p21 and offered an explanation of how the protein switches between its active and passive state and therefore is able to transmit signals. A disadvantage of this approach is the necessity to know the target structure beforehand. Further results of targeted molecular dynamics can be found in [183, 184, 367] and [223, 224].

As numerical example for structural changes of peptides and proteins, we now consider the behavior of a "leucine zipper" in vacuum and in an explicitly modeled aqueous solution. These experiments study how two α-helices bond – they twist into each other – and show a binding mechanism which resembles a zipper. Both strands bond by leucine side chains. This type of bonding can be found for instance as marker of the beginning and the end of a DNA strand. A better understanding of this mechanism can suggest possible points of attack for drugs.

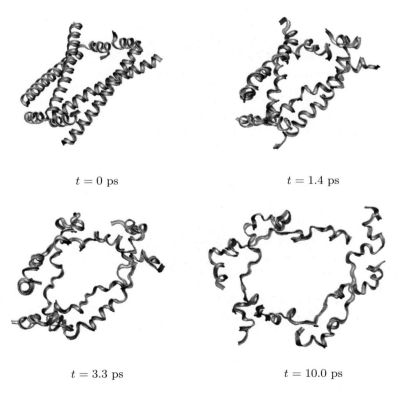

$t = 0$ ps $t = 1.4$ ps

$t = 3.3$ ps $t = 10.0$ ps

Fig. 9.10. Simulation of two leucine zippers in vacuum, evolution over time, ribbon representation.

The numerical simulations start with a configuration which consists of several independent helical strands from the synthetic part of a hepatitis D antigen (PDB entry 1a92) at a temperature of 300 K. Figure 9.10 shows the results of a simulation in vacuum with our higher-order tree method from Chapter 8. One can see how the two pairs of helices move away from each other and how two of the strands start to develop bridge bonds in a screw-like form. A second simulation of leucine zippers, this time in aqueous solution (12129 TIP3P-C molecules) with periodic boundary conditions, can be seen in Figure 9.11. Here, the movements occur substantially slower and the surrounding water has a stabilizing influence on the helices. Again, one can see how the strands start to bond.

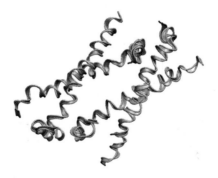

Fig. 9.11. Simulation of two leucine zippers in aqueous solution. Ribbon representation without water molecules, $t = 1.4$ ps.

As another example let us consider the α-amylase inhibitor tendamistat. This protein consists of only 74 amino acids but forms six β-sheets and has two disulfide bridge bonds. Its small size allows to perform a relatively large number of time steps which is enough to observe the complete folding pathway starting from the elongated state of the protein (the amino acid sequence of the PDB entry 3ait) in vacuum. However, in an aqueous solution, the folding process would take about 10 ms even for such a small protein, which can not be reached in a numerical simulation. For our vacuum experiment, we put the 281 atoms of the protein into a cube of dimensions 50 Å× 50 Å× 90 Å at a temperature of 300 K. Then, we cool the system in small successive steps down to 300 K while observing the movement of the protein over time. Some snapshots of the dynamics are shown in Figure 9.12. The different time scales of the different phases of the folding pathway can be seen clearly: The molecule first leaves the elongated, energetically unfavorable state, and then the real folding starts. More extensive experiments for tendamistat can be found in [99].

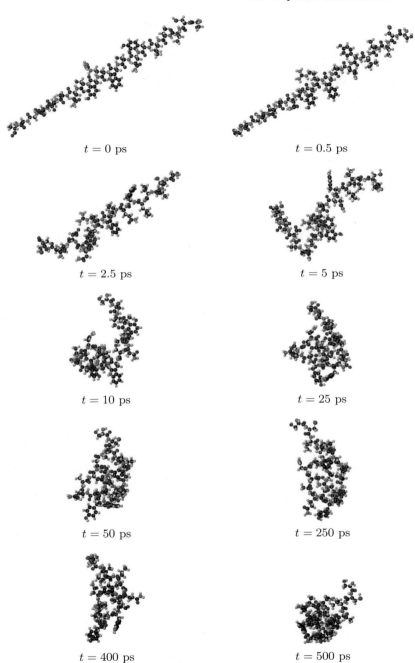

$t = 0$ ps

$t = 0.5$ ps

$t = 2.5$ ps

$t = 5$ ps

$t = 10$ ps

$t = 25$ ps

$t = 50$ ps

$t = 250$ ps

$t = 400$ ps

$t = 500$ ps

Fig. 9.12. Simulation of the folding of tendamistat in vacuum.

9.4 Protein-Ligand Complex and Bonding

In general, the activity of a biopolymer is controlled by small molecules, the so-called ligands. Certain ligands attach to certain parts of the surface of the macromolecule and are thereby "recognized" by it. The efficacy of many drugs and pharmaceuticals depends on this process of ligand recognition. To be able to create new and effective drugs and medicaments, the structure and the mechanism of bonding between the biopolymer and the ligand in the complex has to be known or at least be predictable. This problem is analogous to the task of structure prediction from the last section. The bonding of the ligand to its protein can be understood as a folding problem. Again, the bonding process occurs on a relatively slow time scale.

One example for ligands are enzymes. Enzymes are special proteins that act to synthesize or cleave molecules of some substances inside of the body, such as other proteins or nucleic acids. They work as catalysts for almost all biological systems. The substrate binds to a particular part of the enzyme which is called the active site or binding pocket. The active site is a small three-dimensional part of the enzyme that is shaped like a cave or a crevice. Here, the specificity of the bond depends on the arrangement of the atoms at the active site. Only very few similar substrates fit in the active site.

The determination of the structure of an enzyme and the decoding of the sequence of its amino acids allows to identify the active site. With this information one can then reconstruct the chemical reaction between the enzyme and its substrate. On this basis one may attempt to find new substrates that inhibit the given enzymes. It is of particular interest to construct substrates that fit even better in the active site than the natural substrate. Such substrates could win in competition with the natural substrate and bind instead to the active site. This inhibits the function of the enzyme and prevents or at least delays the enzymatical reaction.

The lock and key model, going back to Fischer [233], makes the assumption that the active site already has its particular shape without the substrate being present. In this model the specificity of the enzyme (lock) for a given substrate (key) results from the geometrically and electrically complementary structure. To say it in a different way: The key fits into the lock. Another model of the bonding mechanism is based on the induced fit hypothesis. In this model it is assumed that the active site of the enzyme, to which the substrate binds, is fully formed only during the bonding process. The mechanisms underlying each of the models are shown in Figure 9.13.

X-ray studies have shown that the binding sites for substrates in enzymes are already preformed, but that there are small conformational changes during bonding. In this sense the induced fit model is more realistic. In the laboratory it is easier to separately produce the components of the enzyme-substrate complex and then to separately determine their structure by X-ray crystallography. However, the results do not allow to directly determine the active site of the enzyme and the surface region of the substrate to which the

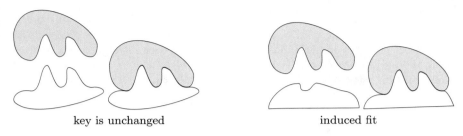

key is unchanged   induced fit

Fig. 9.13. Enzyme bonding models: Lock and key model (left) and induced fit model (right).

enzyme binds. Methods to predict these docking regions are under intensive development. But since such approaches mostly assume rigid geometries of the molecules, they cannot account for the dynamics of induced fit bonding and are often doomed to fail. Certain molecular dynamics methods can make a contribution to the solution of this problem, at least up to some extent. Examples are the techniques already described briefly in the previous chapter, the application of external forces [276, 320, 319], steered molecular dynamics [335, 391, 415], and targeted molecular dynamics [183, 184, 223, 224, 367].

Besides the time complexity, there is another problem in the application of molecular dynamics methods in the study of ligand-protein or enzyme-inhibitor complexes: The recognition and docking process often involves catalytic reactions. It has been shown that quantum mechanical effects occur in those reactions which cannot be neglected. A conventional molecular dynamics method with a given force field, for which the parameters are taken from CHARMM or Amber, for instance, is not flexible enough and produces results that are not sufficiently accurate. Instead, it is necessary to adjust the potential functions in every time step, using a newly computed electron structure. This is of special importance close to the active site. Such an approach can be implemented by coupling the molecular dynamics method to an ab initio method for the approximate computation of the solution of the electronic Schrödinger equation with a fixed configuration of the nuclei, compare Chapter 2. The electronic Schrödinger equation is then treated by local Hartree-Fock or density functional methods [35, 302, 417, 527]. The Hellman-Feynman forces (see Section 2.3) are determined in this way directly from the electron structure. It is therefore not necessary to parametrize the potential function explicitly.

A standard approach to implement the coupling of molecular dynamics methods for the classical treatment of the nuclei with the (local) quantum mechanical treatment of the electrons is the Car-Parinello molecular dynamics method [137, 461]. Only with the use of such ab initio methods in the local neighborhood of the active site, it is possible to reproduce the occurring reactions with sufficient accuracy. One example is HIV, the human immunod-

eficiency virus. This virus is responsible for AIDS, the acquired immunodeficiency syndrome. Since AIDS first occurred about 25 years ago, great efforts
have been made to understand how HIV acts and to develop drugs against
HIV. The structure of the virus has been determined in a relatively short
time, and an understanding of its replication in the cells of the human im-
mune system has been reached. First drugs have been developed to suppress
HIV replication, among them AZT (zidovudine), saquinavir, ritonavir, and
indinavir. However, the virus mutates relatively fast and becomes resistant
to such drugs.

By now, one knows at least three targets for inhibitors against HIV replication: The enzymes protease (PR), reverse transcriptase (RT), and integrase
(IN), see Figure 9.14.

Fig. 9.14. HIV-1: Section of protease (left), reverse transcriptase (center) and
integrase (right).

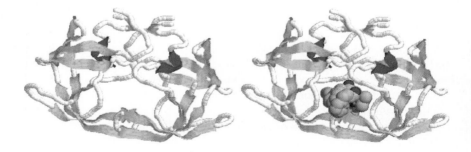

Fig. 9.15. HIV-1 protease (left) together with its inhibitor Ro 31-8959 (right),
PDB entry 1hxb.

Inhibitors for these targets have been and are under intensive development and study. Meanwhile, there exist data bases just for HIV and its inhibitors, see [647]. Great efforts are made to find new drugs. Here, molecular dynamics simulations are an important tool. They have been employed to study the possibility to use zinc ions [673] or C_{60} molecules and variants of fullerenes [431] as inhibitors of HIV protease. Also mutations of protease and their resistance against drugs have been examined in [558, 657] with molecular dynamics methods. The dynamical behavior of protease and reverse transcriptase has been studied in [152, 153, 397, 558] and [402], respectively. Molecular dynamics simulations for integrase and its mutations can be found in [59].

For more accurate studies of the reactions in the HIV-inhibitor complex one needs, as already mentioned, methods that couple quantum mechanical ab initio methods for the electrons with classical molecular dynamics methods for the nuclei. Corresponding result for HIV protease can be found in [388, 637] and for reverse transcriptase in [31].

10 Prospects

In this book we have presented the most important steps in the numerical simulation of molecular dynamics. It was our intention to enable the reader to develop and implement codes for the efficient solution of Newton's equations of motion. We have used the Störmer-Verlet method for the time discretization. For the force evaluation we have discussed, in addition to the linked cell method for short-range potentials, also the SPME method and several tree algorithms for the efficient simulation of long-range Coulomb potentials. A further main theme of this book was the parallelization of the presented algorithms using MPI. This allows to treat problems with large numbers of particles on parallel computers with distributed memory. Finally, we have presented many specific applications of the molecular dynamics method from material sciences, biophysics, and astrophysics. Furthermore, we have given hints and details that will allow readers to implement the presented methods and algorithms on their own.

We have attempted to introduce the most common many-body potentials. However, the limited amount of space has made our list of potentials incomplete. In the following, we would like to address briefly further potentials that are relevant to certain practical applications. The Buckingham potential [129] is often used to model the molecular dynamics of disordered structures and clusters. It is a generalization of the Lennard-Jones potential, where the repulsive $-r_{ij}^{-12}$ term is replaced by the term $A_{ij} \exp(-B_{ij}r_{ij}) - C_{ij}r_{ij}^{-8}$. In the Stockmayer potential [585, 586], the term $1/r_{ij}^3 \cdot \left(\mathbf{p}_i\mathbf{p}_j - 3(\mathbf{p}_i\mathbf{r}_{ij})(\mathbf{p}_j\mathbf{r}_{ij})/r_{ij}^2 \right)$ is added to the Lennard-Jones terms. It models dipole-dipole interaction, where $\mathbf{p}_i$ denotes the dipole moment of particle i. Many-body potentials are substantially more expensive than simple two-body potentials, but allow for a correspondingly more accurate model.[1] They play an important role especially in material sciences. For example, the Stillinger-Weber potential [584] and the Tersoff potential [602] are used for the simulation of silicon; the potentials of Vashishta [640] and Tsuneyuki [618] are used for silicon oxide. It is a further interesting problem how to model the *polarizability* of materials with molecular dynamics methods. Here, core-shell potentials [385, 434, 551] are used. To this end, the atom is conceptually split into its core and its shell.

---

[1] First three-body potentials were introduced by, amongst others, Axilrod and Teller [52].

The charge of the polarized atom is divided analogously between the core and the shell. The mass of the atom is either also split into a fictitious mass of the core and a fictitious mass of the shell (adiabatic model) or the shell is assigned no mass at all (static model). These two models lead to different numerical approaches: Since both parts – core and shell – have a mass in the adiabatic model, they can interact like ordinary particles in a molecular dynamics method. In the static model, the shell is explicitly attached to the core by a harmonic spring. Then, in every time step, one has to first minimize the energy with respect to the harmonic potential of the shell, for instance by the method of conjugate gradients; the cores are then moved classically according to Newton's equations of motion.

Furthermore, we have discussed the computation of a number of different macroscopic parameters such as the energy, the temperature, the pressure, the diffusion coefficient, the radial distribution functions, or cis-trans statistics. We could not discuss the computation of structure factors, more complicated transport coefficients, and correlation functions of higher order, which can be gained with the Green-Kubo relation and the linear response theory of statistical mechanics. Here, we refer to [34, 141, 239, 511] and the literature cited therein. In addition, short-range and long-range order functions play an important role in crystalline and amorphous materials. Waves associated to those order functions, the so-called phonons, can be extracted from the autocorrelation of the velocities. For this purpose, one needs in general very large simulation domains and high numbers of particles to reduce finite size effects in the results.

Finally, we would like to point out that the methods and techniques presented in this book are not only limited to conventional molecular dynamics. They can also be applied directly to ab initio molecular dynamics, such as Ehrenfest molecular dynamics, Born-Oppenheimer molecular dynamics, or Car-Parinello molecular dynamics. With these methods, the potential is no longer given as a fixed parametrized function but is determined in each time step directly from the electronic structure as computed by the Hartree-Fock method, the density functional theory, the configuration interaction method, or the coupled cluster method. Ab initio molecular dynamics methods do not need complex empirical potential functions with parameter sets that have been fitted time-consumingly to results of measurements, since such methods do not have parameters by construction. Since they take quantum mechanical effects into account, they furthermore allow for the simulation of the dynamics of chemical reactions. At the moment, such methods are still limited to systems with a hundred atoms, but they are expected to become more and more powerful, not at least because of the increasing performance and capacity of parallel computers. The molecular dynamics method can also be combined with a Monte-Carlo approach [193]. In hybrid Monte-Carlo simulations, one computes short trajectories by molecular dynamics for a sequence of random initial configurations and performs a statistical averaging over these trajectories in the Monte-Carlo part of the algorithm.[2] Application areas for such methods are solid state physics and polymer physics [236, 307, 423], more recently also biophysics [552, 553] and path integrals [624]. Molecular dynamics methods can moreover be employed for a number of more general ensembles, for open systems, for non-equilibrium phenomena (NEMD: non-equilibrium molecular dynamics) and for quantum mechanical problems (QMD: quantum molecular dynamics). Details and further references can be found in [34] and [282].

Molecular dynamics methods can also be extended without difficulties beyond Newton's equations of motion to other areas in which the motion of particles plays a role.

Vortex methods [144, 156, 380, 408] are used in the simulation of incompressible flows with high Reynolds numbers. The vorticity is discretized in such methods by N Lagrangian particles and the flow equation is thus transformed into a system of $2N$ ordinary differential equations which is similar to Newton's equations of motion. The theorems of Kelvin and Helmholtz, which determine the dynamics of the vorticity of inviscid fluids, are enforced directly in the resulting discrete method. The particles are now propagated over time and interact by potential functions which are derived from the flow equation and the types of the particles used. The velocity of each particle can be computed by the Biot-Savart law. The techniques discussed in this book, especially tree methods, can be used to implement these computations in an efficient way, see also [190, 191, 661].

The *smoothed particle hydrodynamics* method (SPH) [194, 254, 394, 437, 369] is a meshfree Lagrangian particle method for the numerical simulation of flow problems. It simulates the fluid with help of particles which possess a certain size, an interior density distribution, a velocity, and, if needed, also a temperature or a charge. To this end, a kernel function, usually with compact support, is assigned to each particle. The well-known equations of fluid mechanics, such as the Euler equations or the Navier-Stokes equations, can then be formulated approximately as systems of ordinary differential equations in which the particles are convectively propagated over time and interact with each other by potentials. These potentials result from the application of the differential operators of the flow equations to the kernel functions. Altogether, particles move according to a force field as in molecular dynamics methods, where in each time step, the forces acting on each particle have to be computed. Here, the techniques described in this book can be applied with just a few simple modifications. The SPH method is often used for astrophysical flow problems as well as for flow problems with free surfaces.

---

[2] See also the Folding@Home project at http://folding.stanford.edu which carried out the longest protein folding simulations so far. There, the folding of the villin headpiece using the so-called MSM sampling technique was studied up to a time span of 500 μs.

Finally, the described computational methods can also be adopted to the fast evaluation of discretized integral transforms and the fast solution of integral equations. Examples are the fast Gauss transform [267, 268], the fast Radon transform [110, 120], or the fast evaluation of radial basis functions [66, 67, 68].

We hope that, after having worked through this book, readers will find their way quickly into these application areas of particle methods and will be able to apply the techniques learned for molecular dynamics with success also for such problems.

A Appendix

A.1 Newton's, Hamilton's, and Euler-Lagrange's Equations

We consider the equations of classical mechanics for a system in $\mathbb{R}^d$ which consists of N particles with the masses $\{m_1, \ldots, m_N\}$. We denote the position of the ith particle at time t by $\mathbf{x}_i(t) \in \mathbb{R}^d$ and its velocity by $\mathbf{v}_i(t) = \dot{\mathbf{x}}(t) \in \mathbb{R}^d$. The momentum is then given by $\mathbf{p}_i(t) := m_i \mathbf{v}_i(t) = m_i \dot{\mathbf{x}}(t)$.

The motion of the particles obeys Newton's equations of motion

$$m_i \ddot{\mathbf{x}}_i = -\nabla_{\mathbf{x}_i} V(\mathbf{x}), \qquad \text{or} \qquad \begin{aligned} \dot{\mathbf{x}}_i &= \mathbf{v}_i \\ m_i \dot{\mathbf{v}}_i &= -\nabla_{\mathbf{x}_i} V(\mathbf{x}), \end{aligned} \tag{A.1}$$

where the function

$$V : \mathbb{R}^{dN} \to \mathbb{R}$$

denotes the potential energy. Here, we have combined the positions of the particles into a vector $\mathbf{x} := (\mathbf{x}_1, \ldots, \mathbf{x}_N)^T$. Analogously, we combine the velocities and momenta of the particles into vectors $\mathbf{v} := (\mathbf{v}_1, \ldots, \mathbf{v}_N)^T$ and $\mathbf{p} := (\mathbf{p}_1, \ldots, \mathbf{p}_N)^T$.

An alternative description of the motion of the system results from Hamilton's principle, sometimes called the principle of least action. To this end, we consider the Lagrangian

$$\mathcal{L}(\mathbf{x}, \mathbf{v}) := \frac{1}{2} \sum_{i=1}^{N} m_i \mathbf{v}_i^2 - V(\mathbf{x}), \tag{A.2}$$

which depends on the positions $\mathbf{x}$ and velocities $\mathbf{v}$. The motion of the system is determined by the condition that it will make the action integral

$$L(\mathbf{x}) := \int_0^T \mathcal{L}(\mathbf{x}(t), \dot{\mathbf{x}}(t)) \mathrm{d}t$$

stationary, i.e., that the integral's first variation δL vanishes. This means

$$\delta L(\mathbf{x}, \mathbf{y}) := \lim_{\varepsilon \to 0} \frac{L(\mathbf{x} + \varepsilon \mathbf{y}) - L(\mathbf{x})}{\varepsilon}$$

$$= \int_0^T \nabla_{\mathbf{x}} \mathcal{L}(\mathbf{x}, \dot{\mathbf{x}}) \mathbf{y} + \nabla_{\dot{\mathbf{x}}} \mathcal{L}(\mathbf{x}, \dot{\mathbf{x}}) \dot{\mathbf{y}} \, \mathrm{d}t = 0 \qquad \forall \mathbf{y} \in C_c^\infty((0,T); \mathbb{R}^{dN}).$$

This is equivalent to the Euler-Lagrange equation

$$\nabla_{\mathbf{x}} \mathcal{L}(\mathbf{x}, \dot{\mathbf{x}}) - \frac{\mathrm{d}}{\mathrm{d}t} \nabla_{\dot{\mathbf{x}}} \mathcal{L}(\mathbf{x}, \dot{\mathbf{x}}) = 0. \tag{A.3}$$

One can easily see by substituting (A.2) into (A.3) that the equations are equivalent to Newton's equations of motion (A.1).

One obtains a third description of the motion of the system using the Legendre transform $(\mathbf{x}, \dot{\mathbf{x}}, t) \rightarrow (\mathbf{x}, \mathbf{p}, t)$. We define the Hamiltonian, which depends on the positions $\mathbf{x}$ and the momenta $\mathbf{p}$, as the Legendre transform of the Lagrangian,

$$\mathcal{H}(\mathbf{x}, \mathbf{p}) := \sum_{i=1}^{N} \frac{\mathbf{p}_i^2}{m_i} - \mathcal{L}\left(\mathbf{x}, \frac{\mathbf{p}}{\mathbf{m}}\right) = \frac{1}{2} \sum_{i=1}^{N} \frac{\mathbf{p}_i^2}{m_i} + V(\mathbf{x}),$$

with $\frac{\mathbf{p}}{\mathbf{m}} := \left(\frac{\mathbf{p}_1}{m_1}, \frac{\mathbf{p}_2}{m_2}, \ldots, \frac{\mathbf{p}_N}{m_N}\right)^T$. Hamilton's equations for a system $(\mathbf{x}, \mathbf{p})$ are then given by

$$\dot{\mathbf{x}} = \nabla_{\mathbf{p}} \mathcal{H}(\mathbf{x}, \mathbf{p}), \qquad \dot{\mathbf{p}} = -\nabla_{\mathbf{x}} \mathcal{H}(\mathbf{x}, \mathbf{p}). \qquad (A.4)$$

Again, it is easy to see that Newton's equations of motion, together with the definition of the momenta $\mathbf{p}_i := m_i \dot{\mathbf{x}}_i$, are equivalent to Hamilton's equations (A.4). The relation

$$\frac{\mathrm{d}}{\mathrm{d}t} \mathcal{H}(\mathbf{x}, \mathbf{p}) = \nabla_{\mathbf{x}} \mathcal{H}(\mathbf{x}, \mathbf{p}) \dot{\mathbf{x}} + \nabla_{\mathbf{p}} \mathcal{H}(\mathbf{x}, \mathbf{p}) \dot{\mathbf{p}}$$

$$= \nabla_{\mathbf{x}} \mathcal{H}(\mathbf{x}, \mathbf{p}) \nabla_{\mathbf{p}} \mathcal{H}(\mathbf{x}, \mathbf{p}) - \nabla_{\mathbf{p}} \mathcal{H}(\mathbf{x}, \mathbf{p}) \nabla_{\mathbf{x}} \mathcal{H}(\mathbf{x}, \mathbf{p}) = 0$$

implies that the sum of kinetic and potential energy is conserved for any system which satisfies Hamilton's equations.

A.2 Suggestions for Coding and Visualization

The algorithms described in this book can be implemented in many different programming languages. We have used the language C in our examples, see [40, 354].

It is advisable to group the procedures needed in the implementation of the algorithms into different modules (and files) since that makes it easier to obtain a clearly laid out program in C and to avoid redundancies. The different files are then automatically compiled and linked together. It is recommended to split the source code into header files (.h) containing the declaration of procedures and data types and into (.c) files containing the implementation of the procedures. For instance, in the SPME method from Section 7.3, the procedures for the computation of the short-range and long-range force terms can be combined into one single file mesh.c. Dependencies and files are usually automatically managed by integrated development environments. Computers running variants of the UNIX operating systems allow to automatize the compilation and linking process with the program make. Further information can be found in books covering the pertinent features of UNIX, as for instance [107]. The Makefile necessary for the program make could look in our case as given in algorithm A.1.

Algorithm A.1 `Makefile`

```
OBJ = particle.o mesh.o        # the modules
CC = cc                        # the compiler
CFLAGS = -...                  # optional flags for the compiler
CLIBS = -lfft -lmpi -lm        # the libraries used
AOUT = a.out


.c.o:                          # rules for compilation
        $(CC) -c $(CFLAGS) $*.c

$(AOUT): $(OBJ)                # rules for linking
        $(CC) -o $(AOUT) $(OBJ) $(CLIBS)

particle.o: particle.h  mpi.h   # further dependencies
mesh.o:     mesh.h particle.h mpi.h fft.h
```

In the following we give explanations for some of the commands in this file:

- The modules that are supposed to be linked together into a program are collected in the variable `OBJ`. The files with the extension `.o` are modules which are already compiled; they are called object files.
- Rules are marked by a colon. The first rule describes how to compile a module from its source code.
- The second rule describes how the executable program is built from the modules.
- The lines with `$(CC)` have to start with a tabulator character so that `make` can distinguish between rules and actions.
- The `-l` option tells the linker into which libraries it should look for object files. In our case, in addition to the math library, we also link to a library for the fast Fourier transform `fft` and the MPI library for parallel computing with message passing.
- The `-c` options allows the creation of object files without further linking.
- Further rules specify the dependencies of the files among each other. For instance, the object file `particle.o` depends also on the header file `particle.h`. Thus, the object file has to be recompiled when the header file has been changed. These kinds of entries can be created automatically in UNIX using the call `makedepend *.c`. Some compilers are also able to determine dependencies directly from the `include` commands in the files without the programmer specifying them in the `Makefile`. This ensures that the dependencies are complete and up-to-date, which is not necessarily the case if the dependencies are updated manually.

Of course, languages for object-oriented programming such as `C++` and `Java` can be used for the implementation instead of `C`. The data structures struct of our implementation are then promoted to classes class and the procedures are accordingly assigned to those classes. This allows a more detailed structuring and a better encapsulation of data and algorithms. One should at least use general implementations with container classes for linked lists and trees, as suggested by the concepts of generic programming. Time integration and force computation offer themselves as abstract interfaces for entire hierarchies of algorithms.

The particle data structure 3.1 and the three functions updateX, updateV and force in algorithms 3.5 and 3.7 could be combined to a common class in an object-oriented implementation, see code fragment A.1. In this way, for different potentials and time integration schemes, one obtains different types (and classes) of particles with their own methods for time integration and force computation. Of course, one could also structure the program using multiple inheritance or parametrized classes instead.

Code fragment A.1 A Class for the Störmer-Verlet Method

```
class GravitationStoermerVerletParticle: public Particle {
private:
  real F_old[DIM];
public:
  GravitationStoermerVerletParticle();
  void updateX(real delta_t);
  void updateV(real delta_t);
  void force(const GravitationStoermerVerletParticle& j);
};
```

There are other programming languages such as Pascal, Fortran90, and their variants. Implementations of our algorithms in these languages will differ at least syntactically from our example programs in C.

The results of a simulation, including the number, positions, and velocities of the particles as well as relevant computed quantities, such as the kinetic and potential energy, should be written in regular intervals into an output file. The data from this output file can then be used in the further analysis of the results of the simulation. Using an appropriate visualization program, one can display and animate the data from this file. The simplest visualization represents the particles of the system at a given time as points or small spheres which may be colored according to the velocity or the type of the particle. However, if there are too many particles, such a representation reaches its limits and is no longer adequate. Instead, one could use a color representation of the density of the particles for the visualization. The dynamic behavior of the particles can be shown in a movie created from a sequence of representations of the system at different times. There are various programs with different capabilities, as for instance gnuplot, IDL, MS

Visualizer, AVS, OpenDX, VTK, or Explorer. Specialized representations of molecules are possible for instance with RASMOL, ProteinExplorer, CHIME, EXPLOR, KiNG, Jmol, WebMol, MIDAS, and VMD.

A.3 Parallelization by MPI

We addressed the parallelization of algorithms for parallel computers with distributed memory in this book. In such a parallel computer, every processor has its own local memory where all the data needed by local processes must be present. In turn, each process owns a certain part of this local memory. The processes can communicate with each other by messages; one process sends another process data in such messages. Thus, a parallel program does not only consist of a sequence of computational operations as in the sequential case, but additional send and receive operations have to be inserted at the appropriate places by the programmer. We use the SPMD (single program, multiple data) approach for our programs. In this approach the programmer writes *one* program that runs on every processor in parallel. The copies on different processors then communicate by explicit send and receive operations and are synchronized by receive operations.

Different approaches have been used in the course of the development of parallel computers to implement such communication operations. Based on earlier message passing libraries such as PVM, Parmacs, NX/2, Picl, or chip, a uniform standard "MPI" (Message Passing Interface) has been established [7, 271, 272, 273, 458]. Nowadays, there is at least one implementation of MPI available for all existing parallel computers. For testing and debugging purposes one can also use the freely available MPI implementations "MPICH" (from Argonne National Lab) [11] or "LAM" (from Notre Dame University) [12], which run on many parallel computers and are able to simulate a parallel system on a single computer or on a network of workstations.

MPI basically provides a library which allows to start a certain number of processes at the same time on one or several processors. These processes exchange data among each other. The processes are identified by a unique process number. MPI is a very powerful and complex system with more than 120 different functions, for more details see [272, 273, 458]. Fortunately, many programs can be parallelized using only six of these functions. The six functions are:

- `MPI_Init()`:
  Initialization of the MPI library.
- `MPI_Finalize()`:
  Termination of the MPI library.
- `MPI_Comm_size()`:
  Determine the number `numprocs` of processes started.
- `MPI_Comm_rank()`:
  Determine the local process number $\texttt{myrank} \in \{0, \ldots, \texttt{numprocs} - 1\}$.

– MPI_Send() or MPI_Isend():
  Send MPI message.
– MPI_Recv():
  Receive MPI message.

The initialization of a parallel main program is given in Algorithm A.2.

Algorithm A.2 Parallel Main Program a.out

```
#include <mpi.h>
int main(int argc, char *argv[])
{
  int myrank, numprocs;
  MPI_Init(&argc, &argv);
  MPI_Comm_size(MPI_COMM_WORLD, &numprocs);
  MPI_Comm_rank(MPI_COMM_WORLD, &myrank);
  ...                        // main part consisting of the
  ...                        // computational and communication
  ...                        // operations of the parallel program
  MPI_Finalize();
  return 0;
}
```

The MPI library has to be initialized first by calling MPI_Init. In between the calls to MPI_Init and MPI_Finalize, data can be sent to other processes or can be received from other processes. In total, there are numprocs parallel processes available for the parallel program. The value of numprocs can be queried by MPI_Comm_size. The processes are numbered in increasing order starting at 0. To determine the rank of the process, one can use the function MPI_Comm_rank. After the call to this function, the return variable myrank contains a value between 0 and numprocs-1. Using the process number myrank and the total number numprocs of processes, one can control the behavior of the local process.

Parallel programs can be started in different ways, depending on the parallel computer used. Ultimately, the executable program has to be copied to each processor and the desired amount of processes have to be started on each processor. Two possible ways to start the parallel program a.out in two processes are for instance mpirun -np 2 a.out and mpiexec -n 2 a.out.

There are a number of different send and receive functions in MPI for the sending and receiving of data between two processes. We first consider the routines

```
int MPI_Send(void* data, int length, MPI_Datatype, int to,
             int tag, MPI_Comm);
int MPI_Recv(void* data, int length, MPI_Datatype, int from,
             int tag, MPI_Comm, MPI_Status *);
```

in more detail. With their help one can transport a vector of length length between two processes. The data type (for instance MPI_INT or MPI_DOUBLE) and a pointer to the first element data have to be given as arguments. For the send operation, the rank of the target process has to be given, for the corresponding receive operation the process, from which the message originates, has to be given. The field tag can be used as a filter for messages, as only messages with matching tags are received. The communicator MPI_Comm is usually MPI_COMM_WORLD.[1]

The functions just introduced have one flaw, however: They are blocking operations. Thus, the program cannot continue unless the operation has been completed. This sounds harmless. But imagine a situation in which every process sends some data to its neighboring process. If we write a program in which MPI_Send is followed by MPI_Recv, it might work if only a small amount of data is to be transported. But if a lot of data have to be sent, the operation MPI_Send blocks until the operation MPI_Recv is encountered by the receiving process and enough memory is allocated to receive the message. In the worst case, all processes get stuck in the operation MPI_Send and no process ever reaches the receiving operation MPI_Recv which would allow the parallel program to proceed. To eliminate this kind of deadlock among the processes one could change the communication pattern, for instance by having all processes with even numbers send and having all processes with odd numbers receive in the first step, and reversing the pattern in the second step. However, it is simpler to use a nonblocking send routine. The following Algorithm A.3 solves our problem.

Algorithm A.3 Nonblocking Communication

```
void comm(void *send_data, int send_length, int send_to,
          void *recv_data, int recv_length, int recv_from,
          int tag, MPI_Datatype datatype) {
  MPI_Request req;
  MPI_Status status1, status2;
  MPI_Isend(send_data, send_length, datatype, send_to,
            tag, MPI_COMM_WORLD, &req);
  MPI_Recv (recv_data, recv_length, datatype, recv_from,
            tag, MPI_COMM_WORLD, &status1);
  MPI_Wait(&req, &status2);
}
```

The call MPI_Isend (I stands for immediate) does not block. We can therefore use the receive operation MPI_Recv without change. To make sure that all data are sent we now have to insert a further call to MPI_Wait, which

---

[1] One can also use other communicators to combine only some of the processes to a smaller, virtual, parallel computer, see [272].

waits for the completion of the MPI operation `MPI_Request`. Only then we can change the vector of data pointed to by `send_data`.

In principle, one can write complete parallel programs with these point-to-point communication operations. For operations that involve all processes, as for instance a sum or a maximum of values distributed on all processes, it is advantageous to use a few additional specialized routines. They are:

```
int MPI_Allreduce(void* datain, void* dataout, int length,
                  MPI_Datatype, MPI_Op, MPI_Comm);
int MPI_Allgather(void* datain, int lengthin, MPI_Datatype,
                  void* dataout, int lengthout, MPI_Datatype,
                  MPI_Comm);
```

The call to `MPI_Allreduce` computes the data contributed by each process according to `MPI_Op` and returns the final result to all processes. Possible choices for `MPI_Op` are `MPI_SUM`, `MPI_PROD`, `MPI_MIN` or `MPI_MAX`. It is guaranteed that all processes receive exactly the same final result. The precise sequence of operations depends on the MPI implementation. Thus, different MPI implementations might lead to differently rounded values. Usually, all data are first sent to a master process which then *reduces* the data according to `MPI_Op`.

The function `MPI_Allgather` is a simplified variant of such a reduction. It delivers the collected data to all processes without any further processing. Both functions block until all processes have contributed their data.

Finally, we present a small parallel example program that employs the communication operations just introduced. We use the main program from Algorithm A.2 completed by a call to a new routine `solve` which uses the nonblocking communication routine `comm` from Algorithm A.3. In addition, it employs the MPI function `MPI_Wtime()` which returns the value of a local system clock. The implementation of `solve` is given in A.4.

The routine `solve` from Algorithm A.4 solves a one-dimensional Poisson problem $-\Delta x = f$ with periodic boundary conditions. This is implemented with a damped Jacobi iteration.[2] The unknowns are stored for each process in $x[1]$ to $x[n]$, the corresponding right hand side is stored in $f[1]$ to $f[n]$. The data are distributed to the processes so that each process only owns n values from a much longer parallel vector. The neighboring processes for each process are called `left` and `right`. To implement periodic boundary conditions, the left neighbor of process 0 is set to process `numprocs-1` and correspondingly the right neighbor of process `numprocs-1` is set to process 0. A loop over j then executes m Jacobi iterations. First, the "ghost cells" $x[0]$ and $x[n+1]$ are filled with the correct values from the neighboring processes in a nonblocking communication step. A damped Jacobi iteration step is executed next. We use the discretization stencil $[-1 \quad 2 \quad -1]$ for the

---

[2] This program is just intended as a first example for parallelization. The parallel FFT solver discussed in Chapter 7 is substantially faster for our applications.

Algorithm A.4 Parallel Iterative Jacobi-Solver

```
void solve() {
#define n 10
#define m 50
  double x[n+2], y[n+2], f[n+2];
  int myrank, numprocs;
  MPI_Comm_size(MPI_COMM_WORLD, &numprocs); // to determine
  MPI_Comm_rank(MPI_COMM_WORLD, &myrank);   // neighboring processes
  int left  = (myrank + numprocs - 1) % numprocs;
  int right = (myrank + 1) % numprocs;
  for (int i=1; i<=n; i++) {
    x[i] = 0;
    f[i] = sin((i + n * myrank) * 2. * M_PI / (n * numprocs));
  }
  real t = MPI_Wtime();
  for (int j=0; j<m; j++) { // iteration counter
    comm(&x[n], 1, right, &x[0],   1, left,  7, MPI_DOUBLE);
    comm(&x[1], 1, left,  &x[n+1], 1, right, 8, MPI_DOUBLE);
    for (int i=1; i<=n; i++)
      y[i] = 0.5 * x[i] + (f[i] + x[i-1] + x[i+1]) * 0.25;
    double s = 0, c;
    for (int i=1; i<=n; i++) s += y[i];
    MPI_Allreduce(&s, &c, 1, MPI_DOUBLE, MPI_SUM, MPI_COMM_WORLD);
    for (int i=1; i<=n; i++)
      x[i] = y[i] - c / (n * numprocs); // force average to 0
  }
  printf("proc %d: %g sec\n", myrank, MPI_Wtime()-t);
  for (int i=1; i<=n; i++)
    printf("x[%d] = %g\n", i + n * myrank, x[i]);
}
```

Laplace operator. A damping factor of $1/2$ is employed. The solution of our problem is only defined up to an additive constant because of the periodic boundary conditions. We therefore force the solution to have zero average. To this end, we first compute the local sum in s, compute then the global sum with MPI_Allreduce, and finally subtract out the average on all processes in parallel. At the end of the program, we print the time needed for the solution and the solution itself.

A.4 Maxwell-Boltzmann Distribution

The Maxwell-Boltzmann distribution is given for a vector $\mathbf{v} = (v_1, v_2, v_3) \in \mathbb{R}^3$ as an appropriately transformed Gaussian $N(0, 1)$ normal distribution

$$f(\mathbf{v}) := \left(\frac{m}{2\pi k_B T}\right)^{\frac{3}{2}} e^{-\frac{m\mathbf{v}^2}{2k_B T}}. \tag{A.5}$$

Here, T denotes the temperature, m denotes the mass of the particles, and $k_B = 1.380662\cdot 10^{-23} \frac{J}{K}$ denotes the Boltzmann constant. If a velocity vector $\mathbf{v}$ is distributed according to a Maxwell-Boltzmann distribution, its magnitude has the distribution

$$f_0(||\mathbf{v}||) := 4\pi \mathbf{v}^2 f(\mathbf{v}) = 4\pi \left(\frac{m}{2\pi k_B T}\right)^{\frac{3}{2}} \mathbf{v}^2 e^{-\frac{m\mathbf{v}^2}{2k_B T}}. \tag{A.6}$$

The mean square velocity thus satisfies

$$\langle \mathbf{v}^2 \rangle := \int_0^\infty \mathbf{v}^2 f_0(||\mathbf{v}||) d||\mathbf{v}|| = \frac{3k_B T}{m}. \tag{A.7}$$

Every degree of freedom v_d, $d = 1, 2, 3$, contributes a mean kinetic energy of

$$\frac{1}{2} m \langle v_d^2 \rangle = \frac{k_B T}{2} \tag{A.8}$$

to the total kinetic energy. This is the so-called equidistribution theorem of thermodynamics. Here, $\langle v_d^2 \rangle$ is the mean squared velocity for the dth component of $\mathbf{v}$, $d = 1, 2, 3$. In general a system with N particles of equal mass (which has $N_f = 3N$ degrees of freedom) then satisfies

$$E_{kin} = \frac{1}{2} \sum_{i=1}^N m\mathbf{v}_i^2 = \frac{1}{2} \sum_{i=1}^N m \sum_{d=1}^3 (\mathbf{v}_i)_d^2 = \frac{N_f}{2} k_B T. \tag{A.9}$$

To construct a set of particles from a Maxwell-Boltzmann distribution, one first generates random vectors which are distributed according to a multivariate $N(0, 1)$ normal distribution. They are then scaled by a factor of $\sqrt{k_B T/m}$. This factor can also be expressed in terms of the kinetic energy or the mean velocity. (A.8) and (A.9) imply

$$\sqrt{(k_B T)/m} = \sqrt{(2E_{kin})/(N_f m)} = \sqrt{\langle v_d^2 \rangle}. \tag{A.10}$$

Thus, given the kinetic energy or the mean velocity, we easily obtain a corresponding Maxwell-Boltzmann distribution as well.

Algorithm A.5 shows a possible implementation following Marsagila [414], see also [109, 356] for the general Box-Muller method. There, the function GaussDeviate generates random numbers distributed according to a $N(0, 1)$ normal distribution. First, two random numbers a_1 and a_2 from a uniform random distribution are generated (here using the rand function for simplicity[3]) which are interpreted as coordinates of a point in the unit circle. The polar transform [414] then produces two numbers

---

[3] Here, two independent uniformly distributed random numbers are needed. The rand function is not really adequate since the same algorithm with the same seed is used for both numbers. However, the generation of good random numbers is an extensive subject that cannot be covered here in detail, see also [490].

$$b_1 = a_1 \sqrt{(-2\ln r)/r}, \qquad b_2 = a_2 \sqrt{(-2\ln r)/r}$$

from a normal distribution, with $r = a_1^2 + a_2^2$. These values are multiplied with the value `factor` in the function `MaxwellBoltzmann`. The quantities $\sqrt{\langle v_d^2 \rangle}$, E_{kin}, or T were given as parameter in the examples from Chapter 3. These values then must be plugged into (A.10) to obtain the respective value of `factor`.[4]

Algorithm A.5 Maxwell-Boltzmann Distribution

```
void MaxwellBoltzmann(Particle *p, real factor) {
  for (int d=0; d<DIM; d++)
    p->v[d] = factor * GaussDeviate ();
}

real GaussDeviate(void) {
  real a1, a2, s, r, b1;
  static int iset = 0;
  static real b2;
  if (!iset) {
    do {                                      // two uniformly
      a1 = 2.0 * rand () / (RAND_MAX + 1.0) - 1.0; // distributed
      a2 = 2.0 * rand () / (RAND_MAX + 1.0) - 1.0; // random numbers
      r = a1 * a1 + a2 * a2;                  // from (-1,1)
    } while (r>=1.0);            // is (a1,a2) inside the unit circle?
    s = sqrt (-2.0 * log (r) / r);     // polar transform
    b1 = a1 * s;
    b2 = a2 * s;
    iset = 1;
    return b1;
  }
  else {
    iset = 0;
    return b2;
  }
}
```

---

[4] Other algorithms for the generation of normally distributed random variables are implemented for instance in the *GNU Scientific Library* [24] and can be used as library functions.

A.5 Parameters for the Brenner Potential and Data for Initial Configurations

Parameter Table for the Splines in the Brenner Potential

In the following, we give the nodal values for the two-dimensional and three-dimensional cubic Hermite splines H_{CC}, H_{CH} and K needed in the definition of the Brenner potential in the formula (5.23) and (5.25). They smooth the transition from the bound to the unbound state. Further explanations can be found in [122, 123, 602].

Carbon		Hydrogen	Hydrocarbons	
-	-	-	$H_{CC}(1,1)$	-0.0226
-	-	-	$H_{CC}(2,0)$	-0.0061
-	-	-	$H_{CC}(3,0)$	0.0173
-	-	-	$H_{CC}(1,2)$	0.0149
-	-	-	$H_{CC}(2,1)$	0.0160
-	-	-	$H_{CH}(1,0)$	-0.0984
-	-	-	$H_{CH}(2,0)$	-0.2878
-	-	-	$H_{CH}(3,0)$	-0.4507
-	-	-	$H_{CH}(0,1)$	-0.2479
-	-	-	$H_{CH}(0,2)$	-0.3221
-	-	-	$H_{CH}(1,1)$	-0.3344
-	-	-	$H_{CH}(2,1)$	-0.4438
-	-	-	$H_{CH}(0,3)$	-0.4460
-	-	-	$H_{CH}(1,2)$	-0.4449
-	-	-	$\frac{\partial H_{CH}(1,1)}{\partial C}$	-0.17325
-	-	-	$\frac{\partial H_{CH}(2,0)}{\partial C}$	-0.09905
-	-	-	$\frac{\partial H_{CH}(0,2)}{\partial H}$	-0.17615
-	-	-	$\frac{\partial H_{CH}(1,1)}{\partial H}$	-0.09795
$K(2,3,1)$	-0.0363	-	-	
$K(2,3,2)$	-0.0363	-	-	
$K(1,2,2)$	-0.0243	-	-	
-	-	-	$K(1,1,1)$	0.1264
-	-	-	$K(2,2,1)$	0.0605
-	-	-	$K(1,2,1)$	0.0120
-	-	-	$K(1,3,1), K(1,3,2)$	-0.0903
-	-	-	$K(0,3,1), K(0,3,2)$	-0.0904
-	-	-	$K(0,2,2)$	-0.0269
-	-	-	$K(0,2,1)$	0.0427
-	-	-	$K(0,1,1)$	0.0996
-	-	-	$K(1,1,2)$	0.0108
-	-	-	$\frac{\partial K(3,1,1)}{\partial i}$	-0.0950
-	-	-	$\frac{\partial K(3,2,1)}{\partial i}$	-0.10835
-	-	-	$\frac{\partial K(3,1,2)}{\partial i}$	-0.0452
-	-	-	$\frac{\partial K(2,3,2)}{\partial i}$	0.01345
-	-	-	$\frac{\partial K(2,4,2)}{\partial i}$	-0.02705
-	-	-	$\frac{\partial K(3,4,2)}{\partial i}$	0.04515
-	-	-	$\frac{\partial K(3,4,1)}{\partial i}$	0.04515
-	-	-	$\frac{\partial K(3,2,2)}{\partial i}$	-0.08760

Table A.1. Nodal values for the two-dimensional and three-dimensional cubic Hermite splines H_{CC}, H_{CH} and K in the Brenner potential. The following properties hold: $K(i,j,k) = K(j,i,k)$, $K(i,j,k > 2) = K(i,j,2)$, and $\partial K(i,j,k)/\partial i = \partial K(j,i,k)/\partial i$. The partial derivatives are needed for the force computation. Values not listed are equal to zero.

Data for Dehydrobenzene and a C_{60} Molecule

We give the positions of the atoms for a C_{60} molecule and a dehydrobenzene molecule in the Tables A.2 and A.3. They are needed for the initial configuration for the simulations in Section 5.1.3. For further details see also the web site [25].

	$x_1[\text{Å}]$	$x_2[\text{Å}]$	$x_3[\text{Å}]$		$x_1[\text{Å}]$	$x_2[\text{Å}]$	$x_3[\text{Å}]$
C	4.13127	2.04365	0.24220	C	2.72152	1.66445	0.42819
C	1.86478	2.86901	0.34933	C	2.78173	3.98528	0.06360
C	4.13936	3.48140	-0.00753	C	5.78527	0.70439	3.17888
C	4.89453	0.45453	2.02247	C	5.18828	1.44270	0.98960
C	6.25133	2.30427	1.49597	C	6.61994	1.85765	2.84958
C	1.65940	0.48262	3.70735	C	1.26109	0.92194	2.37547
C	2.41952	0.73208	1.46188	C	3.50540	0.16862	2.24948
C	3.04602	0.05591	3.62921	C	-0.05841	4.24502	3.15406
C	0.39476	4.42145	1.75617	C	0.79918	3.08211	1.28696
C	0.45394	2.09464	2.31554	C	-0.06301	2.80125	3.49715
C	3.04828	6.89601	2.18925	C	3.53632	6.16548	1.01429
C	2.45515	5.29269	0.54756	C	1.23755	5.52335	1.37420
C	1.64629	6.46513	2.45815	C	6.69230	4.65624	2.21740
C	6.31185	3.71062	1.17977	C	5.18969	4.31888	0.44867
C	4.87531	5.65965	0.95503	C	5.81036	5.85288	2.08070
C	4.29026	2.99332	7.03625	C	2.90557	3.49798	7.11094
C	2.92756	4.93406	6.76120	C	4.31490	5.30613	6.45220
C	5.18211	4.11225	6.65676	C	3.91401	0.20644	4.75482
C	3.43612	0.85954	5.97599	C	4.55447	1.68667	6.49877
C	5.72588	1.45256	5.58961	C	5.30675	0.52704	4.50972
C	0.30899	2.34704	4.82214	C	0.70633	3.29736	5.88615
C	1.82469	2.70502	6.62986	C	2.10143	1.34401	6.04534
C	1.20806	1.18868	4.87518	C	1.26425	6.21960	3.81811
C	2.13242	6.47272	4.94969	C	1.81030	5.53400	6.06702
C	0.70986	4.69924	5.55349	C	0.37137	5.11396	4.19912
C	5.32320	6.54249	3.22881	C	5.72396	6.08319	4.56802
C	4.54943	6.27301	5.45155	C	3.46748	6.86751	4.67617
C	3.93181	7.01512	3.28944	C	6.99425	2.77353	3.84704
C	6.54811	2.56795	5.22097	C	6.27829	3.90803	5.77491
C	6.57361	4.92334	4.71704	C	7.01550	4.19431	3.52312

Table A.2. Coordinates of the atoms of a C_{60} molecule.

	$x_1[\text{Å}]$	$x_2[\text{Å}]$	$x_3[\text{Å}]$
C	12.99088	3.50000	2.83052
C	12.99088	3.50000	4.21148
C	14.18373	3.50000	2.07479
H	14.14772	3.50000	0.98434
C	15.35883	3.50000	2.81770
H	16.31845	3.50000	2.29236
C	15.35883	3.50000	4.22430
H	16.31845	3.50000	4.74964
C	14.18373	3.50000	4.96721
H	14.14772	3.50000	6.05766

Table A.3. Coordinates of the atoms of a dehydrobenzene molecule.

References

1. http://www.top500.org.
2. http://www.chim.unifi.it/orac.html.
3. http://www.mpa-garching.mpg.de/∼volker/gadget.
4. http://coho.physics.mcmaster.ca/hydra.
5. http://www.ivec.org/GULP.
6. http://www.ccp5.ac.uk/librar.shtml.
7. http://www.mpi-forum.org.
8. http://www.new-npac.org/projects/cdroms/cewes-1999-06-vol1/nhse/hpcc-survey/index.html.
9. http://www.openmp.org.
10. http://www.epm.ornl.gov/pvm.
11. http://www-unix.mcs.anl.gov/mpi/mpich.
12. http://www.lam-mpi.org.
13. http://www.pccluster.org.
14. http://www.hyper.com, (Hyperchem).
15. http://www.cs.sandia.gov/∼sjplimp.
16. http://www.mpa-garching.mpg.de/millennium/.
17. http://www.mpa-garching.mpg.de/galform/press/.
18. http://cosmicweb.uchicago.edu/sims.html.
19. http://www.ifa.hawaii.edu/∼barnes/barnes.html.
20. http://www.psc.edu/general/software/packages/charmm/tutorial/mackerell/membrane.html.
21. http://moose.bio.ucalgary.ca/index.php?page=Peter_Tieleman.
22. http://persweb.wabash.edu/facstaff/fellers/coordinates.html.
23. http://www.lrz-muenchen.de/∼heller/membrane/membrane.html.
24. http://www.gnu.org.
25. http://people.nas.nasa.gov/globus/papers/MGMS_EC1/simulation/data/.
26. *The nanotechnology site.* http://www.pa.msu.edu/cmp/csc/nanotech.
27. *The nanotube site.* http://www.pa.msu.edu/cmp/csc/nanotube.html.
28. G. ABELL, *Empirical chemical pseudopotential theory of molecular and metallic bonding*, Phys. Rev. B, 31 (1985), pp. 6184–6196.
29. J. ADAMS AND S. FOILES, *Development of an embedded atom potential for a bcc metal: Vanadium*, Phys. Rev. B, 41 (1990), pp. 3316–3328.
30. R. ADAMS, *Sobolev spaces*, Academic Press, New York, 1975.
31. F. ALBER AND P. CARLONI, *Ab initio molecular dynamics studies on HIV-1 reverse transcriptase triphosphate binding site: Implications for nucleoside analog drug resistance*, Protein Sci., 9 (2000), pp. 2535–2546.

32. W. ALDA, W. DZWINEL, J. KITOWSKI, J. MOŚCIŃSKI, M. POGODA, AND D. YUEN, *Rayleigh-Taylor instabilities simulated for a large system using molecular dynamics*, Tech. Rep. UMSI 96/104, Supercomputer Institute, University of Minnesota, Minneapolis, 1996.

33. B. ALDER AND T. WAINWRIGHT, *Phase transition for a hard sphere system*, J. Chem. Phys., 27 (1957), pp. 1208–1209.

34. M. ALLEN AND D. TILDESLEY, *Computer simulation of liquids*, Clarendon Press, Oxford, 1987.

35. J. ALMLÖF, *Notes on Hartree-Fock theory and related topics*, in Lecture notes in quantum chemistry II, B. Roos, ed., vol. 64 of Lecture Notes in Chemistry, Springer, Berlin, 1994, pp. 1–90.

36. S. ALURU, *Greengard's N-body algorithm is not order N*, SIAM J. Sci. Comput., 17 (1996), pp. 773–776.

37. A. AMADEI, A. LINSSEN, AND H. BERENDSEN, *Essential dynamics of proteins*, Proteins: Struct. Funct. Genet., 17 (1993), pp. 412–425.

38. J. AMBROSIANO, L. GREENGARD, AND V. ROHKLIN, *The fast multipole method for gridless particle simulations*, Comp. Phys. Comm., 48 (1988), pp. 117–125.

39. G. AMDAHL, *Validity of the single processor approach to achieving large scale computing capabilities*, in Proc. AFIPS Spring Joint Computer Conf., Reston, Va., 1967, AFIPS Press, pp. 483–485.

40. AMERICAN NATIONAL STANDARDS INSTITUTE, *Programming Languages - C*, Washington, DC, 1999. ANSI/ISO/IEC Standard No. 9899-1999.

41. M. AMINI AND R. HOCKNEY, *Computer simulation of melting and glass formation in a potassium chloride microcrystal*, J. Non-Cryst. Solids, 31 (1979), pp. 447–452.

42. H. ANDERSEN, *Molecular dynamics simulation at constant pressure and/or temperature*, J. Chem. Phys., 72 (1980), pp. 2384–2393.

43. ———, *RATTLE: A 'velocity' version of the SHAKE algorithm for molecular dynamics calculations*, J. Comput. Phys., 52 (1983), pp. 24–34.

44. B. ANDREWS, T. ROMO, J. CLARAGE, B. PETTITT, AND G. PHILLIPS, *Characterizing global substates of myoglobin*, Structure Folding Design, 6 (1998), pp. 587–594.

45. J. ANTOSIEWICZ, E. BLACHUT-OKRASINSKA, T. GRYCUK, J. BRIGGS, S. WLODEK, B. LESYNG, AND J. MCCAMMON, *Predictions of pK_as of titratable residues in proteins using a Poisson-Boltzmann model of the solute-solvent system*, in Computational Molecular Dynamics: Challenges, Methods, Ideas, P. Deuflhard, J. Hermans, B. Leimkuhler, A. Mark, S. Reich, and R. Skeel, eds., vol. 4 of Lecture Notes in Computational Science and Engineering, Springer, New York, 1999, pp. 176–196.

46. K. AOKI AND T. AKIYAMA, *Spontaneous wave pattern formation in vibrated granular materials*, Phys. Rev. Lett., 77 (1996), pp. 4166—4169.

47. A. APPEL, *An efficient program for many-body simulation*, SIAM J. Sci. Stat. Comput., 6 (1985), pp. 85–103.

48. V. ARNOLD, *Mathematical methods of classical mechanics*, Springer, New York, 1978.

49. ———, ed., *Dynamical systems III: Mathematical aspects of classical and celestial mechanics*, vol. 3 of Encyclopaedia of Mathematical Sciences, Springer, New York, 1994.

50. N. ARONSZAJN AND K. SMITH, *Theory of Bessel potentials I*, Ann. Inst. Fourier (Grenoble), 11 (1961), pp. 385–475.

51. D. ASIMOV, *Geometry of capped nanocylinders*, 1998. http://www.research. att.com/areas/stat/dan/nano-asimov.ps.gz.

52. B. AXILROD AND E. TELLER, *Interaction of the van der Waals type between three atoms*, J. Chem. Phys., 11 (1943), pp. 299–300.

53. I. BABUSKA AND W. RHEINBOLDT, *Error estimates for adaptive finite element computations*, SIAM J. Numer. Anal., 15 (1978), pp. 736–754.

54. I. BABUSKA AND M. SURI, *The p and h-p versions of the finite element method, basic principles and properties*, SIAM Rev., 36 (1994), pp. 578–632.

55. M. BAINES, *Moving finite elements*, Oxford University Press, Oxford, 1994.

56. R. BALESCU, *Statistical dynamics, matter out of equilibrium*, Imperial College Press, London, 1997.

57. J. BARKER, R. FISCHER, AND R. WATTS, *Liquid argon: Monte Carlo and molecular dynamics calculations*, Mol. Phys., 21 (1971), pp. 657–673.

58. J. BARNES AND P. HUT, *A hierarchical $O(N \log(N))$ force-calculation algorithm*, Nature, 324 (1986), pp. 446–449.

59. M. BARRECA, A. CHIMIRRI, L. DE LUCA, A. MONFORTE, P. MONFORTE, A. RAO, M. ZAPPALA, J. BALZARINI, E. DE CLERCQ, C. PANNECOUQUE, AND M. WITVROUW, *Discovery of 2,3-diaryl-1,3-thiazolidin-4-ones as potent anti-HIV-1 agents*, Bioorg. Med. Chem. Lett., 11 (2001), pp. 1793–1796.

60. E. BARTH, B. LEIMKUHLER, AND S. REICH, *A test set for molecular dynamics*, Tech. Rep. 05, University of Leicester, MCS, 2001.

61. E. BARTH AND T. SCHLICK, *Extrapolation versus impulse in multiple-timestepping schemes: Linear analysis and applications to Newtonian and Langevin dynamics*, J. Chem. Phys., 109 (1998), pp. 1633–1642.

62. ———, *Overcoming stability limitations in biomolecular dynamics: Combining force splitting via extrapolation with Langevin dynamics in LN*, J. Chem. Phys., 109 (1998), pp. 1617–1632.

63. M. BASKES, *Application of the embedded-atom method to covalent materials: A semiempirical potential for silicon*, Phys. Rev. Lett., 59 (1987), pp. 2666–2669.

64. ———, *Modified embedded-atom potentials for cubic materials and impurities*, Phys. Rev. B, 46 (1992), pp. 2727–2742.

65. M. BASKES, J. NELSON, AND A. WRIGHT, *Semiempirical modified embedded-atom potentials for silicon and germanium*, Phys. Rev. B, 40 (1989), pp. 6085–6100.

66. R. BEATSON AND L. GREENGARD, *A short course on fast multipole methods*, in Wavelets, Multilevel Methods and Elliptic PDEs, M. Ainsworth, J. Levesley, W. Light, and M. Marletta, eds., Numerical Mathematics and Scientific Computation, Oxford University Press, Oxford, 1997, pp. 1–37.

67. R. BEATSON AND G. NEWSAM, *Fast evaluation of radial basis functions: I*, Comp. Math. Applic., 24 (1992), pp. 7–19.

68. ———, *Fast evaluation of radial basis functions: Moment based methods*, SIAM J. Sci. Comput., 19 (1998), pp. 1428–1449.

69. D. BEAZLEY AND P. LOMDAHL, *Message-passing multi-cell molecular dynamics on the Connection Machine 5*, Parallel Comp., 20 (1994), pp. 173–195.

70. ———, *Lightweight computational steering of very large scale molecular dynamics simulations*, in Supercomputing '96 Conference Proceedings: November 17–22, Pittsburgh, PA, ACM, ed., New York, 1996, ACM Press and IEEE Computer Society Press.

71. D. BEAZLEY, P. LOMDAHL, N. GRONBECH-JENSEN, R. GILES, AND P. TAMAYO, *Parallel algorithms for short-range molecular dynamics*, in Annual Reviews of Computational Physics, D. Stauffer, ed., vol. 3, World Scientific, 1996, pp. 119–175.

72. M. BEBENDORF, S. RJASANOW, AND E. TYRTYSHNIKOV, *Approximation using diagonal-plus-skeleton matrices*, in Mathematical aspects of boundary element methods, M. Bonnet, A. Sandig, and W. Wendland, eds., Chapman & Hall/CRC Research Notes in Mathematics, 1999, pp. 45–53.

73. A. BEJAN, *Convection heat transfer*, Wiley-Interscience, New York, 1984.

74. H. BEKKER, *Molecular dynamics simulation methods revised*, Proefschrift, Rijksuniversiteit Groningen, 1996.

75. M. BELHADJ, H. ALPER, AND R. LEVY, *Molecular dynamics simulations of water with Ewald summation for the long range electrostatic interactions*, Chem. Phys. Lett., 179 (1991), pp. 13–20.

76. T. BELYTSCHKO, Y. LU, AND L. GU, *Element-free Galerkin methods*, Int. J. Numer. Meth. Eng., 27 (1994), pp. 229–256.

77. A. BEN-NAIM AND F. STILLINGER, *Aspects of the statistical-mechanical theory of water*, in Water and Aqueous Solutions, R. Horne, ed., Wiley, New York, 1972, pp. 295–330.

78. G. BENETTIN AND A. GIORGILLI, *On the Hamiltonian interpolation of near to the identity symplectic mappings*, J. Stat. Phys., 74 (1994), pp. 1117–1143.

79. H. BERENDSEN, J. GRIGERA, AND T. STRAATSMA, *The missing term in effective pair potentials*, J. Phys. Chem., 91 (1987), pp. 6269–6271.

80. H. BERENDSEN, J. POSTMA, W. VAN GUNSTEREN, A. DI NOLA, AND J. HAAK, *Molecular dynamics with coupling to an external bath*, J. Chem. Phys., 81 (1984), pp. 3684–3690.

81. H. BERENDSEN, J. POSTMA, W. VAN GUNSTEREN, AND J. HERMANS, *Interaction models for water in relation to protein hydration*, in Intermolecular Forces, B. Pullman, ed., Reidel Dordrecht, Holland, 1981, pp. 331–342.

82. H. BERENDSEN AND D. TIELEMAN, *Molecular dynamics: Studies of lipid bilayers*, in Encyclopedia of Computational Chemistry, P. von Ragué Schleyer, ed., vol. 3, Wiley, New York, 1998, pp. 1639–1650.

83. H. BERENDSEN, D. VAN DER SPOEL, AND R. VAN DRUNEN, *GROMACS: A message-passing parallel molecular dynamics implementation*, Comp. Phys. Comm., 91 (1995), pp. 43–56.

84. C. BERMAN AND L. GREENGARD, *A renormalization method for the evaluation of lattice sums*, J. Math. Phys., 35 (1994), pp. 6036–6048.

85. H. BERMAN, J. WESTBROOK, Z. FENG, G. GILLILAND, T. BHAT, H. WEISSIG, I. SHINDYALOV, AND P. BOURNE, *The protein data bank*, Nucleic Acids Research, 28 (2000), pp. 235–242. http://www.rcsb.org/pdb.

86. J. BERNAL AND R. FOWLER, *A theory of water and ionic solution, with particular reference to hydrogen and hydroxyl ions*, J. Chem. Phys., 1 (1933), pp. 515–548.

87. E. BERTSCHINGER, *COSMICS: Cosmological initial conditions and microwave anisotropy codes*, Dept. of Physics, MIT, Cambridge. http://arcturus.mit.edu/cosmics.

88. G. BEYLKIN, *On the representation of operators in bases of compactly supported wavelets*, SIAM J. Num. Anal., 29 (1992), pp. 1716–1740.

89. ———, *On the fast Fourier transform of functions with singularities*, PAM report 195, University of Colorado at Boulder, 1994.

90. K. BINDER AND G. CICCOTTI, eds., *Monte Carlo and molecular dynamics of condensed matter systems*, vol. 49 of Conference Proceedings, Italian Physical Society, Bologna, 1995. Euroconference on Computer Simulation in Condensed Matter Physics and Chemistry.

91. G. BIRD, *Molecular gas dynamics and the direct simulation of gas flows*, Oxford University Press, 1994.

92. C. BIZON, M. SHATTUCK, J. SWIFT, W. MCCORMICK, AND H. SWINNEY, *Patterns in 3D vertically oscillated granular layers: Simulation and experiment*, Phys. Rev. Lett., 80 (1997), pp. 57–60.

93. D. BLACKSTON AND T. SUEL, *Highly portable and efficient implementations of parallel adaptive N-body methods*, tech. rep., Computer Science Division, University of California at Berkeley, 1997.

94. G. BLELLOCH AND G. NARLIKAR, *A practical comparison of N-body algorithms*, in Parallel Algorithms, vol. 30 of Series in Discrete Mathematics and Theoretical Computer Science, American Mathematical Society, 1997.

95. J. BOARD, Z. HAKURA, W. ELLIOTT, AND W. RANKIN, *Scalable variants of multipole-accelerated algorithms for molecular dynamics applications*, in Proc. 7. SIAM Conf. Parallel Processing for Scientific Computing, D. Bailey, P. Bjørstad, J. Gilbert, M. Mascagni, R. Schreiber, H. Simon, V. Torczon, and L. Watson, eds., Philadelphia, 1995, SIAM, pp. 295–300.

96. J. BOARD, C. HUMPHRES, C. LAMBERT, W. RANKIN, AND A. TOUKMAJI, *Ewald and multipole methods for periodic N-body problems*, in Lecture Notes in Computational Science and Engineering, Vol. 4, Springer-Verlag, 1998.

97. S. BOGUSZ, T. CHEATHAM, AND B. BROOKS, *Removal of pressure and free energy artefacts in charged periodic systems via net charge corrections to the Ewald potential*, J. Chem. Phys., 108 (1998), pp. 7070–7084.

98. S. BOND, B. LEIMKUHLER, AND B. LAIRD, *The Nosé-Poincaré method for constant temperature molecular dynamics*, J. Comput. Phys., 151 (1999), pp. 114–134.

99. A. BONVIN AND W. VAN GUNSTEREN, *β-hairpin stability and folding: Molecular dynamics studies of the first β-hairpin of tendamistat*, J. Mol. Biol., 296 (2000), pp. 255–268.

100. S. BORESCH AND O. STEINHAUSER, *Presumed versus real artifacts of the Ewald summation technique: The importance of dielectric boundary conditions*, Ber. Bunsenges. Phys. Chem., 101 (1997), pp. 1019–1029.

101. S. BÖRM, $\mathcal{H}^2$-*matrices – multilevel methods for the approximation of integral operators*. Max Planck Institute for Mathematics in the Sciences, Preprint Nr. 7/2003, 2003.

102. S. BÖRM AND W. HACKBUSCH, *Data-sparse approximation by adaptive* $\mathcal{H}^2$-*matrices*, Computing, 69 (2002), pp. 1–35.

103. F. BORNEMANN, *Homogenization in time of singularly perturbed mechanical systems*, no. 1687 in Lecture Notes in Mathematics, Springer, Berlin, 1998.

104. F. BORNEMANN, P. NETTESHEIM, AND C. SCHÜTTE, *Quantum-classical molecular dynamics as an approximation to full quantum dynamics*, J. Chem. Phys., 105 (1996), pp. 1074–1083.

105. F. BORNEMANN AND C. SCHÜTTE, *Homogenization approach to smoothed molecular dynamics*, Nonlinear Analysis, 30 (1997), pp. 1805–1814.

106. F. BORNEMANN AND C. SCHÜTTE, *A mathematical investigation of the Car-Parrinello method*, Numer. Math., 78 (1998), pp. 359–376.

107. S. BOURNE, *The UNIX V Environment*, Addison-Wesley, 1987.

108. J. BOUSSINESQ, *Theorie analytique de la chaleur*, vol. 2, Gauthier-Villars, 1903.

109. G. BOX AND M.MULLER, *A note on the generation of random normal deviates*, Ann. Math. Stat., 29 (1958), pp. 610–611.

110. M. BRADY, *A fast discrete approximation for the Radon transform*, SIAM J. Comput., 27 (1998), pp. 107–119.

111. D. BRAESS, *Finite Elemente*, Springer, Berlin, 1992.

112. J. BRAMBLE, J. PASCIAK, AND J. XU, *Parallel multilevel preconditioners*, Math. Comp., 55 (1990), pp. 1–22.

113. C. BRANDEN AND J. TOOZE, *Introduction to protein structure*, Garland Publishing, 1998.

114. A. BRANDT, *Multi-level adaptive technique (MLAT) for fast numerical solutions to boundary value problems*, in Lecture Notes in Physics 18, H. Cabannes and R. Temam, eds., Heidelberg, 1973, Proc. 3rd Int. Conf. Numerical Methods in Fluid Mechanics, Springer, pp. 82–89.

115. ———, *Multi-level adaptive technique (MLAT). I. The multi-grid method*, IBM Research Report RC-6026, IBM T. Watson Research Center, Yorktown Heights, NY, 1976.

116. ———, *Multi-level adaptive solutions to boundary-value problems*, Math. Comp., 31 (1977), pp. 333–390.

117. ———, *Multigrid techniques: 1984 guide with applications to fluid dynamics*, tech. rep., GMD-Studien Nr. 85, Bonn, 1984.

118. ———, *Multilevel computations of integral transforms and particle interactions with oscillatory kernels*, Comp. Phys. Comm., 65 (1991), pp. 24–38.

119. A. BRANDT AND A. LUBRECHT, *Multilevel matrix multiplication and fast solution of integral equations*, J. Comput. Phys., 90 (1990), pp. 348–370.

120. A. BRANDT, J. MANN, M. BRODSKI, AND M. GALUN, *A fast and accurate multilevel inversion of the Radon transform*, SIAM J. on Applied Math., 60 (1999), pp. 437–462.

121. A. BRANDT AND C. VENNER, *Multilevel evaluation of integral transforms with asymptotically smooth kernel*, SIAM J. Sci. Comput., 12 (1998), pp. 468–492.

122. D. BRENNER, *Empirical potential for hydrocarbons for use in simulating the chemical vapor deposition of diamond films*, Phys. Rev. B, 42 (1990), pp. 9458–9471.

123. D. BRENNER, O. SHENDEROVA, J. HARRISON, S. STUART, B. NI, AND S. SINNOTT, *A second-generation reactive empirical bond order (REBO) potential energy expression for hydrocarbons*, J. Phys.: Condens. Matter, 14 (2002), pp. 783–802.

124. M. BROKATE AND J. SPREKELS, *Hysteresis and phase transitions*, Springer, New York, 1996.

125. B. BROOKS, R. BRUCCOLERI, B. OLAFSON, D. STATES, S. SWAMINATHAN, AND M. KARPLUS, *CHARMM: A program for macromolecular energy, minimization, and dynamics calculations*, J. Comput. Chem., 4 (1983), pp. 187–217.

126. B. BROOKS AND M. HODOSCEK, *Parallelization of CHARMM for MIMD machines*, Chem. Design Automation News, 7 (1992), pp. 16–22.

127. B. BROOKS, D. JANEZIC, AND M. KARPLUS, *Harmonic-analysis of large systems: I. Methodology*, J. Comput. Chem., 16 (1995), pp. 1522–1542.

128. R. BRUNNE, K. BERNDT, P. GÜNTERT, K. WÜTHRICH, AND W. VAN GUN-STEREN, *Structure and internal dynamics of the bovine pancreatic trypsin inhibitor in aqueous solution from long-time molecular dynamics simulations*, Proteins: Struct. Funct. Genet., 23 (1995), pp. 49–62.

129. E. BUCKINGHAM, *The classical equation of state of gaseous helium, neon and argon*, Proc. Roy. Soc. London A, 168 (1938), pp. 264–283.

130. C. BUNGE, J. BARRIENTOS, A. BUNGE, AND J. COGORDAN, *Hartree-Fock and Roothaan-Hartree-Fock energies for the ground states of He through Xe*, Phys. Rev. A, 46 (1992), pp. 3691–3696.

131. A. BUTZ, *Alternative algorithm for Hilbert's space-filling curve*, IEEE Trans. Comput., (1971), pp. 424–426.

132. P. BUTZER AND K. SCHERER, *Approximationsprozesse und Interpolationsmethoden*, Bibliographisches Institut, Mannheim, 1968.

133. A. CAGLAR AND M. GRIEBEL, *On the numerical simulation of Fullerene nanotubes: $C_{100.000.000}$ and beyond!*, in Molecular Dynamics on Parallel Computers, NIC, Jülich 8-10 February 1999, R. Esser, P. Grassberger, J. Grotendorst, and M. Lewerenz, eds., World Scientific, 1999, pp. 1–27.

134. M. CALVO AND J. SANZ-SERNA, *The development of variable-step symplectic integrators with application to the two-body problem*, SIAM J. Sci. Stat. Comput., 14 (1993), pp. 936–952.

135. C. CAMPBELL, *Rapid granular flows*, Annu. Rev. Fluid Mech., 22 (1990), pp. 57–92.

136. B. CANO AND A. DURAN, *An effective technique to construct symmetric variable-stepsizes linear multistep methods for second-order systems*, Tech. Rep. 10, Dpto. Matematica Aplicada y Computación, Universidad de Valladolid, 2000.

137. R. CAR AND M. PARRINELLO, *Unified approach for molecular dynamics and density functional theory*, Phys. Rev. Lett., 55 (1985), pp. 2471–2474.

138. V. CARRAVETTA AND E. CLEMENTI, *Water-water interaction potential: An approximation of the electron correlation contribution by a function of the SCF density matrix*, J. Chem. Phys., 81 (1984), pp. 2646–2651.

139. J. CARRIER, L. GREENGARD, AND V. ROKHLIN, *A fast adaptive multipole algorithm for particle simulations*, SIAM J. Sci. Stat. Comput., 9 (1988), pp. 669–686.

140. S. CHALASANI AND P. RAMANATHAN, *Parallel FFT on ATM-based networks of workstations*, Cluster Comp., 1 (1998), pp. 13–26.

141. D. CHANDLER, *Introduction to modern statistical mechanics*, Oxford University Press, New York, 1987.

142. S. CHARATI AND S. STERN, *Diffusion of gases in silicone polymers. Molecular dynamics simulations*, Macromolecules, 31 (1998), pp. 5529–5535.

143. H. CHENG, L. GREENGARD, AND V. ROKHLIN, *A fast adaptive multipole algorithm in three dimensions*, J. Comput. Phys., 155 (1999), pp. 468–498.

144. J. CHRISTIANSEN, *Vortex methods for flow simulations*, J. Comput. Phys., 13 (1973), pp. 363–379.

145. C. CHUI, *An introduction to wavelets*, Academic Press, Boston, 1992.

146. P. CIARLET, *The finite element method for elliptic problems*, North-Holland, Amsterdam, 1978.

147. G. CICCOTTI, D. FRENKEL, AND I. MCDONALD, eds., *Simulation of liquids and solids*, North Holland, Amsterdam, 1987.

148. G. CICCOTTI AND W. HOOVER, eds., *Molecular dynamics simulations of statistical-mechanical systems*, Proc. of international School of Physics, "Enrico Fermi" Course XCVII, Varenna, Italy, 1985, North Holland, Amsterdam, 1986.

149. T. CLARK, R. HANXLEDEN, J. MCCAMMON, AND L. SCOTT, *Parallelizing molecular dynamics using spatial decomposition*, in Proc. Scalable High Performance Computing Conference-94, IEEE Computer Society Press, 1994, pp. 95–102.

150. T. CLARK, J. MCCAMMON, AND L. SCOTT, *Parallel molecular dynamics*, in Proc. 5. SIAM Conf. Parallel Processing for Scientific Computing, 1992, pp. 338–344.

151. E. CLEMENTI AND C. ROETTI, *Atomic data and nuclear data tables*, vol. 14, Academic Press, New York, 1974.

152. J. COLLINS, S. BURT, AND J. ERICKSON, *Activated dynamics of flap opening in HIV-1 protease*, Adv. Exp. Med. Bio., 362 (1995), pp. 455–460.

153. ———, *Flap opening in HIV-1 protease simulated by 'activated' molecular dynamics*, Nat. Struct. Biol., 2 (1995), pp. 334–338.

154. M. COOKE, D. STEPHENS, AND J. BRIDGEWATER, *Powder mixing – a literature survey*, Powder Tech., 15 (1976), pp. 1–20.

155. J. COOLEY AND J. TUKEY, *An algorithm for the machine computation of complex Fourier series*, Math. Comp., 19 (1965), pp. 297–301.

156. G. COTTET AND P. KOUMOUTSAKOS, *Vortex methods: Theory and practice*, Cambridge University Press, Cambridge, 2000.

157. H. COUCHMAN, *Mesh-refined P^3M: A fast adaptive N-body algorithm*, Astrophys. J., 368 (1991), pp. 23–26.

158. H. COUCHMAN, P. THOMAS, AND F. PEARCE, *Hydra: An adaptive-mesh implementation of P^3M-SPH*, Astrophys. J., 452 (1995), pp. 797–813.

159. T. CREIGHTON, *Proteins, structures and molecular properties*, Freeman, New York, 1992.

160. M. CROWLEY, T. DARDEN, T. CHEATHAM, AND D. DEERFIELD, *Adventures in improving the scaling and accuracy of a parallel molecular dynamics program*, J. Supercomp., 11 (1997), pp. 255–278.

161. M. DA GAMA AND K. GUBBINS, *Adsorption and orientation of amphiphilic molecules at a liquid-liquid interface*, Mol. Phys., 59 (1986), pp. 227–239.

162. V. DAGGETT AND M. LEVINE, *A model of the molten globule state from molecular dynamics simulations*, Proc. Natl. Acad. Sci., 89 (1992), pp. 5142–5146.

163. V. DAGGETT AND M. LEVITT, *Protein unfolding pathways explored through molecular dynamics simulations*, J. Mol. Biol., 232 (1993), pp. 600–618.

164. W. DAHMEN AND A. KUNOTH, *Multilevel preconditioning*, Numer. Math., 63 (1992), pp. 315–344.

165. W. DAHMEN, S. PRÖSSDORF, AND R. SCHNEIDER, *Wavelet approximation methods for pseudo-differential equations II: Matrix compression and fast solution*, Adv. Comp. Math., 1 (1993), pp. 259–335.

166. J. DANBY, *Fundamentals of celestial mechanics*, Willmann-Bell, Richmond, 2 ed., 1988.

167. T. DARDEN, A. TOUKMAJI, AND L. PEDERSEN, *Long-range electrostatic effects in biomolecular simulations*, J. Chimie Physique Physico-Chimie Biologique, 94 (1997), pp. 1346–1364.

168. T. DARDEN, D. YORK, AND L. PEDERSEN, *Particle mesh Ewald: An $N\log(N)$ method for Ewald sums in large systems*, J. Chem. Phys., 98 (1993), pp. 10089–10092.

169. X. DAURA, K. GADEMANN, H. SCHÄFER, B. JAUN, D. SEEBACH, AND W. VAN GUNSTEREN, *The β-peptide hairpin in solution: Conformational study of a β-hexapeptide in methanol by NMR spectroscopy and MD simulation*, J. Amer. Chem. Soc., 123 (2001), pp. 2393–2404.

170. X. DAURA, B. JAUN, D. SEEBACH, W. VAN GUNSTEREN, AND A. MARK, *Reversible peptide folding in solution by molecular dynamics simulation*, J. Mol. Biol., 280 (1998), pp. 925–932.

171. X. DAURA, W. VAN GUNSTEREN, AND A. MARK, *Folding-unfolding thermodynamics of a β-heptapeptide from equilibrium simulations*, Proteins: Struct. Funct. Genet., 34 (1999), pp. 269–280.

172. X. DAURA, W. VAN GUNSTEREN, D. RIGO, B. JAUN, AND D. SEEBACH, *Studying the stability of a helical β-heptapeptide by molecular dynamics simulations*, Chem. Europ. J., 3 (1997), pp. 1410–1417.

173. M. DAW AND M. BASKES, *Semiempirical, quantum mechanical calculation of hydrogen embrittlement in metals*, Phys. Rev. Lett., 50 (1983), pp. 1285–1288.

174. ———, *Embedded-atom method: Derivation and application to impurities, surfaces, and other defects in metals*, Phys. Rev. B, 29 (1984), pp. 6443–6453.

175. P. DE BAKKER, P. HÜNENBERGER, AND J. MCCAMMON, *Molecular dynamics simulations of the hyperthermophilic protein Sac7d from Sulfolobus acidocaldaricus: Contribution of salt bridges to thermostability*, J. Mol. Biol., 285 (1999), pp. 1811–1830.

176. C. DE BOOR, *Practical guide to splines*, Springer, New York, 1978.

177. S. DE LEEUW, J. PERRAM, AND E. SMITH, *Simulation of electrostatic systems in periodic boundary conditions. I. Lattice sums and dielectric constants*, Proc. Roy. Soc. London A, 373 (1980), pp. 27–56.

178. S. DEBOLT AND P. KOLLMAN, *AMBERCUBE MD, Parallelization of AMBER's molecular dynamics module for distributed-memory hypercube computers*, J. Comput. Chem., 14 (1993), pp. 312–329.

179. M. DESERNO AND C. HOLM, *How to mesh up Ewald sums. I. A theoretical and numerical comparison of various particle mesh routines*, J. Chem. Phys., 109 (1998), pp. 7678–7693.

180. P. DEUFLHARD AND A. HOHMANN, *Numerical analysis in modern scientific computing. An introduction.*, vol. 43 of Texts in Applied Mathematics, Springer, Berlin, Heidelberg, New York, 2003.

181. E. DEUMENS, A. DIZ, R. LONGO, AND Y. ÖHRN, *Time-dependent theoretical treatments of the dynamics of electrons and nuclei in molecular systems*, Rev. Modern Phys., 66 (1994), pp. 917–983.

182. B. DEY, A. ASKAR, AND H. RABITZ, *Multidimensional wave packet dynamics within the fluid dynamical formulation of the Schrödinger equation*, J. Chem. Phys., 109 (1998), pp. 8770–8782.

183. J. DIAZ, M. ESCALONA, S. KUPPENS, AND Y. ENGELBORGHS, *Role of the switch II region in the conformational transition of activation of Ha-ras-p21*, Protein Sci., 9 (2000), pp. 361–368.

184. J. DIAZ, B. WROBLOWSKI, J. SCHLITTER, AND Y. ENGELBORGHS, *Calculation of pathways for the conformational transition between the GTP- and GDP-bound states of the Ha-ras-p21 protein: Calculations with explicit solvent simulations and comparison with calculations in vacuum*, Proteins: Struct. Funct. Genet., 28 (1997), pp. 434–451.

185. H. DING, N. KARASAWA, AND W. GODDARD, *The reduced cell multipole method for Coulomb interactions in periodic systems with million-atom unit cells*, Chem. Phys. Lett., 196 (1992), pp. 6–10.

186. P. DIRAC, *Note on exchange phenomena in the Thomas atom*, Proc. Cambridge Phil. Soc., 26 (1930), pp. 376–385.

187. ———, *The principles of quantum mechanics*, Oxford University Press, Oxford, 1947, ch. V.

188. S. DODELSON, *Modern cosmology*, Academic Press, New York, 2003.

189. S. DOUADY, S. FAUVE, AND C. LAROCHE, *Subharmonic instabilities and defects in a granular layer under vertical vibrations*, Europhys. Lett., 8 (1989), pp. 621–627.

190. C. DRAGHICESCU, *An efficient implementation of particle methods for the incompressible Euler equations*, SIAM J. Num. Anal., 31 (1994), pp. 1090–1108.

191. C. DRAGHICESCU AND M. DRAGHICESCU, *A fast algorithm for vortex blob interactions*, J. Comput. Phys., 1 (1995), pp. 69–78.

192. Y. DUAN AND P. KOLLMAN, *Pathways to a protein folding intermediate observed in a 1-microsecond simulation in aqueous solution*, Science, 282 (1998), pp. 740–744.

193. S. DUANE, A. KENNEDY, B. PENDLETON, AND D. ROWETH, *Hybrid Monte Carlo*, Phys. Lett. B, 195 (1987), pp. 216–222.

194. C. DUARTE, *A review of some meshless methods to solve partial differential equations*, Tech. Rep. 95-06, TICAM, University of Texas at Austin, 1995.

195. C. DUARTE AND J. ODEN, *Hp-clouds – A meshless method to solve boundary-value problems*, Tech. Rep. 95-05, TICAM, University of Texas at Austin, 1995.

196. J. DUBINSKI, *A parallel tree code*, New Astronomy, 1 (1996), pp. 133–147.

197. B. DÜNWEG, G. GREST, AND K. KREMER, *Molecular dynamics simulations of polymer systems*, in Numerical Methods for Polymeric Systems, S. Whittington, ed., Berlin, 1998, Springer, pp. 159–196.

198. W. DZWINEL, W. ALDA, J. KITOWSKI, J. MOŚCIŃSKI, M. POGODA, AND D. YUEN, *Rayleigh-Taylor instability - complex or simple dynamical system?*, Tech. Rep. UMSI 97/71, Supercomputer Institute, University of Minnesota, Minneapolis, 1997.

199. W. DZWINEL, W. ALDA, J. KITOWSKI, J. MOŚCIŃSKI, R. WCISLO, AND D. YUEN, *Macro-scale simulations using molecular dynamics method*, Tech. Rep. UMSI 95/103, Supercomputer Inst., University Minnesota, 1995.

200. W. DZWINEL, J. KITOWSKI, J. MOŚCIŃSKI, AND D. YUEN, *Molecular dynamics as a natural solver*, Tech. Rep. UMSI 98/99, Supercomputer Institute, University of Minnesota, Minneapolis, 1998.

201. D. EARN AND S. TREMAINE, *Exact numerical studies of Hamiltonian maps: Iterating without roundoff error*, Physica D, 56 (1992), pp. 1–22.

202. J. EASTWOOD, *Optimal particle-mesh algorithms*, J. Comput. Phys., 18 (1975), pp. 1–20.

203. O. EDHOLM, O. BERGER, AND F. JÄHNIG, *Structure and fluctuations of bacteriorhodopsin in the purple membrane; A molecular dynamics study*, J. Mol. Biol., 250 (1995), pp. 94–111.

204. P. EHRENFEST, *Bemerkung über die angenäherte Gültigkeit der klassischen Mechanik innerhalb der Quantenmechanik*, Z. Phys., 45 (1927), pp. 455–457.

205. M. EICHINGER, H. HELLER, AND H. GRUBMÜLLER, *EGO - an efficient molecular dynamics program and its application to protein dynamics simulations*, in Workshop on Molecular Dynamics on Parallel Computers, R. Esser, P. Grassberger, J. Grotendorst, and M. Lewerenz, eds., World Scientific, Singapore, 2000, pp. 154–174.

206. W. ELLIOTT, *Multipole algorithms for molecular dynamics simulations on high performance computers*, Ph.D. thesis, Duke University, May 1995.

207. W. ELLIOTT AND J. BOARD, *Fast multipole algorithm for the Lennard-Jones potential*, Techn. Report 94-005, Duke University, Department of Electrical Engineering, 1994.

208. F. ERCOLESSI AND J. ADAMS, *Interatomic potentials from first-principles calculations: The force-matching method*, Europhys. Lett., 26 (1994), pp. 583–588.

209. F. ERCOLESSI, M. PARRINELLO, AND E. TOSATTI, *Simulation of gold in the glue model*, Phil. Mag. A, 58 (1988), pp. 213–226.

210. R. ERNST AND G. GREST, *Search for a correlation length in a simulation of the glass transition*, Phys. Rev. B, 43 (1991), pp. 8070–8080.

211. K. ESSELINK, *A comparison of algorithms for long-range interactions*, Comp. Phys. Comm., 87 (1995), pp. 375–395.

212. K. ESSELINK AND P. HILBERS, *Efficient parallel implementation of molecular dynamics on a toroidal network: II. Multi-particle potentials*, J. Comput. Phys., 106 (1993), pp. 108–114.

213. K. ESSELINK, P. HILBERS, N. VAN OS, B. SMIT, AND S. KARABORNI, *Molecular dynamics simulations of model oil/water/surfactant systems*, Colloid Surface A, 91 (1994), pp. 155–167.

214. U. ESSMANN, L. PERERA, AND M. BERKOWITZ, *The origin of the hydration interaction of lipid bilayers from MD simulation of dipalmitoylphosphatidylcholine membrane in gel and cristalline phases*, Langmuir, 11 (1995), pp. 4519–4531.

215. U. ESSMANN, L. PERERA, M. BERKOWITZ, T. DARDEN, H. LEE, AND L. PEDERSEN, *A smooth particle mesh Ewald method*, J. Chem. Phys., 103 (1995), pp. 8577–8593.

216. P. EWALD, *Die Berechnung optischer und elektrostatischer Gitterpotentiale*, Ann. Phys., 64 (1921), pp. 253–287.

217. R. FAROUKI AND S. HAMAGUCHI, *Spline approximation of "effective" potentials under periodic boundary conditions*, J. Comput. Phys., 115 (1994), pp. 276–287.

218. M. FEIT, J. FLECK, AND A. STEIGER, *Solution of the Schrödinger equation by a spectral method*, J. Comput. Phys., 47 (1982), pp. 412–433.

219. S. FELLER, R. PASTOR, A. ROJNUCKARIN, S. BOGUSZ, AND B. BROOKS, *Effect of electrostatic force truncation on interfacial and transport properties of water*, J. Phys. Chem., 100 (1996), pp. 17011–17020.

220. S. FELLER, R. VENABLE, AND R. PASTOR, *Computer simulation of a DPPC phospholipid bilayer: Structural changes as a function of molecular surface area*, Langmuir, 13 (1997), pp. 6555–6561.

221. R. FERELL AND E. BERTSCHINGER, *Particle-mesh methods on the Connection Machine*, Int. J. Modern Physics C, 5 (1994), pp. 933–956.

222. E. FERMI, J. PASTA, AND S. ULAM, *Studies of non-linear problems*, Tech. Rep. LA-1940, Los Alamos, LASL, 1955.

223. P. FERRARA, J. APOSTOLAKIS, AND A. CAFLISH, *Computer simulation of protein folding by targeted molecular dynamics*, Proteins: Struct. Funct. Genet., 39 (2000), pp. 252–260.

224. ——, *Targeted molecular dynamics simulations of protein folding*, J. Phys. Chem. B, 104 (2000), pp. 4511–4518.

225. R. FEYNMAN, *Forces in molecules*, Phys. Rev., 56 (1939), pp. 340–343.

226. F. FIGUEIRIDO, R. LEVY, R. ZHOU, AND B. BERNE, *Large scale simulation of macromolecules in solution: Combining the periodic fast multipole method with multiple time step integrators*, J. Chem. Phys., 106 (1997), pp. 9835–9849.

227. D. FINCHAM, *Parallel computers and molecular simulation*, Mol. Sim., 1 (1987), pp. 1–45.

228. ——, *Optimisation of the Ewald sum for large systems*, Mol. Sim., 13 (1994), pp. 1–9.

229. J. FINEBERG, S. GROSS, M. MARDER, AND H. SWINNEY, *Instability in dynamic fracture*, Phys. Rev. Lett., 67 (1991), pp. 457–460.

230. ——, *Instability in the propagation of fast cracks*, Phys. Rev. B, 45 (1992), pp. 5146–5154.

231. J. FINNEY, *Long-range forces in molecular dynamics calculations on water*, J. Comput. Phys., 28 (1978), pp. 92–102.

232. M. FINNIS AND J. SINCLAIR, *A simple empirical N-body potential for transition metals*, Phil. Mag. A, 50 (1984), pp. 45–55.

233. E. FISCHER, *Einfluss der Configuration auf die Wirkung der Enzyme*, Ber. Dt. Chem. Ges., 27 (1894), pp. 2985–2993.

234. M. FLYNN, *Very high-speed computing systems*, Proc. IEEE, 54 (1966), pp. 1901–1909.

235. E. FOREST AND R. RUTH, *Fourth order symplectic integration*, Physica D, 43 (1990), pp. 105–117.

236. B. FORREST AND U. SUTER, *Hybrid Monte Carlo simulations of dense polymer systems*, J. Chem. Phys., 101 (1994), pp. 2616–2629.

237. J. FOX AND H. ANDERSEN, *Molecular dynamics simulations of a supercooled monoatomic liquid and glass*, J. Phys. Chem., 88 (1984), pp. 4019–4027.

238. D. FRENKEL AND J. MCTAGUE, *Computer simulations of freezing and supercooled liquids*, Annu. Rev. Phys. Chem., 31 (1980), pp. 491–521.

239. D. FRENKEL AND B. SMIT, *Understanding molecular simulation: From algorithms to applications*, Academic Press, New York, 1996.

240. M. FRIGO AND S. JOHNSON, *FFTW: An adaptive software architecture for the FFT*, in ICASSP conf. proceeding, vol. 3, 1998, pp. 1381–1384.

241. F. FUMI AND M. TOSI, *Ionic sizes and Born repulsive parameters in the NaCl-type alkali halides I. The Huggins-Mayer and Pauling forms*, J. Phys. Chem. Solids, 25 (1964), pp. 31–43.

242. R. GABDOULLINE AND R. WADE, *Brownian dynamics simulation of protein-protein diffusional encounter*, Methods Comp. Phys., 14 (1998), pp. 329–341.

243. ——, *Protein-protein association: Investigation of factors influencing association rates by Brownian dynamics simulations*, J. Mol. Biol., 306 (2001), pp. 1139–1155.

244. J. GALE, *GULP - a computer program for the symmetry adapted simulation of solids*, J. Chem. Soc. Faraday Trans., 93 (1997), pp. 629–637.

245. B. GARCIA-ARCHILLA, J. SANZ-SERNA, AND R. SKEEL, *The mollified impulse method for oscillatory differential equations*, SIAM J. Sci. Comput., 20 (1998), pp. 930–963.

246. R. GAREMYR AND A. ELOFSSON, *A study of the electrostatic treatment in molecular dynamics simulations*, Proteins: Struct. Funct. Genet., 37 (1999), pp. 417–428.

247. C. GEAR, *Numerical initial value problems in ordinary differential equations*, Prentice-Hall, Englewood Cliffs, NJ, 1971.

248. A. GEIST, A. BEGUELIN, J. DONGERRA, W. JIANG, R. MANCHEK, AND V. SUNDERAM, *PVM: Parallel virtual machine*, MIT Press, Cambridge, MA, 1994.

249. S. GELATO, D. CHERNOFF, AND I. WASSERMANN, *An adaptive hierarchical particle-mesh code with isolated boundary conditions*, Astrophys. J., 480 (1997), pp. 115–131.

250. W. GELBART, A. BEN-SHAUL, AND D. ROUX, eds., *Micelles, membranes, microemulsions, and monolayers*, Springer, New York, 1994.

251. M. GERSTEIN AND M. LEVIN, *Simulating water and the molecules of life*, Scientific American, (1998), pp. 100–105.

252. J. GIBSON, A. GOLAND, M. MILGRAM, AND G. VINEYARD, *Dynamics of radiation damage*, Phys. Rev., 120 (1960), pp. 1229–1253.

253. K. GIEBERMANN, *Multilevel approximation of boundary integral operators*, Computing, 67 (2001), pp. 183–207.

254. R. GINGOLD AND J. MONAGHAN, *Kernel estimates as a basis for general particle methods in hydrodynamics*, J. Comput. Phys., 46 (1982), pp. 429–453.

255. G. GOLUB AND C. V. LOAN, *Matrix computations*, The Johns Hopkins University Press, Baltimore, 1996.

256. S. GOREINOV, E. TYRTYSHNIKOV, E. YEREMIN, AND A. YU, *Matrix-free iterative solution strategies for large dense linear systems*, Linear Algebra Appl., 4 (1997), pp. 273–294.

257. S. GOREINOV, E. TYRTYSHNIKOV, AND N. ZAMARASHKIN, *A theory of pseudoskeleton approximations*, Linear Algebra Appl., 261 (1997), pp. 1–21.

258. D. GOTTLIEB AND S. ORSZAG, *Numerical analysis of spectral methods: Theory and applications*, SIAM, CMBS, Philadelphia, 1977.

259. D. GRAY AND A. GIORGINI, *On the validity of the Boussinesq approximation for liquids and gases*, J. Heat Mass Transfer, 19 (1976), pp. 545–551.

260. L. GREENGARD, *The rapid evaluation of potential fields in particle systems*, PhD thesis, Yale University, 1987.

261. L. GREENGARD AND W. GROPP, *A parallel version of the fast multipole method*, Comp. Math. Applic., 20 (1990), pp. 63–71.

262. L. GREENGARD AND J. LEE, *A direct adaptive Poisson solver of arbitrary order accuracy*, J. Comput. Phys., 125 (1996), pp. 415–424.

263. L. GREENGARD AND V. ROKHLIN, *A fast algorithm for particle simulations*, J. Comput. Phys., 73 (1987), pp. 325–348.

264. ———, *On the efficient implementation of the fast multipole method*, Research Report YALEU/DCS/RR-602, Yale University, Department of Computer Science, New Haven, Conneticut, 1988.

265. ———, *On the evaluation of electrostatic interactions in molecular modeling*, Chemica Scripta, 29A (1989), pp. 139–144.

444   References

266. ———, *A new version of the fast multipole method for the Laplace equation in three dimensions*, Acta Numerica, 6 (1997), pp. 229–269.

267. L. GREENGARD AND J. STRAIN, *The fast Gauss transform*, SIAM J. Sci. Comput., 12 (1991), pp. 79–94.

268. L. GREENGARD AND X. SUN, *A new version of the fast Gauss transform*, Doc. Math., Extra Volume ICM, III (1999), pp. 575–584.

269. G. GREST AND S. NAGEL, *Frequency-dependent specific heat in a simulation of the glass transition*, J. Phys. Chem., 91 (1987), pp. 4916–4922.

270. M. GRIEBEL AND M. SCHWEITZER, *A particle-partition of unity method for the solution of elliptic, parabolic and hyperbolic PDEs*, SIAM J. Sci. Comput., 22 (2000), pp. 853–690.

271. W. GROPP, S. HUSS-LEDERMAN, A. LUMSDAINE, E. LUSK, B. NITZBERG, W. SAPHIR, AND M. SNIR, *MPI: The complete reference*, vol. 2, MIT Press, Cambridge, MA, 1996.

272. W. GROPP, E. LUSK, AND A. SKJELLUM, *Using MPI*, MIT Press, Cambridge, MA, 1994.

273. ———, *Portable parallel programming with the Message-Passing Interface*, MIT Press, Cambridge, MA, 1999.

274. H. GRUBMÜLLER, *Dynamiksimulationen sehr grosser Makromoleküle auf einem Parallelrechner*, Diplomarbeit, TU München, 1989.

275. H. GRUBMÜLLER, H. HELLER, A. WINDEMUTH, AND K. SCHULTEN, *Generalized Verlet algorithm for efficient molecular dynamics simulation with long range interactions*, Mol. Sim., 6 (1991), pp. 121–142.

276. H. GRUBMÜLLER, B. HEYMANN, AND P. TAVAN, *Ligand binding: Molecular dynamics calculation of the streptavidin-biotin rupture force*, Science, 271 (1996), pp. 997–999.

277. S. GUATTERY AND G. MILLER, *On the quality of spectral separators*, SIAM J. Matrix Anal. Appl., 19 (1998), pp. 701–719.

278. P. GUMBSCH AND G. BELTZ, *On the continuum versus atomistic descriptions of dislocation nucleation versus cleavage in nickel*, Model. Simul. Mater. Sci. Eng., 3 (1995), pp. 597–613.

279. T. GUO, P. NIKOLAEV, A. RINZLER, D. TOMANEK, D. COLBERT, AND R. SMALLEY, *Self assembly of tubular Fullerene*, J. Phys. Chem., 99 (1995), pp. 10694–10697.

280. S. GUPTA, *Computing aspects of molecular dynamics simulations*, Comp. Phys. Comm., 70 (1992), pp. 243–270.

281. H. HABERLAND, Z. INSEPOV, AND M. MOSELER, *Molecular-dynamics simulations of thin-film growth by energetic cluster impact*, Phys. Rev. B, 51 (1995), pp. 11061–11067.

282. R. HABERLANDT, S. FRITZSCHE, G. PEINELA, AND K. HEINZINGER, *Molekulardynamik, Grundlagen und Anwendungen*, Vieweg Lehrbuch Physik, Vieweg, Braunschweig, 1995.

283. W. HACKBUSCH, *Multi-Grid methods and applications*, Springer Series in Computational Mathematics 4, Springer, Berlin, 1985.

284. ———, *Elliptic differential equations. Theory and numerical treatment*, Springer, Berlin, 1992.

285. ———, *Iterative solution of large sparse systems*, Springer, New York, 1994.

286. ———, *A sparse matrix arithmetic based on $\mathcal{H}$-matrices. Part I: Introduction to $\mathcal{H}$-matrices*, Computing, 62 (1999), pp. 89–108.

287. W. HACKBUSCH AND S. BÖRM, *Approximation of boundary element operators by adaptive $\mathcal{H}^2$ matrices.* Max Planck Institute for Mathematics in the Sciences, Preprint Nr. 5/2003, 2003.

288. W. HACKBUSCH, B. KHOROMSKIJ, AND S. SAUTER, *On $\mathcal{H}^2$-matrices,* in Lectures on Applied Mathematics, Springer Berlin, 2000, pp. 9–29.

289. W. HACKBUSCH AND Z. NOWAK, *On the fast matrix multiplication in the boundary element method by panel clustering,* Numer. Math., 54 (1989), pp. 463–491.

290. W. HACKBUSCH AND S. SAUTER, *On the efficient use of the Galerkin-method to solve Fredholm integral equations,* Appl. Math., 38 (1993), pp. 301–322.

291. J. HAILE AND S. GUPTA, *Extension of molecular dynamics simulation method. III. Isothermal systems,* J. Chem. Phys., 70 (1983), pp. 3067–3076.

292. E. HAIRER, *Backward analysis of numerical integrators and symplectic methods,* Ann. Numer. Math., 1 (1994), pp. 107–132.

293. ———, *Backward error analysis for multistep methods,* Numer. Math., 84 (1999), pp. 199–232.

294. E. HAIRER AND P. LEONE, *Order barriers for symplectic multi-value methods,* in Numerical analysis 1997, Proceedings of the 17th Dundee Biennial Conference, D. Griffiths, D. Higham, and G. Watson, eds., vol. 380 of Research Notes in Mathematics, Pitman, 1998, pp. 133–149.

295. E. HAIRER AND C. LUBICH, *The life-span of backward error analysis for numerical integrators,* Numer. Math., 76 (1997), pp. 441–462.

296. E. HAIRER, C. LUBICH, AND G. WANNER, *Geometric numerical integration. Structure-preserving algorithms for ordinary differential equations,* vol. 31 of Series in Computational Mathematics, Springer, Berlin, 2002.

297. E. HAIRER, S. NØRSETT, AND G. WANNER, *Solving ordinary differential equations I, Nonstiff problems,* Springer, Berlin, 1993.

298. E. HAIRER AND D. STOFFER, *Reversible long-term integration with variable step sizes,* SIAM J. Sci. Stat. Comput., 18 (1997), pp. 257–269.

299. B. HALPERIN AND D. NELSON, *Theory of two-dimensional melting,* Phys. Rev. Lett., 41 (1978), pp. 121–124.

300. S. HAMMES-SCHIFFER AND J. TULLY, *Proton transfer in solution: Molecular dynamics with quantum transitions,* J. Chem. Phys., 101 (1994), pp. 4657–4667.

301. J. HAN, A. GLOBUS, R. JAFFE, AND G. DEARDORFF, *Molecular dynamics simulations of carbon nanotube based gears,* Nanotech., (1997), p. 103.

302. N. HANDY, *Density functional theory,* in Lecture notes in quantum chemistry II, B. Roos, ed., vol. 64 of Lecture Notes in Chemistry, Springer, Berlin, 1994, pp. 91–124.

303. E. HARRISON, *Cosmology: The science of the universe,* Cambridge University Press, Cambridge, 2000.

304. B. HARTKE AND E. CARTER, *Ab initio molecular dynamics with correlated molecular wave functions: Generalized valence bond molecular dynamics and simulated annealing,* J. Chem. Phys., 97 (1992), pp. 6569–6578.

305. S. HAYWARD, A. KITAO, F. HIRATA, AND N. GO, *Effect of solvent on collective motions in globular protein,* J. Mol. Biol., 234 (1993), pp. 1207–1217.

306. T. HEAD-GORDON AND F. STILLINGER, *An orientational perturbation theory for pure liquid water,* J. Chem. Phys., 98 (1993), pp. 3313–3327.

307. D. HEERMANN AND L. YIXUE, *A global-update simulation method for polymer systems,* Macromol. Chem. Theor. Simul., 2 (1993), pp. 299–308.

308. H. HELLER, *Simulation einer Lipidmembran auf einem Parallelrechner*, Dissertation, TU München, 1993.

309. H. HELLER, H. GRUBMÜLLER, AND K. SCHULTEN, *Molecular dynamics simulation on a parallel computer*, Mol. Sim., 5 (1990), pp. 133–165.

310. H. HELLER, M. SCHÄFER, AND K. SCHULTEN, *Molecular dynamics simulations of a bilayer of 200 lipids in the gel and in the liquid-crystal phases*, J. Phys. Chem., 97 (1993), pp. 8343–8360.

311. H. HELLMANN, *Zur Rolle der kinetischen Elektronenenergie für die zwischenatomaren Kräfte*, Z. Phys., 85 (1933), pp. 180–190.

312. V. HELMS AND J. MCCAMMON, *Conformational transitions of proteins from atomistic simulations*, in Computational Molecular Dynamics: Challenges, Methods, Ideas, P. Deuflhard, J. Hermans, B. Leimkuhler, A. Mark, S. Reich, and R. Skeel, eds., vol. 4 of Lecture Notes in Computational Science and Engineering, Springer, Berlin, 1999, pp. 66–77.

313. P. HENRICI, *Fast Fourier methods in computational complex analysis*, SIAM Review, 21 (1979), pp. 481–527.

314. J. HERMANS, R. YUN, J. LEECH, AND D. CAVANAUGH, *SIGMA: SI-mulations of MA-cromolecules*, Department of Biochemistry and Biophysics, Univ. of North Carolina. http://hekto.med.unc.edu:8080/HERMANS/software/SIGMA/index.html.

315. L. HERNQUIST, *Performance characteristics of tree codes*, Astrophys. J. supp. series, 64 (1987), pp. 715–734.

316. ———, *An analytical model for spherical galaxies and bulges*, The Astrophysical Journal, 356 (1990), pp. 359–364.

317. H. HERRMANN AND S. LUDING, *Modeling granular media on the computer*, Contin. Mech. Thermodyn., 10 (1998), pp. 189–231.

318. B. HESS, H. BEKKER, H. BERENDSEN, AND J. FRAAIJE, *LINCS: A linear constraint solver for molecular simulations*, J. Comput. Chem., 18 (1997), pp. 1463–1472.

319. B. HEYMANN, *Beschreibung der Streptavidin-Biotin-Bindung mit Hilfe von Molekulardynamiksimulationen*, Diplomarbeit, Fakultät für Physik, LMU München, 1996.

320. B. HEYMANN AND H. GRUBMÜLLER, *AN02/DNP unbinding forces studied by molecular dynamics AFM simulations*, Chem. Phys. Lett., 303 (1999), pp. 1–9.

321. D. HILBERT, *Über die stetige Abbildung einer Linie auf ein Flächenstück*, Mathematische Annalen, 38 (1891), pp. 459–460.

322. R. HOAGLAND, M. DAW, S. FOILES, AND M. BASKES, *An atomic model of crack tip deformation in aluminum using an embedded atom potential*, J. Mater. Res., 5 (1990), pp. 313–324.

323. R. HOCKNEY, *The potential calculation and some applications*, Methods Comp. Phys., 9 (1970), pp. 136–211.

324. R. HOCKNEY AND J. EASTWOOD, *Computer simulation using particles*, IOP Publising Ltd., London, 1988.

325. P. HOHENBERG AND W. KOHN, *Inhomogeneous electron gas*, Phys. Rev. B, 136 (1964), pp. 864–871.

326. B. HOLIAN, P. LOMDAHL, AND S. ZHOU, *Fracture simulations via large-scale nonequilibrium molecular dynamics*, Physica A, 240 (1997), pp. 340–348.

327. W. HOOVER, *Canonical dynamics: Equilibrium phase-space distributions*, Phys. Rev. A, 31 (1985), pp. 1695–1697.

328. ———, *Molecular dynamics*, vol. 258 of Lecture Notes in Physics, Springer, Berlin, 1986.

329. T. HRYCAK AND V. ROKHLIN, *An improved fast multipole algorithm for potential fields*, Research Report YALEU/DCS/RR-1089, Yale University, Department of Computer Science, New Haven, Conneticut, 1995.

330. L. HUA, H. RAFII-TABAR, AND M. CROSS, *Molecular dynamics simulation of fractures using an N-body potential*, Phil. Mag. Lett., 75 (1997), pp. 237–244.

331. P. HUANG, J. PEREZ, AND G. LOEW, *Molecular-dynamics simulations of phospholipid-bilayers*, J. Biomol. Struct. Dyn., 11 (1994), pp. 927–956.

332. W. HUMPHREY, A. DALKE, AND K. SCHULTEN, *VMD - visual molecular dynamics*, J. Molec. Graphics, 14 (1996), pp. 33–38.

333. S. IIJIMA, *Helical microtubules of graphitic carbon*, Nature, 354 (1991), pp. 56–58.

334. J. IZAGUIRRE, *Longer time steps for molecular dynamics*, PhD thesis, University of Illinois at Urbana-Champaign, 1999.

335. S. IZRAILEV, S. STEPANIANTS, M. BALSERA, Y. OONA, AND K. SCHULTEN, *Molecular dynamics study of the unbinding of the avidin-biotin complex*, Biophys. J., 72 (1997), pp. 1568–1581.

336. K. JACOBSEN, J. NORSKOV, AND M. PUSKA, *Interatomic interactions in the effective-medium theory*, Phys. Rev. B, 35 (1987), pp. 7423–7442.

337. H. JAEGER, S. NAGEL, AND R. BEHRINGER, *Granular solids, liquids, and gases*, Rev. Modern Phys., 68 (1996), pp. 1259–1273.

338. W. JAFFE, *A simple model for the distribution of light in spherical galaxies*, Monthly Notices of the Royal Astronomical Society, 202 (1983), pp. 995–999.

339. J. JANAK AND P. PATTNAIK, *Protein calculations on parallel processors: II. Parallel algorithm for forces and molecular dynamics*, J. Comput. Chem., 13 (1992), pp. 1098–1102.

340. D. JANEZIC AND F. MERZEL, *Split integration symplectic method for molecular dynamics integration*, J. Chem. Inf. Comp. Sci., 37 (1997), pp. 1048–1054.

341. L. JAY, *Runge-Kutta type methods for index three differential-algebraic equations with applications to Hamiltonian systems*, Dissertation, Section de mathématiques, Université de Genève, 1994.

342. J. JERNIGAN AND D. PORTER, *A tree code with logarithmic reduction of force terms, hierarchical regularization of all variables, and explicit accuracy controls*, Astrophys. J. supp. series, 71 (1989), pp. 871–893.

343. C. JESSOP, M. DUNCAN, AND W. CHAU, *Multigrid methods for n-body gravitational systems*, J. Comput. Phys., 115 (1994), pp. 339–351.

344. C. JOHNSON AND L. SCOTT, *An analysis of quadrature errors in second-kind boundary integral methods*, SIAM J. Num. Anal., 26 (1989), pp. 1356–1382.

345. W. JORGENSEN, *Transferable intermolecular potential functions of water, alcohols and ethers*, J. Amer. Chem. Soc., 103 (1981), pp. 335–340.

346. ———, *Revised TIPS model for simulations of liquid water and aqueous solutions*, J. Chem. Phys., 77 (1982), pp. 4156–4163.

347. W. JORGENSEN, J. CHANDRASEKHAR, J. MADURA, R. IMPLEY, AND M. KLEIN, *Comparison of simple potential functions for simulating liquid water*, J. Chem. Phys., 79 (1983), pp. 926–935.

348. W. JORGENSEN AND J. TIRADO-RIVES, *Optimized potentials for liquid simulations: "The OPLS potential functions for proteins. Energy minimizations for crystals of cyclic peptides and crambin"*, J. Amer. Chem. Soc., 110 (1988), pp. 1657–1666.

349. B. JUNG, H. LENHOF, R. MÜLLER, AND C. RÜB, *Parallel algorithms for MD-simulations of synthetic polymers*, Tech. Rep. MPI-I-7-1-003, Max Planck Institut für Informatik, Saarbrücken, 1997.

350. K. KADAU, *Molekulardynamik-Simulationen von strukturellen Phasenumwandlungen in Festkörpern, Nanopartikeln und ultradünnen Filmen*, Dissertation, Universität Duisburg, Fachbereich Physik – Technologie, 2001.

351. M. KARPLUS AND J. MCCAMMON, *Protein structural fluctuations during a period of 100 ps*, Nature, 277 (1979), pp. 578–579.

352. ———, *The internal dynamics of globular proteins*, CRC Crit. Revs. Biochem., 9 (1981), pp. 293–349.

353. W. KAUFMANN AND L. SMARR, *Supercomputing and the transformation of science*, Scientific American Library, New York, 1993.

354. B. KERNIGHAN AND D. RITCHIE, *The C programming language*, Prentice-Hall, Englewood Cliffs, NJ, 1988.

355. R. KESSLER, *Nonlinear transition in three-dimensional convection*, J. Fluid Mech., 174 (1987), pp. 357–379.

356. D. KNUTH, *The art of computer programming, seminumerical algorithms*, vol. 2, Addison-Wesley, 1997.

357. ———, *The art of computer programming, sorting and searching*, vol. 3, Addison-Wesley, 1998.

358. W. KOB AND H. ANDERSEN, *Testing mode-coupling theory for a supercooled binary Lennard-Jones mixture: The van Hove correlation function*, Phys. Rev. E, 51 (1995), pp. 4626–4641.

359. W. KOŁOS, *Adiabatic approximation and its accuracy*, Adv. Quant. Chem., 5 (1970), pp. 99–133.

360. R. KOSLOFF, *Time-dependent quantum-mechanical methods for molecular dynamics*, J. Phys. Chem., 92 (1988), pp. 2087–2100.

361. ———, *Propagation methods for quantum molecular dynamics*, Annu. Rev. Phys. Chem., 45 (1994), pp. 145–178.

362. A. KOX, J. MICHELS, AND F. WIEGEL, *Simulation of a lipid monolayer using molecular dynamics*, Nature, 287 (1980), pp. 317–319.

363. K. KREMER, *Computer simulation methods for polymer physics*, in Monte Carlo and Molecular Dynamics of Condensed Matter Systems, K. Binder and G. Ciccotti, eds., vol. 49 of Conference Proceedings, Italian Physical Society SFI, Bologna, 1995.

364. H. KROTO, J. HEATH, S. O'BRIEN, R. CURL, AND R. SMALLEY, *C60: Buckminsterfullerene*, Nature, 318 (1985), pp. 162–163.

365. K. KRYNICKI, C. GREEN, AND D. SAWYER, *Pressure and temperature dependence of self-diffusion in water*, Disc. Faraday Soc., 66 (1980), p. 199.

366. R. KUBO, *Statistical mechanics*, Elsevier, Amsterdam, 1965.

367. S. KUPPENS, J. DIAZ, AND Y. ENGELBORGHS, *Characterization of the hinges of the effector loop in the reaction pathway of the activation of ras-proteins. Kinetics of binding of beryllium trifluoride to V29G and I36G mutants of Ha-ras-p21*, Protein Sci., 8 (1999), pp. 1860–1866.

368. S. KURZ, O. RAIN, AND S. RJASANOW, *The adaptive cross approximation technique for the 3D boundary element method*, IEEE Transaction on Magnetics, 38 (2002), pp. 421 – 424.

369. P. LAGUNA, *Smoothed particle interpolation*, Astrophys. J., 439 (1995), pp. 814–821.

370. C. LAMBERT, T. DARDEN, AND J. BOARD, *A multipole-based algorithm for efficient calculation of forces and potentials in macroscopic periodic assemblies of particles*, J. Comput. Phys., 126 (1996), pp. 274–285.

371. L. LANDAU AND E. LIFSCHITZ, *Mechanics*, Course of Theoretical Physics, Vol. 1, Pergamon Press, Oxford, 1976.

372. ———, *Quantum mechanics*, Course of Theoretical Physics, Vol. 3, Pergamon Press, Oxford, 1977.

373. A. LEACH, *Molecular modelling: Principles and applications*, Addison-Wesley, 1996.

374. H. LEE, T. DARDEN, AND L. PEDERSEN, *Accurate crystal molecular dynamics simulations using particle-mesh-Ewald: RNA dinucleotides – ApU and GbC*, Chem. Phys. Lett., 243 (1995), pp. 229–235.

375. C. LEFORESTIER, R. BISSELING, C. CERJAN, M. FEIT, R. FRIESNER, A. GULDBERG, A. HAMMERICH, G. JOLICARD, W. KARRLEIN, H. MEYER, N. LIPKIN, O. RONCERO, AND R. KOSLOFF, *A comparison of different propagation schemes for the time dependent Schrödinger equation*, J. Comput. Phys., 94 (1991), pp. 59–80.

376. B. LEIMKUHLER, *Reversible adaptive regularization: Perturbed Kepler motion and classical atomic trajectories*, Phil. Trans. Royal Soc. A, 357 (1999), pp. 1101–1133.

377. B. LEIMKUHLER AND S. REICH, *Geometric numerical methods for Hamiltonian mechanics*, Cambridge Monographs on Applied and Computational Mathematics, Cambridge University Press, 2004.

378. B. LEIMKUHLER AND R. SKEEL, *Symplectic numerical integrators in constrained Hamiltonian systems*, J. Comput. Phys., 112 (1994), pp. 117–125.

379. H. LEISTER, *Numerische Simulation dreidimensionaler, zeitabhängiger Strömungen unter dem Einfluß von Auftriebs- und Trägheitskräften*, Dissertation, Universität Erlangen-Nürnberg, 1994.

380. A. LEONARD, *Vortex methods for flow simulations*, J. Comput. Phys., 37 (1980), pp. 289–335.

381. I. LEVINE, *Quantum chemistry*, Prentice-Hall, Englewood Cliffs, NJ, 2000.

382. M. LEVITT, *Molecular dynamics of native protein, I. Computer simulation of trajectories*, J. Mol. Biol., 168 (1983), pp. 595–620.

383. J. LEWIS AND K. SINGER, *Thermodynamic properties and self-diffusion of molten sodium chloride*, J. Chem. Soc. Faraday Trans. 2, 71 (1975), pp. 41–53.

384. S. LIN, J. MELLOR-CRUMMEY, B. PETTIT, AND G. PHILLIPS, *Molecular dynamics on a distributed-memory multiprocessor*, J. Comput. Chem., 13 (1992), pp. 1022–1035.

385. P. LINDAN AND M. GILLAN, *Shell-model molecular dynamics simulation of superionic conduction in CaF_2*, J. Phys. Condens. Matter, 5 (1993), pp. 1019–1030.

386. K. LINDSAY, *A three-dimensional cartesian tree-code and applications to vortex sheet roll-up*, PhD thesis, Dept. of Math., University of Michigan, 1997.

387. C. LINEWEAVER AND T. DAVIS, *Misconceptions about the big bang*, Scientific American, 3 (2005), pp. 36–45.

388. H. LIU, F. MÜLLER-PLATHE, AND W. VAN GUNSTEREN, *Combined quantum/classical molecular dynamics study of the catalytic mechanism of HIV-protease*, J. Mol. Biol., 261 (1996), pp. 454–469.

389. H. LIU AND K. SCHULTEN, *Steered molecular dynamics simulations of force-induced protein domain folding*, Proteins: Struct. Funct. Genet., 35 (1999), pp. 453–463.

390. P. LOMDAHL, P. TAMAYO, N. GRØNBECH-JENSEN, AND D. BEAZLEY, *50 GFlops molecular dynamics on the Connection Machine 5*, in Proc. of the 1993 Conf. on Supercomputing, ACM Press, 1993, pp. 520–527.

391. H. LU, B. ISRALEWITZ, A. KRAMMER, V. VOGEL, AND K. SCHULTEN, *Unfolding of titin immunoglobulin domains by steered molecular dynamics*, Biophys. J., 75 (1998), pp. 662–671.

392. H. LU, A. KRAMMER, B. ISRALEWITZ, V. VOGEL, AND K. SCHULTEN, *Computer modeling of force-induced titin domain unfolding*, Adv. Exp. Med. Bio., 481 (2000), pp. 143–160.

393. H. LU AND K. SCHULTEN, *Steered molecular dynamics simulation of conformational changes of immunoglobulin domain I27 interpret atomic force microscopy observations*, J. Chem. Phys., 247 (1999), pp. 141–153.

394. L. LUCY, *A numerical approach to the testing of the fission hypothesis*, Astronom. J., 82 (1977), pp. 1013–1024.

395. S. LUDING, *Granular materials under vibration: Simulations of rotating species*, Phys. Rev. B, 52 (1995), pp. 4442–4457.

396. S. LUDING, H. HERRMANN, AND A. BLUMEN, *Simulations of two-dimensional arrays of beads under external vibrations: Scaling behavior*, Phys. Rev. E, 50 (1994), pp. 3100–3108.

397. X. LUO, R. KATO, AND J. COLLINS, *Dynamic flexibility of protein-inhibitor complexes: A study of the HIV-1 protease/KNI-272 complex*, J. Amer. Chem. Soc., 120 (1998), pp. 12410–12418.

398. S. LUSTIG, S. RASTOGI, AND N. WAGNER, *Telescoping fast multipole methods using Chebyshev economization*, J. Comput. Phys., 122 (1995), pp. 317–322.

399. B. LUTY, I. TIRONI, AND W. VAN GUNSTEREN, *Lattice-sum methods for calculating electrostatic interactions in molecular simulations*, J. Chem. Phys., 103 (1995), pp. 3014–3021.

400. B. LUTY AND W. VAN GUNSTEREN, *Calculating electrostatic interactions using the particle-particle particle-mesh method with nonperiodic boundary long-range interactions*, J. Phys. Chem., 100 (1996), pp. 2581–2587.

401. A. MACKERELL, B. BROOKS, C. BROOKS, L. NILSSON, B. ROUX, Y. WON, AND M. KARPLUS, *CHARMM: The energy function and its parameterization with an overview of the program*, in The Encyclopedia of Computational Chemistry, P. von Ragué Schleyer, ed., vol. 1, John Wiley & Sons, Chichester, 1998, pp. 271–277.

402. M. MADRID, A. JACOBO-MOLINA, J. DING, AND E. ARNOLD, *Major subdomain rearrangement in HIV-1 reverse transcriptase simulated by molecular dynamics*, Proteins: Struct. Funct. Genet., 35 (1999), pp. 332–337.

403. J. MADURA, J. BRIGGS, R. WADE, AND R. GABDOULLINE, *Brownian dynamics*, in Encyclopedia of Computational Chemistry, P. von Ragué Schleyer, ed., vol. 1, Wiley, New York, 1998, pp. 141–154.

404. M. MAHONEY AND W. JORGENSEN, *A five-site model for liquid water and the reproduction of the density anomaly by rigid, nonpolarizable potential functions*, J. Chem. Phys., 112 (2000), pp. 8910–8922.

405. ——, *Diffusion constant of the TIP5P model of liquid water*, J. Chem. Phys., 114 (2001), pp. 363–366.

406. Y. MAO, M. RATNER, AND M. JARROLD, *One water molecule stiffens a protein*, J. Amer. Chem. Soc., 122 (2000), pp. 2950–2951.

407. Z. MAO, A. GARG, AND S. SINNOTT, *Molecular dynamics simulations of the filling and decorating of carbon nanotubules*, Nanotechnology, 10 (1999), pp. 273–277.

408. C. MARCHIORO AND M. PULVIRENTI, *Vortex methods in two-dimensional fluid dynamics*, vol. 203 of Lecture Notes in Physics, Springer, Berlin, 1984.

409. M. MARDER, *Molecular dynamics of cracks*, Comp. Sci. Eng., 1 (1999), pp. 48–55.

410. M. MARESCHAL AND E. KESTEMONT, *Order and fluctuations in nonequilibrium molecular dynamics simulations of two-dimensional fluids*, J. Stat. Phys., 48 (1987), pp. 1187–1201.

411. S. MARRINK AND H. BERENDSEN, *Simulation of water transport through a lipid membrane*, J. Phys. Chem., 98 (1994), pp. 4155–4168.

412. S. MARRINK, O. BERGER, D. TIELEMAN, AND F. JÄHNIG, *Adhesion forces of lipids in a phospholipid membrane studied by molecular dynamics simulations*, Biophys. J., 74 (1998), pp. 931–943.

413. S. MARRINK AND A. MARK, *Effect of undulations on surface tension in simulated bilayers*, J. Phys. Chem. B, 105 (2001), pp. 6122–6127.

414. G. MARSAGLIA, *Random numbers fall mainly in the plane*, Proc. Nat. Acad. Sci., 61 (1968), pp. 25–28.

415. P. MARSZALEK, H. LU, H. LI, M. CARRION-VASQUEZ, A. OBERHAUSER, K. SCHULTEN, AND J. M. FERNANDEZ, *Mechanical unfolding intermediates in titin modules*, Nature, 402 (1999), pp. 100–103.

416. G. MARTYNA, M. KLEIN, AND M. TUCKERMAN, *Nosé-Hoover chains – the canonical ensemble via continuous dynamics*, J. Chem. Phys., 97 (1992), pp. 2635–2643.

417. D. MARX AND J. HUTTER, *Ab initio molecular dynamics: Theory and implementation*, in Modern Methods and Algorithms of Quantum Chemistry, J. Grotendorst, ed., vol. 1 of NIC series, John von Neumann Institute for Computing, Jülich, 2000, pp. 301–449.

418. H. MATUTTIS, S. LUDING, AND H. HERRMANN, *Discrete element simulations of dense packings and heaps made of spherical and non-sperical particles*, Powder Tech., 109 (2000), pp. 278–292.

419. J. MCCAMMON, B. GELIN, AND M. KARPLUS, *Dynamics of folded proteins*, Nature, 267 (1977), pp. 585–590.

420. J. MCCAMMON AND S. HARVEY, *Dynamics of proteins and nucleic acids*, Cambridge University Press, 1987.

421. R. MCLACHLAN, *On the numerical integration of ordinary differential equations by symmetric composition methods*, SIAM J. Sci. Comput., 16 (1995), pp. 151–168.

422. R. MCLACHLAN, G. REINOUT, AND W. QUISPEL, *Splitting methods*, in Acta Numerica, A. Iserles, ed., vol. 11, Cambridge Univ. Press, Cambridge, 2002, pp. 341–434.

423. B. MEHLIG, D. HEERMANN, AND B. FORREST, *Hybrid Monte Carlo method for condensed-matter systems*, Phys. Lett. B, 45 (1992), pp. 679–685.

424. M. MELENK, S. BÖRM, AND M. LÖHNDORF, *Approximation of integral operators by variable-order interpolation.* Max Planck Institute for Mathematics in the Sciences, Preprint Nr. 82/2002, 2002.

425. F. MELO, P. UMBANHOWAR, AND H. SWINNEY, *Transition to parametric wave patterns in a vertically oscillated granular layer*, Phys. Rev. Lett., 72 (1994), pp. 172–175.

426. ———, *Hexagons, kinks, and disorder in oscillated granular layers*, Phys. Rev. Lett., 75 (1995), pp. 3838–3841.

427. A. MESSIAH, *Quantum mechanics*, vol. 1 & 2, North-Holland, Amsterdam, 1961/62.

428. N. METROPOLIS, A. ROSENBLUTH, M. ROSENBLUTH, A. TELLER, AND E. TELLER, *Equation of state calculations by fast computing machines*, J. Chem. Phys., 21 (1953), pp. 1087–1092.

429. R. MEYER, *Computersimulationen martensitischer Phasenübergänge in Eisen-Nickel- und Nickel-Aluminium-Legierungen*, Dissertation, Universität Duisburg, Fachbereich Physik – Technologie, 1998.

430. R. MEYER AND P. ENTEL, *Molecular dynamics study of iron-nickel alloys*, in IV European Symposium on Martensitic Transformations, A. Planes, J. Ortín, and L. Mañosa, eds., Les editions de physique, 1995, pp. 123–128.

431. H. MI, M. TUCKERMAN, D. SCHUSTER, AND S. WILSON, *A molecular dynamics study of HIV-1 protease complexes with C_{60} and fullerene-based anti-viral agents*, Proc. Electrochem. Soc., 99 (1999), pp. 256–269.

432. R. MIKULLA, J. STADLER, P. GUMBSCH, AND H. TREBIN, *Molecular dynamics simulations of crack propagation in quasicrystals*, Phil. Mag. Lett., 78 (1998), pp. 369–376.

433. G. MILLER, S. TENG, W. THURSTON, AND S. VAVASIS, *Geometric separators for finite-element meshes*, SIAM J. Sci. Comput., 19 (1998), pp. 364–386.

434. P. MITCHELL AND D. FINCHAM, *Shell-model simulations by adiabatic dynamics*, J. Phys. Condens. Matter, 5 (1993), pp. 1019–1030.

435. M. MIYAMOTO AND R. NAGAY, *Three-dimensional models for the distribution of mass in galaxies*, PASJ, Publication of the Astronomical Society of Japan, 27 (1975), pp. 533–543.

436. M. MOLLER, D. TILDESLEY, K. KIM, AND N. QUIRKE, *Molecular dynamics simulation of a Langmuir-Blodgett film*, J. Chem. Phys., 94 (1991), pp. 8390–8401.

437. J. MONAGHAN, *Simulating free surface flows with SPH*, J. Comput. Phys., 52 (1994), pp. 393–406.

438. R. MOUNTAIN AND D. THIRUMALAI, *Measures of effective ergodic convergence in liquids*, J. Phys. Chem., 93 (1989), pp. 6975–6979.

439. M. MÜLLER AND H. HERRMANN, *DSMC - a stochastic algorithm for granular matter*, in Physics of dry granular media, H. Herrmann, J. Hovi, and S. Luding, eds., NATO ASI Series, Kluwer, Dordrecht, 1998, pp. 413–420.

440. M. MÜLLER, S. LUDING, AND H. HERRMANN, *Simulations of vibrated granular media in two and three dimensional systems*, in Friction, arching and contact dynamics, D. Wolf and P. Grassberger, eds., World Scientific, Singapore, 1997, pp. 335–341.

441. F. MÜLLER-PLATHE, *YASP: A molecular simulation package*, Comp. Phys. Comm., 78 (1993), pp. 77–94.

442. F. MÜLLER-PLATHE, H. SCHMITZ, AND R. FALLER, *Molecular simulation in polymer science: Understanding experiments better*, Prog. Theor. Phys. (Kyoto), Supplements, 138 (2000), pp. 311–319.

443. D. MYERS, *Surfactant science and technology*, VCH Publishers, New York, 1992.

444. A. NAKANO, R. KALIA, AND P. VASHISHTA, *Scalable molecular dynamics, visualization, and data-management algorithms for material simulations*, Comp. Sci. Eng., 1 (1999), pp. 39–47.

445. B. NAYROLES, G. TOUZOT, AND P. VILLON, *Generalizing the finite element method: Diffusive approximation and diffusive elements*, Comput. Mech., 10 (1992), pp. 307–318.

446. M. NELSON, W. HUMPHREY, A. GURSOY, A. DALKE, L. KALÉ, R. SKEEL, AND K. SCHULTEN, *NAMD - A parallel, object-oriented molecular dynamics program*, Int. J. Supercomp. Appl. High Perf. Comp., 10 (1996), pp. 251–268.

447. P. NETTESHEIM, *Mixed quantum-classical dynamics: A unified approach to mathematical modeling and numerical simulation*, Dissertation, Freie Universität Berlin, Fachbereich Mathematik und Informatik, 2000.

448. H. NEUNZERT, A. KLAR, AND J. STRUCKMEIER, *Particle methods: Theory and applications*, Tech. Rep. 95-113, Fachbereich Mathematik, Universität Kaiserslautern, 1995.

449. H. NEUNZERT AND J. STRUCKMEIER, *Boltzmann simulation by particle methods*, Tech. Rep. 112, Fachbereich Mathematik, Universität Kaiserslautern, 1994.

450. Z. NISHIYAMA, *Martensitic transformation*, Academic Press, New York, 1978.

451. S. NOSÉ, *A molecular dynamics method for simulations in the canonical ensemble*, Mol. Phys., 53 (1984), pp. 255–268.

452. ——, *A unified formulation of the constant temperature molecular dynamics method*, J. Chem. Phys., 81 (1984), pp. 511–519.

453. S. NOSÉ AND M. KLEIN, *Constant pressure molecular dynamics for molecular systems*, Mol. Phys., 50 (1983), pp. 1055–1076.

454. L. NYLAND, J. PRINS, AND J. REIF, *A data-parallel implementation of the adaptive fast multipole algorithm*, in Proceedings of the 1993 DAGS/PC Symposium, Dartmouth College, Hanover, NH, 1993, pp. 111–123.

455. A. OBERBECK, *Über die Wärmeleitung der Flüssigkeiten bei Berücksichtigung der Strömungen infolge von Temperaturdifferenzen*, Ann. Phys. Chem., 7 (1879), pp. 271–292.

456. P. OSWALD, *Multilevel finite element approximation: Theory and applications*, Teubner, Stuttgart, 1994.

457. H. OZAL AND T. HARA, *Numerical analysis for oscillatory natural convection of low Prandtl number fluid heated from below*, Numer. Heat Trans. A, 27 (1995), pp. 307–318.

458. P. PACHECO, *Parallel programming with MPI*, Morgan Kaufmann Publishers, 1997.

459. T. PADMANABHAN, *Structure formation in the universe*, Cambridge University Press, Cambridge, 1993.

460. V. PANDE AND D. ROKHSAR, *Molecular dynamics simulations of unfolding and refolding of a hairpin fragment of protein G*, Proc. Natl. Acad. Sci., 96 (1999), pp. 9062–9067.

461. M. PARRINELLO, *Simulating complex systems without adjustable parameters*, Comp. Sci. Eng., 2 (2000), pp. 22–27.

462. M. PARRINELLO AND R. RAHMAN, *Crystal structure and pair potentials: A molecular-dynamics study*, Phys. Rev. Lett., 45 (1980), pp. 1196–1199.

463. D. PASCHEK AND A. GEIGER, *Moscito.* http://ganter.chemie.uni-dortmund.de/MOSCITO/index.shtml.

464. A. PASKIN, A. GOHAR, AND G. DIENES, *Computer simulation of crack propagation*, Phys. Rev. Lett., 44 (1980), pp. 940–943.

465. G. PASTORE, *Car-Parrinello methods and adiabatic invariants*, in Monte Carlo and Molecular Dynamics of Condensed Matter Systems, K. Binder and G. Ciccotti, eds., Italian Physical Society SIF, Bologna, 1996, ch. 24, pp. 635–647.

466. G. PASTORE, E. SMARGIASSI, AND F. BUDA, *Theory of ab initio molecular-dynamics calculations*, Phys. Rev. A, 44 (1991), pp. 6334–6347.

467. D. PATTERSON AND J. HENNESSY, *Computer architecture. A quantitive approach*, Morgan Kaufmann Publishers, San Francisco, 2 ed., 1996.

468. L. PAULING, *The nature of the chemical bonding*, Oxford University Press, London, 1950.

469. M. PAYNE, M. TETER, AND D. ALLAN, *Car-Parrinello methods*, J. Chem. Soc. Faraday Trans., 86 (1990), pp. 1221–1226.

470. G. PEANO, *Sur une courbe qui remplit toute une aire plaine*, Mathematische Annalen, 36 (1890), pp. 157–160.

471. D. PEARLMAN, D. CASE, J. CALDWELL, W. ROSS, T. CHEATHAM, S. DE-BOLT, D. FERGUSON, G. SEIBEL, AND P. KOLLMAN, *AMBER, a computer program for applying molecular mechanics, normal mode analysis, molecular dynamics and free energy calculations to elucidate the structures and energies of molecules*, Comp. Phys. Comm., 91 (1995), pp. 1–41.

472. P. PEEBLES, *The large-scale structure of the universe*, Princeton University Press, Princeton, 1980.

473. E. PEN, *A linear moving adaptive particle-mesh N-body algorithm*, Astrophys. J. supp. series, 100 (1995), pp. 269–280.

474. J. PÉREZ-JORDÁ AND W. YANG, *A concise redefinition of the solid spherical harmonics and its use in fast multipole methods*, J. Chem. Phys., 104 (1996), pp. 8003–8006.

475. L. PERONDI, P. SZELESTEY, AND K. KASKI, *Atomic structure of a dissociated edge dislocation in copper*, in Multiscale phenomena in materials - experiments and modeling, V. Bulatov, T. de la Rubia, N. Ghoniem, E. Kaxiras, and R. Phillips, eds., vol. 578, Materials Research Society, Pittsburgh, 1999, pp. 223–228.

476. C. PETER, X. DAURA, AND W. VAN GUNSTEREN, *Peptides of aminoxy acids: a molecular dynamics simulation study of conformational equilibria under various conditions*, J. Amer. Chem. Soc., 122 (2000), pp. 7461–7466.

477. H. PETERSEN, *Accuracy and efficiency of the particle mesh Ewald method*, J. Chem. Phys., 103 (1995), pp. 3668–3678.

478. H. PETERSEN, D. SOELVASON, J. PERRAM, AND E. SMITH, *Error estimates for the fast multipole method. I. The two-dimensional case*, Proc. R. Soc. Lond. A, 448 (1995), pp. 389–400.

479. ———, *Error estimates for the fast multipole method. II. The three-dimensional case*, Proc. R. Soc. Lond. A, 448 (1995), pp. 401–418.

480. S. PFALZNER AND P. GIBBON, *A 3D hierarchical tree code for dense plasma simulation*, Comp. Phys. Comm., 79 (1994), pp. 24–38.

481. ———, *Many-body tree methods in physics*, Cambridge University Press, 1996.

482. S. PLIMPTON, *LAMMPS - large-scale atomic/molecular massively parallel simulator.* http://lammps.sandia.gov/.

483. ———, *Fast parallel algorithms for short-range molecular dynamics*, J. Comput. Phys., 117 (1995), pp. 1–19.

484. S. PLIMPTON AND B. HENDRICKSON, *Parallel molecular dynamics algorithms for simulation of molecular systems*, in Parallel Computing in Computational Chemistry, T. Mattson, ed., ACS Symposium Series 592, American Chemical Society, 1995, pp. 114–132.

485. S. PLIMPTON, R. POLLOCK, AND M. STEVENS, *Particle-mesh Ewald and rRESPA for parallel molecular dynamics simulations*, in Proceedings of the eighth SIAM conference on parallel processing for scientific computing, Minneapolis, SIAM, 1997.

486. H. PLUMMER, *On the problem of distribution in globular star clusters*, Monthly notices of the Royal Astronomical Society, 71 (1911), pp. 460–470.

487. P. POHL AND G. HEFFELFINGER, *Massively parallel molecular dynamics simulation of gas permeation across porous silica membranes*, J. Membrane Sci., 155 (1999), pp. 1–7.

488. E. POLLOCK AND J. GLOSLI, *Comments on P3M, FMM and the Ewald method for large periodic Coulomb systems*, Comp. Phys. Comm., 95 (1996), pp. 93–110.

489. T. PÖSCHEL AND H. HERRMANN, *Size segregation and convection*, Europhys. Lett., 29 (1995), pp. 123–128.

490. W. PRESS, B. FLANNERY, S. TEUKOLSKY, AND W. VETTERLING, *Numerical recipes in C – the art of scientific computing*, Cambridge University Press, Cambridge, 1988, ch. 7.2.

491. P. PROCACCI, T. DARDEN, E. PACI, AND M. MARCHI, *ORAC: A molecular dynamics program to simulate complex molecular systems with realistic electrostatic interactions*, J. Comput. Chem., 18 (1997), pp. 1848–1862.

492. P. PROCACCI AND M. MARCHI, *Taming the Ewald sum in molecular dynamics simulations of solvated proteins via a multiple time step algorithm*, J. Chem. Phys., 104 (1996), pp. 3003–3012.

493. P. PROCACCI, M. MARCHI, AND G. MARTYNA, *Electrostatic calculations and multiple time scales in molecular dynamics simulation of flexible molecular systems*, J. Chem. Phys., 108 (1996), pp. 8799–8803.

494. A. PUHL, M. MANSOUR, AND M. MARESCHAL, *Quantitative comparison of molecular dynamics with hydrodynamics in Rayleigh-Benard convection*, Phys. Rev. A, 40 (1989), pp. 1999–2011.

495. X. QIAN AND T. SCHLICK, *Efficient multiple timestep integrators with distance-based force splitting for particle-mesh-Ewald molecular dynamics simulations*, J. Chem. Phys., 116 (2002), pp. 5971–5983.

496. H. RAFII-TABAR, L. HUA, AND M. CROSS, *A multi-scale atomistic-continuum modelling of crack propagation in a 2-D macroscopic plate*, J. Phys. Condens. Matter, 10 (1998), pp. 2375–2387.

497. H. RAFII-TABAR AND A. SUTTON, *Long-range Finnis-Sinclair potentials for f.c.c. metallic alloys*, Phil. Mag. Lett., 63 (1991), pp. 217–224.

498. A. RAHMAN, *Correlations in the motion of atoms in liquid Argon*, Phys. Rev. A, 136 (1964), pp. 405–411.

499. A. RAHMAN AND F. STILLINGER, *Molecular dynamics study of liquid water*, J. Chem. Phys., 55 (1971), pp. 3336–3359.

500. W. RANKIN, *Efficient parallel implementations of multipole based N-body algorithms*, Ph.D. thesis, Duke University, 1999.

501. W. RANKIN AND J. BOARD, *A portable distributed implementation of the parallel multipole tree algorithm*, in Proceedings of the 1995 IEEE Symposium on High Performance Distributed Computing, 1995.

502. D. RAPAPORT, *Microscale hydrodynamics: Discrete-particle simulation of evolving flow patterns*, Phys. Rev. A, 36 (1987), pp. 3288–3299.

503. ——, *Large-scale molecular dynamics simulation using vector and parallel computers*, Comp. Phys. Reports, 9 (1988), pp. 1–53.

504. ——, *Molecular-dynamics study of Rayleigh-Benard convection*, Phys. Rev. Lett., 60 (1988), pp. 2480–2483.

505. ——, *Unpredictable convection in a small box: Molecular-dynamics experiments*, Phys. Rev. A, 46 (1992), pp. 1971–1984.

506. ——, *Subharmonic surface waves in vibrated granular media*, Physica A, 249 (1998), pp. 232–238.

507. K. REFSON, *Molecular dynamics simulation of solid n-butane*, Physica B, 131 (1985), pp. 256–266.

508. S. REICH, *Symplectic integration of constrained Hamiltonian systems by Runge-Kutta methods*, Tech. Rep. 93-13, Univ. British Columbia, CS, Vancouver, 1993.

509. ——, *Modified potential energy functions for constrained molecular dynamics*, Numer. Algo., 19 (1998), pp. 213–221.

510. ——, *Backward error analysis for numerical integrators*, SIAM J. Num. Anal., 36 (1999), pp. 1549–1570.

511. L. REICHL, *A modern course in statistical physics*, Edward Arnold Ltd, 1980.

512. ——, *Equilibrium statistical mechanics*, Prentice-Hall, Englewood Cliffs, NJ, 1989.

513. M. REIF, M. GAUTEL, F. OSTERHELT, J. FERNANDEZ, AND H. GAUB, *Reversible folding of individual titin immunoglobulin domains by AFM*, Science, 276 (1997), pp. 1109–1112.

514. J. REINHOLD, *Quantentheorie der Moleküle*, Teubner, Braunschweig, 1994.

515. D. REMLER AND P. MADDEN, *Molecular dynamics without effective potentials via the Car-Parrinello approach*, Mol. Phys., 70 (1990), pp. 921–966.

516. G. RHODES, *Crystallography made crystal clear*, Academic Press, New York, 2000.

517. R. RICHERT AND A. BLUMEN, eds., *Disorder effects on relaxational processes*, Springer, 1994.

518. S. RICK, S. STUART, AND B. BERNE, *Dynamical fluctuating charge force fields: Application to liquid water*, J. Chem. Phys., 101 (1994), pp. 6141–6156.

519. M. RIEF AND H. GRUBMÜLLER, *Kraftspektroskopie von einzelnen Biomolekülen*, Phys. Blätter, 57 (2001), pp. 55–61.

520. D. ROBERTSON, D. BRENNER, AND J. MINTMIRE, *Energetics of nanoscale graphitic tubules*, Phys. Rev. B, 45 (1992), pp. 12592–12595.

521. A. ROBINSON, W. RICHARDS, P. THOMAS, AND M. HANN, *Behavior of cholesterol and its effect on head group and chain conformations in lipid bilayers: A molecular dynamics study*, Biophys. J., 68 (1995), pp. 164–170.

522. D. ROCCATANO, A. AMADEI, A. DI NOLA, AND H. BERENDSEN, *A molecular dynamics study of the 41-56 β-hairpin from B1 domain of protein G*, Protein Sci., 8 (1999), pp. 2130–2143.

523. D. ROCCATANO, R. BIZZARRI, G. CHILLEMI, AND N. SANNA, *Development of a parallel molecular dynamics code on SIMD computers: Algorithm for use of pair list criterion*, J. Comput. Chem., 19 (1998), pp. 685–694.

524. C. RÖHR, *Intermetallische Phasen.* http://ruby.chemie.uni-freiburg.de/Vorlesung/intermetallische_2_3.html.

525. V. ROKHLIN, *Rapid solution of integral equations of classical potential theory*, J. Comput. Phys., 60 (1985), pp. 187–207.

526. B. ROOS, ed., *Lecture notes in quantum chemistry I*, vol. 58 of Lecture Notes in Chemistry, Springer, Berlin, 1992.

527. ——, *The multiconfigurational (MC) self-consistent field (SCF) theory*, in Lecture notes in quantum chemistry, B. Roos, ed., vol. 58 of Lecture Notes in Chemistry, Springer, Berlin, 1992, pp. 177–254.

528. ——, ed., *Lecture notes in quantum chemistry II*, vol. 64 of Lecture Notes in Chemistry, Springer, Berlin, 1994.

529. V. ROSATO, M. GUILLOPE, AND B. LEGRAND, *Thermodynamical and structural-properties of fcc transition-metals using a simple tight-binding model*, Phil. Mag. A, 59 (1989), pp. 321–336.

530. J. ROTH, *IMD - A molecular dynamics program and application*, in Molecular dynamics on parallel computers, R. Esser, P. Grassberger, J. Grotendorst, and M. Lewerenz, eds., Workshop on Molecular dynamics on parallel computers, Jülich, 1999, John von Neumann Institute for Computing (NIC), Research Center Jülich, Germany, World Scientific, 2000, pp. 83–94.

531. S. RUBINI AND P. BALLONE, *Quasiharmonic and molecular-dynamics study of the martensitic transformation in Ni-Al alloys*, Phys. Rev. B, 48 (1993), pp. 99–111.

532. R. RUTH, *A canonical integration technique*, IEEE Trans. Nucl. Sci., NS-30 (1983), pp. 2669–2671.

533. J. RYCKAERT AND A. BELLMANS, *Molecular dynamics of liquid n-butane near its boiling point*, Chem. Phys. Lett., 30 (1975), pp. 123–125.

534. J. RYCKAERT, G. CICCOTTI, AND H. BERENDSEN, *Numerical integration of the Cartesian equation of motion of a system with constraints: Molecular dynamics of N-alkanes*, J. Comput. Phys., 23 (1977), pp. 327–341.

535. H. SAGAN, *Space-filling curves*, Springer, New York, 1994.

536. J. SAKURAI, *Modern quantum mechanics*, Addison-Wesley, 1985, ch. 2.4.

537. J. SALMON, *Parallel hierarchical N-body methods*, PhD thesis, California Institute of Technology, 1990.

538. J. SALMON AND M. WARREN, *Skeletons from the treecode closet*, J. Comput. Phys., 111 (1994), pp. 136–155.

539. J. SALMON, M. WARREN, AND G. WINCKELMANS, *Fast parallel tree codes for gravitational and fluid dynamical N-body problems*, Int. J. Supercomp. Appl. High Perf. Comp., 8 (1994), pp. 129–142.

540. H. SAMET, *Design and analysis of spatial data structures*, Addison-Wesley, 1990.

541. S. SAMUELSON AND G. MARTYNA, *Two dimensional umbrella sampling techniques for the computer simulation study of helical peptides at thermal equilibrium: The 3K(i) peptide in vacuo and solution*, J. Chem. Phys., 109 (1998), pp. 11061–11073.

542. B. SANDAK AND A. BRANDT, *Multiscale fast summation of long range charge and dipolar interactions*, in Multiscale Computational Methods in Chemistry and Physics, A. Brandt, J. Bernholc, and K. Binder, eds., IOS Press, 2001.

543. M. SANGSTER AND M. DIXON, *Interionic potentials in alkali halides and their use in simulation of molten salts*, Adv. Phys., 25 (1976), pp. 247–342.

544. H. SATO, Y. TANAKA, H. IWAMA, S. KAWAKIKA, M. SAITO, K. MORIKAMI, T. YAO, AND S. TSUTSUMI, *Parallelization of AMBER molecular dynamics program for the AP1000 highly parallel computer*, in Proc. Scalable High

Performance Computing Conference-92, IEEE Computer Society Press, 1992, pp. 113–120.

545. S. SAUTER, *Variable order panel clustering*, Computing, 64 (2000), pp. 223–261.

546. U. SCHERZ, *Quantenmechanik: Eine Einführung mit Anwendungen auf Atome, Moleküle und Festkörper*, Teubner, Stuttgart, Leipzig, 1999.

547. C. SCHIFFER AND W. VAN GUNSTEREN, *Structural stability of disulfide mutants of BPTI: A molecular dynamics study*, Proteins: Struct. Funct. Genet., 26 (1996), pp. 66–71.

548. T. SCHLICK, *Molecular modeling and simulation*, Springer, New York, 2002.

549. R. SCHNEIDER, *Multiskalen- und Wavelet-Kompression: Analysisbasierte Methoden zur effizienten Lösung grosser vollbesetzter Gleichungssysteme*, Advances in Numerical Analysis, Teubner, Stuttgart, 1998.

550. I. SCHOENBERG, *Cardinal spline interpolation*, SIAM, Philadelphia, 1973.

551. K. SCHRÖDER AND J. SAUER, *Potential functions for silica and zeolite catalysts based on ab inito calculations. 3. A shell model ion pair potential for silica and aluminosilicates*, J. Phys. Chem., 100 (1996), pp. 11034–11049.

552. C. SCHÜTTE, A. FISCHER, W. HUISINGA, AND P. DEUFLHARD, *A hybrid Monte Carlo method for essential dynamics*, Research report SC 98-04, Konrad Zuse Institut Berlin, 1998.

553. ———, *A direct approach to conformal dynamics based on hybrid Monte Carlo*, J. Comput. Phys., 151 (1999), pp. 146–169.

554. F. SCHWABL, *Quantenmechanik*, Springer, Berlin, 1990.

555. E. SCHWEGLER, G. GALLI, AND F. GYGI, *Water under pressure*, Phys. Rev. Lett., 84 (2000), pp. 2429–2432.

556. M. SCHWEITZER, *A parallel multilevel partition of unity method for ellicptic partial differential equations*, Lecture Notes in Computational Science and Engineering, Vol. 29, Springer, 2003.

557. M. SCHWEITZER, G. ZUMBUSCH, AND M. GRIEBEL, *Parnass2: A cluster of dual-processor PCs*, Tech. Rep. CSR-99-02, TU Chemnitz, 1999. Proceedings of the 2nd Workshop Cluster-Computing, Karlsruhe, Chemnitzer Informatik Berichte.

558. W. SCOTT AND C. SCHIFFER, *Curling of flap tips in HIV-1 protease as a mechanism for substrate entry and tolerance of drug resistance*, Structure Folding Design, 8 (2000), pp. 1259–1265.

559. R. SEDGEWICK, *Algorithms in C*, Addison-Wesley, 1990.

560. R. SEDGEWICK AND P. FLAJOLET, *An introduction to the analysis of algorithms*, Addison-Wesley, 1996.

561. J. SHIMADA, H. KANEKO, AND T. TAKADA, *Efficient calculations of Coulombic interactions in biomolecular simulations with periodic boundary conditions*, J. Comput. Chem., 14 (1993), pp. 867–878.

562. ———, *Performance of fast multipole methods for calculating electrostatic interactions in biomacromolecular simulations with periodic boundary conditions*, J. Comput. Chem., 15 (1994), pp. 28–43.

563. D. SHOLL, *Predicting single-component permeance through macroscopic zeolite membranes from atomistic simulations*, Eng. Chem. Res., 39 (2000), pp. 3737–3746.

564. J. SINGH, C. HOLT, J. HENNESSY, AND A. GUPTA, *A parallel adaptive fast multipole method*, in Proceedings of the Supercomputing '93 Conference, 1993.

565. R. SKEEL, *Symplectic integration with floating-point arithmetic and other approximations*, Appl. Numer. Math., 29 (1999), pp. 3–18.

566. R. SKEEL, G. ZHANG, AND T. SCHLICK, *A family of symplectic integrators: Stability, accuracy, and molecular dynamics applications*, SIAM J. Sci. Comput., 18 (1997), pp. 203–222.

567. B. SMIT, P. HILBERS, K. ESSELINK, L. RUPERT, N. VAN OS, AND G. SCHLIJPER, *Structure of a water/oil interface in the presence of micelles: A computer simulation study*, J. Phys. Chem., 95 (1991), pp. 6361–6368.

568. B. SMITH, W. GROPP, AND P. BJØRSTAD, *Domain decomposition: Parallel multilevel methods for elliptic partial differential equations*, Cambridge University Press, Cambridge, 1996.

569. L. SMITH, C. DOBSON, AND W. VAN GUNSTEREN, *Molecular dynamics simulations of human α-lactalbumin. Changes to the structural and dynamical properties of the protein at low pH*, Proteins: Struct. Funct. Genet., 36 (1999), pp. 77–86.

570. P. SMITH AND B. PETTITT, *Ewald artefacts in liquid state molecular dynamics simulations*, J. Chem. Phys., 105 (1996), pp. 4289–4293.

571. W. SMITH AND T. FORESTER, *Parallel macromolecular simulations and the replicated data strategy: I. The computation of atomic forces*, Comp. Phys. Comm., 79 (1994), pp. 52–62.

572. A. SMONDYREV AND M. BERKOWITZ, *Structure of dipalmitoylphosphatidylcholine/cholesterol bilayer at low and high cholesterol concentrations: Molecular dynamics simulation*, Biophys. J., 77 (1999), pp. 2075–2079.

573. ———, *Molecular dynamics simulation of dipalmitoylphosphatidylcholine membrane with cholesterol sulfate*, Biophys. J., 78 (2000), pp. 1672–1680.

574. D. SOLVASON, J. KOLAFA, H. PETERSON, AND J. PERRAM, *A rigorous comparison of the Ewald method and the fast multipole method in two dimensions*, Comp. Phys. Comm., 87 (1995), pp. 307–318.

575. J. SORENSON, G. HURA, R. GLAESER, AND T. HEAD-GORDON, *What can x-ray scattering tell us about the radial distribution functions of water?*, J. Chem. Phys., 113 (2000), pp. 9149–9161.

576. M. SPECTOR, E. NAJANJO, S. CHIRUVOLU, AND J. ZASADZINSKI, *Conformations of a tethered membrane: Crumpling in graphite oxide*, Phys. Rev. Lett., 73 (1994), pp. 2867–2870.

577. R. SPLINTER, *A nested-grid particle-mesh code for high-resolution simulations of gravitational instability in cosmology*, Mon. Not. R. Astron. Soc., 281 (1996), pp. 281–293.

578. M. SPRIK, J. HUTTER, AND M. PARRINELLO, *Ab initio molecular dynamics simulation of liquid water: Comparison of three gradient-corrected density functionals*, J. Chem. Phys., 105 (1996), pp. 1142–1152.

579. D. SRIVASTAVA AND S. BARNARD, *Molecular dynamics simulation of large-scale carbon nanotubes on a shared-memory architecture*, in Proc. SuperComputing 97, NASA Ames Research Center, 1997.

580. S. STEPANIANTS, S. IZRAILEV, AND K. SCHULTEN, *Extraction of lipids from phospholipid membranes by steered molecular dynamics*, J. Mol. Modeling, 3 (1997), pp. 473–475.

581. T. STERLING, J. SALMON, D. BECKER, AND D. SAVARESE, *How to build a Beowulf*, MIT Press, Cambridge, MA, 1999.

582. H. STERN, F. RITTNER, B. BERNE, AND R. FRIESNER, *Combined fluctuating charge and polarizable dipole models: Application to a five-site water potential function*, J. Chem. Phys., 115 (2001), pp. 2237–2251.

583. F. STILLINGER AND A. RAHMAN, *Improved simulation of liquid water by molecular dynamics*, J. Chem. Phys., 60 (1974), pp. 1545–1557.

584. F. STILLINGER AND T. WEBER, *Computer simulation of local order in condensed phases of silicon*, Phys. Rev. B, 31 (1985), pp. 5262–5271. errata: Phys. Rev. B 33, 1451 (1986).

585. W. STOCKMAYER, *Second virial coefficients of polar gases*, J. Chem. Phys., 9 (1941), pp. 389–402.

586. ——, *Theory of molecular size distribution and gel formation in branched chain polymers*, J. Chem. Phys., 11 (1943), pp. 45–55.

587. J. STOER AND R. BULIRSCH, *Introduction to numerical analysis*, Springer, Berlin, Heidelberg, New York, 2002.

588. C. STÖRMER, *Sur les trajectoires des corpuscles életrisés dans l'espace sous l'action du magnetisme terrestre avec application aux aurores boréales*, Arch. Sci. Phys. Nat., 24 (1907), pp. 221–247.

589. T. STOUCH, *Permeation of lipid membranes: Molecular dynamics simulations*, in Encyclopedia of Computational Chemistry, P. von Ragué Schleyer, ed., vol. 3, Wiley, New York, 1998, pp. 2038–2045.

590. G. STRANG, *On the construction and comparison of difference schemes*, SIAM J. Numer. Anal., 5 (1968), pp. 506–517.

591. W. STREETT, D. TILDESLEY, AND G. SAVILLE, *Multiple time-step methods in molecular dynamics*, Mol. Phys., 35 (1978), pp. 639–648.

592. A. SUENAGA, Y. KOMEIJI, M. UEBAYASI, T. MEGURO, AND I. YAMATO, *Molecular dynamics simulation of unfolding of histidine-containing phospho-carrier protein in water*, J. Chem. Software, 4 (1998), pp. 127–142.

593. A. SUTTON AND J. CHEN, *Long-range Finnis-Sinclair potentials*, Phil. Mag. Lett., 61 (1990), pp. 139–146.

594. M. SUZUKI, *Fractal decomposition of exponential operators with applications to many-body theories and Monte Carlo simulations*, Phys. Lett. A, 146 (1990), pp. 319–323.

595. P. SWARZTRAUBER, *Multiprocessor FFTs*, Parallel Comp., 5 (1987), pp. 197–210.

596. W. SWOPE, H. ANDERSEN, P. BERENS, AND K. WILSON, *A computer simulation method for the calculation of equilibrium constants for the formation of physical clusters of molecules: Application to small water clusters*, J. Chem. Phys., 76 (1982), pp. 637–649.

597. B. SZABO AND I. BABUSKA, *Finite element analysis*, John Wiley, New York, 1991.

598. T. TANAKA, S. YONEMURA, K. YASHI, AND Y. TSUJI, *Cluster formation and particle-induced instability in gas-solid flows predicted by the DSMC method*, JSME Int. J. B, 39 (1996), pp. 239–245.

599. J. TAUSCH, *The fast multipole method for arbitrary Green's functions*, in Current Trends in Scientific Computing, Z. Chen, R. Glowinski, and K. Li, eds., American Mathematical Society, 2003.

600. ——, *The variable order fast multipole method for boundary integral equations of the second kind*, Computing, 72 (2004), pp. 267–291.

601. M. TEODORO, G. PHILLIPS, AND L. KAVRAKI, *Singular value decomposition of protein conformational motions: Application to HIV-1 protease*, in Currents in Computational Molecular Biology, M. Satoru, R. Shamir, and T. Tagaki, eds., Tokyo, 2000, Universal Academy Press Inc., pp. 198–199. The Third ACM International Conference on Computational Biology (RECOMB).

602. J. TERSOFF, *New empirical model for the structural properties of silicon*, Phys. Rev. Lett., 56 (1986), pp. 632–635.

603. ———, *New empirical approach for the structure and energy of covalent systems*, Phys. Rev. B, 37 (1988), pp. 6991–7000.

604. P. TEUBEN, *The stellar dynamics toolbox NEMO*, in Astronomical Data Analysis Software and Systems IV, R. Shaw, H. Payne, and J. Hayes, eds., vol. 77 of PASP Conf Series, 1995, pp. 398–401.

605. J. THEILHABER, *Ab initio simulations of sodium using time-dependent density-functional theory*, Phys. Rev. B, 46 (1992), pp. 12990–13003.

606. T. THEUNS, *Parallel P3M with exact calculation of short range forces*, Comp. Phys. Comm., 78 (1994), pp. 238–246.

607. J. THIJSSEN, *Computational physics*, Cambridge University Press, 1999.

608. D. TIELEMAN AND H. BERENDSEN, *A molecular dynamics study of the pores formed by Escherichia coli OmpF porin in a fully hydrated palmitoyl-olelyl-phosphatidylcholine bilayer*, Biophys. J., 74 (1998), pp. 2786–2801.

609. D. TIELEMAN, S. MARRINK, AND H. BERENDSEN, *A computer perspective of membranes: Molecular dynamics studies of lipid bilayer systems*, Biochem. Biophys. Acta, 1331 (1997), pp. 235–270.

610. D. TIELEMAN, M. SANSOM, AND H. BERENDSEN, *Alamethicin helices in a bilayer and in solution: Molecular dynamics simulations*, Biophys. J., 76 (1999), pp. 40–49.

611. D. TOBIAS, K. TU, AND M. KLEIN, *Atomic scale molecular dynamics simulations of lipid membranes*, Curr. Opin. Coll. Int. Sci., 2 (1997), pp. 15–26.

612. A. TOUKMAJI AND J. BOARD, *Ewald summation techniques in perspective: A survey*, Comp. Phys. Comm., 95 (1996), pp. 73–92.

613. A. TOUKMAJI, D. PAUL, AND J. BOARD, *Distributed particle-mesh Ewald: A parallel Ewald summation method*, in Proc. of International Conference on Parallel and Distributed Processing Techniques and Applications (PDPTA'96), CSREA Press, Athens, 1996, pp. 33–43.

614. S. TOXVAERD, *Comment on constrained molecular dynamics of macromolecules*, J. Chem. Phys., 87 (1987), pp. 6140–6143.

615. J. TRAUB, G. WASILKOWSKI, AND H. WOZNIAKOWSKI, *Information-based complexity*, Academic Press, New York, 1988.

616. H. TRIEBEL, *Interpolation theory, function spaces, differential operators*, Dt. Verlag Wiss., Berlin, 1978.

617. H. TROTTER, *On the product of semi-groups of operators*, Proc. Am. Math. Soc., 10 (1959), pp. 545–551.

618. S. TSUNEYUKI, M. TSUKADA, H. AOKI, AND Y. MATSUI, *First-principles interatomic potential of silica applied to molecular dynamics*, Phys. Rev. Lett., 61 (1988), pp. 869–872.

619. K. TU, D. TOBIAS, K. BLASIE, AND M. KLEIN, *Molecular dynamics investigation of the structure of a fully hydrated gel-phase dipalmitoylphosphatidylcholine bilayer*, Biophys. J., 70 (1996), pp. 598–608.

620. M. TUCKERMAN, B. BERNE, AND G. MARTYNA, *Molecular dynamics algorithm for multiple time scales: Systems with long range forces*, J. Chem. Phys., 94 (1991), pp. 6811–6815.

621. ——, *Reversible multiple time scale molecular dynamics*, J. Chem. Phys., 97 (1992), pp. 1990–2001.

622. M. TUCKERMAN, B. BERNE, AND A. ROSSI, *Molecular dynamics algorithm for multiple time scales: Systems with disparate masses*, J. Chem. Phys., 94 (1991), pp. 1465–1469.

623. M. TUCKERMAN, G. MARTYNA, AND B. BERNE, *Molecular dynamics algorithm for condensed systems with multiple time scales*, J. Chem. Phys., 93 (1990), pp. 1287–1291.

624. M. TUCKERMAN, G. MARTYNA, M. KLEIN, AND B. BERNE, *Efficient molecular dynamics and hybrid Monte Carlo algorithms for path integrals*, J. Chem. Phys., 99 (1993), pp. 2796–2808.

625. J. TULLY, *Molecular dynamics with electronic transitions*, J. Chem. Phys., 93 (1990), pp. 1061–1071.

626. ——, *Mixed quantum-classical dynamics: Mean-field and surface-hopping*, in Classical and Quantum Dynamics in Condensed Phase Simulations, B. Berne, G. Ciccotti, and D. Coker, eds., World Scientific, Singapore, 1998, ch. 21, pp. 489–509.

627. ——, *Nonadiabatic dynamics*, in Modern Methods for Multidimensional Dynamics Computations in Chemistry, D. Thompson, ed., World Scientific, Singapore, 1998, pp. 34–79.

628. D. TURNER AND J. MORRIS, *AL_CMD*, Condensed Matter Physics Group, Ames Laboratory. http://cmp.ameslab.gov/cmp/CMP_Theory/cmd/alcmd_source.html.

629. P. UMBANHOWAR, F. MELO, AND H. SWINNEY, *Localized excitations in a vertically vibrated granular layer*, Nature, 382 (1996), pp. 793–796.

630. P. VAN DER PLOEG AND H. BERENDSEN, *Molecular dynamics of model membranes*, Biophys. Struct. Mechanism. Suppl., 6 (1980), pp. 106–108.

631. ——, *Molecular dynamics simulation of a bilayer membrane*, J. Chem. Phys., 76 (1982), pp. 3271–3276. Note a typo: σ should be 0.374 nm.

632. D. VAN DER POEL, B. DE GROOT, S. HAYWARD, H. BERENDSEN, AND H. VOGEL, *Bending of the calmodulin central helix: A theoretical study*, Protein Sci., 5 (1996), pp. 2044–2053.

633. D. VAN DER SPOEL, P. VAN MAAREN, AND H. BERENDSEN, *A systematic study of water models for molecular dynamics simulation: Derivation of water models optimized for use with a reaction field*, J. Chem. Phys., 108 (1998), pp. 10220–10230.

634. W. VAN GUNSTEREN AND H. BERENDSEN, *Computer simulation of molecular dynamics: Methodology, applications and perspectives in chemistry*, Angew. Chem. Int. Ed. Engl., 29 (1990), pp. 992–1023.

635. W. VAN GUNSTEREN AND M. KARPLUS, *Effect of constrains, solvent and cristal environment on protein dynamics*, Nature, 293 (1981), pp. 677–678.

636. ——, *Protein dynamics in solution and in a crystalline environment: A molecular dynamics study*, Biochem., 21 (1982), pp. 2259–2274.

637. W. VAN GUNSTEREN, H. LIU, AND F. MÜLLER-PLATHE, *The elucidation of enzymatic reaction mechanisms by computer simulation: Human immunodeficiency virus protease catalysis*, J. Mol. Struct. (Theochem), 432 (1998), pp. 9–14.

638. P. VAN MAAREN AND D. VAN DER SPOEL, *Molecular dynamics simulations of water with novel shell-model potentials*, J. Phys. Chem. B, 105 (2001), pp. 2618–2626.

639. P. VASHISHTA, R. KALIA, AND A. NAKANO, *Large-scale atomistic simulations of dynamic fracture*, Comp. Sci. Eng., 1 (1999), pp. 56–65.

640. P. VASHISHTA, R. KALIA, J. RINO, AND I. EBBSJO, *Interaction potential for SiO_2: a molecular-dynamics study of structural correlations*, Phys. Rev. B, 41 (1990), pp. 12197–12209.

641. P. VASHISHTA, A. NAKANO, R. KALIA, AND I. EBBSJO, *Crack propagation and fracture in ceramic films – million atom molecular dynamics simulation on parallel computers.*, Mater. Sci. Eng. B, 37 (1996), pp. 56–71.

642. R. VENABLE, B. BROOKS, AND R. PASTOR, *Molecular dynamics simulations of gel (L-beta I) phase lipid bilayers in constant pressure and constant surface area ensembles*, J. Chem. Phys., 112 (2000), pp. 4822–4832.

643. R. VENABLE, Y. ZHANG, B. HARDY, AND R. PASTOR, *Molecular dynamics simulations of a lipid bilayer and of hexadecane – an investigation of membrane fluidity*, Science, 262 (1993), pp. 223–226.

644. R. VERFÜHRT, *A review of a posteriori error estimation and adaptive mesh-refinement techniques*, J. Wiley & Teubner, Chichester, 1996.

645. L. VERLET, *Computer "experiments" on classical fluids. I. Thermodynamical properties of Lennard-Jones molecules*, Phys. Rev., 159 (1967), pp. 98–103.

646. J. VILLUMSEN, *A new hierachical particle-mesh code for very large scale cosmological N-body simulations*, Astrophys. J. supp. series, 71 (1989), pp. 407–431.

647. J. VONDRASEK AND A. WLODAWER, *Database of HIV proteinase structures*, Trends Biochem. Sci., 22 (1997), pp. 183–187.

648. A. VOTER, *The embedded atom method*, in Intermetallic Compounds: Principles and Practice, J. Westbrook and R. Fleischer, eds., J. Wiley, Chichester, 1994, pp. 77–90.

649. A. VOTER AND S. CHEN, *Accurate interatomic potentials for Ni, Al, and Ni_3Al*, Mat. Res. Soc. Symp. Proc., 82 (1987), pp. 175–180.

650. H. WANG AND R. LESAR, *An efficient fast-multipole algorithm based on an expansion in the solid harmonics*, J. Chem. Phys., 104 (1996), pp. 4173–4179.

651. M. WARREN AND J. SALMON, *Astrophysical N-body simulations using hierarchical tree data structures*, in Supercomputing '92, Los Alamitos, 1992, IEEE Comp. Soc., pp. 570–576.

652. ———, *A parallel hashed oct-tree N-body algorithm*, in Supercomputing '93, Los Alamitos, 1993, IEEE Comp. Soc., pp. 12–21.

653. ———, *A parallel, portable and versatile treecode*, in Proc. 7. SIAM Conf. Parallel Processing for Scientific Computing, D. Bailey, P. Bjørstad, J. Gilbert, M. Mascagni, R. Schreiber, H. Simon, V. Torczon, and L. Watson, eds., Philadelphia, 1995, SIAM, pp. 319–324.

654. ———, *A portable parallel particle program*, Comp. Phys. Comm., 87 (1995), pp. 266–290.

655. K. WATANABE AND M. KLEIN, *Effective pair potentials and the properties of water*, J. Chem. Phys., 131 (1989), pp. 157–167.

656. J. WATSON AND F. CRICK, *A structure for deoxyribose nucleic acid*, Nature, 171 (1953), pp. 737–738.

657. I. WEBER AND R. HARRISON, *Molecular mechanics analysis of drug resistant mutations of HIV protease*, Protein Eng., 12 (1999), pp. 469–474.

658. W. WEBER, P. HÜNENBERGER, AND J. MCCAMMON, *Molecular dynamics simulations of a polyalanine octapeptide under Ewald boundary conditions: Influence of artificial periodicity on peptide conformation*, J. Phys. Chem. B, 104 (2000), pp. 3668–3675.

659. C. WHITE AND M. HEAD-GORDON, *Derivation and efficient implementation of the fast multipole method*, J. Chem. Phys., 8 (1994), pp. 6593–6605.

660. J. WILLIAMSON, *Four cheap improvements to the particle-mesh code*, J. Comput. Phys., 41 (1981), pp. 256–269.

661. G. WINCKELMANS, J. SALMON, M. WARREN, AND A. LEONARD, *The fast solution of three-dimensional fluid dynamical N-body problems using parallel tree codes: Vortex element method and boundary element method*, in Proc. 7. SIAM Conf. Parallel Processing for Scientific Computing, D. Bailey, P. Bjørstad, J. Gilbert, M. Mascagni, R. Schreiber, H. Simon, V. Torczon, and L. Watson, eds., Philadelphia, 1995, SIAM, pp. 301–306.

662. A. WINDEMUTH, *Advanced algorithms for molecular dynamics simulation: The program PMD*, in Parallel Computing in Computational Chemistry, T. Mattson, ed., vol. 592 of ACS Symposium Series, American Chemical Society, 1995, pp. 151–169.

663. J. WISDOM, *The origin of the Kirkwood gaps: A mapping for asteroidal motion near the 1/3 commensurability*, Astronom. J., 87 (1982), pp. 577–593.

664. S. WLODEK, J. ANTOSIEWICZ, AND J. MCCAMMON, *Prediction of titration properties of structures of a protein derived from molecular dynamics trajectories*, Protein Sci., 6 (1997), pp. 373–382.

665. J. WOENCKHAUS, R. HUDGINS, AND M. HARROLD, *Hydration of gas-phase proteins: A special hydration site on gas-phase BPTI*, J. Amer. Chem. Soc., 119 (1997), pp. 9586–9587.

666. T. WOOLF AND B. ROUX, *Structure, energetics, and dynamics of lipid-protein interactions: A molecular dynamics study of the gramicidin A channel in a DMPC bilayer*, Proteins: Struct. Funct. Genet., 24 (1996), pp. 92–114.

667. T. WUNG AND F. TSENG, *A color-coded particle tracking velocimeter with application to natural convection*, Experiments Fluids, 13 (1992), pp. 217–223.

668. R. WYATT, *Quantum dynamics with trajectories, Introduction to quantum hydrodynamics*, Interdisciplinary Applied Mathematics, Vol. 28, Springer, New York, 2005.

669. D. XIE, A. TROPSHA, AND T. SCHLICK, *An efficient projection protocol for chemical databases: The singular value decomposition combined with truncated-Newton minimization*, J. Chem. Inf. Comp. Sci., 40 (2000), pp. 167–177.

670. B. YAKOBSON AND R. SMALLEY, *Fullerene nanotubes: $C_{1,000,000}$ and beyond*, American Scientist, 85 (1997), pp. 324–337.

671. J. YEOMANS, *Equilibrium statistical mechanics*, Oxford University Press, Oxford, 1989.

672. D. YORK, T. DARDEN, AND L. PEDERSEN, *The effect of long-range electrostatic interactions in simulations of macromolecular crystals: A comparison of the Ewald and truncated list methods*, J. Chem. Phys., 99 (1993), pp. 8345–8348.

673. D. YORK, T. DARDEN, L. PEDERSEN, AND M. ANDERSON, *Molecular modeling studies suggest that zinc ions inhibit HIV-1 protease by binding at catalytic aspartates*, Environmental Health Perspectives, 101 (1993), pp. 246–250.

674. D. YORK, A. WLODAWER, L. PEDERSEN, AND T. DARDEN, *Atomic level accuracy in simulations of large protein crystals*, Proc. Natl. Acad. Sci., 91 (1994), pp. 8715–8718.

675. D. YORK AND W. YANG, *The fast Fourier Poisson method for calculating Ewald sums*, J. Chem. Phys., 101 (1994), pp. 3298–3300.

676. H. YOSHIDA, *Construction of higher order symplectic integrators*, Phys. Lett. A, 150 (1990), pp. 262–268.

677. H. YSERENTANT, *A new class of particle methods*, Numer. Math., 76 (1997), pp. 87–109.

678. ——, *A particle model of compressible fluids*, Numer. Math., 76 (1997), pp. 111–142.

679. X. YUAN, C. SALISBURY, D. BALSARA, AND R. MELHEM, *A load balance package on distributed memory systems and its application to particle-particle particle-mesh (P^3M) methods*, Parallel Comp., 23 (1997), pp. 1525–1544.

680. Y. YUAN AND P. BANERJEE, *A parallel implementation of a fast multipole based 3-D capacitance extraction program on distributed memory multicomputers*, Journal of Parallel and Distributed Computing, 61 (2001), pp. 1751–1774.

681. Y. ZELDOVICH, *The evolution of radio sources at large redshifts*, Mon. Not. R. Astron. Soc., 147 (1970), pp. 139–148.

682. G. ZHANG AND T. SCHLICK, *LIN: A new algorithm combining implicit integration and normal mode techniques for molecular dynamics*, J. Comput. Chem., 14 (1993), pp. 1212–1233.

683. ——, *The Langevin/implicit-Euler/normal-mode scheme (LIN) for molecular dynamics at large timesteps*, J. Chem. Phys., 101 (1994), pp. 4995–5012.

684. ——, *Implicit discretization schemes for Langevin dynamics*, Mol. Phys., 84 (1995), pp. 1077–1098.

685. Y. ZHANG, T. WANG, AND Q. TANG, *Brittle and ductile fracture at the atomistic crack-tip in copper-crystals*, Scripta Metal. Mater., 33 (1995), pp. 267–274.

686. F. ZHAO AND S. JOHNSSON, *The parallel multipole method on the Connection Machine*, SIAM J. Sci. Stat. Comput., 12 (1991), pp. 1420–1437.

687. F. ZHOU AND K. SCHULTEN, *Molecular dynamics study of a membrane water interface*, J. Phys. Chem., 100 (1995), pp. 2194–2207.

688. ——, *Molecular dynamics study of phospholipase A2 on a membrane surface*, Proteins: Struct. Funct. Genet., 25 (1996), pp. 12–27.

689. R. ZHOU AND B. BERNE, *A new molecular dynamics method combining the reference system propagator algorithm with a fast multipole method for simulating proteins and other complex systems*, J. Chem. Phys., 103 (1995), pp. 9444–9459.

690. R. ZHOU, E. HARDER, H. XU, AND B. BERNE, *Efficient multiple time step method for use with Ewald and particle mesh Ewald for large biomolecular systems*, J. Chem. Phys., 115 (2001), pp. 2348–2358.

691. S. ZHOU, D. BEAZLEY, P. LOMDAHL, AND B. HOLIAN, *Large-scale molecular dynamics simulations of three-dimensional ductile fracture*, Phys. Rev. Lett., 78 (1997), pp. 479–482.

692. S. ZHOU, D. BEAZLEY, P. LOMDAHL, A. VOTER, AND B. HOLIAN, *Dislocation emission from a three-dimensional crack – a large-scale molecular dynamics study*, in Advances in Fracture Research - 97, B. Karihaloo, Y. Mai, M. Ripley, and R. Ritchie, eds., vol. 6, Pergamon Press, Amsterdam, 1997, p. 3085.

693. G. ZUMBUSCH, *Parallel multilevel methods. Adaptive mesh refinement and loadbalancing*, Teubner, 2003.

Index

Editorial Policy

§1. Textbooks on topics in the field of computational science and engineering will be considered. They should be written for courses in CSE education. Both graduate and undergraduate textbooks will be published in TCSE. Multidisciplinary topics and multidisciplinary teams of authors are especially welcome.

§2. Format: Only works in English will be considered. They should be submitted in camera-ready form according to Springer-Verlag's specifications.
Electronic material can be included if appropriate. Please contact the publisher.
Technical instructions and/or TeX macros are available via
http://www.springer.com/sgw/cda/frontpage/0,11855,5-40017-2-71391-0,00.html

§3. Those considering a book which might be suitable for the series are strongly advised to contact the publisher or the series editors at an early stage.

General Remarks

TCSE books are printed by photo-offset from the master-copy delivered in camera-ready form by the authors. For this purpose Springer-Verlag provides technical instructions for the preparation of manuscripts. See also *Editorial Policy*.

Careful preparation of manuscripts will help keep production time short and ensure a satisfactory appearance of the finished book.

The following terms and conditions hold:

Regarding free copies and royalties, the standard terms for Springer mathematics monographs and textbooks hold. Please write to martin.peters@springer.com for details.

Authors are entitled to purchase further copies of their book and other Springer books for their personal use, at a discount of 33,3 % directly from Springer-Verlag.

Series Editors

Timothy J. Barth
NASA Ames Research Center
NAS Division
Moffett Field, CA 94035, USA
e-mail: barth@nas.nasa.gov

Michael Griebel
Institut für Numerische Simulation
der Universität Bonn
Wegelerstr. 6
53115 Bonn, Germany
e-mail: griebel@ins.uni-bonn.de

David E. Keyes
Department of Applied Physics
and Applied Mathematics
Columbia University
200 S. W. Mudd Building
500 W. 120th Street
New York, NY 10027, USA
e-mail: david.keyes@columbia.edu

Risto M. Nieminen
Laboratory of Physics
Helsinki University of Technology
02150 Espoo, Finland
e-mail: rni@fyslab.hut.fi

Dirk Roose
Department of Computer Science
Katholieke Universiteit Leuven
Celestijnenlaan 200A
3001 Leuven-Heverlee, Belgium
e-mail: dirk.roose@cs.kuleuven.ac.be

Tamar Schlick
Department of Chemistry
Courant Institute of Mathematical
Sciences
New York University
and Howard Hughes Medical Institute
251 Mercer Street
New York, NY 10012, USA
e-mail: schlick@nyu.edu

Editor at Springer: Martin Peters
Springer-Verlag, Mathematics Editorial IV
Tiergartenstrasse 17
D-69121 Heidelberg, Germany
Tel.: *49 (6221) 487-8185
Fax: *49 (6221) 487-8355
e-mail: martin.peters@springer.com

Texts
in Computational Science
and Engineering

1. H. P. Langtangen, *Computational Partial Differential Equations*. Numerical Methods and Diffpack Programming. 2nd Edition
2. A. Quarteroni, F. Saleri, *Scientific Computing with MATLAB and Octave*. 2nd Edition
3. H. P. Langtangen, *Python Scripting for Computational Science*. 2nd Edition
4. H. Gardner, G. Manduchi, *Design Patterns for e-Science*.
5. M. Griebel, S. Knapek, G. Zumbusch, *Numerical Simulation in Molecular Dynamics*.

For further information on these books please have a look at our mathematics catalogue at the following URL: www.springer.com/series/5151

Monographs
in Computational Science
and Engineering

1. J. Sundnes, G.T. Lines, X. Cai, B. F. Nielsen, K.-A. Mardal, A. Tveito, *Computing the Electrical Activity in the Heart*.

For further information on these books please have a look at our mathematics catalogue at the following URL: www.springer.com/series/7417

Lecture Notes
in Computational Science
and Engineering

1. D. Funaro, *Spectral Elements for Transport-Dominated Equations*.
2. H. P. Langtangen, *Computational Partial Differential Equations*. Numerical Methods and Diffpack Programming.
3. W. Hackbusch, G. Wittum (eds.), *Multigrid Methods V*.
4. P. Deuflhard, J. Hermans, B. Leimkuhler, A. E. Mark, S. Reich, R. D. Skeel (eds.), *Computational Molecular Dynamics: Challenges, Methods, Ideas*.
5. D. Kröner, M. Ohlberger, C. Rohde (eds.), *An Introduction to Recent Developments in Theory and Numerics for Conservation Laws*.

6. S. Turek, *Efficient Solvers for Incompressible Flow Problems.* An Algorithmic and Computational Approach.

7. R. von Schwerin, *Multi Body System SIMulation.* Numerical Methods, Algorithms, and Software.

8. H.-J. Bungartz, F. Durst, C. Zenger (eds.), *High Performance Scientific and Engineering Computing.*

9. T. J. Barth, H. Deconinck (eds.), *High-Order Methods for Computational Physics.*

10. H. P. Langtangen, A. M. Bruaset, E. Quak (eds.), *Advances in Software Tools for Scientific Computing.*

11. B. Cockburn, G. E. Karniadakis, C.-W. Shu (eds.), *Discontinuous Galerkin Methods.* Theory, Computation and Applications.

12. U. van Rienen, *Numerical Methods in Computational Electrodynamics.* Linear Systems in Practical Applications.

13. B. Engquist, L. Johnsson, M. Hammill, F. Short (eds.), *Simulation and Visualization on the Grid.*

14. E. Dick, K. Riemslagh, J. Vierendeels (eds.), *Multigrid Methods VI.*

15. A. Frommer, T. Lippert, B. Medeke, K. Schilling (eds.), *Numerical Challenges in Lattice Quantum Chromodynamics.*

16. J. Lang, *Adaptive Multilevel Solution of Nonlinear Parabolic PDE Systems.* Theory, Algorithm, and Applications.

17. B. I. Wohlmuth, *Discretization Methods and Iterative Solvers Based on Domain Decomposition.*

18. U. van Rienen, M. Günther, D. Hecht (eds.), *Scientific Computing in Electrical Engineering.*

19. I. Babuška, P. G. Ciarlet, T. Miyoshi (eds.), *Mathematical Modeling and Numerical Simulation in Continuum Mechanics.*

20. T. J. Barth, T. Chan, R. Haimes (eds.), *Multiscale and Multiresolution Methods.* Theory and Applications.

21. M. Breuer, F. Durst, C. Zenger (eds.), *High Performance Scientific and Engineering Computing.*

22. K. Urban, *Wavelets in Numerical Simulation.* Problem Adapted Construction and Applications.

23. L. F. Pavarino, A. Toselli (eds.), *Recent Developments in Domain Decomposition Methods.*

24. T. Schlick, H. H. Gan (eds.), *Computational Methods for Macromolecules: Challenges and Applications.*

25. T. J. Barth, H. Deconinck (eds.), *Error Estimation and Adaptive Discretization Methods in Computational Fluid Dynamics.*

26. M. Griebel, M. A. Schweitzer (eds.), *Meshfree Methods for Partial Differential Equations.*

27. S. Müller, *Adaptive Multiscale Schemes for Conservation Laws.*

28. C. Carstensen, S. Funken, W. Hackbusch, R. H. W. Hoppe, P. Monk (eds.), *Computational Electromagnetics.*

29. M. A. Schweitzer, *A Parallel Multilevel Partition of Unity Method for Elliptic Partial Differential Equations.*

30. T. Biegler, O. Ghattas, M. Heinkenschloss, B. van Bloemen Waanders (eds.), *Large-Scale PDE-Constrained Optimization.*

31. M. Ainsworth, P. Davies, D. Duncan, P. Martin, B. Rynne (eds.), *Topics in Computational Wave Propagation.* Direct and Inverse Problems.

32. H. Emmerich, B. Nestler, M. Schreckenberg (eds.), *Interface and Transport Dynamics.* Computational Modelling.

33. H. P. Langtangen, A. Tveito (eds.), *Advanced Topics in Computational Partial Differential Equations.* Numerical Methods and Diffpack Programming.

34. V. John, *Large Eddy Simulation of Turbulent Incompressible Flows.* Analytical and Numerical Results for a Class of LES Models.

35. E. Bänsch (ed.), *Challenges in Scientific Computing – CISC 2002.*

36. B. N. Khoromskij, G. Wittum, *Numerical Solution of Elliptic Differential Equations by Reduction to the Interface.*

37. A. Iske, *Multiresolution Methods in Scattered Data Modelling.*

38. S.-I. Niculescu, K. Gu (eds.), *Advances in Time-Delay Systems.*

39. S. Attinger, P. Koumoutsakos (eds.), *Multiscale Modelling and Simulation.*

40. R. Kornhuber, R. Hoppe, J. Périaux, O. Pironneau, O. Widlund, J. Xu (eds.), *Domain Decomposition Methods in Science and Engineering.*

41. T. Plewa, T. Linde, V.G. Weirs (eds.), *Adaptive Mesh Refinement – Theory and Applications.*

42. A. Schmidt, K. G. Siebert, *Design of Adaptive Finite Element Software.* The Finite Element Toolbox ALBERTA.

43. M. Griebel, M.A. Schweitzer (eds.), *Meshfree Methods for Partial Differential Equations II.*

44. B. Engquist, P. Lötstedt, O. Runborg (eds.), *Multiscale Methods in Science and Engineering.*

45. P. Benner, V. Mehrmann, D.C. Sorensen (eds.), *Dimension Reduction of Large-Scale Systems.*

46. D. Kressner (ed.), *Numerical Methods for General and Structured Eigenvalue Problems.*

47. A. Boriçi, A. Frommer, B. Joó, A. Kennedy, B. Pendleton (eds.), *QCD and Numerical Analysis III.*

48. F. Graziani (ed.), *Computational Methods in Transport.*

49. B. Leimkuhler, C. Chipot, R. Elber, A. Laaksonen, A. Mark, T. Schlick, C. Schütte, R. Skeel (eds.), *New Algorithms for Macromolecular Simulation.*

50. M. Bücker, G. Corliss, P. Hovland, U. Naumann, B. Norris (eds.), *Automatic Differentiation: Applications, Theory, and Implementations.*

51. A. M. Bruaset, A. Tveito (eds.), *Numerical Solution of Partial Differential Equations on Parallel Computers.*

52. K. H. Hoffmann, A. Meyer (eds.), *Parallel Algorithms and Cluster Computing.*

53. H.-J. Bungartz, M. Schäfer (eds.), *Fluid-Structure Interaction.*

54. J. Behrens, *Adaptive Atmospheric Modeling.*

55. O. Widlund, D. Keyes (eds.), *Domain Decomposition Methods in Science and Engineering XVI.*

56. S. Kassinos, C. Langer, G. Iaccarino, P. Moin (eds.), *Complex Effects in Large Eddy Simulations.*

57. M. Griebel, M. A. Schweitzer (eds.), *Meshfree Methods for Partial Differential Equations III.*

For further information on these books please have a look at our mathematics catalogue at the following URL: www.springer.com/series/3527